NATO ASI Series
Advanced Science Institutes Series

A series presenting the results of activities sponsored by the NATO Science Committee, which aims at the dissemination of advanced scientific and technological knowledge, with a view to strengthening links between scientific communities.

The Series is published by an international board of publishers in conjunction with the NATO Scientific Affairs Division

A	Life Sciences	Plenum Publishing Corporation
B	Physics	London and New York
C	Mathematical and Physical Sciences	Kluwer Academic Publishers
D	Behavioural and Social Sciences	Dordrecht, Boston and London
E	Applied Sciences	
F	Computer and Systems Sciences	Springer-Verlag
G	Ecological Sciences	Berlin Heidelberg New York
H	Cell Biology	London Paris Tokyo Hong Kong
I	Global Environmental Change	Barcelona Budapest

PARTNERSHIP SUB-SERIES

1. Disarmament Technologies	Kluwer Academic Publishers
2. Environment	Springer-Verlag/Kluwer Acad. Publishers
3. High Technology	Kluwer Academic Publishers
4. Science and Technology Policy	Kluwer Academic Publishers
5. Computer Networking	Kluwer Academic Publishers

The Partnership Sub-Series incorporates activities undertaken in collaboration with NATO's Cooperation Partners, the countries of the CIS and Central and Eastern Europe, in Priority Areas of concern to those countries.

NATO-PCO DATABASE

The electronic index to the NATO ASI Series provides full bibliographical references (with keywords and/or abstracts) to about 50 000 contributions from international scientists published in all sections of the NATO ASI Series. Access to the NATO-PCO DATABASE is possible via a CD-ROM "NATO Science & Technology Disk" with user-friendly retrieval software in English, French and German (© WTV GmbH and DATAWARE Technologies Inc. 1992).

The CD-ROM can be ordered through any member of the Board of Publishers or through NATO-PCO, Overijse, Belgium.

Series I: Global Environmental Change, Vol. 55

Springer
*Berlin
Heidelberg
New York
Barcelona
Budapest
Hong Kong
London
Milan
Paris
Santa Clara
Singapore
Tokyo*

Modelling Soil Erosion by Water

Edited by

John Boardman
David Favis-Mortlock

University of Oxford
Environmental Change Unit
5 South Parks Road
OX1 3UB Oxford, U.K.

With 122 Figures and 83 Tables

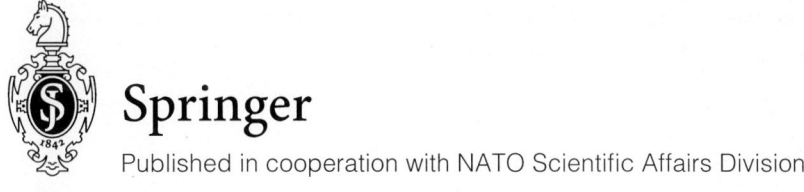

Published in cooperation with NATO Scientific Affairs Division

Proceedings of the NATO Advanced Research Workshop "Global Change: Modelling Soil Erosion by Water", held at the University of Oxford, September 11–14, 1995

Cataloging-in-Publication Data applied for

Die Deutsche Bibliothek - CIP-Einheitsaufnahme

Modelling soil erosion by water : [proceedings of the NATO Advanced Research Workshop "Global Change: Modelling Soil Erosion by Water ", held at the University of Oxford, September 11 - 14, 1995] / ed. by John Boardman ; David Favis-Mortlock. Publ. in cooperation with NATO Scientific Affairs Division. - Berlin ; Heidelberg ; New York ; Barcelona ; Budapest ; Hong Kong ; London ; Milan ; Paris ; Santa Clara ; Singapore ; Tokyo : Springer, 1998
  (NATO ASI series : Ser. I, Global environmental change ; Vol. 55)
  ISBN 3-540-64034-7

ISSN 1431-7125
ISBN 3-540-64034-7 Springer-Verlag Berlin Heidelberg New York

This work is subject to copyright. All rights are reserved, whether the whole or part of the material is concerned, specifically the rights of translation, reprinting, reuse of illustrations, recitation, broadcasting, reproduction on microfilm or in any other way, and storage in data banks. Duplication of this publication or parts thereof is permitted only under the provisions of the German Copyright Law of September 9, 1965, in its current version, and permission for use must always be obtained from Springer-Verlag. Violations are liable for prosecution under the German Copyright Law.

© Springer-Verlag Berlin Heidelberg 1998
Printed in Germany

Typesetting: Camera ready by the authors/editors
Printed on acid-free paper
SPIN: 10514491   31/3137 - 5 4 3 2 1 0

CONTENTS

SECTION 1. BACKGROUND TO THE MODEL EVALUATION
1. Modelling Soil Erosion by Water
   (J. Boardman and D.T. Favis-Mortlock) 3
2. Towards an Improved Predictive Capability for Soil Erosion under
   Global Change (C. Valentin) 7
3. Modelling Soil Erosion in Real Landscapes: a Western European
   Perspective (J. Boardman) 17

SECTION 2. MODEL EVALUATION WITH COMMON DATASETS
4. Evaluation of Plot Runoff and Erosion Forecasts using the CSEP and
   MEDRUSH Models (M.J. Kirkby) 33
5. Evaluation of the Water Erosion Prediction Project (WEPP) Model
   for Hillslopes (M.A. Nearing and A.D. Nicks) 43
6. GLEAMS Model Evaluation — Hydrology and Erosion Components
   (A.D. Nicks) 55
7. EUROSEM: an Evaluation with Single Event Data from the C5
   Watershed, Oklahoma, USA (J.N. Quinton and R.P.C. Morgan) 65
8. Comparison of Simulated and Observed Runoff and Soil Loss on
   Three Small United States Watersheds (T.S. Ramanarayanan,
   M.V. Padmanabhan, G.N. Gajanan and J.R. Williams) 75
9. Validation of Field-Scale Soil Erosion Models using Common Datasets
   (D.T. Favis-Mortlock) 89

SECTION 3. MODEL EVALUATION WITH USER-SUPPLIED DATASETS
10. Predicting Runoff in Semiarid Woodlands: Evaluation of the WEPP
    Model (B.P. Wilcox and J.R. Simanton) 131
11. Evaluation of Field-Scale Erosion Models on the UK South Downs
    (D.T. Favis-Mortlock) 141

SECTION 4. MODELLING ISSUES
12. Modelling Across Scales: the MEDALUS Family of Models
    (M.J. Kirkby) 161
13. Problems Regarding the Use of Soil Erosion Models
    (R.P. Rudra, W.T. Dickinson and G.J. Wall) 175
14. Cross-Scale Aspects of EPA Erosion Studies
    (J.J. Lee) 191
15. Scale Issues and a Scale Transfer Method for Erosion Modelling
    (D. King, D.M. Fox, Y. Le Bissonnais and V. Danneels) 201
16. Infiltration for Soil Erosion Models: Some Temporal and Spatial
    Complications (T.P. Burt) 213
17. Saturation Overland Flow on Loess Soils in the Netherlands
    (F.J.P.M. Kwaad) 225
18. Incorporating Crusting Processes in Erosion Models
    (Y. Le Bissonnais, D.M. Fox and L.-M. Bresson) 237

19. The Role of Soil Aggregates in Soil Erosion Processes
    (D. Torri, R. Ciampalini and P.A. Gil) 247
20. Process-Based Approaches to Modelling Soil Erosion
    (C.W. Rose and W.L. Hogarth) 259
21. Sensitivity of Sediment-Transport Equations to Errors in Hydraulic
    Models of Overland Flow (J. Wainwright and A.J. Parsons) 271
22. Gully Erosion: Importance and Model Implications
    (J.W.A. Poesen, K. Vandaele and B. van Wesemael) 285
23. Field Data and Erosion Models
    (R. Evans) 313
24. Effects of Agricultural Land Use on Spatial and Temporal Distribution
    of Soil Erosion in Small Catchments: Implications for Modelling
    (A.-V. Auzet, J. Boiffin, B. Ludwig and J. Guérif) 329
25. Sensitivity of the Model LISEM to Variables Related to Agriculture
    (V. Jetten, A.P.J. de Roo and J. Guérif) 339
26. Applying GIS to Catchment-Scale Soil Erosion Modelling
    (R. McDonnell) 351
27. Snowmelt and Frozen Soils in Simulation Models
    (P.F. Botterweg) 365
28. The Use of USLE Components in Models
    (A.D. Nicks) 377

SECTION 5. MODEL DESCRIPTIONS
29. The EUROSEM Model (R.P.C. Morgan, J.N. Quinton, R.E. Smith,
    G. Govers, J.W.A. Poesen, G. Chisci and D. Torri) 389
30. Griffith University Erosion System Template (GUEST)
    (C.W. Rose, K.J. Coughlan and B. Fentie) 399
31. A Continuous Catchment-Scale Erosion Model
    (J.G. Arnold and R. Srinivasan) 413
32. LISEM: a Physically-Based Hydrologic and Soil Erosion Catchment
    Model (A.P.J. de Roo, V. Jetten, C. Wesseling and C. Ritsema) 429
33. APEX: a New Tool for Predicting the Effects of Climate and
    CO_2 Changes on Erosion and Water Quality (J.R. Williams, J.G. Arnold,
    R. Srinivasan and T.S. Ramanarayanan) 441
34. A Dynamic Model of Gully Erosion
    (A. Sidorchuk) 451
35. Alternative Approaches to Soil Erosion Prediction and Conservation
    Using Expert Systems and Neural Networks (T.M. Harris and J. Boardman) 461

SECTION 6. MODEL APPLICATIONS: ACTUAL AND POTENTIAL
36. Soil Erosion Modelling in Hungary
    (D. Lóczy, A. Kertész and T. Huszár) 481
37. Definition and Mapping of Desertification Units in Mediterranean Areas
    Under Rainfed Cereals (C.S. Kosmas and N.G. Danalatos) 491
38. Hydrological and Erosion Processes in the Research Catchments of
    Vallcebre (Pyrenees) (F. Gallart, J. Latron and D. Regüés) 503

SECTION 7. CONCLUSIONS
39. Modelling Soil Erosion by Water: Some Conclusions
 (J. Boardman and D.T. Favis-Mortlock) 515

APPENDICES
A. List of Acronyms Used 521
B. Participants at 'Global Change: Modelling Soil Erosion by Water' 523

Index 527

Attendees at the NATO Advanced Research Workshop 'Global Change: Modelling Soil Erosion by Water', 11-14th September 1995, University of Oxford, UK

Front row (left to right): Favis-Mortlock, Evans, Skidmore, Boardman, Valentin, Guérif, Ingram. Second row: Jansky, Bilge, Lee, Auzet, Kwaad, Morgan, Harris, Walling, Parsons, Wainwright, de Roo. Third row: Tomás, Kirkby, H. Fox, Moore, Botterweg, Quinton. Fourth row: D. Fox, King, Srinivasan, Arnold, Kosmas. Fifth row: Baade, Lóczy, Govers, Stromberg, Poesen, Gallart, Nicks. Sixth row: Wilcox, Rose, Imeson, Jetten, Nearing, Burt, Coutinho, Torri, Sidorchuk, Mirtskhulava. Missing: Le Bissonnais, Fix, Gregory, McDonnell, Tinker.

SECTION 1. BACKGROUND
TO THE MODEL EVALUATION

1. MODELLING SOIL EROSION BY WATER

John Boardman[1] and David Favis-Mortlock[2]

[1] School of Geography and Environmental Change Unit
Mansfield Road
University of Oxford
Oxford OX1 3TB
UK

[2] Environmental Change Unit
University of Oxford
5 South Parks Road
Oxford OX1 3UB
UK

Introduction

This volume is the Proceedings of the NATO Advanced Research Workshop 'Global Change: Modelling Soil Erosion by Water', which was held on 11-14th September 1995, at the University of Oxford, UK. The meeting was also one of a series organised by the IGBP-GCTE[1] Soil Erosion Network, which is a component of GCTE's Land Degradation Task (3.3.2) (Ingram *et al.*, 1996; Valentin, this volume).

One aim of the GCTE Soil Erosion Network is to evaluate the suitability of existing soil erosion models for predicting the possible impacts of global change upon soil erosion. Due to the wide range of erosion models currently in use or under development, it was decided to evaluate models in the following sequence (Favis-Mortlock *et al.*, 1996):

- field-scale water erosion models
- catchment[2]-scale water erosion models
- wind erosion models
- models with a landscape-scale and larger focus.

As part of this strategy, the first stage of the GCTE validation of field-scale erosion models was carried out at the Oxford NATO-ARW.

---

[1] A list of Acronyms forms Appendix A.
[2] Throughout, the European 'catchment' is used in preference to the US 'watershed'. Similarly, European spellings (e.g. 'modelling') are used, except in references. The spelling 'talweg' is used in preference to 'thalweg'. 'evaluation' and 'validation' are considered to be equivalent throughout.

The structure of this book

Following the introductory first section, the second and third of this book's seven sections deal with the results of the GCTE model evaluation exercise. It was recognised at an early stage that the use of common — i.e. shared — datasets offers a systematic and objective approach to model comparison. A 'minimum dataset' (Ingram, 1994) was drawn up, and this was used as the basis for dataset selection (Favis-Mortlock, this volume a). However, there are some problems with exclusive reliance on the common-dataset approach: the datasets may not include all variables required by all models; also the model users are likely to be unfamiliar with the datasets, and so experience some difficulty in choosing the best values of variables for their model (cf. Quinton and Morgan, this volume). Accordingly, a parallel approach was adopted for the GCTE exercise: models were run with both common datasets and user-supplied data (Table 1).

Table 1. Design of the GCTE field-scale erosion model evaluation (adapted from Favis-Mortlock *et al.*, 1996, with permission). Ch = Chapter in this book

Model	Originator	Ch.	GCTE field-scale model evaluation			
			With common datasets		With user-supplied datasets	
			Model user	Ch.	Model user	Ch
GLEAMS	USDA	-	Arlin Nicks, USDA-ARS	6, 9	David Favis-Mortlock, University of Oxford, UK	11
EUROSEM	Silsoe College, UK	29	John Quinton, Silsoe College, UK	7, 9	-	-
GUEST	Griffith University, Australia	30	Calvin Rose, Griffith University, Australia	-	-	-
EPIC	USDA	-	Jimmy Williams et al., USDA-ARS	8, 9	David Favis-Mortlock, University of Oxford, UK	11
CSEP	MEDALUS	12	Mike Kirkby, University of Leeds, UK	4, 9	-	-
MEDRUSH	MEDALUS	12	Mike Kirkby, University of Leeds, UK	4, 9	-	-
WEPP	USDA	-	Mark Nearing, NSERL, USDA-ARS	5, 9	(1) Brad Wilcox, Los Alamos National Laboratory, US	10
					(2) David Favis-Mortlock, University of Oxford, UK	11

Erosion modelling is not static: while evaluating present-day models for their usefulness in global change modelling, it is of equal importance to look to future models. The chapters of the fourth section of this book ('Modelling Issues') consider weaknesses or omissions in current modelling approaches, and suggest ways in which these may be remedied. A range of themes, central to the future development of erosion models, are covered. Fundamental issues of scale form the focus of the first group of chapters; hydrology (infiltration in particular) is a second theme, which then moves on via crusting to sediment detachment and transport. A move to more field-scale features follows, with a consideration of gully erosion and its implications for modelling. The field-scale perspective continues, with discussions of the implications of agricultural activity on erosion at this scale. This section ends with presentations of the role of GIS; approaches to modelling snowmelt (essential for the evaluation of global change impacts in higher latitudes); and — since it appears that the USLE's heritage will not die off in future approaches to erosion modelling, but instead will probably just fade away — the foundations and subsequent development of the Universal Soil Loss Equation.

The book's fifth section presents descriptions of specific erosion models or modelling approaches. These include both field-scale models and models or methodologies with a larger focus. The sixth section presents potential or actual applications of models in areas where modelling is new, or has not yet been applied. A seventh section draws some conclusions.

Acknowledgements

We wish to thank all who refereed chapters; in addition, our thanks to Dr Tony Guerra for careful and thorough proof-reading, and to Anna Winton for typing. We would like to acknowledge funding from the NATO Office for Scientific Affairs for the NATO Advanced Research Workshop 'Global Change: Modelling Soil Erosion by Water'. This book is a contribution to the Soil Erosion Network of the GCTE, which is a Core Research Project of the International Geosphere-Biosphere Programme.

Finally, as well as the pleasure which the preparation of this book has brought to all involved, there has however been good reason for sadness: two of the contributors, Jeff Lee and Arlin Nicks, died while the book was being prepared. We hope that this book serves in some small way as a memorial to them.

References

Favis-Mortlock, D.T., Quinton, J.N. and Dickinson, W.T. (1996). The GCTE validation of soil erosion models for global change studies. *Journal of Soil and Water Conservation* **51**(5), 397-403.

Ingram, J.S.I. (ed.) (1994). *Report of the GCTE Workshop 'Soil Erosion under Global Change'*, Paris, France, 29-31 March 1994, GCTE Focus 3 Associate Office, Oxford, UK. 22 pp.

Ingram, J.S.I., Lee, J.J. and Valentin, C. (1996). The GCTE Soil Erosion Network: a multi-participatory research program. *Journal of Soil and Water Conservation* **51**(5), 377-380.

2. TOWARDS AN IMPROVED PREDICTIVE CAPABILITY FOR SOIL EROSION UNDER GLOBAL CHANGE

Christian Valentin

ORSTOM
B.P. 11416
Niamey
Niger

Abstract

Global change encompasses not only changes in atmospheric composition and climate but also change in land use. The interaction of these 'drivers' is likely to have a major influence on the fate of terrestrial ecosystems over the next few decades. One aspect intimately associated with such change is soil erosion. The Soil Erosion Network launched by the Global Change and Terrestrial Ecosystems (GCTE) Core Project of the International Geosphere-Biosphere Programme (IGBP) aims at an improved predictive capacity of soil erosion under global change. Its objectives are: (i) to design and undertake experiments and monitoring programmes to provide a predictive understanding of the impacts of changes in climate and land-use on soil erosion; (ii) to refine and adapt current soil erosion models for use in global change studies from plot to regional scales. Three major themes have been identified: (i) linking erosion processes across temporal and spatial scales; (ii) identification and quantification of key thresholds for soil erosion; (iii) soil erosion feedbacks to Global Change. A planning workshop for collaborative experiments in West Africa has taken place and a workshop of model comparison at the catchment scale is planned.

Introduction

Although soil is an essential natural resource it is being degraded at an unprecedented scale, both in rate and geographical extent. Soil degradation ranges from soil loss via erosion, through chemical depletion, to solute accumulation like salinisation. These can be locally very severe, but soil erosion is a widespread problem and has been identified as the major type of human-induced land degradation from a global perspective (Oldeman *et al.*, 1991). Soil erosion not only hinders sustainable land management, but it can also cause off-site environmental problems such as siltation of lakes and reservoirs (Ottichilo *et al.*, 1991) and increased dust in the air over long distances (Prospero and Nees, 1977; Mann, 1987).

Global change is expected to exacerbate current problems in influencing soil erosion, both directly through changes in wind and rainfall regimes, and indirectly through changes in soil moisture, soil organic matter, soil fauna, soil structure and vegetation cover. However, global change encompasses far more than just climate change alone. It includes changes in atmospheric composition, such as the concentration of CO_2 and other greenhouse gases which have direct impacts on vegetation, and also change in land use, as driven by demographic, economic, technological and social pressures. Over the next few decades, this human dimension of global change will probably have a more profound influence on the fate of terrestrial ecosystems than will changes in climate and atmospheric composition alone. With regard to soil erosion by wind and water, the effects of land use change are more immediate and extensive (Valentin *et al.*, 1994).

This potential for global change to modify erosion hazard requires a relevant planning response. In order to improve predictive capability for soil erosion under global change, the Global Change and Terrestrial Ecosystems Core Project (Steffen *et al.*, 1992) of the International Geosphere-Biosphere Programme (IGBP) has established a specific international research Task including the GCTE Soil Erosion Network (Ingram and Gregory, 1996). The aims of this paper are to present: (i) the background and the objectives, (ii) the main components, (iii) the implementation plan of this network.

The GCTE Soil Erosion Network

To gain an improved predictive capacity of soil erosion under global change, an international collaboration between research teams and disciplines is crucial. The Soil Erosion Network of the GCTE Soil Degradation Task has been established to provide the necessary international co-ordination and linkages. It was launched at a workshop held in Paris, 29-31 March 1994 (Ingram, 1994), where it was agreed that the Network will rely heavily on collaboration with many national and international programmes or institutions addressing some GCTE aspects, e.g. the Environmental Protection Agency in the USA (Lee *et al.*, 1993) and the IGBP Transects (Koch *et al.*, 1995) among others.

In Paris, the participants agreed on dual objectives:
- To design and undertake experiments and monitoring programmes to provide a predictive understanding of the impacts of changes in climate and land-use on soil erosion.
- To refine and adapt current soil erosion models for use in global change studies from plot to regional scales.

It was emphasised that both modellers and experimentalists will provide equal input to the network. It was also agreed that the network should concentrate on the humid tropics and semi-arid regions, not only because they are likely to be very sensitive to global climate change, but also because these are areas of actual and potential food security problems related to erosion. Following this meeting, questionnaires were designed for the purpose of obtaining 'metadata' (data describing a piece of research, as distinct from the data/model code itself) on models, monitoring and experimental work. An invitation to apply to join the Network was mailed widely. A first set of 32 applications was received in the GCTE office in Oxford.

The criteria for formal membership acceptance were discussed during the workshop in February 1995 in Corvallis, Oregon, sponsored by the US EPA. The submitted work must meet some general requirements: it must address the stated objectives and be of high scientific quality. If an application is rejected, it must be for good scientific reasons unless there is an administrative reason for rejecting it, such as an inability to agree to the GCTE Data and Model Sharing Policy. For experiments, it was decided that they must be multi-spatial and/or multi-temporal scale. This requirement might be less strict for work conducted in tropical regions or any region where the availability of suitable data remains scarce. It was decided also that the submitted model must have been validated and encoded. Monitoring projects must be over two years and preferably multi-scale. Based on the above criteria four applications were rejected, nine soil erosion models were included in the network, complemented by ten experimental datasets and nine monitoring projects. They have been endorsed by the Scientific Steering Committee of GCTE as formal contributions to the GCTE Core Research Programme. Further applications have been sent to Oxford since then, and similarly screened by the Soil Erosion Working Group during the Oxford NATO workshop.

Network components
The network comprises three closely interrelated components.

1. Long term monitoring
Infrequent climatic events, such as heavy storms, typhoons, etc. can trigger severe erosion, that would be unpredictable from short-term records. Long-term erosion monitoring is therefore essential to observe possible transient and non-equilibrium responses to climatic and

land use changes. There are still relatively few studies which investigate current erosion rates in the tropics, especially over the long term.

2. Experimentation

Most experiments have been carried out using land cover and tillage as key variables. The network will build on these base-line data for land-use change and will also examine the impact of other global change driving forces such as changed variance and mean for key climate parameters. It will also investigate the effect of soil organic matter upon soil erosion, via its impact on aggregation and aggregate stability, due to possible effects of elevated CO_2.

3. Modelling

The broad range of data covering the space-time domain flowing from the many monitoring and experimental programmes world-wide will be used to calibrate and validate soil erosion simulation models. It must be remembered that simulation models provide an aid in identifying gaps in understanding, and provide a unique forum for bringing together disciplines. Models can also be used for policy formulation. For these two purposes, science development and decision support, we need models that are both highly sensitive to global change and sufficiently flexible to be relevant under the widest range of conditions.

While the main thrust of the GCTE Erosion Network will be to establish new research programmes, a considerable body of information concerning soil erosion in the last few decades has already been collected, much of it stored in substantial archives of air-photos and satellite images. These archives could be further used for developing, on a retrospective basis, predictive models. Another relatively new technique uses ^{137}Cs measurements to provide information on erosion rates since the period of A-bomb testing. Additional information on longer time-span erosion, related to climatic changes, could be derived from the numerous data on the Late Quaternary period. Lake sediments preserve historical records of material export from watersheds and can be analysed using palaeolimnological methods. Pollen, charcoal, chemical, physical, magnetic mineral and radiocarbon dating (^{14}C) along with archaeological records, and local documentary evidence could be interpreted in the light of models of climatic change. However, helpful as they are, the use of palaeoclimatic analogues should be pursued with care, as it is not yet clear whether they can by used as surrogates for contemporary or future climates.

The Network implementation plan

Core Research Projects

Four Core Research Projects (CRP) were developed at the 1995 Corvallis workshop to meet the objectives of the Task. In the four CRPs the experimental and modelling aspects will be closely integrated, leading to the development of the models necessary for predicting erosion from plot to regional scales. The four Core Research Projects are:

- CRP1: GCTE Soil Erosion Network
- CRP2: Linking erosion processes across temporal and spatial scales
- CRP3: Identification and quantification of key thresholds for soil resource erosion
- CRP4: Soil erosion feedbacks to global change

The first CRP aims at co-ordinating the activities of the three others which are strongly interrelated. Whilst CRP1 and CRP2 constitute the bulk of the Task activities in terms of genuine soil erosion research, CRP3 and CRP4 are more product-oriented, the former towards the issue of sustainable development, the latter to the IGBP programmes dealing with atmospheric and climate change.

CRP1 - Soil Erosion Network

Objective

- To develop research co-ordination and enhance communication to identify the best experimental and monitoring datasets and the most robust models for global change studies.

Specific questions

- Which erosion models are most robust for global change studies? Given the wide variety of available models, a phased strategy is essential. The approach is to group them by the main erosive agent (water or wind) and by spatial scale categories, field scale and catchment scale (<10 km²). Tested models should concurrently be highly sensitive to climatic and land use changes and sufficiently flexible to be relevant under the widest range of conditions. A first workshop on Global Change: Modelling Soil Erosion by Water, was held in Oxford, UK, 11-14 September 1995 to compare soil erosion models at the field-scale and to plan model comparison at the catchment scale (this volume).

- Which experimental and monitored data sets will best help in model development? Experimental studies test a clearly defined hypothesis, whilst monitoring studies involve the measurement of baseline variables (e.g. rainfall and wind intensity, soil moisture, surface roughness, soil erosion rates, etc.) over several years with or without a clear initial hypothesis. Both experimental and monitoring data are essential to calibrate, validate and improve soil erosion models. Common datasets for model comparison should represent a wide range of present climatic and land use conditions, ideally for temperate, semi-arid and humid tropical sites.

CRP2 - Linking erosion processes across temporal and spatial scales

Objective
- To make more realistic predictions of the impact of land use and climate change on soil erosion by monitoring, experimenting and modelling soil erosion processes across temporal and spatial scales.

Specific questions
- What are the most significant erosion processes for the various hierarchical spatial and temporal levels in various environments (intensively cultivated, rangeland, semi-natural, and natural land) and what are their spatial and temporal boundary conditions?
- How do soil erosion processes at different scales interact with each other, with other processes of soil degradation and with hydrological processes (such as infiltration and runoff generation)?
- What are the interactions between patterns and erosion processes operating at various temporal and spatial scales? In other words, how do these patterns control soil erosion processes at various scales? How may environmental change lead to loss of patterns (including biodiversity loss) and, therefore, affect water and sediment movement?

The first two questions have been addressed in the Oxford workshop (Lee, 1998; Kirkby, 1998). The third question is crucial because change of land use is often associated with change in land patterns. This question was partly addressed in a symposium held in Paris, in April 1996, on banding vegetation patterning.

CRP3 - Identification and quantification of key thresholds for soil erosion

Objectives

- To identify and understand critical thresholds in the landscape and in the soil profile, which lead to irreversible changes in the rate or style of soil degradation at a site.
- To develop indicators which forecast irreversible change, using an appropriate combination of available data, remotely sensed imagery and ground monitoring programmes.

Specific questions

- At what level and when is loss of a given function critical?
- What are the processes and properties responsible for soil resilience, and what are the predicative indicators? Since erosion processes commonly only start beyond a certain threshold set of conditions, emphasis will be put on the determination of such thresholds, on reversibility of processes, and on soil resilience. This point is crucial in the context of sustainable land management.

CRP4 - Soil erosion feedbacks to global change

Objective

- To provide data and understanding for the programmes dealing with atmosphere and climate change

Specific question

- What are the direct and indirect atmospheric and climatic feedbacks from erosion, and how do they operate?

CRP4 meets the second objective of GCTE on the feedbacks to the atmosphere. Erosion is affected by, and feeds back to: (i) climate change, and (ii) land use change, but within this Task only the first feedback is addressed. Soil erosion feeds back to climate change primarily in two ways: (i) through effects on the concentration of greenhouse gases (CO_2, CH_4, N_2O, etc.) in the atmosphere; (ii) through changes in surface albedo. This can occur directly through crusting of the soil surface or through removal of upper soil horizons and exposure of lower horizons with different albedos. Arid and semi-arid regions are recognised as sources for dust that is transported by the atmosphere. Satellite images of dust storms demonstrate that large wind erosion events produce an atmospheric aerosol that reflects more of the incoming solar radiation. Warming generally occurs in the dust layer and cooling beneath, reducing the formation of clouds and changing their spatial distribution, thus affecting the climate (Bergametti, 1992).

Future action

1996

SALT erosion workshop

A planning workshop for collaborative experimental and monitoring programmes in West Africa was organised for late January 1996. It included a field trip to the three main semi-arid sites from Senegal to Niger, along a decreasing gradient of human pressure of the GCTE SALT transect (SAvannahs in the Long Term). West Africa has been experiencing a definite downward trend in rainfall amount for twenty-five years, whilst the near doubling of population over the same period has severely exacerbated erosion problems. Moreover, West Africa constitutes a large continuous mass, with a fairly homogeneous terrain, distributed across a strong and clear moisture gradient. Such conditions seem therefore most propitious to examine the possible impacts of global change on soil erosion and its potential feedbacks on climate in tropical regions. The SALT project, part of the GCTE Core Research Programme, is based on process studies conducted on eight major sites in Côte d'Ivoire, Burkina Faso, Mali, Niger, and Senegal, and on a number of secondary sites, spanning a 1,000-km long climatic transect from Côte d'Ivoire to Mali, and a perpendicular human density transect of nearly 2,000 km from Senegal to Niger (Menaut *et al.*, 1993). Remote-sensing data are used to extrapolate the results of the site studies. The processes studied include primary productivity, organic matter and nutrient cycling; soil/vegetation/atmosphere interactions; soil surface structure and erosion processes; vegetation structure and dynamics; ecosystem response to disturbances (e.g. fire, grazing, and cultivation); and changes in the hydrology of small catchments. As a follow-up to the workshop, a proposal of concerted action is being drafted to be submitted to the European Commission.

1997

Catchment-scale erosion model sensitivity and comparison workshop

This meeting will be held in Utrecht in April.

Wind erosion model sensitivity and comparison workshop

This is likely to take place before or during the 50th anniversary of Agricultural Research wind erosion programme in Kansas (USA), in June.

Network experimental, monitoring and modelling review and synthesis workshop

This will be held in the arid south-eastern part of Spain, in September.

1998

Specific symposium of GCTE Activity on Soils and Global Change

To be held during the International Soil Science Society Congress in Montpellier (France), August.

Conclusions

Because soil erosion results from a complex web of interactions, interdisciplinarity is essential to achieve the objectives of the GCTE-Soil Degradation Task. Close links should be developed therefore with not only other branches of physical geography, soil science, and hydrology, but also with landscape ecology, agronomy and atmosphere sciences. Moreover, an excellent integration of long-term monitoring, experimental and modelling programmes at different scales both in time and space needs to be achieved. Links must also be strengthened with other Tasks of GCTE, most notably with that on Global Change Impact on Soil Organic Matter (Task 3.3.1).

Both water and wind erosion are commonly accelerated by land-use change (especially the clearing of vegetation cover), presently the main manifestation of global change. Therefore, strong links must be made with the International Human Dimensions Programme (IHDP), notably with the joint IGBP-IHDP Core Project 'Land Use and Cover Change' (LUCC). Collaboration with PAGES on Past Global Changes will be extended to maximise the benefit of palaeodata. Closer links should be explored with IGAC on atmospheric chemistry and other programmes involved in climate change simulation (BAHC, GEWEX).

Acknowledgements

This paper is a contribution to the Soil Erosion Network of GCTE, a Core Research Project of the IGBP.

References

Bergametti, G. (1992). Atmospheric cycle of desert dust. *Encyclopaedia of Earth Systems Science* **1**, 171-182.

Ingram, J.S.I. (ed.) (1994). *Report of the GCTE Workshop 'Soil Erosion under Global Change'*, Paris, 29-31 March 1994. GCTE, Oxford, 22 pp.

Ingram, J.S.I. and Gregory, P.J. (1996). *Effects of Global Change on Soils: Implementation Plan*. GCTE Report **12**, GCTE Focus 3 Office, Wallingford. 56 pp.

Kirkby, M.J. (1998). Modelling across scales: the MEDALUS family of models. In, Boardman, J. and Favis-Mortlock, D.T. (eds), *Modelling Soil Erosion by Water*, Springer-Verlag NATO-ASI Global Change Series, Heidelberg.

Koch, G.W., Scholes, R.J., Steffen, W.L., Vitousek, P.M. and Walker, B.H. (eds) (1995). *The IGBP Terrestrial Transects: Science Plan*. IGBP Report **36**, Stockholm.

Lee, J.J., Phillips, D.L. and Liu R. (1993). The effect of trends in tillage practices on erosion and carbon content of soils in the US corn belt. *Water, Air, and Soil Pollution* **70**, 389-401.

Lee, J.J. (1998). Cross-scale aspects of EPA erosion studies. In, Boardman, J. and Favis-Mortlock, D.T. (eds), *Modelling Soil Erosion by Water*, Springer-Verlag NATO-ASI Global Change Series, Heidelberg.

Mann, R. (1987). Development and the Sahel disaster: the case of Gambia. *The Ecologist* **17**, 84-90.

Menaut, J.C., Saint, G. and Valentin, C. (1993). SALT. Les Savanes à Long Terme. Analyse de la dynamique des savanes d'Afrique de l'Ouest : mécanismes sous-jacents et spatialisation des processus. CNRS, Paris, *Lettre du Programme Environnement*, **10**, 34-36.

Oldeman, L.R., Hakkeling, R.T.A. and Sombroek, W.G. (1991). *World Map of the Status of Human-Induced Soil Degradation: an Explanatory Note*. ISRIC, Wageningen, UNEP, Nairobi, 34p.

Ottichilo, W. K.; Kinuthia, J. H.; Ratego, P. O. and Nasubo, G. (1991). *Weathering the Storm: Climate Change and Investment in Kenya*. Regional Centre for Services in Surveying, Mapping and Remote Sensing, Nairobi, Kenya. ACTS Press; Stockholm, Sweden; Stockholm Environment Institute. ACTS Environmental Policy Series **3**, 90 pp.

Prospero, J.M. and Nees, R.T. (1977). Dust concentration in the atmosphere of the equatorial North Atlantic: possible relationship to the Sahelian Drought. *Science*, **196**, 1196-1198.

Steffen, W.L., Walker, B.H., Ingram, J.S.I. and Koch, G.W. (eds) (1992). *Global Change and Terrestrial Ecosystems: The Operational Plan*. IGBP Report **21**, Stockholm. 95 pp.

Valentin, C., Collinet, J. and Albergel, J. (1994). Assessing erosion in West African savannas under global change: overview and research needs. *XVth International Congress of Soil Science*, Acapulco, Mexico, Vol.7a, pp. 253-274.

3. MODELLING SOIL EROSION IN REAL LANDSCAPES: A WESTERN EUROPEAN PERSPECTIVE

John Boardman

School of Geography and Environmental Change Unit
Mansfield Road
University of Oxford
Oxford OX1 3TB
UK

Abstract

The challenge to soil erosion modellers is to take into account information from a variety of sources, the laboratory, the experimental plot and the field and to integrate it into predictive models that take account of processes in real landscapes at the correct spatial and temporal scales. While substantial progress has been made in recent years many issues remain unresolved.

More information is required about present day rates: are they high enough to be of concern? Is the impact of erosion on the sustainability of agricultural production an issue, and if so in what areas? Monitoring offers some answers to these questions.

Little is known about erosion and its role in the history of our soils. This raises questions about major controls on erosional processes: land use and farming practice, soil character, and climate. Research is now concentrating on several previously neglected processes which present a formidable challenge to modellers e.g. crusting, stoniness and ephemeral gullying. Much research is experimental and small scale in character: integrating it into field and catchment-scale models is not easy.

The old geomorphological question of the role of extreme events is still of relevance. However, a knowledge of thresholds, and the sensitivity of arable systems to high frequency, low magnitude events is of equal importance. The threat (or promise?) of climate change raises questions for erosion modellers; not least because the most profound changes may be in terms of land use as new crops move into new areas. Finally, we are apt to forget that the reason why erosion occurs is largely due to economic, social and political factors that our models tend to neglect. This is understandably due to the difficulty of including such factors; however, their inclusion presents a challenge for future generations of models.

Introduction

Two major aims for erosion studies may be identified: to control erosion and preserve the soil resource, and to contribute to the understanding of the long-term evolution of landscape. To achieve these aims it is necessary to have some knowledge of processes and to be able to predict the rate, extent and frequency of erosion in the field.

It is important to emphasise these aims in a volume largely devoted to the development and testing of soil erosion models. The models *contribute* to much wider aims: we are some way from understanding, let alone replicating, the infinite complexity of real landscapes.

To emphasise this point: it is easy to agree to the assertion that we know enough about erosion and the challenge is to apply that knowledge in real (and difficult) social, economic and political contexts. Indeed, in general terms, the amount of technical knowledge is impressive. But it is only complete and satisfactory in certain well ordered and rather simplified environments, such as the laboratory or the experimental plot. On a topographically complex hillslope, with typical soil variability, a growing crop, a changing microtopography produced by tillage and varying rainfall, we have difficulty predicting even to within an order of magnitude, a rate of erosion. Therefore, some degree of humility is appropriate and advances in modelling are desperately needed to predict erosion on real landscapes.

The aim of this paper is to take a broad view of the main issues in erosion modelling, from a western European perspective.

The quality of spatial data

An obvious preliminary question to ask is do we know where erosion is occurring or is likely to occur? The maps available are at a regional scale and are able only to identify broad areas where the risk of erosion is high, e.g. the CORINE project maps (Commission of the European Communities, 1992); the Soil Erosion Map of Western Europe (De Ploey, 1989a); and the National Soil Map of England and Wales (Mackney *et al.*, 1983). The latter map is of most value because soil associations at risk of erosion are mapped at a scale of 1:250 000. However, none of the maps are at a farm or field scale and are therefore of very limited use in predicting erosion.

If we do not have maps that can be used for specific areas at a farm scale, do we at least know what rates of erosion are occurring on average, on fields, or on the scale of river catchments?

There have been dramatic claims for rates of erosion in Europe. Pimentel (1993 p. 2) for example suggests that soil loss rates in Europe range between 10 and 20 $t\,ha^{-1}yr^{-1}$, and suggests an average figure of about 17 t ha^{-1} yr^{-1} (Pimentel et al., 1995). Many published figures are based on either sediment yields in rivers or experimental plot estimates; both have severe disadvantages. Monitoring schemes suggest that figures such as those of Pimentel are inflated and a consensus around 1-5 t ha^{-1} yr^{-1} for eroding fields would seem to exist (Table 1). However, rates from monitoring schemes are not strictly comparable because of different methodologies; and the proportion of eroding fields in the landscape may not be stated.

Table 1. Some results from erosion monitoring schemes in western Europe

Country and monitored area	Period	Erosion rate and source
Central Belgium: 86 fields	1981-85	Mean = 3.6t ha^{-1} (Govers, 1991)
North-east Scotland	1984-86	Mean = 6.7 m^3 ha^{-1} Median = 2.5 m^3 ha^{-1} (Watson and Evans, 1991)
UK Soil Survey of England and Wales: 17 localities in England and Wales (c. 826km²)	1982-86	Mean = 2.3 m^3 ha^{-1} (Evans, 1995)
UK: South Downs	1982-91	Median = 0.5 - 5.0 m^3 ha^{-1} yr^{-1} (Boardman and Favis-Mortlock, 1993)
Southern Sweden: 90 km²	1986-88	Median = 0.8 t ha^{-1} yr^{-1} (Alstrom and Bergman-Akerman, 1992)
Northern France: 33 small catchments (3-95 ha)	1988-91	0 - 11.7 m^3 ha^{-1} yr^{-1} (Ludwig et al., 1995)

Even if, for the moment, we are unable to answer satisfactorily the questions raised above the issue of sustainability has still to be addressed. Are present day rates of erosion so high as to be of concern? Can we sustain agricultural production on soils of declining thickness, and for how long? Or perhaps can we at least identify areas of special concern? Using a very simple model, Morgan (1987) attempts to quantify the life of soils given different assumptions about erosion rates (Table 2).

rates over a decade (Table 3), and provides information on spatial variability with respect to high-magnitude storm events. It also may show that rates on individual fields or catchment areas can be well in excess of mean regional rates. For example, in 1987-88 in the monitored area of the South Downs, rates on 18 fields exceeded 10 m³ ha^{-1} compared to the mean rates on all eroded fields of 3.3 m³ ha^{-1} (Boardman, 1988). However, the short time period of most monitoring schemes means that low-frequency weather conditions are not sampled and case studies of such events are necessary, e.g. the storm with a return period in some locations in excess of one thousand years which affected southern England in May 1993 (Boardman *et al.*, 1996). When such events are sampled in short-term monitoring studies the results are dominated by the event!

Table 3. Erosion rates in the monitored area, South Downs, 1982-1991

Year	Median soil loss (m³ ha^{-1})	Total soil loss (m³)	Number of sites
82-83	1.7	1816	68
83-84	0.6	27	7
84-85	1.1	182	25
85-86	0.7	541	49
86-87	0.7	211	34
87-88	5.0	13529	97
88-89	0.5	2	1
89-90	1.4	940	51
90-91	2.3	1527	43
91-92	1.2	112	14

The concentration of such studies in UK is noteworthy (Table 4). This may result from a tradition of empiricism and the number of practising geomorphologists. There seem to be few such studies from other areas of western Europe.

There is a further reason why extreme events should be part of the concern of the modeller. In western Europe much of the interest in erosion has focused on off-site effects, particularly the flooding of property by runoff from agricultural land (Boardman *et al.*, 1994). The more serious cases have been associated with extreme meteorological events.

Table 4. Extreme events causing erosion on agricultural land in Britain

Day/period	Precipitation (mm)	Location	Reference
22.6.1941	75	Blaydon, Tyne valley	Morris (1942)
24-26.9.1976	83	Worfield, Shropshire	Reed (1979)
30.9 to 3.10.1976	90	Yorkshire Wolds	Foster (1978)
7.10.87	63	South Downs	Boardman (1988)
31.3 to 1.4.1992	110	Kelso, SE Scotland	Davidson and Harrison (1995)
	>50	Angus, Scotland	Kirkbride and Reeves (1993)
January 1993	e.g. 440	Strath Earn, Scotland	Davidson and Harrison (1995)
26.5.1993	>100	Oxfordshire and Berkshire	Boardman et al. (1996)

Finally, and in partial contradiction to the last point, enthusiasm for extreme events must not blind us to the fact that the thresholds at which runoff and rilling begin are low on bare arable land (Table 5). There are also many examples of serious erosion occurring as a result of relatively modest rainfall amounts where land use, soil and farming practice factors dominate (Evans and Nortcliff, 1978; Boardman, 1983; Frost and Speirs, 1984; Boardman and Spivey, 1987). It is also the case that frequent, low-magnitude runoff events associated with low rates of erosion are responsible for transfer of agricultural pollutants to watercourses, e.g. Alstrom and Bergman-Akerman (1992); Harrod (1994).

Table 5. Some suggested rainfall thresholds for rilling in Britain

>7.5 mm day^{-1}	Evans and Nortcliff (1978) based on Evans and Morgan (1974)
>1mm h^{-1} during rainfalls of 10 mm on compacted soils	Reed (1979)
KE>10 index based on >10 mm h^{-1} for >10 min	Morgan (1980)
Successive rainfalls in winter >20-25 mm	Evans (1980)
>10 mm in summer, >20 mm in 3 days in winter	Evans (1981)
15-20 mm rainfall events, especially when soil at field capacity	Speirs and Frost (1985)
30 mm in 2 days	Boardman (1990)
15 mm in 1 day with maximum intensity >4 mm h^{-1}	Chambers et al. (1992)

Climate change and erosion

There have been few model-based studies of the effects of climate change on erosion in western Europe (Boardman et al., 1990). Monitoring of present-day erosion indicates that year to year changes in mean regional rates can be as much as an order of magnitude (Table 3) but are generally much less (e.g. Evans, 1993). Considerable variation in rates between crop types (Evans, 1988) suggests that the main effect of climate change will be a change to more at-risk crops particularly if these are grown on silty and loamy soils. An example of this is the substitution of winter cereals by maize on sandy soils in southern England, which could lead to a trebling of erosion rates (Boardman and Favis-Mortlock, 1996).

The use of modelling for climate change/land use change impact studies is essential as there is no other practicable way of prediction. It is, however, incumbent on the modeller to show that reasonable results can be achieved for present-day situations before attempting to predict the future.

Beyond the physical landscape

Modellers are apt to forget that the reason why erosion occurs is rooted in socio-economic factors: crops at risk of erosion are grown on erodible soils because it is profitable to do so. Unless modellers are able to take such factors into account their explanations and predictions will be incomplete. The 'real landscape' is not farmed, or allowed to erode, by experimenters or modellers but by people with concerns about productivity and (hopefully) sustainability.

Conclusion

This short review has attempted to emphasise both the strengths and weaknesses of current western European knowledge and approaches as they affect the modelling of erosion. The need for integrated approaches encompassing the field and laboratory worker and the modeller is essential if progress is to be maintained.

Acknowledgement

I thank Dr David Favis-Mortlock and a referee for helpful comments on a draft of this paper which is a contribution to the Soil Erosion Network of GCTE, a Core Research Project of the IGBP.

References

Allen, M. J. (1992). Products of erosion and the Prehistoric land-use of the Wessex Chalk. In, Bell, M. and Boardman, J. (eds) *Past and Present Soil Erosion: Archaeological and Geographical Perspectives*, Oxbow Monograph 22, Oxbow Books, Oxford, 37-52.

Alstrom, K. and Bergman-Akerman, A. (1992). Contemporary soil erosion rates on arable land in southern Sweden. *Geografiska Annaler* **74A**(2-3), 101-108.

Auzet, A.V., Boiffin, J., Papy, F., Ludwig, B. and Maucorps, J. (1993). Rill erosion as a function of the characteristics of cultivated catchments in the north of France. *Catena* **20**, 41-62.

Bell, M. and Boardman, J. (1992). *Past and Present Soil Erosion: Archaeological and Geographical Perspectives*, Oxbow Monograph 22, Oxbow Books, Oxford.

Bell, M. and Walker, M.J.C. (1992). *Late Quaternary Environmental Change: Physical and Human Perspectives*, Longman.

Boardman, J. (1983). Soil erosion at Albourne, West Sussex, England. *Applied Geography* **3**, 317-329.

Boardman, J. (1988). Severe erosion on agricultural land in East Sussex, UK, October 1987. *Soil Technology* **1**, 333-348.

Boardman, J. (1990). Soil erosion on the South Downs: a review. In, Boardman, J., Foster, I.D.L. and Dearing, J.A. (eds) *Soil Erosion on Agricultural Land*, Wiley, Chichester, 87-105.

Boardman, J., Burt, T.P., Evans, R., Slattery, M.C. and Shuttleworth, H. (1996). Soil erosion and flooding as a result of a summer thunderstorm in Oxfordshire and Berkshire, May 1993. *Applied Geography* **16**(1), 21-34.

Boardman, J., Evans, R., Favis-Mortlock, D.T. and Harris, T.M. (1990). Climate change and soil erosion on agricultural land in England and Wales. *Land Degradation and Rehabilitation* **2**(2), 95-106.

Boardman, J. and Favis-Mortlock, D.T. (1993). Simple methods of characterizing erosive rainfall with reference to the South Downs, southern England. In, Wicherek, S. (ed.), *Farm Land Erosion: In Temperate Plains Environment and Hills*, Elsevier, 17-29.

Boardman, J. and Favis-Mortlock, D.T. (1996). Implications of climate change and land use change for soil erosion in the UK. *Paper presented at IAG Regional Conference, Hungary 1996*.

Boardman, J., Ligneau, L., de Roo, A. and Vandaele, K. (1994). Flooding of property by runoff from agricultural land in northwestern Europe. *Geomorphology* **10**, 183-196.

Boardman, J. and Robinson, D.A. (1985). Soil erosion, climatic vagary and agricultural change on the Downs around Lewes and Brighton, autumn 1982. *Applied Geography* **5**, 243-258.

Boardman, J. and Spivey, D. (1987). Flooding and erosion in west Derbyshire, April 1983. *East Midland Geographer* **10**(2), 36-44.

Chambers, B.J., Davies, D.B. and Holmes, S. (1992). Monitoring of water erosion on arable farms in England and Wales, 1989-90. *Soil Use and Management* **8**(4), 163-170.

Commission of the European Communities (1992). *CORINE Soil Erosion Risk and Important Land Resources in the Southern Regions of the European Community*, EUR 13233 EN, Directorate General Environment, Consumer Protection and Nuclear Safety, EC, Brussels.

Davidson, D.A. and Harrison, D.J. (1995). The nature, causes and implications of water erosion on arable land in Scotland. *Soil Use and Management* **11**, 63-68.

De Ploey, J. (1989a). *Soil Erosion Map of Western Europe*, Catena-Verlag, Cremlingen-Destedt, Germany.

De Ploey, J. (1989b). Erosional systems and perspectives for erosion control in European loess areas. *Catena* **1**, 93-102.

Dearing, J.A., Elner, J.K., and Happey-Wood, C.M. (1981). Recent sediment flux and erosional processes in a Welsh upland-lake catchment based on magnetic susceptibility measurements. *Quaternary Research* **16**, 356-372.

Edwards, K. and Rowntree, K.M. (1980). Radiocarbon and palaeoenvironmental evidence for changing rates of erosion at a Flandrian stage site in Scotland. In, Cullingford, R.A., Davidson, D.A. and Lewin, J. (eds), *Timescales in Geomorphology*, Wiley, Chichester, 207-233.

Evans, R. (1980). Characteristics of water-eroded fields in lowland England. In, De Boodt, M. and Gabriels, D. (eds), *Assessment of Erosion*, Wiley, Chichester, 77-87.

Evans, R. (1981). Assesments of soil erosion and peat wastage for parts of East Anglia, England. A field visit. In, Morgan, R.P.C. (ed.), *Soil Conservation: Problems and Prospects*, Wiley, Chichester, 521-530.

Evans, R. (1988). *Water Erosion in England and Wales 1982-1984*, Report for Soil Survey and Land Research Centre, Silsoe.

Evans, R. (1990). Soil erosion: its impact on the English and Welsh landscape since woodland clearance. In, Boardman, J. Foster, I.D.L. and Dearing, J.A. (eds), *Soil Erosion on Agricultural Land*, Wiley, Chichester, 231-254.

Evans, R. (1993). Extent, frequency and rates of rilling of arable land in localities in England and Wales. In, Wicherek, S. (ed.), *Farm land Erosion: In Temperate Plains Environment and Hills*, Elsevier, 177-190.

Evans, R. (1995). Some methods of directly assessing water erosion of cultivated land - a comparison of measurements made on plots and in fields. *Progress in Physical Geography* **19**(1), 115-129.

Evans, R. and Morgan, R.P.C. (1974). Water erosion of arable land. *Area* **6**(3), 221-225.

Evans, R. and Nortcliff, S. (1978). Soil erosion in north Norfolk. *Journal Agricultural Science, Cambridge* **90**, 185-192.

Favis-Mortlock, D.T. (1995). The use of synthetic weather for soil erosion modelling. In, McGregor, D.F.M. and Thompson, D.A. (eds), *Geomorphology and Land Management in a Changing Environment*, Wiley, Chichester, 265-282.

Favis-Mortlock, D.M., Boardman, J. and Bell, M. (1995). Modelling long-term anthropogenic erosion of a loess cover. Paper presented at colloquium '*Floods, Slopes and River Beds*', Paris, March 1995.

Foster, S. (1978). An example of gullying on arable land on the Yorkshire Wolds. *Naturalist* **103**, 157-161.

Frost, C.A. and Speirs, R.B. (1984). Water erosion of soils in south-east Scotland - a case study. *Research and Development in Agriculture* **1**(3), 145-152.

Govers, G. (1991). Rill erosion on arable land in central Belgium: rates, controls and predictability. *Catena* **18**, 133-155.

Harrod, T. (1994). Runoff, soil erosion and pesticide pollution in Cornwall. In, Rickson, R.J. (ed.), *Conserving Soil Resources: European Perspectives*, CAB International, Wallingford, UK, 105-115.

Kirkbride, M.P. and Reeves, A.D. (1993). Soil erosion caused by low-intensity rainfall in Angus, Scotland. *Applied Geography* **13**, 299-311.

Ludwig, B., Boiffin, J., Chadœuf, J. and Auzet, A.V. (1995). Hydrological structure and erosion damage caused by concentrated flow in cultivated catchments. *Catena* **25**, 227-252.

Mackney, D., Hodgson, J.M., Hollis, J.M. and Staines, S.J. (1983). *Legend for the 1:250,000 Soil Map of England and Wales*, Soil Survey of England and Wales, Harpenden.

Morgan, R.P.C. (1980). Soil erosion and conservation in Britain. *Progress in Physical Geography* **4**, 24-47.

Morgan, R.P.C. (1987). Sensitivity of European soils to ultimate physical degradation. In, Barth, H. and L'Hermite, P. (eds), *Scientific Basis for Soil Protection in the European Community*, Elsevier, 147-157.

Morris, F.G. (1942). Severe erosion near Blaydon, county Durham. *Geographical Journal* **100**, 256-261.

Pimentel, D. (1993). Overview. In, Pimentel, D. (ed.), *World Soil Erosion and Conservation*, Cambridge University Press, Cambridge, UK, 1-5.

Pimentel, D., Harvey, C. Resosudarmo, P. Sinclair, K., Kurz, D., McNair, M., Crist, S., Shpritz, L., Fitton, L., Saffouri, R. and Blair, R. (1995). Environmental and economic costs of soil erosion and conservation benefits. *Science* **267**, 1117-1123.

Poesen, J., Vandaele, K. and van Wesemael, B. (1998). Gully erosion: importance and model implications. In, Boardman, J. and Favis-Mortlock, D.T. (eds), *Modelling Soil Erosion by Water*. Springer-Verlag NATO-ASI Global Change Series, Heidelberg.

Reed, A.H. (1979). Accelerated erosion of arable soils in the United Kingdom by rainfall and run-off. *Outlook on Agriculture* **10**, 41-48.

Richter, G. (1986). Investigation of soil erosion in central Europe. *SEESOIL* **3**, 14-27.

Shotton, F.W. (1978). Archaeological inferences from the study of alluvium in the lower Severn-Avon valleys. In, Limbrey, S. and Evans, J.G. (eds), *The Effect of Man on the Landscape: the Lowland Zone*. CBA Research Report 21, Council for British Archaeology, 27-32.

Speirs, R.B. and Frost, C.A. (1985). The increasing incidence of accelerated soil water erosion on arable land in the east of Scotland. *Research and Development in Agriculture* **2**(3), 161-167.

Watson, A. and Evans, R. (1991). A comparison of estimates of soil erosion made in the field and from photographs. *Soil and Tillage Research* **19**, 17-27.

SECTION 2. MODEL EVALUATION WITH COMMON DATASETS

4. EVALUATION OF PLOT RUNOFF AND EROSION FORECASTS USING THE CSEP AND MEDRUSH MODELS

Mike Kirkby

School of Geography
University of Leeds
Leeds LS2 9TJ
UK

Abstract

Three models have been applied to the GCTE common data sets. The simplest CEP-EROS model has been applied to five areas (US-C5, W12 and R5; Port-10 and 13) with substantial calibration, a modified version which includes subsurface flow (CEP-TOPER) has also been applied to five catchments (US-C5, W12 and R5; Canada Gu4 and Gu5) and the MEDRUSH model has been applied to US-R5 only, with no calibration owing to time constraints. For the R5 catchment, a forecast has also been made for the distribution of erosion over the catchment. Results for the US site are broadly consistent with expectations, with lower erodibility in clays and for uncultivated areas. The Portuguese sites, however, show higher runoff from the natural vegetation (site 13), but dramatically lower erodibility. Lower runoff thresholds would not be consistent with the observed runoff rates, but would bring erodibilities closer to values which might be compared to the US sites. The results from MEDRUSH on the US-R5 catchment have been obtained with prior estimation of parameter values, but no calibration. They show good agreement on the number of runoff events, but suggest that the overland flow velocities used are too low, with runoff events lasting too long.

The CEP-EROS model

The CEP-EROS model has three main components, which are currently run as distinct computer programs. Climatically based runoff thresholds are obtained from a global database of monthly precipitation, temperature and potential evapotranspiration, which are used to grow uncultivated vegetation to equilibrium (De Ploey *et al.*, 1991; Kirkby and Cox, 1995). The CSEP program provides estimates of climate-based potential runoff thresholds month by month, on a regional basis. The thresholds obtained do not contain specific factors for soil type, land use or land management, so that they have not proved suitable to make detailed comparisons, although they have provided a starting value for optimising parameter values for each individual site.

The second component is a hydrologically based model working from daily rainfalls. There is a threshold rainfall, below which there is no runoff forecast. Runoff is estimated as a constant fraction of rainfall in excess of this threshold. Sediment transport for each day is then estimated as proportional to the square of the daily runoff, so that the total for an event at a point in the erosion plot is estimated by the three expressions:

Unit Runoff = (Daily rainfall - Runoff threshold) × Runoff proportion (1)

Runoff volume = Unit Runoff × Contributing Plot Area (2)

Sediment Transport = Erodibility × (Runoff volume)^2 × (Local gradient) (3)

Specific Sediment Yield = Sediment Transport ÷ Contributing Plot Area

The three key parameters of the model are thus:
1. the threshold rainfall, below which there is no runoff,
2. the proportion of additional rainfall which runs off (Equation 1), and
3. the erodibility constant in Equation 2, which converts the runoff to sediment transport.

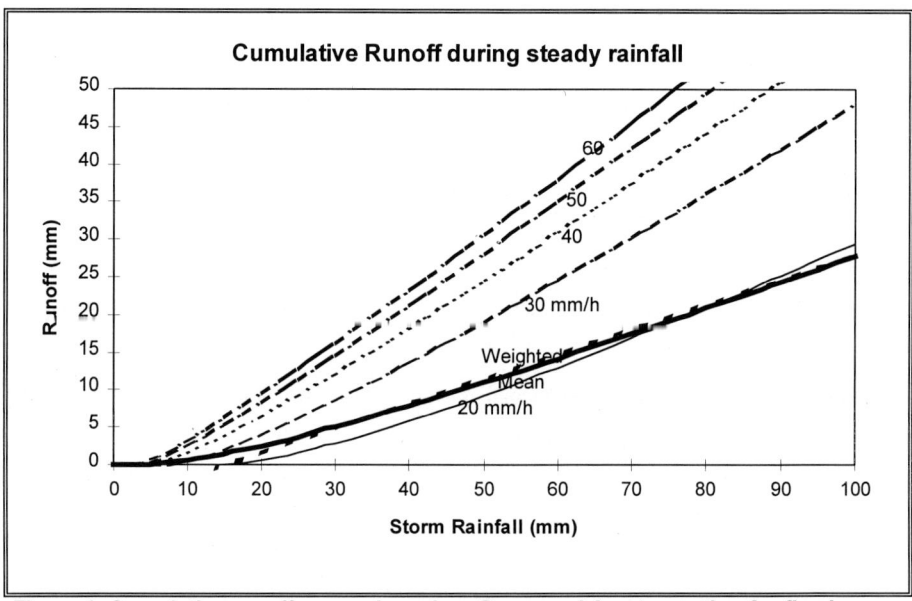

Figure 1. Cumulative runoff curves based on Green and Ampt equation for fixed rainfall intensities, and for the weighted mean over an exponential distribution of intensities (with mean of 20 mm hr^{-1}). The heavy broken line shows a straight line approximation of the type used in the CEP-EROS model, with a threshold of 15mm and 33 per cent of the excess as runoff

This approach differs from the USLE/ RUSLE approach in several significant ways. First the model is essentially a hydrological model for estimating runoff, and sediment transport is derived from the runoff, rather than directly from the rainfall kinetic energy or a related function. Second, soil properties are not contained within a single erodibility parameter but are distributed between the three parameters above, which makes it easier to understand the distinct roles of the soil in promoting runoff and in resisting erosion. A third important difference is that sediment transport is, in principle, calculable at every point down the length of an erosion plot or natural flow line, so that local sediment loss may be calculated by applying the continuity equation for the sediment budget for each section of the slope.

The type of linear runoff model used in CEP-EROS is compared with calculations based on the Green and Ampt (1911) equation in Figure 1. It may be seen that the relation between rainfall and runoff varies with intensity, but that, in the absence of detailed intensity data, a mean curve may be used to provide an average across the distribution of intensities. The runoff threshold and proportion of runoff can therefore be derived from measured soil infiltration curves, combined with the distribution of intensities where this is available. It may be seen, for example, that an increase in mean intensity tends to increase the average proportion of runoff, while having a lesser effect on the runoff threshold. For this system, the accumulated runoff for a constant rainfall intensity is given by the expression:

$$R = 0 \qquad \text{for } P < \frac{iB}{(i-A)^2}$$

$$R = (i-A)t - \left[2Bt - \left(\frac{B}{i-A}\right)^2\right]^{0.5} \qquad \text{otherwise}$$

where i is the net rainfall rate,
$\quad R$ is the accumulated total runoff after time t,
and A, B are constants in the Green Ampt equation:
Infiltration Rate, $f = A + B/S$
where S is the storage term $\int_0^t (f - A)dt$

This simple model is used here for individual daily rainfalls, for the purpose of this comparison. It was originally designed for use with an explicit integration over the daily rainfall distribution, to give estimates of total average annual runoff and sediment yield, together with their frequency distributions. Given this difference in emphasis, the three main parameters have been obtained in this study by calibration against the period for which data was provided. The optimised parameters have then been used to run the hydrological model, with a daily time step, in order to forecast the total event runoff and sediment yield at the catchment outlet. Clearly this approach leads to errors associated with the unknown distribution of rainfall within events, and its variability. It may also underestimate the importance of soil moisture conditions at the start of each event, a factor which becomes increasingly important at higher temporal resolution. Soil moisture has been estimated in CEP-EROS using a simple linear one store model, with decay rates of 30-50 per cent per day, and the stored moisture is subtracted from the runoff threshold, although this correction only modestly improves the runoff forecasts.

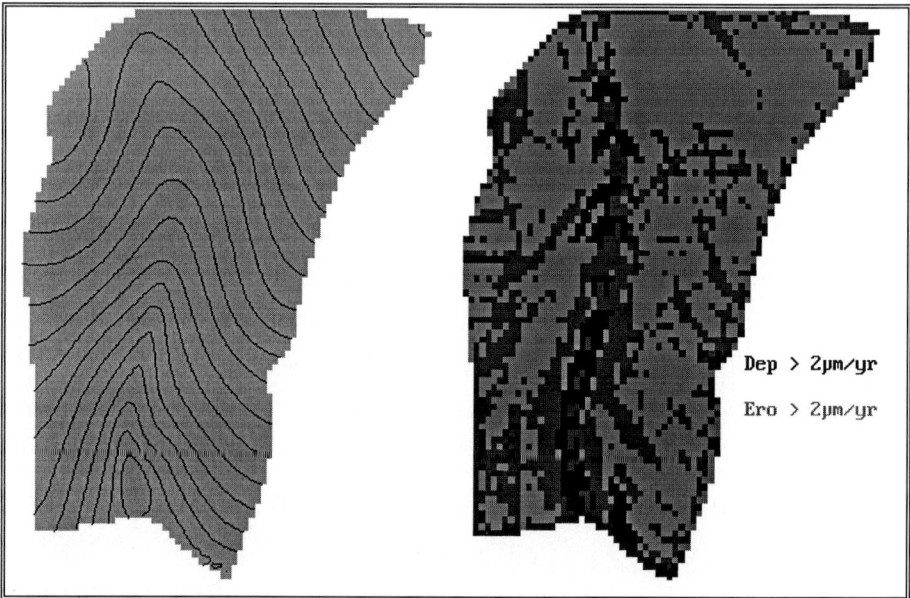

Figure 2. Simulated distribution of erosion and deposition for the US-R5 catchment over the test period:
a) Re-contoured map based on 5m DEM derived from digitised contour map
b) Pattern of erosion and deposition assuming splash at diffusivity of 5 $cm^2 a^{-1}$

The third component of the CEP-EROS model is an estimate of the areal distribution of erosion or deposition over the erosion plot. This has been implemented only for US-R5, which has the greatest relief of the test areas. An additional parameter required is an estimate of the rate of rainsplash transport (assumed proportional to gradient), which is not obtainable from the plot sediment yield data, since sediment yield at the output is very insensitive to it. Figure

2(a) shows the re-contoured DEM created by scanning the scanned contour map and rasterising on a 5 m grid. It may be seen that some significant detail may be lost close to the stream channel. Figure 2(b) shows the simulated spatial pattern of erosion and deposition, assuming a rather low splash diffusivity of 5 $cm^2 a^{-1}$. For this rate, the area close to the stream is mainly depositional (although with patches of erosion interspersed), and the main area of erosion is on the steepest gradients along each side of the valley bottom. However, it should be noted that this pattern is very dependent on the diffusivity value chosen, so that the pattern is only illustrative.

Table 1. Comparison of runoff thresholds and specific erodibilities for the CEP-EROS model

Site	Soil	Land Use	Runoff Threshold (mm)	Per cent Runoff above Threshold	Specific Erodibility (T Ha^{-1} mm^{-2})
US-C5	Silt loam	Wheat	30	50	1.5×10^{-3}
US-W12	Clay	Corn/sorghum/ wheat	30	45	2.4×10^{-4}
US-R5	Silt loam	Virgin grass	32	59	3.3×10^{-5}
Ca-Gu3	loam	Corn, minimum tillage	#	#	# 4.0×10^{-2}
Ca-Gu4	loam	Corn, ploughed etc.	#	#	# 8.0×10^{-2}
Port-10	Med'n Red	Wheat/ fallow	120	35	1.3×10^{-5}*
Port-13	Med'n Red	Natural vegetation	58	40	3.0×10^{-8}*

* Assuming that original sediment yields provided for Portugal were 1000 x too large!
# CEP-EROS not applied. Erodibility values obtained from CEP-TOPER model.

Results from the CEP-EROS model are summarised in terms of the key parameter values in Table 1. The US site results are broadly consistent with expectations, with lower erodibility in heavier soils and for uncultivated areas. The Portuguese sites, however, show generally low rates of runoff, and, unexpectedly, more runoff from the natural vegetation (site 13) than from the wheat/ fallow site (10). However the erodibility for the natural site is exceedingly low, so that total rates of erosion remain very low.

The CEP-TOPER model

Recognising the shortcomings of the CEP-EROS hydrology for use as a daily hydrological model, particularly because of the limitations in its treatment of soil moisture, a variant CEP-TOPER has been developed to take better account of saturated and unsaturated subsurface moisture and flow. This was thought particularly relevant for the Canadian catchments, where the hydrological balance is relatively moist. The hydrological model has been extended to take account of sub-surface flow, using the semi-distributed model TOPMODEL (Beven and Kirkby, 1979). This model simulates the subsurface saturated downslope drainage at any point in a catchment, with percolation into the saturated zone from an unsaturated store. Here, the saturated zone is represented as a modified linear store, with the time constant depending on the relative levels of saturated and unsaturated soil moisture (Kirkby, 1986). This type of model is better able to simulate the seasonal variations in soil water levels, and the dynamic response between storms, but is still not able to take account of the variations in storm intensity within the time unit of the single day. Although this is recognised as a significant limitation of any model using daily time steps, high resolution rainfall data is not widely enough available to provide a basis for widespread application of a model which requires it.

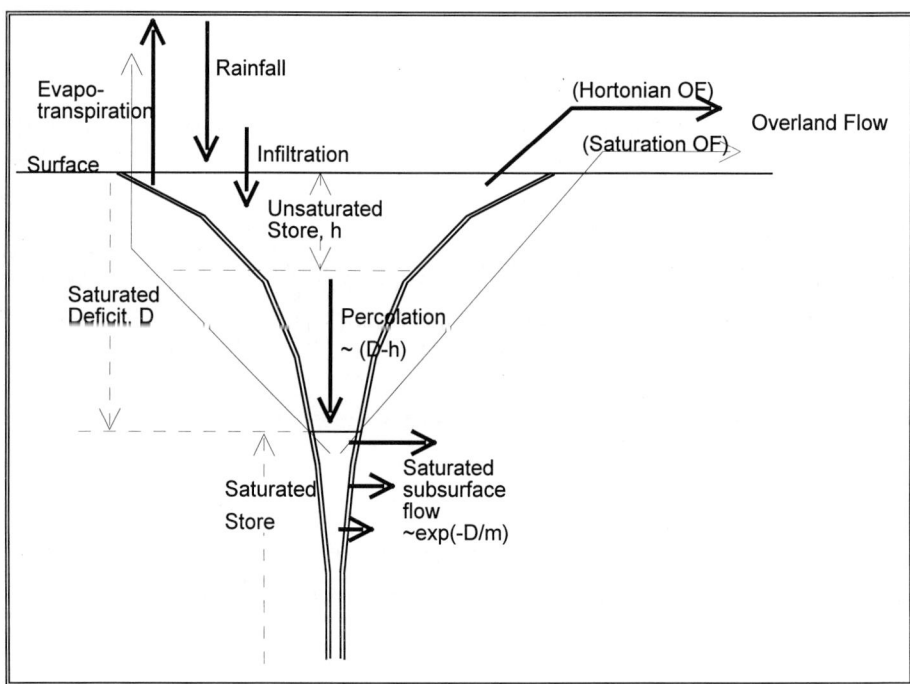

Figure 3. Flow diagram for components of hydrological model in CEP-TOPER

Figure 3 shows the main hydrological components of the model. The soil water store is shown as funnel shaped to indicate its decreasing transmission and storage capacity with depth, so that, although in principle saturation extends indefinitely downwards, the saturated subsurface flow remains finite. The main effects of including the saturated store are to allow saturation overland flow during long wet periods, even at low rainfall intensities, and to allow the percolation rate to vary dynamically with the state of both saturated and unsaturated stores. Evapotranspiration is able to take place at potential rate from the unsaturated store, and at a reduced rate from the saturated store.

Figure 4 shows a time series for rainfall and runoff for the American C5 Wheat Catchment, near Chickasha, Oklahoma. The figure shows observed rainfall and runoff for the calibration period (from 1971), together with the forecast daily soil water deficits and runoff for the entire period. It may be seen that, although the major runoff events are generally forecast in a reasonable way, there are a number of lesser actual runoff events which are not forecast, and some forecast events which did not occur. This discrepancy is thought to be related to the

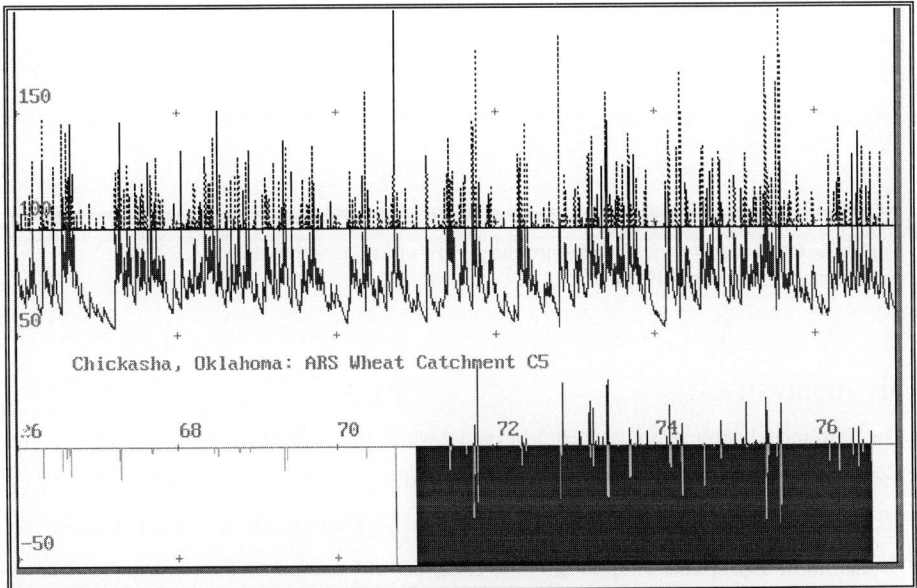

Figure 4. Rainfall, runoff and erosion for the ARS Wheat catchment C5, near Chickasha, Oklahoma. Measured values for rainfall and for runoff (above lower line in shaded calibration time period only). Forecast values for soil water storage and runoff for full period (below lower line). Scale shows years

use of a daily rainfall step, with the consequent loss of resolution. Figure 5 shows a direct comparison for the calibration period. The erodibility has been calibrated to optimise the mean values. The problem of spurious and absent forecasts is even more marked for the sediment than for the runoff, but there appears to be moderate agreement for the largest sediment yield events.

In common with other erosion models, it is possible to obtain better forecasts of sediment yield if forecasts of storm runoff are not used as an explicit intermediate stage. The view taken by the author is that, although the statistical fit may be improved, the loss of understanding can only reduce any prospect for transferring the results to other areas.

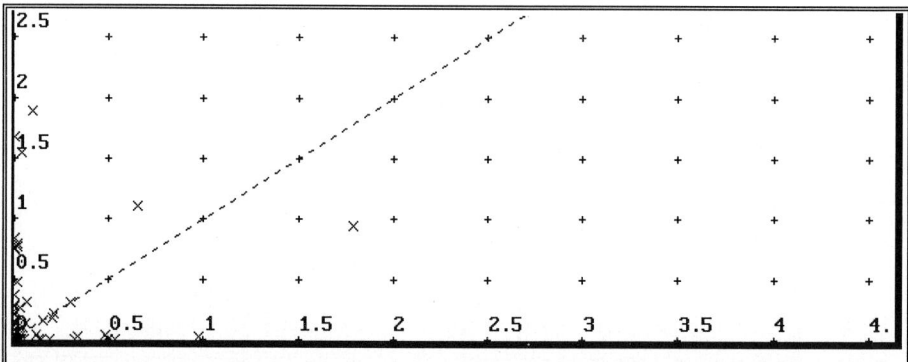

Figure 5: Observed (x-axis) and Forecast (y-axis) Sediment yield for US C-5 catchment. Units are T Ha^{-1}

The MEDRUSH model

The third model applied was a catchment model which has been developed for areas of up to 2500 km^2 within the MEDALUS project on desertification in the Mediterranean (Kirkby *et al.*, 1995). It is primarily suitable for larger areas and for uncultivated land. It therefore seemed most appropriate to apply it to the American R5 grassed sub-catchment. Because of the large number of parameters and their somewhat different style from the other models, it was not practicable to do a full calibration run in the time available. However, a single uncalibrated run gave reasonable results for runoff, which have been included in the model comparisons (Favis-Mortlock, 1998). Sediment yield values were somewhat too high, but the distribution between storms appeared reasonable. It is hoped that there will be a fuller opportunity to test the MEDRUSH model in the next proposed series of tests, targeted specifically on catchment scale models.

Conclusion

The internal consistency of the CEP based models can be evaluated from the summary in Table 1. First there is a clear reduction in erodibility of at least two orders of magnitude for fully vegetated sites (PORT-10 and US-R5) in comparison with similar cropped sites. Second there is a clear difference in erodibility according to soil types, with falling erodibility across the soil types loam, silt loam, clay, and Mediterranean Red (stony) soils, which is in broad agreement with other experience. These categories each differ by about one order of magnitude. The main anomaly observed is in the relative runoff rates between the two Portuguese sites. The wheat/fallow site (PORT-10) has a much higher threshold for runoff than the site under natural vegetation (PORT-13), even though the difference in erodibilities ensures that the naturally vegetated site is eroded less. The other sites all show lower runoff thresholds, but without marked differences in runoff thresholds related to either soil or vegetation differences. The differences in runoff do not therefore appear to follow the expected pattern of sensitivity to vegetation cover. The observed disparities are thought to be associated with:

1. the use of fixed parameters throughout the year, which ignores changes in surface and cover associated with the cultivation cycle;
2. a lack of information on crusting: well developed crusts are able to provide a combination of low runoff threshold and low erodibility;
3. the probable absence of well developed organic soil components for any of the test sites: organic matter generally raises runoff thresholds and reduces erodibility.

It appears that models with a small number of physically based parameters are able to reproduce many of the predictive features of more complex models, particularly in contexts where there is only limited scope for prior parameterisation. At present, the inheritance of the USLE and its developments, such as RUSLE, may act as an impediment to the development of new approaches. It is essential to retain the level of conceptual simplicity associated with these well established tools, while providing a better physical basis for maintaining steepland soils, particularly to meet the expanding demand for sustainable development of tropical areas. The CEP approach is thought to offer a physical basis for such a conceptual advance within a simple framework, although it currently lacks the wealth of experience in parameter selection. There is perhaps an opportunity for joint development, drawing on the extensive data banks which already exist world-wide.

References

Beven, K.J. and Kirkby, M.J. (1979). Towards a simple, physically based, variable contributing area model of catchment hydrology. *International Association of Hydrological Sciences Bulletin* **24**, 43-69.

De Ploey, J., Kirkby, M.J. and Ahnert, F. (1991). Hillslope erosion by rainstorms - a magnitude-frequency analysis. *Earth Surface Processes and Landforms* **16**, 399-409.

Favis-Mortlock, D.T. (1998). Validation of field-scale soil erosion models using common datasets. In, Boardman, J. and Favis-Mortlock, D.T. (eds), *Modelling Soil Erosion by Water*, Springer-Verlag NATO-ASI Global Change Series, Heidelberg.

Green, W.H. and Ampt, G.A., (1911). Studies in Soil Physics I: The flow of air and water through soils. *Journal of Agricultural Soils* **4,** 1-24.

Kirkby, M.J. (1986). A runoff simulation model based on hillslope topography. In, Gupta, V.K. *et al.* (eds), *Scale problems in Hydrology: Runoff Generation and Response.* Reidel, Dordrecht, 39-56.

Kirkby M.J. and N.J. Cox, (1995). A climatic index for soil erosion potential (CSEP) including seasonal and vegetation factors. *Catena* **25**, 333-52.

Kirkby, M.J., McMahon, M.L. and Abrahart, R.J. (1995). *MEDALUS II, Project 2, Unpublished Final Report: The MEDRUSH model.* Submitted to E.C. Environment Program on Desertification.

5. EVALUATION OF THE WATER EROSION PREDICTION PROJECT (WEPP) MODEL FOR HILLSLOPES

M.A. Nearing[1] and A.D. Nicks[2,a]

[1] National Soil Erosion Research Laboratory
USDA-Agricultural Research Service
West Lafayette
IN 47907-1196
USA

[2] National Agricultural Water Quality Laboratory
USDA-Agricultural Research Service
Durant
OK 74702
USA

Abstract

The USDA-Water Erosion Prediction Project (WEPP) computer model is process-based soil erosion prediction technology. Process-based erosion models provide several advantages over empirically based erosion prediction technology, including most notably: 1) capabilities for estimating spatial and temporal distributions of net soil loss, 2) more reliable extrapolation to ungauged areas, and 3) the ability to better predict off-site delivery of sediment, including particle size information. The purpose of this paper is to present an evaluation of the WEPP erosion model as applied to the case of soil loss on hillslopes. Cases where deposition is active were not considered. WEPP was compared with the Universal Soil Loss Equation and its revision, RUSLE, relative to soil erodibility and selected agricultural land use effects on erosion. This paper also presents data from natural runoff plots at nine locations evaluated relative to the WEPP model predictions. Plot data were taken from the National Repository of Soil Erosion Data located at the Agricultural Research Service's National Soil Erosion Research Laboratory. Historical climate and management information, as well as representative slope and soil information, was used to build input files for the model. A total of 544 erosion years of data from 64 plots were used. These data are among the most reliable from the current data base of soil erosion in the US. Results indicated that WEPP followed the trends in erodibility and cropping factors as represented by USLE and RUSLE. Comparisons of predictions with measured data were reasonable relative to the expected degree of accuracy for erosion prediction.

Introduction

Thorough evaluation and testing of the WEPP model (Foster and Lane, 1987; Nearing *et al.*, 1989; Flanagan and Nearing, 1995) is critical to acceptance of the technology. An important aspect of the evaluation process involves determining how well the model results compare to existing data and information on rates of erosion, including existing empirical models of

---

[a] Deceased

erosion for conditions under which the empirical models were developed (Nearing *et al.*, 1990). Evaluation and testing are also important to the land owner and the conservation planner. WEPP is a conservation planning tool: this implies that land management decisions, which always have associated monetary and social costs, will be based in part on the results of the model. It is important that the WEPP model results be critically and thoroughly evaluated relative to the best existing information on rates of soil erosion. The purpose of this study was to evaluate the WEPP model as applied to the case of uniformly sloped, single cropped conditions with no toe-slope deposition.

WEPP and USLE/RUSLE comparisons
Soil erodibility
A direct comparison of erodibility values between WEPP and USLE or RUSLE is complicated by the fact that WEPP uses three 'erodibility' parameters: rill erodibility (K_r), interrill erodibility (K_i), and critical hydraulic shear stress, (τ_c). The issue is complicated further since the RUSLE erodibility (K) factor incorporates infiltration differences between soils, whereas the WEPP erodibilities are essentially independent of infiltration. In WEPP calculations, soil detachment and transport are functions of excess rainfall, i.e., runoff. Also, the WEPP erodibility and infiltration parameters are adjusted within the model on a daily basis as a function of several inter-dependent mechanisms.

In order to compare 'soil erodibilities' we used WEPP to calculate average annual erosion rates, A_{wf}, on a clean-tilled, fallow, unit plot of 9 percent slope and 22.1 m length. We then divided the result by the RUSLE erosivity factor, R, for the location to obtain an equivalent WEPP K factor, i.e.,

$$\text{WEPP K} = A_{wf} / R. \qquad [1]$$

Forty-five soils of widely varying characteristics were tested at four different locations. Erosion was simulated for 100 years. The stations were selected based on their compatibility with soil loss gradients across the states in which they are located, so as to reduce the possibility of using non-representative stations (Baffaut *et al.*, 1996). Stations used were: Salina, KS; Greensboro, AL; Delphi, IN; and Richmond, VA. RUSLE 'R' factors at each of these locations were 160, 390, 155, and 200, respectively. The WEPP equivalent 'K' value for each soil was taken as the average of the calculated equivalent 'K' values from the 4 locations.

The equivalent WEPP 'K' values correlated well with the RUSLE K values (Figure 1). The difference between WEPP equivalent 'K' and RUSLE K was not statistically correlated to soil properties. The slope of the regression line was not significantly different from one, therefore the bias in erodibilities between the two models was not considered significant.

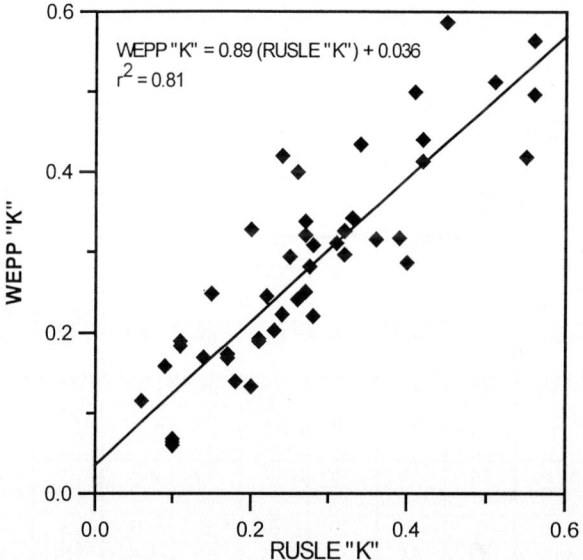

Figure 1. Comparison between WEPP and USLE/RUSLE erodibility factors. Equation 1 was used to calculate the WEPP 'K'

Management factors

We also compared WEPP response to RUSLE and the USLE relative to the effects of cropping and management on erosion (Table 1). All comparisons were for simulations, with no attempt at this point to compare to measured data. WEPP was tested at three locations using input data for a silt loam soil (Providence) on a unit plot (9 per cent slope, 22.1 m length). Crop management practices were: conventional corn with spring turn-plough with both residue removed and residue left; conventional corn with fall turn plough with both residue removed and left; conservation corn; no-till corn; and conventional cotton. Corn yield was set in all simulations to be 80 bu/acre and lint cotton was 500 lbs/acre. Above-ground biomass at harvest was about 4000 lbs/acre for corn and 2300 lbs/acre for cotton. The annual average 'C' factors for both USLE and RUSLE were obtained by weighting soil loss ratios at different crop stages by their EI values. The analogous WEPP 'C' factor is given by the ratio:

$$\text{WEPP 'C'} = A_{wc} / A_{wf} \qquad [2]$$

where A_{wc} and A_{wf} are the WEPP predicted average annual soil loss for cropped and fallow plots, respectively. All of these results are based on the use of estimated (uncalibrated) input parameters, using procedures recommended in the WEPP User Summary.

Table 1. Comparison of WEPP (W), USLE (U), and RUSLE (R) annual C factor by crop management and locations

Cropping and Management	Hollysprings, MS			Jefferson City, MO			Morris, MN		
	W	U	R	W	U	R	W	U	R
Corn, Res. left, spring TP	29	29	30	36	34	32	38	39	29
Corn, Res. Removed spr TP	54	55	49	51	53	42	47	49	35
Corn, Res. left, fall TP	47	45	50	42	45	38	38	38	30
Corn, Res. Removed fall TP	57	63	58	51	59	45	47	53	36
Corn, Conservation till.	14	18	25	17	19	26	15	20	23
Corn, No-till	6	6	9	5	5	5	3	5	2
Cotton, Conv. Tillage	49	51	27	46	53	29	NA	NA	NA

WEPP comparisons with natural runoff plot data

Comparisons were made between WEPP predictions and measured data from natural runoff and erosion plots located at nine stations in the United States (Table 2, Figures. 2, 3, 4, and 5). These data were distributed on the WEPP95 CD released in August, 1995. Input data files were constructed for soils, topography, climate, and cropping and management using information from the original field data sheets. Again, all of these results are based on the use of estimated (uncalibrated) input parameters, using procedures recommended in the WEPP User Summary.

Table 2. Site, cropping and management, soil, data period, numbers of replicates and selected events used in this study

Site	Crop Management Systems	Soil	Reps.	Years	Events Used
Holly Springs, MS	1. fallow	Providence sil	2	61-68	208
	2. conv. corn		2	"	163
	3. meadow/corn		2	62-68	127
	4. corn/bean		2	70-80	406
	5. no-till corn/bean		2	"	406
	6. no-till corn/bean		2	"	405
	7. no-till corn/bean		2	70-76	267
Madison, SD	1. fallow	Egan sicl	3	62-70	59
	2. conv. corn		3	"	48
	3. cons. corn		3	"	50
	4. oats		3	62-64	15
Morris, MN	1. fallow	Barnes l	3	62-71	67
	2. conv. corn		3	"	67
	3. meadow/corn/oats		3	"	41
Presque Isle, ME	1. fallow	Caribou gr sil	3	61-65	65
	2. potato		3	"	64
	3. potato-oats-meadow		3	"	46
Watkinsville GA	1. fallow	Cecil scl	2	61-67	147
	2. conv. corn		2	"	97
	3. conv. cotton		2	"	112
	4. corn/meadow		2	"	83
Bethany, MO	1. alfalfa	Shelby sil	1	31-40	83
	2. brome grass		1	"	79
Geneva, NY	1. fallow	Ontario l	1	37-46	97
	2. winter rye		1	"	77
	3. conv. soybean		1	"	45
	4. red clover		1	37-41	19
	5. brome grass		1	37-46	30
Guthrie, OK	1. fallow	Stephensv. fsl	1	42-56	170
	2. conv. cotton		1	"	140
	3. meadow		1	"	96
	4. wheat-clover-cotton		1	"	124
Castana, IA	1. fallow	Monona sil	2	60-71	90

Table 3 presents some overall statistics of the fit of the WEPP model to the data outlined in Table 2. Also presented in Table 3 are results of two previous studies (Risse et al., 1993; Rapp, 1994) conducted on the USLE and RUSLE models for a larger natural runoff plot data set, from which the data for the WEPP comparisons are a subset. WEPP appears from this limited analysis to provide soil loss predictions with essentially the same relative degree of accuracy as do the USLE and RUSLE. These data are not independent from the models used.

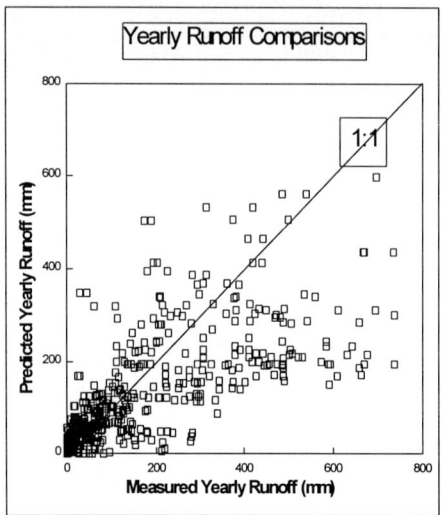

Figure 2. Yearly values of measured and WEPP-predicted runoff for data from Table 2

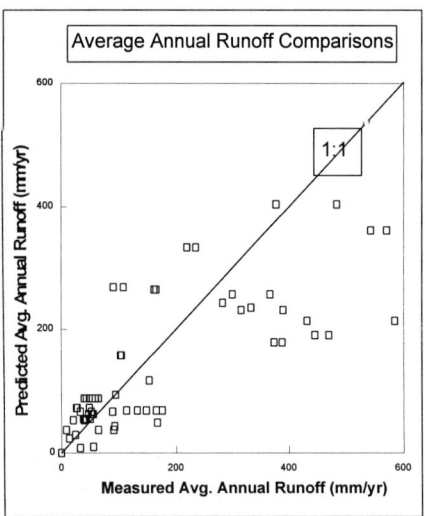

Figure 3. Average annual values of measured and WEPP-predicted runoff for data from Table 2

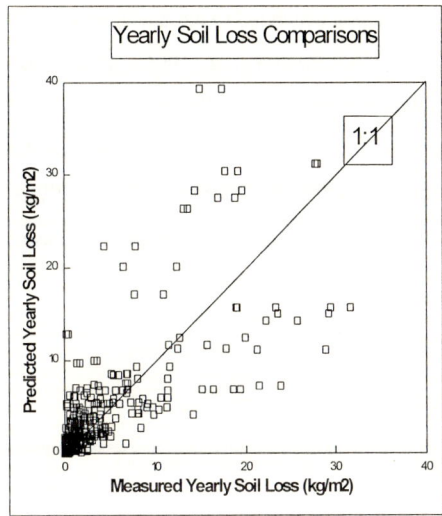

Figure 4. Yearly values of measured and WEPP-predicted soil loss for data from Table 2

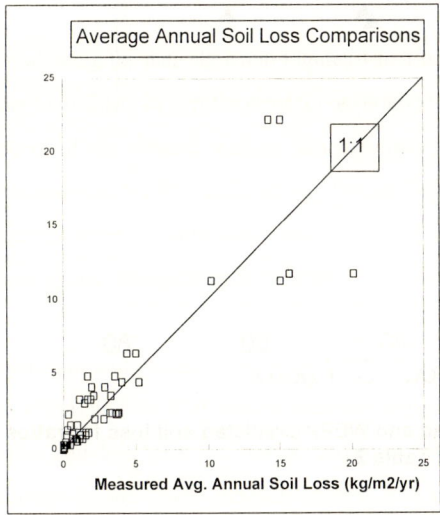

Figure 5. Average annual values of measured and WEPP-predicted soil loss for data from Table 2

6. GLEAMS MODEL EVALUATION — HYDROLOGY AND EROSION COMPONENTS

A.D. Nicks[a]

USDA-Agricultural Research Service
National Agricultural Water Quality Laboratory
Durant
OK 74702
USA

Abstract
The seven watershed data sets provided were evaluated using the hydrology and erosion model components of GLEAMS. Overall, the model did a better job of estimating the hydrology and erosion for annual periods than for monthly or storm by storm comparisons with the observed training data sets. CREAMS/GLEAMS originally developed as a planning tool and not an absolute prediction model, is used to evaluate proposed agricultural management systems against the in place conventional farming practices. The use of the NRCS curve number method for runoff estimation and the USLE technology for erosion simulation is an improvement over USLE procedures, but the model is limited in the prediction of storm events.

Introduction

The CREAMS (Chemicals, Runoff and Erosion, from Agricultural Management Systems) model (Knisel, 1980) was developed to address the issues raised by guidelines established by the Clean Water Act Amendments of 1972 (Public Law No. 92-500) by providing a tool to evaluate non-point sources of pollution. The GLEAMS (Groundwater Loading Effects of Agricultural Management Systems) model (Knisel, 1993; Leonard et al., 1987) is an extension of the CREAMS model with improved hydrology and nutrient components and an added component for vertical flux of pesticides. The GLEAMS model is made up of four separate component sub-models: hydrology, erosion, nutrients and pesticides. The hydrology and erosion models can be run separately.

The hydrology component of GLEAMS has been modified to allow for two evapotranspiration options; the Priestly-Taylor method which uses daily air temperature and solar radiation for estimating potential evapotranspiration (ET), and the Penman-Monteith method which utilises additional daily wind movement and relative humidity inputs. Another modification to the hydrology is the division of the plant root zone into a minimum of three and a maximum of 12

---

[a] Deceased

computational layers. Soil physical properties required for each layer for hydrologic, erosion and pesticide computations are included in the hydrology parameters. The erosion component of the GLEAMS model remains unchanged from that of the CREAMS model.

Hydrology component

The Natural Resource Conservation Service (NRCS, formerly SCS) curve number method was used in the hydrology component of GLEAMS. This method, modified by Williams and Nicks (1982) partitions storm rainfall into runoff and infiltration as lumped sums. An advantage of this method was that it uses readily available daily rainfall totals. Peak flow rate are estimated with the modified rational formula. Soil water routing and storage used a weighting technique for distribution of infiltrated water. Two methods are used for calculating ET, the Priestly-Taylor (1972) and the Penman-Monteith (Monteith, 1965). Both evaporation from bare soil and transpiration from plants are calculated. Parameters required by the hydrology component for runoff, evaporation, soil water storage, percolation, and peak flow calculation were derived from the soil, management, and plot configuration data.

Erosion component

The erosion component of GLEAMS was based on Universal Soil Loss Equation (USLE) (Wischmeier and Smith, 1978) technology. Erosion calculations requires storm runoff volume and peak flow rate passed from the hydrology component. Erosion by detachment is calculated for both interrill and rill areas using two modifications to USLE (Foster et al., 1980). Rainfall erosivity is calculated from rainfall volume. The Yalin sediment transport equation (Yalin, 1963) is used to describe transport capacity. Sediment transport and deposition in two channels can be calculated with the model. Also, an impoundment, such as a small farm pond or drop-inlet terrace may be included in the flow regime.

Parameters for the erosion model consists of particle sizes (up to five particles classes); a representative slope profile; the USLE soil erosivity value for each soil along the slope; channel profiles; the crop planting, tillage operation, and harvest dates for each crop in the rotation and corresponding Manning's n values, USLE C factor and USLE practices value for each operation date. As many as ten field operation dates may be included.

Evaluation data

Evaluation data were provided for seven plot and watershed locations in the US (three sites); Canada (two sites) and Portugal (two sites). The size of the erosion areas range from 0.0016 to 9.6 ha, and the land use ranged from natural cover to continuous crop rotations. Table 1 lists some of the characteristics for each site.

Table 1. Evaluation site characteristics

Site	Area (ha)	Slope (%)	Length (m)	Land Use
US-R5	5.2	1	550	Continuous wheat
US-C5	9.6	3	380	Virgin native grass land
US-W12	4.0	2	50	Corn-sorghum-wheat
Port-P10	0.016	15	20	Wheat-fallow
Port-P13	0.016	18	20	Natural vegetation
Canada-C3	0.027	8	44	Continuous corn, minimum-tillage
Canada-C4	0.027	8	44	Continuous corn, conventional tillage

Also provided with the test data sets were precipitation, meteorological data (maximum and minimum air temperature, and solar radiation), soil physical characteristics, tillage operations and dates, observed runoff, and sediment load for 'training periods'. Not all the data were of the same quality. Some data sets were more complete than others and were of longer record length. Observed data for runoff and erosion was only provided for the 'training periods' to allow a blind evaluation of the models. The 'training period' data was provided to allow calibration and tuning of the models. Precipitation and tillage operation were provided for the entire record length to allow a comparison or 'testing period'. Table 2 lists the periods of record and the 'training periods'.

Table 2. Period of record for the evaluation data sets

Site	Observed Precipitation	Training Runoff and Erosion
US-C5	1966-1976	1971-1976
US-R5	1966-1978	1973-1978
US-W12	1985-1989	1985-1986
Port-P10	1961-1994	1978-1994
Port-P13	1988-1994	1988-1990
Canada-C3	1971-1975	1971-1973
Canada-C4	1971-1975	1971-1973

Model parameters and results

The GLEAMS hydrology and erosion component model parameter sets were constructed for the training periods for each of the plots and catchments. The procedure for parameterisation of each component consists of using the interactive parameter program for the component, the soil, crop, and tillage operations provided for each site and the supporting database information provided with the model.

In the case of the hydrology model, the interactive program allows the user to enter parameters directly or, in the case of crops, to select leaf area index from stored tables. However, the most useful part of the program is the pre-programmed format sequence of parameters and associated help text files with acceptable ranges for each entry.

Similarly, a parameter program is available for the erosion model component. Required data calculated in the hydrology model is passed to the erosion component, consequently, the number of erosion parameters is reduced. Soil, crop rotation, tillage operations and representative slope profile and channel topography are the major data needed for parameterisation of the erosion component.

Results from the training period model runs are listed in Table 3. The results are grouped by simulated and observed mean monthly and annual values during the training period with precipitation given in the simulated groups. Table 4 lists the results for the entire period of record for all sites. These results are given here for informational purposes: they are evaluated further by Favis-Mortlock (1998). Shown in Figures 1 and 2 are the observed and simulated average annual runoff and soil loss for each site respectively.

Table 3. Summary of observed and simulated runoff and erosion for the training periods

	J	F	M	A	M	J	J	A	S	O	N	D	ANN
					Simulated US-C5								
P[1]	18.12	20.79	59.22	65.96	64.35	126.20	55.46	90.89	82.47	74.59	72.64	28.07	758.74
Q[2]	.00	.05	4.86	4.76	6.35	20.61	5.23	14.36	10.93	12.69	15.10	.29	95.25
S[3]	.00	.00	.03	.03	.04	.11	.14	.36	.21	.22	.20	.00	1.36
					Observed US-C5								
Q	2.96	4.74	14.77	5.24	18.32	7.31	9.19	.58	4.95	18.11	2.38	.03	88.58
S	.07	.09	.21	.03	.01	.06	.44	.00	.06	.44	.02	.00	1.42
					Simulated US-R5								
P	27.28	31.14	66.75	78.49	149.30	79.71	63.45	55.88	80.77	63.70	35.92	17.07	749.45
Q	.00	.00	1.59	3.72	18.04	2.98	4.40	3.33	2.38	2.03	1.70	.00	40.16
S	.00	.00	.00	.00	.03	.00	.01	.00	.00	.00	.00	.00	.05
					Observed US-R5								
Q	.00	.62	1.47	3.93	22.89	3.88	3.87	2.95	1.80	1.70	1.44	.00	44.55
S	.00	.00	.00	.00	.04	.00	.00	.00	.00	.00	.00	.00	.06
					Simulated US-W12								
P	15.75	144.15	42.80	70.10	108.33	85.73	14.60	16.51	86.23	149.73	115.57	121.16	970.66
Q	.00	67.52	1.23	4.41	12.92	13.41	.00	.00	13.31	20.49	10.52	28.41	172.22
S	.00	1.99	.06	.06	.27	.69	.00	.00	.22	.26	.16	.38	4.09
					Observed US-W12								
Q	.00	79.00	.50	10.50	23.00	1.00	.00	.00	.00	8.50	22.50	54.50	199.50
S	.00	2.48	.00	.11	.46	.00	.00	.00	.00	.03	.03	.35	3.46
					Simulated Port-P10								
P	24.65	44.18	23.43	40.04	31.03	6.56	2.21	2.32	19.26	56.94	79.47	67.41	397.51
Q	.12	4.24	.04	1.33	.31	.00	.00	.00	.08	4.56	7.96	18.00	36.64
S	.01	.45	.00	.11	.02	.00	.00	.00	.01	.40	.72	1.67	3.38
					Observed Port-P10								
Q	1.14	1.34	1.02	.52	.09	.01	.00	.00	.32	5.76	5.41	19.25	34.85
S	.01	.00	.00	.02	.01	.00	.00	.00	.04	.07	.27	.06	.48
					Simulated Port-P13								
P	24.13	9.91	17.36	87.12	55.20	.00	.00	.00	21.59	110.49	78.15	173.40	577.34
Q	.39	.00	.48	23.96	10.66	.00	.00	.00	2.06	39.22	21.34	114.22	212.33
S	.00	.00	.00	.25	.10	.00	.00	.00	.02	.51	.21	1.80	2.90
					Observed Port-P13								
Q	.21	.65	.11	8.52	2.70	.00	.00	.00	1.12	20.62	6.10	12.30	52.32
S	.00	.00	.00	.00	.00	.00	.00	.00	.00	.01	.03	.01	.05
					Simulated Canada-C3								
P	.00	.00	.00	.00	190.58	71.71	98.04	66.72	51.48	10.58	.00	.00	489.12
Q	.00	.00	.00	.00	55.80	.00	.51	.37	.00	.00	.00	.00	56.69
S	.00	.00	.00	.00	3.10	.00	.05	.03	.00	.00	.00	.00	3.18
					Observed Canada-C3								
Q	.00	.00	.00	.00	.00	4.88	3.81	2.59	.00	.00	.00	.00	11.27
S	.00	.00	.00	.00	.00	2.24	.93	.11	.00	.00	.00	.00	3.28
					Simulated Canada-C4								
P	.00	.00	.00	.00	190.58	71.71	98.04	66.72	51.48	10.58	.00	.00	489.12
Q	.00	.00	.00	.00	48.88	.00	.00	.00	.00	.00	.00	.00	48.88
S	.00	.00	.00	.00	18.65	.00	.00	.00	.00	.00	.00	.00	18.65
					Observed Canada-C4								
Q	.00	.00	.00	.00	.00	6.37	12.50	2.93	.00	.14	.00	.00	21.95
S	.00	.00	.00	.00	.00	7.95	6.87	1.22	.00	.02	.00	.00	16.06

[1] P - Precipitation (mm)
[2] Q - Runoff (mm)
[3] S - Erosion (tonne/ha)

Table 4. Summary of simulated runoff and erosion for training and test periods

	J	F	M	A	M	J	J	A	S	O	N	D	ANN
US-C5													
P[1]	19.47	28.79	57.87	77.08	99.80	58.47	70.17	67.24	96.40	76.80	29.46	22.88	704.43
Q[2]	.00	.59	3.69	7.39	13.22	5.44	8.71	6.20	15.28	13.01	.65	.15	74.34
S[3]	.00	.00	.02	.06	.12	.03	.28	.11	.30	.19	.01	.00	1.13
US-R5													
P	18.76	29.37	48.22	70.22	120.75	70.51	58.26	57.99	85.30	71.82	32.10	22.35	685.66
Q	.11	.40	.68	3.52	12.00	2.72	4.33	2.76	7.68	8.08	.84	.04	43.16
S	.00	.00	.00	.00	.02	.00	.01	.00	.01	.01	.00	.00	.06
US-W12													
P	27.43	99.97	57.45	52.58	114.55	128.07	41.00	39.67	56.59	86.00	84.02	89.41	876.76
Q	.00	28.59	2.56	3.23	14.04	17.09	.84	4.64	5.81	9.31	5.83	15.46	107.40
S	.00	.81	.07	.05	.21	.40	.02	.08	.11	.11	.14	.46	2.45
Port-P10													
P	41.16	50.70	35.58	32.06	22.94	14.54	1.64	1.16	16.50	57.30	70.34	69.38	413.30
Q	2.33	3.75	1.03	1.08	.17	.00	.00	.00	.02	3.93	4.70	15.19	32.19
S	.22	.38	.10	.09	.01	.00	.00	.00	.00	.35	.41	1.36	2.92
Port-P13													
P	16.36	26.92	27.32	58.86	48.73	8.02	.00	2.14	12.30	71.27	73.12	82.11	427.16
Q	.00	.00	.76	6.11	2.45	.00	.00	.00	.20	11.91	15.51	41.22	78.16
S	.00	.00	.01	.07	.03	.00	.00	.00	.00	.15	.17	.63	1.07
Canada-C3													
P	.00	.00	.00	.00	192.48	87.53	80.47	83.87	50.55	7.87	.00	.00	502.77
Q	.00	.00	.00	.00	67.79	1.14	.13	1.17	.00	.00	.00	.00	70.24
S	.00	.00	.00	.00	1.40	.08	.03	.28	.00	.00	.00	.00	1.80
Canada-C4													
P	.00	.00	.00	.00	192.48	87.53	80.47	83.87	50.55	7.87	.00	.00	502.77
Q	.00	.00	.00	.00	75.42	2.99	.39	1.57	.00	.00	.00	.00	80.36
S	.00	.00	.00	.00	36.70	.13	.05	.32	.00	.00	.00	.00	37.21

[1] P - Precipitation (mm)
[2] Q - Runoff (mm)
[3] S - Erosion (tonne/ha)

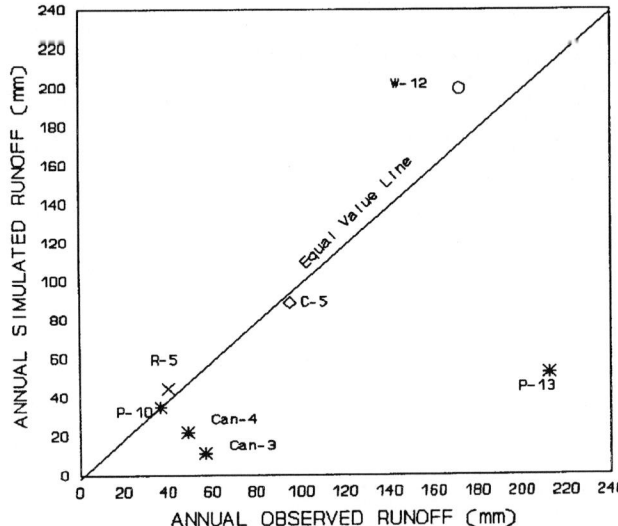

Figure 1. Simulated and observed average runoff during training periods

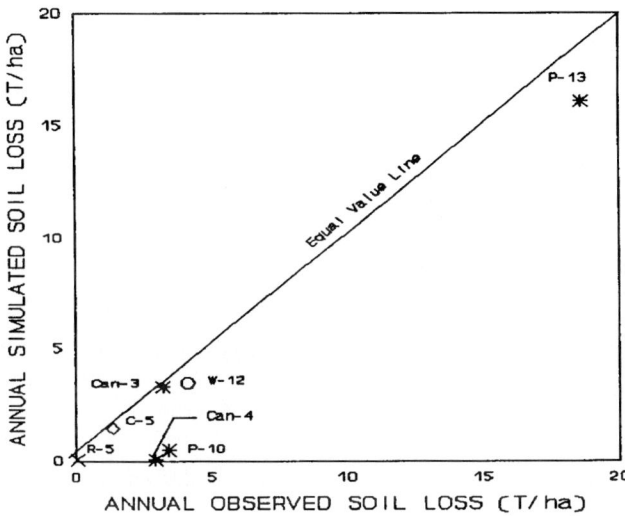

Figure 2. Simulated and observed soil loss during training period

Discussion

The output from the model, both for runoff and erosion prediction, given in Table 3 revealed mixed results for the performance of the model. Annual results for watersheds US-C5, US-R5, and US-W12 were in close agreement with the observed annual values for the training period. Simulated monthly values differ in most cases for watershed US-C5, with an over-prediction of annual runoff and close agreement with observed annual erosion. Results for watershed US-R5 were very close for monthly and annual runoff and erosion. However, this virgin native grassland watershed had very low runoff and soil loss and a less complex vegetative regime than the cropped watersheds and plots.

The more complex the cropping pattern, then the more variation in runoff and erosion should be expected; this is because of the inability of the model parameters to define the tillage and plant growth for day to day changes. For instance, watershed US-W12 under predicted runoff and over predicted erosion on an annual basis, but monthly values varied from under prediction in the first months of the year to an over prediction in the latter months. This watershed had the most complex crop rotation with a contoured tillage pattern.

There appeared to be problems with parameterisation of the model and the data provided for the Portuguese and Canadian plots. For the Canadian plots there was no precipitation data for the first four months or the last two months of the training period. The first month listed, May, had a large amount of precipitation indicating, perhaps, an accumulation for the winter months. The Portuguese P13 plot had a large monthly total precipitation for December as well. The model performed poorly on the Portuguese and Canadian plots and fairly well on the US watersheds; this is probably due to the difference in completeness and detail of the data sets provided. Accumulated precipitation amounts present in the data may have caused over prediction of runoff and erosion because the same total precipitation distributed correctly for each day would produce lower runoff and soil loss.

In general, the GLEAMS model performed well for the purposes it was intended, i.e. the prediction of long-term runoff and erosion, and the response or effects on runoff and erosion due to change in management. Because there were no changes in treatments on individual sites, no test of the response of the model results due to change in management practices could be done.

Conclusions

The GLEAMS model performed as expected on the US watersheds giving close results for the long-term (annual) prediction of runoff and erosion. However, short-term (monthly) were not as close in agreement with the training period data. The model was originally developed to be a response model and not an absolute prediction model. The use of daily precipitation limits somewhat the ability of the model to predict the distribution of storm amounts correctly. Multiple storms during a day, or storm runoff continuing from one day to the next will not be defined and parameters will not be updated correctly by the model. Correspondingly, if the runoff is not correct, then the soil loss calculation will be influenced proportionally.

The model performed poorly on the Portugal and Canadian plots. This was due partly to incorrect parameterisation of the model components and partly to the data supplied. Accumulated precipitation may have caused some of the discrepancies in the results.

Acknowledgement

This paper is a contribution to the Soil Erosion Network of GCTE, which is a Core Research Project of IGBP.

References

Favis-Mortlock, D.T. (1998). Validation of field-scale soil erosion models using common datasets. In, Boardman, J. and Favos-Mortlock, D.T. (eds), *Modelling Soil Erosion by Water*, Springer-Verlag NATO-ASI Global Change Series, Heidelberg.

Foster, G.R., Lane, L.J., Nowlin, J.D., Laflen, J.M., and Young, R.A. (1980). A model to estimate sediment yield from field size areas: Development of model. In, Knisel, W.G. (ed.), *CREAMS: a Field-Scale Model for Chemicals, Runoff, and Erosion from Agricultural Management Systems*. U.S. Department of Agriculture, Science and Education Administration, Conservation Research Report No.26. pp. 36-64.

Knisel, W.G. (ed.) (1980) *CREAMS: A Field-Scale Model for Chemicals, Runoff, and Erosion from Agricultural Management Systems*. U.S. Department of Agriculture, Science and Education Administration, Conservation Report No. 26. 640 pp.

Knisel, W. G., (ed.) (1993). *GLEAMS: Groundwater Loading Effects of Agricultural Management Systems*. University of Georgia, Coastal Plains Experiment Station, Biological and Agricultural Engineering Department Publication No. 5. 260 pp.

Leonard, R.A., Knisel, W.G. and Still, D.A. (1987). GLEAMS: Groundwater Loading Effects of Agricultural Management Systems. *Transactions of the American Society of Agricultural Engineers* **30**(5), 1403-1418.

Monteith, J.L. (1965). Evaporation and the environment. In, *The State and Movement of Water in Living Organisms, XIXth Symposium.* Society for Experimental Biology, Swansea, Cambridge University Press. pp. 205-234.

Priestly, C.H.B. and Taylor, R.J. (1972). On the assessment of surface heat flux and evaporation using large-scale parameters. *Monthly Weather Review* **100**, 81-92.

Wischmeier, W.H., and Smith, D.D. (1978). *Predicting Rainfall Erosion Losses - A Guide to Conservation Planning.* U.S. Department of Agriculture, Agricultural Handbook No. 527. U.S. Government Printing Office, Washington. DC.

Williams, J.R. and Nicks, A.D. (1982). CREAMS hydrology model — Option one. In, Singh, V.P.(ed.), *Applied Modeling in Catchment Hydrology,* Proceedings of the International Symposium on Rainfall-Runoff Modeling, Water Resource Publications, Littleton, CO. pp. 69-86.

Yalin, Y.S. (1963). An expression for bedload transportation. *Journal of the Hydraulics Division,* Proceeding of the American Society of Civil Engineers **89**(HY3), 221-250.

7. EUROSEM: AN EVALUATION WITH SINGLE EVENT DATA FROM THE C5 WATERSHED, OKLAHOMA, USA

John N. Quinton[1] and Roy P.C. Morgan[2]

[1]Department of Water Management and [2]Department of Rural Land Use
School of Agriculture, Food and the Environment
Cranfield University
Silsoe
Bedford MK45 4DT
UK

Abstract

For the GTCE model comparison EUROSEM was evaluated against data for the C5 watershed, Oklahoma, USA. Model calibration was carried out using data for events from the training period with similar rainfall patterns and soil conditions to that of the test events. Calibration was conducted by modifying the saturated hydraulic conductivity and initial moisture content of the soil until broad agreement was reached between the runoff simulated by the model and the observed data. Once this had been achieved the model was independently calibrated to simulate soil loss by modifying the parameters describing the cohesion and erodibility of the soil, and rill density. During the calibration phase parameter values were maintained within realistic limits for the watershed concerned. Once calibrated, the model was applied to four test events. EUROSEM simulated the total runoff and soil loss from three of these events quite well, but failed to reproduce the hydrographs and sedigraphs. Observed hydrographs and sedigraphs must be supplied at the calibration phase if catchment response is to be successfully simulated.

Introduction

The European Soil Erosion Model (EUROSEM) represents a state-of-the-art dynamic process-based model which simulates erosion on an event basis (Morgan et al., 1998). It relies on physical descriptions of the erosion processes which occur at the field and small catchment scale and operates for short timesteps, in the region of one minute.

Most of the work to date on evaluating EUROSEM has been concentrated within Europe. Perhaps the most comprehensive study was that of Quinton (1994a) who applied the model to data collected from the Woburn Erosion Reference Experiment (Catt et al., 1994). He found that the model performed reasonably well in simulating both runoff and soil loss, but that the simulation results were subject to considerable uncertainty as a result of difficulties in the model parameterisation. Other studies with earlier releases of EUROSEM have compared simulations in a deterministic fashion, relying on single outputs from the model, for example

Quinton (1994b) again using the Woburn data, and Albaledejo *et al.* (1994) who evaluated the model against data from erosion plots under semi-natural conditions in Spain. These comparisons illustrated that EUROSEM could simulate both soil loss and runoff well, but was highly sensitive to certain model input parameters.

Choice of test data

Due to its process-based nature, EUROSEM requires a substantial amount of information for its parameterisation. In particular it needs:
- high resolution rainfall data
- soil hydrological information
- detailed surface geometry
- soil mechanical characteristics
- vegetation characteristics

The requirement of the model for breakpoint rainfall data meant that of the data supplied by GTCE only those for the C-5 and R-5 watersheds supplied by the USDA were suitable. Since the R5 watershed had a low mean annual erosion rate (0.06 t ha^{-1} yr^{-1}) compared to that of C-5 (0.85 t ha^{-1} yr^{-1}), it was decided to concentrate on the C5 watershed, in the belief that it would be more important to demonstrate that EUROSEM was able to simulate high magnitude events (Chisci and Morgan, 1988).

The C5 watershed

The C5 watershed is located about 5 km east of Chickasha, Oklahoma in the Washita River basin. It comprises 5.2 ha of gently sloping (0 to 1 per cent) agricultural land (Figure 1). The soils are alluvial in origin, formed on terrace deposits, and are a mixture of silty clay loams (76 per cent) and silt loams (29 per cent). The watershed is under continuous wheat and is representative of cropland within the Central Great Plains.

Rainfall was recorded close to the site using a recording raingauge. Discharge was calculated using water level measurements and stage discharge relationships for the concrete V-notch weir installed at the site.

Model parameterisation

Catchment geometry

The representation of the catchment geometry used in the simulations (Figure 1) was based on the topographical map supplied as part of the GTCE model evaluation data set. Simplifying the watershed was hindered by the poor detail in the topographical map and the lack of information upon the spatial distribution of soil types. This meant that several assumptions had to be made. These were that there was no contribution from western boundary; that the site drained from east to west and north to south; and that there was an ephemeral channel running along the western boundary. On this basis, and using the information provided, the watershed was divided into four plane and four channel elements. Their position and linkage is illustrated in Figure 1.

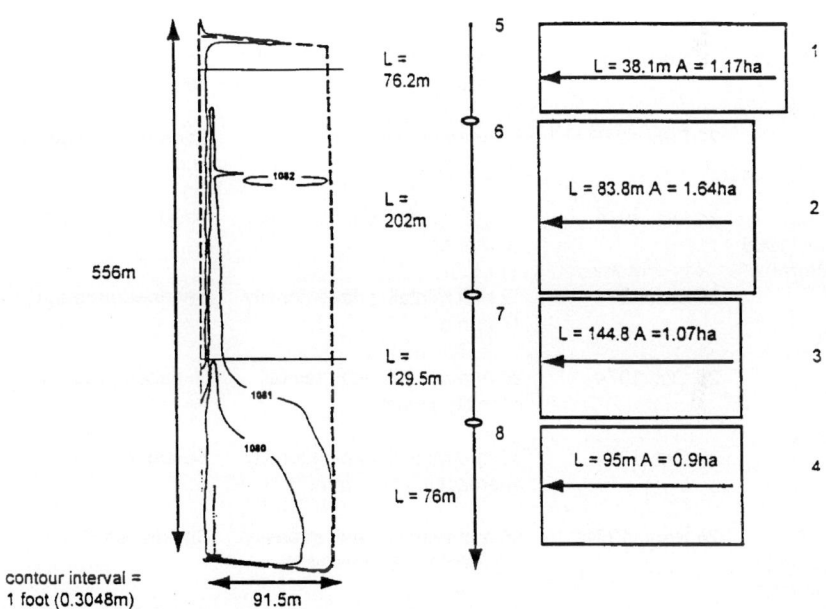

Figure 1. Map and simplified geometry for the watershed. The total channel area was assumed to be 0.42 ha

Selection of training storms

Before continuing the parameterisation of the model it was important to establish the conditions under which the simulations were to take place. The two test storms selected by

GTCE for model evaluation occurred in the autumn, after the harvest of the crop. In order for EUROSEM to simulate these storms well it was decided to make use of storms in the training data set which occurred at the same time of year and with similar soil conditions within the training data set. Two further storms were selected by the authors from the GTCE data set to increase the number of test storms to four. The properties of the test and training storms are given in Table 1.

Table 1. Storm and watershed characteristics of the testing and training events used for the EUROSEM evaluation

Storm type	Date	Storm characteristics	Watershed characteristics
Test storms (GCTE selected)	26-28 Nov. 1968	15 mm rainfall, peak intensity approximately 10 mm hr^{-1}	no information on tillage conditions, seedbed assumed
	22-23 Sept. 1970	90 mm of rainfall, peak rainfall intensity approaching 140 mm hr^{-1}	crop harvested, soil disked three times
Test storms (self selected)	29 Aug. 1966	46 mm of rainfall, peak rainfall, intensity approaching 100 mm hr^{-1}	Soil ploughed after harvest
	22 Feb. 1975	31.5mm of rainfall, peak rainfall intensity of 105 mm hr^{-1}	Seed bed prepared and crop sown
Training storms (1968 test storm)	24 Oct. 1974	10 mm rainfall, peak intensity 10 mm hr^{-1}	Seedbed prepared
	25 Oct. 1974	12 mm rainfall, peak intensity 17 mm hr^{-1}	Seedbed prepared
	28 Oct. 1974	20 mm rainfall, peak rainfall intensity 25mm hr^{-1}	Seedbed prepared
	30 Oct. 1974	35 mm rainfall, peak intensity approaching 60 mm hr^{-1}	Seedbed prepared
Training storms (1970 test storm)	24 Sept. 1971	50 mm rainfall, peak intensity approaching 80 mm hr^{-1}	Soil disked
	2 Oct. 1971	80 mm rainfall, peak intensity 145 mm hr^{-1}	Soil disked
	31 Oct. 1971	55 mm rainfall, peak intensity 45 mm hr^{-1}	Soil disked

Basic parameterisation

Parameterisation of the model was carried out using the limited information supplied and on the basis of the judgement of the authors. Values for the parameters were largely selected from EUROSEM's User Guide (Morgan *et al.*, 1993). The values used are given in Table 2.

Table 2. Parameter values used in the training simulations

Parameter		Value	Unit
Splash erodibility		1.5	g J^{-1}
Cohesion		5	KPa
Saturation water content		0.45	
Capillary drive		500	mm
Saturated hydraulic conductivity		5	mm hr^{-1}
Slope		0-1	%
Maximum slope length		550	m
Manning's 'n'	after disking	0.03	
	after harrowing	0.02	
Surface roughness	after disking	32	
	after harrowing once	25	
	after harrowing twice (drilled surface)	20	

Model calibration

Model calibration was performed by trial and error. To begin with the model was calibrated to simulate the total volume of runoff from the watershed for the seven training storms. This was done by manipulating the values of saturated hydraulic conductivity and initial moisture content, which both invoke a sensitive response from the model, until an acceptable agreement was obtained. Both parameters were held within physically realistic limits. Since no within-storm time series data on runoff were supplied it was not possible to calibrate the hydrograph.

The calibration of the soil loss component of the model was based on the manipulation of the rill density, rill size and cohesion parameters. Since no information was supplied on the number, size or shape of concentrated flow-paths within the watershed, and as, in mechanised agriculture, they are likely to be present, the rill parameters made an ideal group of unknowns upon which to calibrate the model. Cohesion was used due to sensitivity of soil loss to changes in its value.

By carrying out the calibration exercise it was possible to identify a trend in the values of some parameters, which could be used when parameterising the test events. In particular, saturated hydraulic conductivity was seen to decline with degree of cultivation and with the number of days following cultivation. This was presumably due to soil consolidation during this period.

The results of the calibration exercise are given in Table 3. In general it was possible to modify the parameter set used to simulate the observed data. Only in the case of the 2 October 1971 event did difficulties arise. In this case EUROSEM was able to match the runoff, but underestimated the observed soil loss by more than 5 t. Since, in this instance, the average sediment concentration is very different from those of the other storms it is possible that an error may have occurred with the recording equipment at the field site.

Table 3. Model results for the training storms

Storm	Simulated runoff (mm)	Observed runoff (mm)	Simulated soil loss (t ha^{-1})	Observed soil loss (t ha^{-1})
First training set				
24 Oct. 1974	0	0.07	0	1.9×10^{-7}
25 Oct. 1974	0.34	0.36	1.9×10^{-7}	9.6×10^{-4}
28 Oct. 1974	5.21	4.95	3.25×10^{-2}	3.28×10^{-2}
30 Oct. 1974	11.65	10.35	0.1227	0.1148
Second training set				
24 Sept. 1971	12.16	9.43	0.300	0.2942
2 Oct. 1971	36.5	35.35	1.063	1.73×10^{-3}
31 Oct. 1971	26.65	45.18	0.4462	0.3231

Table 4. Model results for test storms

Event date	Observed runoff (mm)	Simulated runoff (mm)	Observed soil loss (t ha^{-1})	Simulated soil loss (t ha^{-1})
29 Aug. 1966	5.41	16.301	0.1105	0.49
28 Nov. 1968	0.84	0	0.0021	0
22 Sept. 1970	14.85	13.91	0.461	0.445
22 Feb. 1975	18.71	13.055	0.4986	0.375

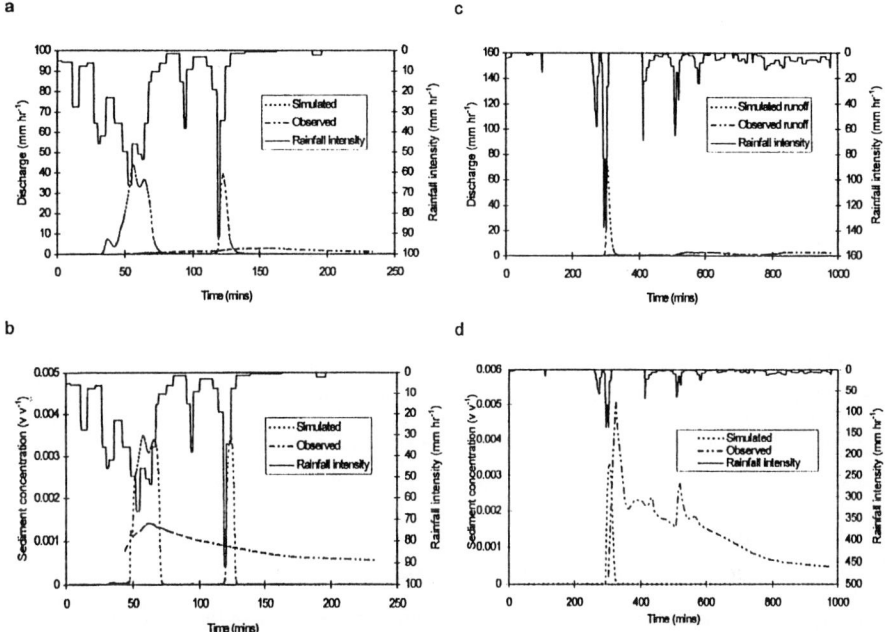

Figure 2. Comparison of the EUROSEM simulation with the observed hydrograph (a) and sedigraph (b) for the event of 29th August 1966, and with the observed hydrograph (c) and sedigraph (d) for the 22nd September 1970

Results for test storms

Table 4 gives details of EUROSEM's performance on simulating the observed runoff and soil loss in the test events. EUROSEM simulates three of the four events well, but overestimates the runoff for the event of 25 February 1975, which in turn leads to an overestimation of the soil loss. The other events appear to have been simulated well. Before concluding that EUROSEM has performed satisfactorily, however, it is instructive to compare the simulated and observed hydrographs and sedigraphs. To illustrate, one simulation is chosen as an example of where EUROSEM did badly (Figure 2a and b) in simulating event totals, and one where it performed well (Figure 2c and d). It is immediately apparent in both cases that the simulated response follows more closely that of the rainfall than does the observed response. This is particularly clear in Figure 2a and b, where the two pulses of the storm are mimicked by the simulation while the observed response shows no relationship with the rainfall. In the case of the storm of 22 September 1970 (Figure 2c), where EUROSEM was particularly accurate in simulating event totals, a similar picture emerges. EUROSEM simulates a single runoff peak in response to a high intensity rainfall pulse, yet the observed data show only a lower magnitude and severely damped hydrograph. EUROSEM produces a sharp response in the sedigraph while the observed data are more muted. However, the simulated and observed sediment concentrations are of a similar order.

Discussion

The results from the simulations of the C5 watershed illustrate one of the problems that face physically based models — it is quite possible to get the right results for the wrong reasons. EUROSEM produced good simulations of event totals yet failed to simulate the dynamic catchment response. The reasons for this are linked to the parameterisation of the model. Firstly EUROSEM was calibrated on the event totals and not on the hydrographs and sedigraphs (these were not provided until after the simulations had been made). A parameter set was selected which adequately simulated a series of training events and gave good simulations for three of the test events. However, several other parameter sets could likely have produced the same result, as illustrated by Beven (1993) and Quinton (1994a). It is, therefore, quite possible, if not probable, that the parameter set used in this study was not the correct one to describe the watershed (if such a parameter set is possible at all!). It is clearly not possible to simulate catchment behaviour with confidence unless the model is first parameterised using data sets which express that behaviour. In the case of C5, it appears that water is being stored within the catchment and then released, giving rise to a hydrograph which appears to be both damped and lagged. Without hydrographs and sedigraphs which exhibit such a response to calibrate the model with, or detailed catchment information, e.g. the positions and properties of rill channels and depressions, the model cannot be expected to simulate hydrographs and sedigraphs successfully.

Even when high quality data sets, such as that used in this study, are made available to modellers it is difficult for models to be accurately parameterised. Certainly the opportunity for the modellers to visit the field site prior to running the model would help to improve the parameterisation — in this case it might have identified reasons why the catchment responded so slowly to high intensity rainfall; but although such a visit might have produced a better interpretation of the catchment, uncertainties would have still remained over the values of many model parameters. To collect information on their spatial and temporal nature is expensive and time consuming, and cannot be done retrospectively. Yet if we are to validate models for use in global change studies, the collection of data sets, from which it can be illustrated that process based models are not only able to simulate event totals, but able to simulate accurately the processes which lead to them, should be of a high priority.

Conclusions

Using the training and test data sets for the GTCE model comparison, EUROSEM was calibrated successfully to produce simulations of total runoff and soil loss for the C5 watershed. Simulation of catchment response over time is not possible unless hydrographs and sedigraphs are available in the training data for calibration, or detailed catchment information is made available for parameterisation.

Acknowledgements

The authors would like to thank the NATO workshop team for their work in producing the data sets, Arlin Nicks of the USDA for supplying the incremental data for the comparisons of hydrographs and sedigraphs and NATO for sponsoring the GTCE model comparison workshop. EUROSEM was developed with funding from the European Commission's 3rd Environment and STEP programmes.

This paper is a contribution to the Soil Erosion Network of the GCTE, which is a Core Research Project of the IGBP.

References

Albaledejo, J. Castillo, V. and Martinez-Mena, M. (1994). EUROSEM: preliminary validation on non-agricultural soils. In, Rickson, R.J. (ed.), *Conserving Soil Resources: European Perspectives*, CAB International, Wallingford. pp 314-325.

Beven, K. (1993). Prophecy, reality and uncertainty in distributed hydrological modelling. *Advances in Water Resources* **16,** 41-51

Catt, J.A., Quinton, J.N., Rickson R.J. and Styles, P.D.R. (1994). Nutrient losses and crop yields in the Woburn Erosion Reference Experiment. In, Rickson, R.J. (ed.), *Conserving Soil Resources: European Perspectives*, CAB International, Wallingford. pp 94-104.

Chisci, G and Morgan, R.P.C. (1988). Modelling soil erosion by water: why and how. In,. Morgan, R.P.C. and Rickson, R.J. (eds), *Erosion Assessment and Modelling*, Commission of European Communities Report EUR 10860EN, pp 121-146.

Morgan, R.P.C., Quinton, J.N. and Rickson, R.J. (1993). *EUROSEM version 3.1 A User Guide*. Silsoe College, Cranfield University, Silsoe, Beds, UK. 83 pp.

Morgan, R.P.C., Quinton, J.N., Smith, R.E., Govers, G., Poesen, J.W.A., Chisci, G. and Torri, D. (1998). The EUROSEM Model. In, Boardman, J. and Favis-Mortlock, D.T. (eds), *Modelling Soil Erosion by Water*, Springer-Verlag NATO-ASI Global Change Series, Heidelberg.

Quinton, J.N. (1994a). *The Validation of Physically Based Erosion Models — with particular reference to EUROSEM.* Unpublished PhD. Thesis, Cranfield University, Silsoe, Beds, UK.

Quinton, J.N. (1994b). The validation of physically-based erosion models, with particular reference to EUROSEM. In, Rickson, R.J. (ed.), *Conserving Soil Resources: European Perspectives*, CAB International, Wallingford. pp 300-313.

8. COMPARISON OF SIMULATED AND OBSERVED RUNOFF AND SOIL LOSS ON THREE SMALL UNITED STATES WATERSHEDS

Tharacad S. Ramanarayanan[1], M.V. Padmanabhan[2], G.N. Gajanan[3], and Jimmy R. Williams[4]

[1]Texas Agricultural Experiment Station
808 East Blackland Road
Temple, TX 76502
USA

[2]Central Research Institute for Dryland Agriculture
Hyderabad
India 500 059

[3]University of Agricultural Sciences (GKVK)
Bangalore
India 560 065

[4]USDA-Agricultural Research Service
808 East Blackland Road
Temple, TX 76502
USA

Abstract

The use of mathematical models has been proven to be cost-effective in determining soil productivity changes due to various agricultural management practices. Nevertheless, many models need to be validated prior to their application. Though many models are physically based, calibration is often needed to better explain weather and land use change effects. In this study, runoff and soil loss simulated by the EPIC (Environmental Policy Integrated Climate — previously known as Erosion Productivity Impact Calculator) model are presented for three small watersheds in the United States, which had different land use and cropping systems. Each data set was divided temporally into testing and training periods. The EPIC model was calibrated using the observed weather, runoff, and soil loss data. Outputs from the model (runoff and soil loss) from the three watersheds for the testing period are presented.

In general, the long term predictions of EPIC were close to the observed values. Furthermore, the model was found to perform satisfactorily for fields under conventional crops like wheat, corn, and sorghum, but the soil loss predicted for a watershed under native grass land was not satisfactory. Information on observed biomass and residue coverage might improve the model performance.

Introduction

Usage of models for predicting runoff and soil loss from agricultural ecosystems is becoming increasingly popular. Since most computer based hydrologic models are just mathematical representations of the soil-water-plant-atmosphere continuum, testing and validating them are essential before application. The Global Change and Terrestrial Ecosystems (GCTE) Soil Erosion Network has undertaken the task of identifying suitable models for simulating soil erosion under global change. Observed data (weather, soil, crop, tillage, runoff and soil loss) from common data sets obtained from three countries (United States, Portugal and Canada) were used. Each data set was temporally divided into a *testing* period and a *training* period. The observed runoff and soil loss during the training period was used to calibrate the models and the testing period was used for validation. This study describes the results from runoff and soil loss simulation using the Environmental Policy Integrated Climate (EPIC) model on three watersheds in the United States.

The EPIC model

The EPIC model was developed by Williams *et al.* (1984) to assess the effect of soil erosion on soil productivity. EPIC, since the time it was used in the 1985 Resources Conservation Act (RCA) for soil productivity estimation, has been expanded and refined to allow simulation of many important processes in agricultural management (Sharpley and Williams, 1990; Williams, 1995).

The components of EPIC can be placed into nine major divisions — hydrology, weather, erosion, nutrients, plant growth, soil temperature, tillage, economics and plant environmental control. The hydrology and erosion components are briefly explained here. Descriptions of other components are given by Sharpley and Williams (1990) and Williams (1995).

Hydrology

The runoff model simulates surface runoff volumes and peak runoff rates, given daily rainfall amounts. Runoff volume is estimated by using a modification of the Soil Conservation Service (SCS) curve number technique (USDA-SCS, 1972). There are two options for estimating the peak runoff rate: the modified Rational formula and the SCS TR-55 method (USDA-SCS, 1986). A stochastic element is included in the Rational equation to allow realistic simulation of peak runoff rates, given only daily rainfall and monthly rainfall intensity. The latest version of EPIC (Version 5300) offers the option of using the Green-Ampt infiltration equation to estimate infiltration. However, in this study the curve number method was used to estimate initial abstraction. The model offers four options for estimating potential evapotranspiration (PET): Penman (1948), Priestley and Taylor (1972), Penman-Monteith (Monteith, 1977), and Hargreaves and Samani (1985).

Erosion

The EPIC component for water-induced erosion simulates erosion caused by rainfall and runoff and by irrigation (sprinkler and furrow). Six options that may be chosen to simulate rainfall/runoff erosion process are: the USLE (Wischmeier and Smith, 1978), the Onstad-Foster modification of the USLE (Onstad and Foster, 1975), the MUSLE (Williams, 1975), two recently developed variations of MUSLE, and a MUSLE structure that accepts input coefficients. The six equations are identical except for their energy components. The USLE depends strictly upon rainfall as an indicator of erosive energy (EI). The MUSLE and its variations use only runoff variables to simulate erosion and sediment yield. The Onstad-Foster equation contains a combination of the USLE and MUSLE energy factors. In this study one of the recently developed variations of MUSLE (MUSLE for small watersheds) was used.

Description of watersheds

Watersheds 1 and 2 are located in Chickasha, Oklahoma, and Watershed 3 is located in Reisel, Texas. Description of site and data sets are summarised in Table 1.

Table 1. Description of watersheds used in the study

Description	Watershed 1	Watershed 2	Watershed 3
Name	Chickasha - C5	Chickasha - R5	Reisel - W12
Location	Oklahoma	Oklahoma	Texas
Area (ha)	5.2	9.6	4.0
Mean Slope (%)	0 - 1 (1)*	0 - 8 (4)*	0 - 2 (2)*
Slope Length (m)	550	380	50
Soil Type	Mclain silty clay loam	Renfrew silt loam	Houston black clay
Crop/Land Use	Winter Wheat	Native Grassland	Corn-Sorghum-Wheat rotation
USLE 'P' Factor	1.0	1.0 (0.4)*	0.4

The number in parentheses was used for simulation

Materials and methods

EPIC input data files were formulated based on the information given by GCTE (1996). The model was run for the entire period of given observed weather data. The model was calibrated based on monthly model outputs during the training period. In the calibration process, the model parameters were adjusted to get a close correspondence between the simulated output and observed data. The calibration procedures adopted for each watershed are explained below.

Watershed 1: Chickasha-C5

Observed runoff and sediment yield data are available for the period August 8, 1965 to September 13, 1976. Observed weather data is available for the period January 1, 1966 to December 31, 1976. The simulations for this watershed was conducted from January 1, 1966 to December 31, 1975 (10 years). The first five years were used as the testing period and second five years as training period. The tillage operation schedule for 1966 to 1968 were missing. Therefore, the tillage operations for this period was substituted with 1969 to 1971 tillage schedules. The Penman-Monteith method was used for estimating PET. The following parameters were changed during the calibration processes: (a) curve number was adjusted and the final curve number used was 91; (b) miscellaneous parameter 6 which affects winter dormancy simulation was changed from a default value of 0.0 to 0.5, which will extend the winter dormancy; (c) the potential heat units accumulation was adjusted and the final value is 1500; and (d) the minimum USLE C factor for wheat was changed from 0.01 to 0.07. It should be noted that the curve number specified in the input file of EPIC is just an initial estimate. During the simulation, the curve number will be updated inside the model based on the prevailing moisture conditions.

Watershed 2: Chickasha-R5

Daily weather data for the period January 26, 1967 to June 21, 1978 were available for this watershed. Some of the weather input were missing during certain parts of this period and in those periods, synthetic weather data generated by EPIC, were used. Though the period January 1, 1973 to June 6, 1978 was designated as the training period for this watershed (GCTE, 1996), January 1, 1973 to December 31, 1977 (5 year period) was used for calibrating the model. The period July 24, 1966 to December 31, 1972 was designated as the testing period (GCTE, 1996). The final EPIC run was made for the period January 1, 1965 to

December 31 1978 using synthetic weather for entire 1965 and 1966, and for the periods January 1 to 26, 1967 and June 22, 1978 to December 31, 1978. The reason for adding two years before the actual testing period is to *prime* the model and allow the pasture to establish. Summer pasture was used to simulate the native virgin grass grown in the watershed. The following calibration adjustments were done for this watershed: (a) the curve number was adjusted and the final number used for this watershed is 89; (b) the wilting point of the top soil layer (0 to 0.33 m) was reduced from 0.147 to 0.127; and (c) the P factor was changed to 0.4 (although this cannot be justified and did not improve the model predictions significantly).

Watershed 3: Riesel-W12

Observed runoff and soil loss data were available only on a monthly basis for the period January 1985 to December 1989. Daily observed weather data was available from January 1, 1985 to December 31, 1989. The period January 1, 1985 to December 31, 1986 and January 1, 1987 to December 1, 1989 were designated as training and testing periods, respectively (GCTE, 1996). The parameters adjusted for this watershed were: (a) the curve number was changed to 91; and (b) the minimum C factor for corn was changed from 0.2 to 0.1.

Results and discussion

For Watersheds 1 and 2 daily, monthly and average annual simulation outputs were compared to the observed data. For Watershed 3, only monthly and average annual outputs were compared. For daily comparisons, the observed and simulated data were visually evaluated by plotting them as time series. For monthly output comparison, the simulated data was regressed over the observed data and the linear regression parameters were used to evaluate the model performance. The absolute and relative difference between the average annual simulated and observed data was calculated to evaluate the annual outputs from the model. The annual averages were calculated for the training periods only. The relative difference in average annual outputs was calculated by:

$$\Delta_r = \left(\frac{O-P}{O}\right) * 100 \qquad (1)$$

where Δ_r is the relative difference (per cent), O is the observed value and P is the predicted value.

Watershed 1: Chickasha-C5

The observed and simulated average annual runoff for the period 1971 to 1975 from Watershed 1 are 110 and 112 mm. The relative difference between the observed and simulated runoff is -2 per cent indicating that EPIC overestimated average annual runoff. The average annual soil loss of 1.7 t ha^{-1} estimated by EPIC matches very well with the observed 1.7 t ha^{-1}. The relative difference between the observed and predicted average annual soil loss is also about 2 per cent.

Table 2. Output from linear regression between observed and predicted monthly runoff and soil loss from three US watersheds

Watershed	Output	r^2	Samples (n)	Intercept (a)	Slope (b)	t$_a$	t$_b$
C5	Runoff	0.47	60	3.39	0.65	1.91	-3.95
	Soil Loss	0.69	60	-0.00	1.03	-0.06	0.33
R5	Runoff	0.91	60	1.18	0.78	3.06	-7.44
	Soil Loss	0.04	60	0.03	2.58	1.91	0.90
W12	Runoff	0.88	24	0.73	0.67	0.39	-6.22
	Soil Loss	0.96	24	-0.00	0.82	-0.17	-5.14

t$_a$ - student's t (t$_{calc}$) for H$_o$: a = 0.0   t$_b$ - t$_{calc}$ for H$_o$: b = 1.0
t$_{0.975,59}$ = 2.00 and t$_{0.975,23}$ = 2.07   If | t$_{calc}$ | ≤ t$_{0.975,n-1}$ accept H$_o$

Table 2 shows the linear regression parameters obtained from regressing simulated monthly runoff during the training period against the observed runoff. The coefficient of determination (r^2) for the linear relationship is 0.47. This indicates that only 47 per cent of the variation in the observed monthly runoff is explained by the linear relationship. A slope of 0.65 of the regression line suggests that the model underestimated monthly runoff. The intercept is not significantly different from zero, but the slope is significantly different from 1.0 at the 95 per cent confidence level. The linear regression results used to evaluate the soil loss simulations are shown in Table 2. The r^2 value of 0.69 indicates a strong linear relationship between the observed and simulated monthly soil loss. The slope and intercept of the regression line are 1.03 and 0.0 and are not significantly different from 1.0 and 0.0, respectively at the 95 per cent confidence level.

Figures 1 and 2 show the time series plot of observed and simulated runoff and soil loss from Watershed 1 for the training period. EPIC simulations are reasonably close to the observed results, although there are some marginal discrepancies. Figures 1 and 2 show no apparent anomalies.

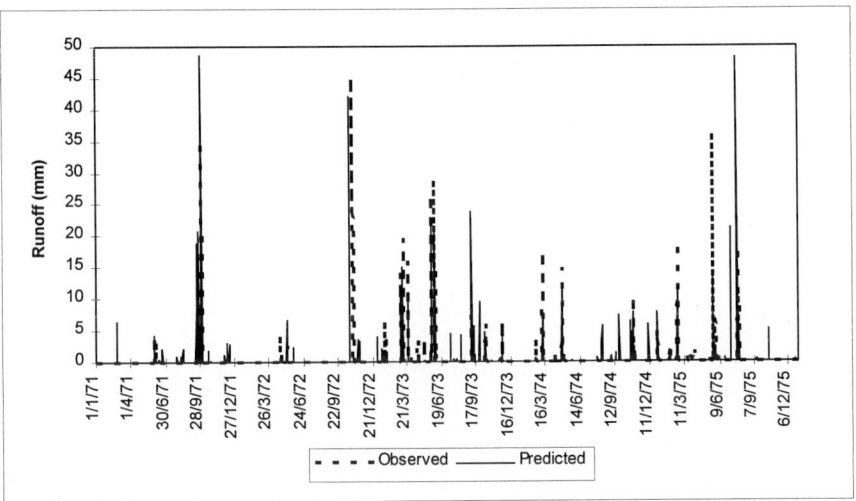

Figure 1. Time series of observed and simulated daily runoff from Watershed 1 for the training period

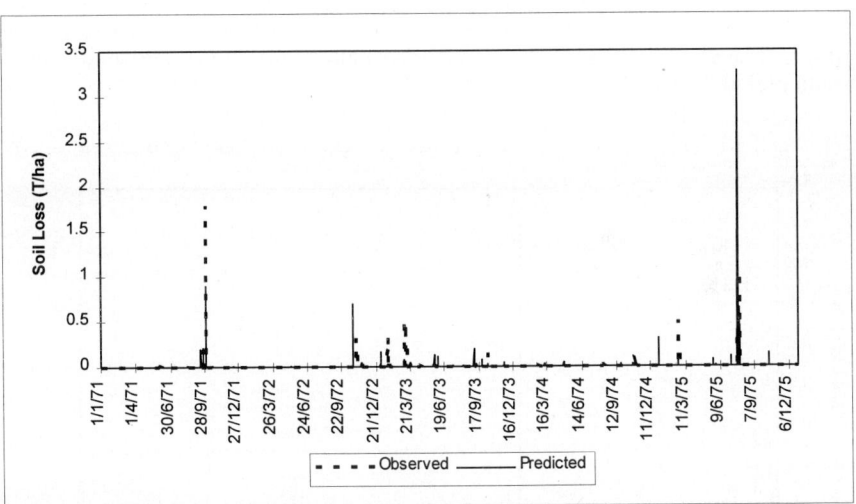

Figure 2. Time series of observed and simulated daily soil loss from Watershed 1 for the training period

From the calibration procedures it was found that runoff and soil loss estimations were sensitive to the crop parameters such as PHU, the potential heat units accumulated from planting to physiological maturity. Observed data for crop yield or biomass yield were not available. Therefore crop growth simulated by EPIC could not be tested or improved, which could have improved the runoff and soil loss estimation. Also the crop residue left on the field after harvest affects runoff and erosion. These are probably some of the reasons for the long-term estimates being closer to the observed values than the monthly and daily estimates.

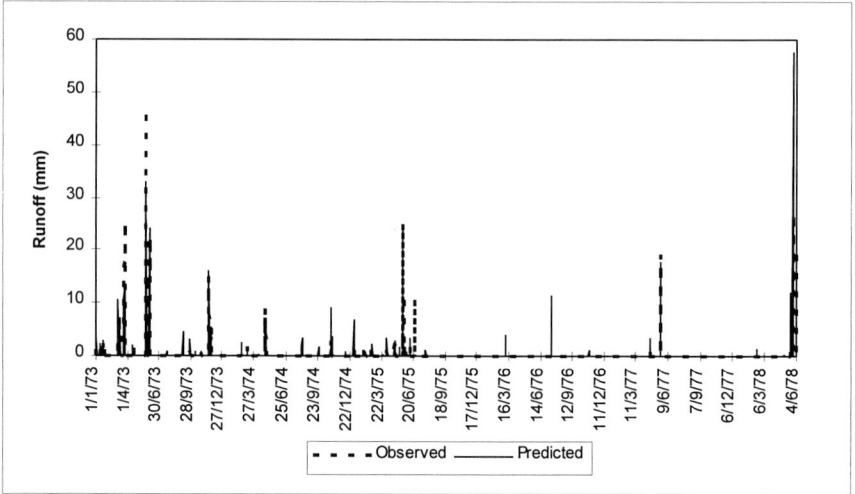

Figure 3. Time series of observed and simulated daily runoff from Watershed 2 for the training period

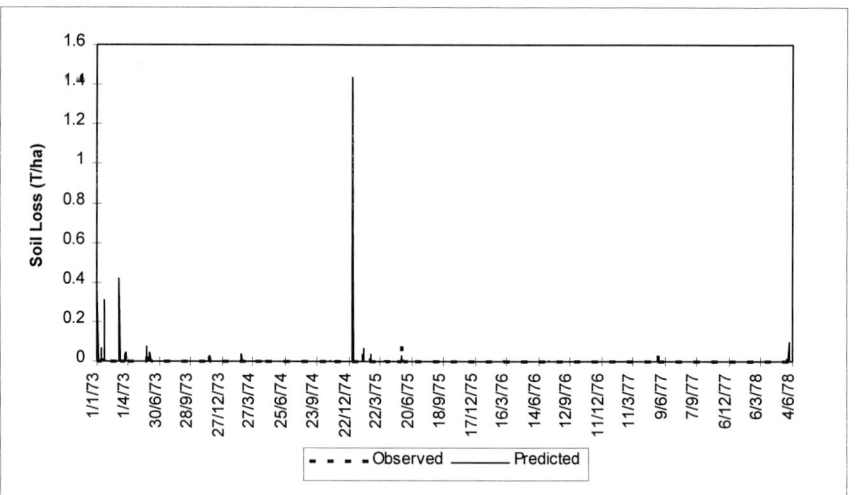

Figure 4. Time series of observed and simulated daily soil loss from Watershed 2 for the training period

Watershed 2: Chickasha-R5

The five year average annual runoff (1973 to 1977) estimated by EPIC is 58 mm as compared to an observed value of 57 mm during the same period; however EPIC overestimated mean annual soil loss from the watershed by about eight times the observed value. One apparent reason for this may be that EPIC failed to simulate the residue and biomass cover on the watershed properly. The watershed had virgin native grass. For evaluating the model performance summer pasture was grown from 1965. The establishment period allowed may not have been enough for crop establishment of good biomass cover and residue. Hence, the existing input parameters for EPIC is not suitable for simulating soil loss from this watershed.

Table 2 presents the linear regression results of the observed and simulated runoff. A coefficient of determination of 0.91 was obtained, indicating a strong linear relationship between the observed and predicted monthly runoff. The slope of the regression line is 0.78 and the intercept is 1.2. The slope and intercept are significantly different from 1.0 and 0.0, respectively. A slope of less than 1.0 indicates that EPIC under estimated monthly runoff values. The regression results for the soil loss shown in Table 2 failed to establish any significant linear relationship between the observed and predicted soil loss values.

Figures 3 and 4 show the time series plot of observed and predicted daily runoff and erosion for the period January 1, 1973 to December 31, 1977. The daily runoff predicted by EPIC shows acceptable correspondence with the observed values. However, soil loss predictions are markedly higher than the observed values.

Watershed 3: Riesel-W12

The average annual runoff estimated by EPIC for 1985 and 1986 is 131 mm, compared to an observed value of 200 mm. The average annual soil loss predicted by EPIC for the same period is 1.6 t ha^{-1}, whereas the observed value is 3.5 t ha^{-1}. EPIC underestimated both runoff and soil loss and the relative difference between observed and predicted values were 34 per cent and 55 per cent for runoff and soil loss, respectively.

Monthly runoff and soil loss predicted by EPIC follow the trend of the observed data very well. The r^2 value for the linear regression between observed and predicted monthly runoff is 0.86, indicating a strong linear relationship. The intercept of the regression line is not significantly different from zero, but the slope is significantly different from 1.0 at 95 per cent confidence level. Furthermore, the slope of the regression line 0.67 indicating an under estimation of monthly runoff by EPIC. The r^2 value for the linear regression between observed and predicted monthly soil loss is 0.96. By analysing the linear regression parameters it is evident that the intercept is not significantly different from zero, but the slope differs from 1.0 at the 95 per cent confidence level. A slope of 0.82 indicates that EPIC underestimated soil loss from the watershed.

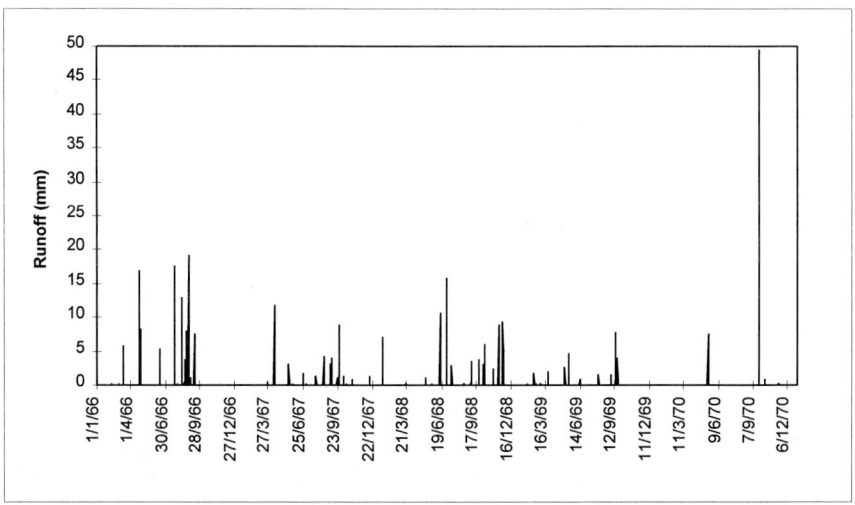

Figure 5. Time series of daily runoff from Watershed 1 for the testing period

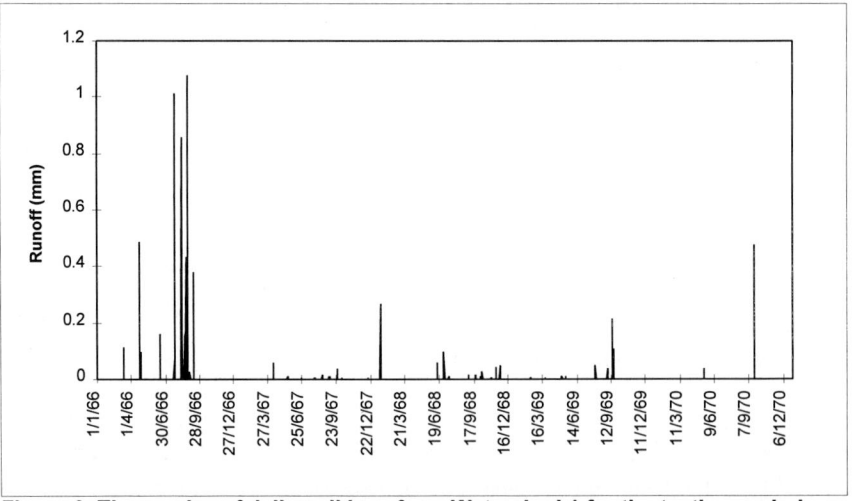

Figure 6. Time series of daily soil loss from Watershed 1 for the testing period

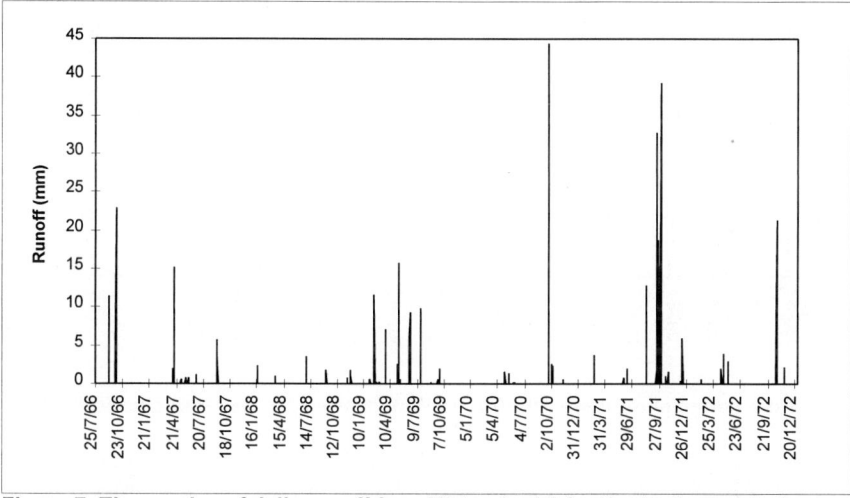
Figure 7. Time series of daily runoff from Watershed 2 for the testing period

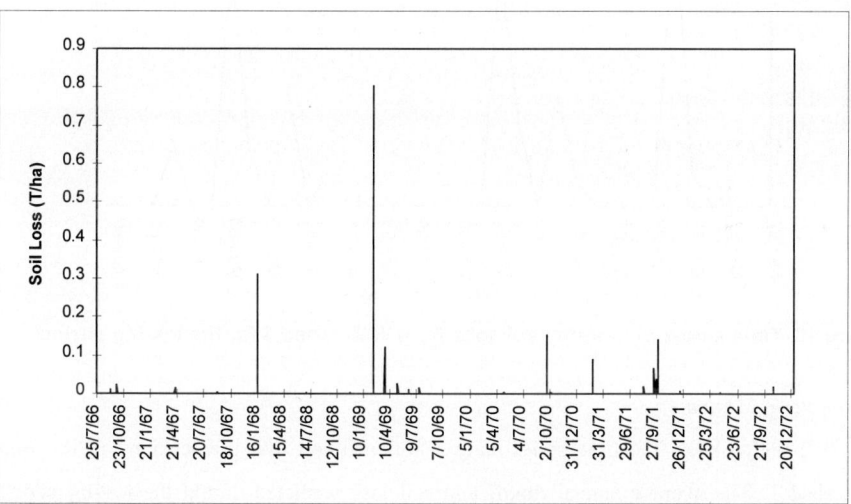
Figure 8. Time series of daily soil loss from Watershed 2 for the testing period

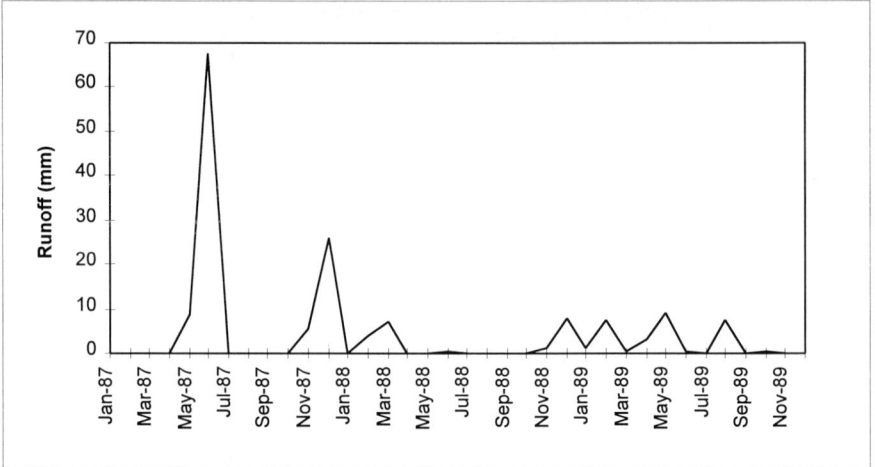

Figure 9. Time series of monthly runoff from Watershed 3 for the testing period

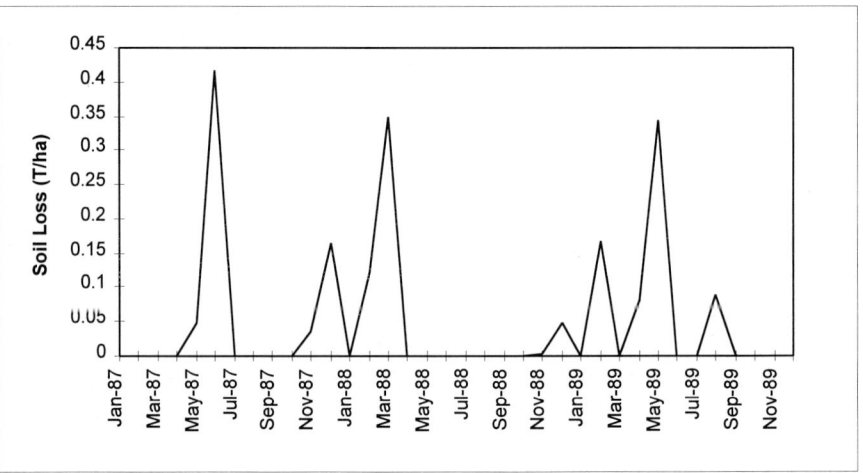

Figure 10. Time series of monthly soil loss from Watershed 3 for the testing period

Testing Period Outputs

Figures 5 and 6 show the time series daily runoff and soil loss predicted by EPIC from Watershed 1. The average annual runoff and soil loss predicted during the testing period (1966 to 1970) is 66 mm and 1.3 t ha^{-1}, respectively. The daily time series of runoff and soil loss predicted from Watershed 2 during the testing period is shown in Figure 7 and the soil loss time series is shown in Figure 8. The average annual runoff and soil loss predicted for the period 1967 to 1972 is 52 mm and the average annual soil loss for the same period is 0.3 t ha^{-1}. For Watershed 3 instead of daily time series monthly time series are presented because daily observed data for this watershed is not available. The monthly time series of runoff and soil loss predicted for the testing period (1987 to 1989) are shown in Figures 9 and 10. The average annual runoff and soil loss predicted during the testing period are 52 mm and 0.6 t ha^{-1}, respectively.

Conclusions

The EPIC model was evaluated for its ability to predict runoff and soil loss from small watersheds. Observed data from three watershed in the United States were used for this purpose. From this study the following conclusions were drawn:

1. Of the three watersheds tested, EPIC's performance was satisfactory in predicting runoff, but soil loss predicted from Watershed 2 did not compare well with the observed data.
2. EPIC performed well under conventional agricultural crops like wheat, corn and sorghum, but did not predict the soil loss under native virgin grassland. The reason for this is that the biomass growth above the ground was not simulated correctly. In other words, native virgin grasslands would have been well established over a long period of time. But in the EPIC simulations the establishment time allowed was not sufficient. But this could be improved only if observed above ground biomass data is available.
3. In general EPIC performed well on an annual basis even though the monthly and daily predictions were less acceptable in most cases in this study. Spatial variability of hydrological parameters may be one of the reasons for this.
4. The runoff predictions are sensitive to curve number, available water capacity of the first soil layer, and potential heat units. Minimum crop cover factor, miscellaneous parameter that governed winter dormancy, and potential heat units affect soil loss predictions. Apart from these the methods of estimation of potential evapotranspiration and soil loss estimation will affect the runoff and soil loss predictions from marginal to significant levels.

References

GCTE (1996). *GCTE Evaluation of Field-Scale Water Erosion Models: Common Data Sets.* An unpublished communication describing the common data sets.

Hargreaves, G.H. and Samani, Z.A. (1985). Reference crop evapotranspiration from temperature. *Applied Engineering in Agriculture* **1**, 96-99.

Monteith, J.L. (1977). Climate and the efficiency of crop production in Britain. *Philosophical Transactions of the Royal Society of London Series B* **281**, 277-329.

Onstad, C.A. and Foster, G.R. (1975). Erosion modeling on a watershed. *Transactions of the American Society of Agricultural Engineers* **18**, 288-292..

Priestley, C.H.B. and Taylor, R.J. (1972). On the assessment of surface heat flux and evaporation using large-scale parameters. *Monthly Weather Review* **100**, 81-92.

Penman, H.L. (1948). Natural evaporation from open, bare soil and grass. *Proceedings of the Royal Society of London Series A* **193**, 120-145.

Sharpley, A.N. and Williams, J.R., (eds) (1990). *EPIC — Erosion/Productivity Impact Calculator: 1. Model Documentation.* U.S. Department of Agriculture Technical Bulletin No. **1768**.

USDA-SCS (1972). *National Engineering Handbook, Hydrology Section* **4**, Chapters 4-10.

USDA-SCS (1986). *Urban Hydrology for Small Watersheds.* U.S. Department of Agriculture Technical Release No. **55**.

Williams, J.R. (1975). Sediment routing for agricultural watersheds. *Water Resources Bulletin* **11**, 965-974.

Williams, J.R. (1995). The EPIC model. In, Singh, V.P. (ed.). *Computer Models of Watershed Hydrology*, Water Resources Publication, Highlands Ranch, Colorado. pp 909-1000.

Williams, J.R., Jones, C.A., and Dyke, P.T. (1984). A modeling approach to determining the relationship between erosion and productivity, *Translations of the American Society of Agricultural Engineers* **27**(1), 129-144.

Wischmeier, W.H. and Smith, D.D. (1978). *Predicting Rainfall Erosion Losses: a Guide to Conservation Planning.* U.S. Department of Agriculture Handbook No. **537**.

9. VALIDATION OF FIELD-SCALE SOIL EROSION MODELS USING COMMON DATASETS

David Favis-Mortlock

Environmental Change Unit
University of Oxford
5 South Parks Road
Oxford OX1 3UB
UK

Abstract
As a first step toward evaluating the suitability of erosion models for global change studies, common datasets (representing 73 site-years of data from seven sites in three countries) were prepared for use with six field-scale erosion models. Five of these are continuous-simulation types (GLEAMS, EPIC, CSEP, MEDRUSH and WEPP); the other is event-based (EUROSEM). Each dataset was split into a 'training set' and a 'testing set'. Measured values for runoff and erosion from the testing set were withheld from the modellers.

An analysis of results from the continuous simulation models is presented here. A comparison of simulated and measured results for runoff and erosion for the training and testing sets, and at several timescales, reveals several model-specific responses as well as a number of more general conclusions:
- calibration is desirable for many models
- runoff is always better simulated than soil loss
- long-term average results are generally best simulated
- there is some evidence that relative results are more reliable than absolute.

Introduction

The choice of an erosion model for global change studies is not straightforward (Favis-Mortlock *et al.*, 1996). A major reason for this is erosion's variability in space and time (e.g. King *et al.*, 1998; Kirkby, 1998b; Lee, 1998; Rudra *et al.*, 1998). Even if only those models which operate at the spatial scale of fields are considered (Boardman and Favis-Mortlock, 1998), the model user must make several choices.

- Is the primary aim of the study to forecast erosion rates over the long-term, or for individual events (cf. Favis-Mortlock and Boardman, 1995; Favis-Mortlock and Savabi, 1996; Favis-Mortlock *et al.*, 1997)?
- Is estimation of absolute amounts of erosion a requirement, or will relative forecasts suffice (e.g. Boardman *et al.*, 1990; Barfield *et al.*, 1991; Favis-Mortlock *et al.*, 1991)?
- Are estimates of future runoff volumes also needed? This might be the case for studies of future flooding risk (cf. Boardman, 1994; 1995; Boardman *et al.*, 1994; 1995).

- Is information on deposition required (cf. Favis-Mortlock and Savabi, 1996)?
- Must the model categorise sediment loss into size classes (e.g. if fertiliser or pesticide loss were being modelled)?

Table 1. Participants in this model evaluation

Model	Version	Type[1]	Model user	Further discussion of results in Chapter:
GLEAMS	2.3	C	Arlin Nicks, USDA-ARS	6
EUROSEM		E	John Quinton, Silsoe, UK	7
GUEST		E	Calvin Rose, Griffith University, Australia	-
EPIC	5300	C	Jimmy Williams, USDA-ARS	8
CSEP		E/C	Mike Kirkby, Leeds, UK	4
MEDRUSH		C	Mike Kirkby, Leeds, UK	4
WEPP	95.7	C	Mark Nearing, NSERL, USDA-ARS	5

[1] as run in this evaluation; C=continuous, E=event

This paper presents a synthesis of results from the main part of the GCTE evaluation of field-scale erosion models, that which uses common datasets (Table 1). Of necessity, it does not attempt to evaluate model suitability with respect to all the above criteria, but concentrates only on those aspects of model performance which are relevant to the first three questions:

- results at different timescales
- relative vs. absolute results
- results for runoff and erosion.

Thus only results from the continuous simulation models are evaluated here.

Table 2. Location and characteristics of the sites chosen for the common datasets

Source	Country	Location	Site	Area (ha)	Slope (%)	Soil	Land use
USDA	USA	Chickasha, Oklahoma	C-5	5.2	0 - 1	McLain silty clay loam and Reinach silt loam	continuous wheat
		Chickasha, Oklahoma	R-5	9.6	0 - 8	Renfrew, Grant and Kingfisher silt loams	virgin native grassland
		Riesel, Texas	W-12	4.0	2	Houston Black clay	maize - sorghum - wheat rotation
MEDALUS[1]	Portugal	Vale Formoso, Mértola, Alentejo	Plot 10	0.016	15	Mediterranean red soil	wheat-fallow rotation
			Plot 13A	0.016	18	Mediterranean red soil	natural vegetation
Guelph University	Canada	Guelph, Ontario	Plot 3	0.027	8	Guelph loam	continuous maize: minimum tillage
			Plot 4	0.027	8	Guelph loam	continuous maize: conv. tillage

[1] supplemented by data from Tomás and Coutinho (1995)

Data

Data was donated (see Acknowledgements) for 29 plots/small catchments in three countries (Table 2): the USA (two locations), Portugal, and Canada (one location each). Seven datasets — representing a total of 73 site-years — were selected for use in the evaluation, with at least one dataset drawn from each of the four locations. The sites to which these datasets refer range from plots of 0.01 ha to small catchments of just under 10 ha. Slopes vary from almost flat to 18 per cent. A range of soil types is also represented, as are both agricultural and natural vegetation covers. Details of climate, runoff and soil loss at these seven sites are given in Table 3. Note that all soil loss measurements have been converted to volumetric units, assuming a bulk density of 1.3 t m^{-3}. In all cases, it is assumed that measured sediment yield is equivalent to soil loss from the area in question (i.e. that storage on this area is minimal).

Table 3. Climate, runoff and soil loss at the sites

Site	Period of record[a]	Number of years of record[a]	Mean January temp.[b] (°C)	Mean July temp.[b] (°C)	Mean annual rainfall (mm)	Mean annual runoff (mm)	Mean annual soil loss (m³ha^{-1})	Median annual soil loss (m³ha^{-1})
US C-5	1966-1975	10	2.9	8.3	700.9	58.6	0.70	0.26
US R-5	1967-1977	10	2.2	7.1	716.1	40.4	0.05	0.04
US W-12	1985-1989	5	4.8	13.0	876.2	116.4	1.25	0.90
Port. Plot 10	1962-1993	32	7.0	18.7	428.3	28.6	0.44	0.23
Port. Plot 13A	1989-1993	5	7.0	18.7	529.2	75.4	0.02	0.00
Can. Plot 3	1971-1975	5	n/a	n/a	509.7	13.5	2.18	0.68
Can. Plot 4	1971-1975	5	n/a	n/a	509.7	26.4	16.32	16.50

[a] for rainfall, runoff and soil loss; complete years only
[b] in some cases for fewer years than rainfall record

The daily meteorological, runoff and sediment yield data supplied for each site largely meet the requirements of a 'minimum dataset' (Ingram, 1994). However, not all datasets are of equal quality. Upon receipt, several sites were discovered to have some missing or erroneous values. Only monthly totals of runoff and sediment yield were recorded for the US W-12 site; and at both Portuguese sites, rainfall, runoff and sediment yield were recorded only at the end of each rainfall event (i.e. on the last day of multi-day events). No rainfall data with a temporal resolution higher than one day was available for the US W-12 site. At the other sites, the length of record for high-resolution rainfall data (in breakpoint form for the US C-5 and R-5 sites; with hourly resolution for the Portuguese and Canadian sites) was always shorter than for daily rainfall data.

Table 4. Training and testing sets for the continuous simulation models

Site	Training set					Testing set				
	Period of record	N (yr)	Mean annual rainfall (mm)	Mean annual runoff (mm)	Mean annual soil loss (m^3ha^{-1})	Period of record	N (yr)	Mean annual rainfall (mm)	Mean annual runoff (mm)	Mean annual soil loss (m^3ha^{-1})
US C-5	1971-1975	5	768.4	98.8	1.23	1966-1970	5	633.4	18.3	0.17
US R-5	1973-1977	5	749.5	57.1	0.06	1967-1972	6	688.3	26.6	0.04
US W-12	1985-1986	2	970.8	61.0	0.31	1987-1989	3	813.2	199.5	2.66
Port. Plot 10	1978-1993	16	412.4	36.9	0.38	1962-1977	16	444.1	20.4	0.51
Port. Plot 13A	1989-1990	2	773.7	161.0	0.05	1991-1993	3	366.2	18.4	0.01
Can. Plot 3	1971-1973	3	512.0	11.3	2.52	1974-1975	2	506.2	16.8	1.65
Can. Plot 4	1971-1973	3	512.0	22.0	12.35	1974-1975	2	506.2	33.0	22.27

Where possible, missing or erroneous values were corrected by consulting the donors of the datasets. Following the methodology described by Favis-Mortlock *et al.* (1996), each time series of runoff and soil loss was then partitioned into a training set and a testing set. Values of runoff and soil loss were removed from the testing set before distribution to modellers (Table 4). The continuous simulation models were required to reproduce the 'missing' values of runoff and soil loss for the whole of the testing set; event models to reproduce these only for specific events within the training set. Note that, due to the presence of temporal trends, mean values for rainfall, runoff and soil loss in the testing and training sets differed markedly in some cases (e.g. US W-12 and Portuguese Plot 13A) from the averages for the whole dataset (Table 5).

Table 5. Annual average values for the training and testing sets as a percentage of annual averages for the whole dataset

Site	Training set			Testing set		
	Rain	Runoff	Soil loss	Rain	Runoff	Soil loss
US C-5	110	169	176	90	31	24
US R-5	105	141	118	96	66	85
US W-12	111	171	213	93	52	25
Port. Plot 10	96	129	85	104	71	115
Port. Plot 13A	146	214	191	69	24	40
Can. Plot 3	100	84	116	99	125	76
Can. Plot 4	100	83	76	99	125	136

Datasets were distributed to modellers in a standard format (Favis-Mortlock, 1995) which was developed for the GCTE erosion model evaluation. This is based on that proposed by Hunt *et al.* (1994), which is in turn a development of the widely used IBSNAT (1989) file format for crop models. The models were then run with the datasets. Modellers were asked to describe which variables, if any, were used for calibration.

Table 6. Models and datasets: shaded areas indicate which model was run with which dataset

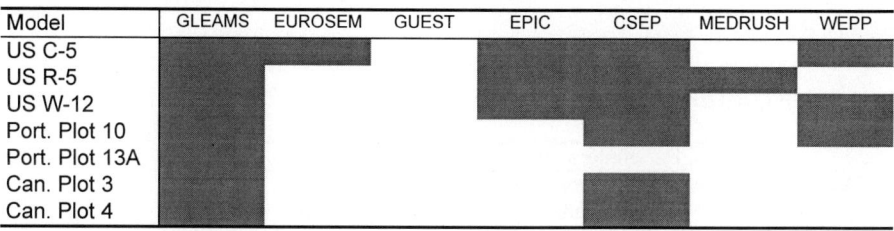

Results

Models and calibration

Not all models were able to work with all datasets (Table 6). Alone of the continuous simulation models, GLEAMS ran simulations with all datasets (Nicks, 1998); details of model calibration are given in Table 7. Due to time constraints (Tharacad Ramanarayanan, personal communication, 1995) EPIC ran only with the US data. These runs were calibrated as described in Table 8 (Ramanarayanan *et al.*, 1998). The CSEP (CEP-EROS in Kirkby, 1998a) model ran all datasets (apart from Portuguese Plot 13A) with substantial calibration (see Kirkby, 1998a for details). The MEDRUSH model ran only with US R-5 and was not calibrated due to time constraints (Kirkby, 1998a).

Table 7. GLEAMS calibration details

Site	Parameters adjusted	Comments
US C-5 US R-5 US W-12	For all: • SCS curve number	'little adjustment needed'[a]
Port. Plot 10 Port. Plot 13A Can. Plot 3 Can. Plot 4	• C factor • Manning's n	'considerable adjustment needed'[a]

[a] Arlin Nicks, personal communication, 1995

Also due to time constraints, the WEPP model ran only the US C-5 and W-12 datasets, and the Portuguese Plot 10 (Nearing and Nicks, 1998). WEPP runs for the US sites were not calibrated; thus in one sense no distinction was made between the testing and training sets. However, no simulations were carried out for the years covered by the testing set at US C-5 (and one year was omitted from the training set). At US W-12, no simulations were run for the years allocated to the training set. Calibration was carried out for WEPP runs with data from Portuguese Plot 10, although the hourly resolution of the Portuguese short-period rainfall data was felt to be too coarse for satisfactory results (Mark Nearing, personal communication, 1995). WEPP was not run for the testing set for this site (also two years were omitted from the training set).

Table 8. EPIC calibration details

Site	Parameters adjusted
US C-5	- SCS curve number - miscellaneous parameter # 6 - total potential heat units for wheat - minimum USLE C factor for wheat crop
US R-5	- SCS curve number - USLE P factor - wilting point for soil top layer
US W-12	- SCS curve number - minimum C factor for maize crop

Of the two single event models, only EUROSEM was able to take part in the evaluation. EUROSEM simulations were carried out for the US C-5 data (Quinton and Morgan, 1998). The GUEST model was not able to run any of the datasets due to lack of appropriate input data (Calvin Rose, personal communication, 1995).

Results for training and testing sets

In the following sections, the convention of using the mean as the measure of distributional central tendency has been followed, even though the median is more appropriate due to the skewness of distributions of runoff and soil loss (Evans, 1990).

Average annual values

When measured and simulated mean annual runoff and soil loss for the training and testing sets (not shown) are compared, results for soil loss show both a rather poorer fit to measured data and a wider divergence between models. For some sites (e.g. US C-5 and W-12), there is no great difference between training and testing sets with regard to their fit to measured data. At other sites there is greater inter-model disagreement in the results from the testing sets. This is almost certainly due to calibration. Particularly good fits to measured data in training set results (again presumably due to calibration) can be seen in the EPIC runs for US C-5 (runoff and soil loss) and US R-5 (runoff only). The 'sign' of the bias in testing set results (i.e. either over- or underestimation) does not necessarily correspond to the 'sign' of the bias in training and testing set averages (Table 5). In general, testing set results both for runoff and soil loss do not consistently underestimate measured values: annual average testing set results are either broadly correct, or overestimated.

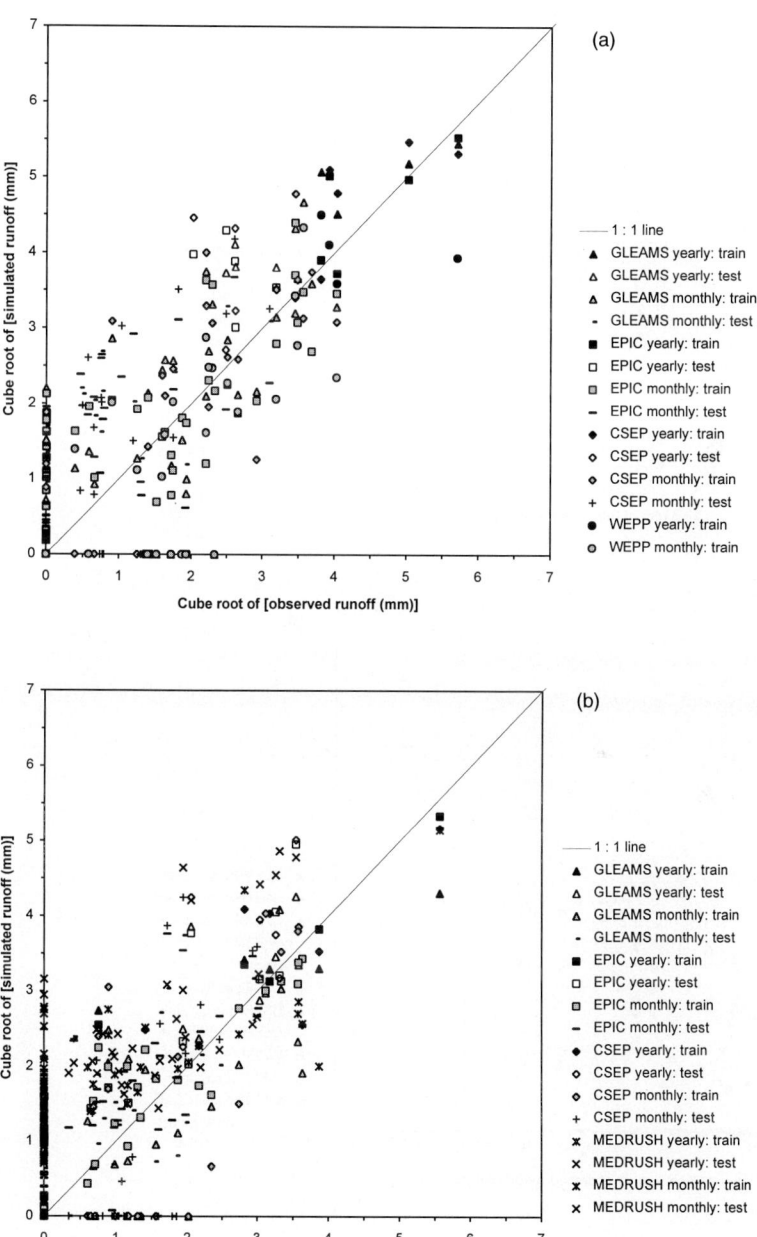

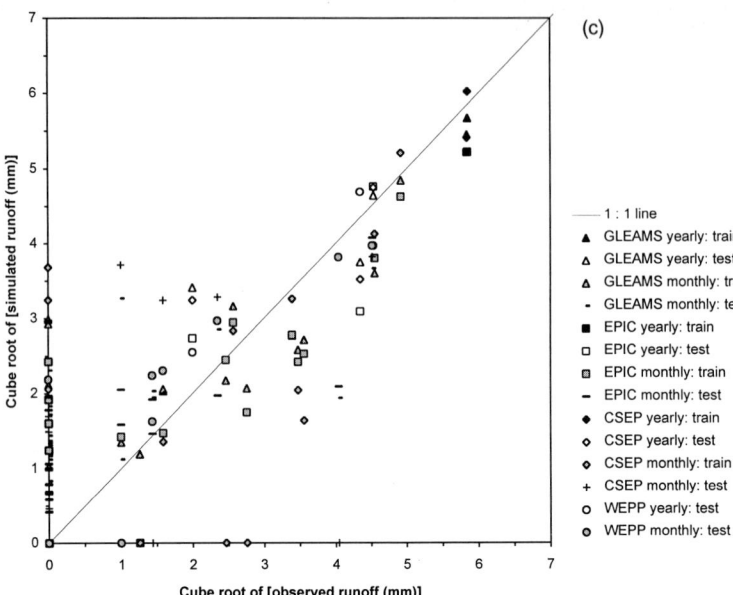

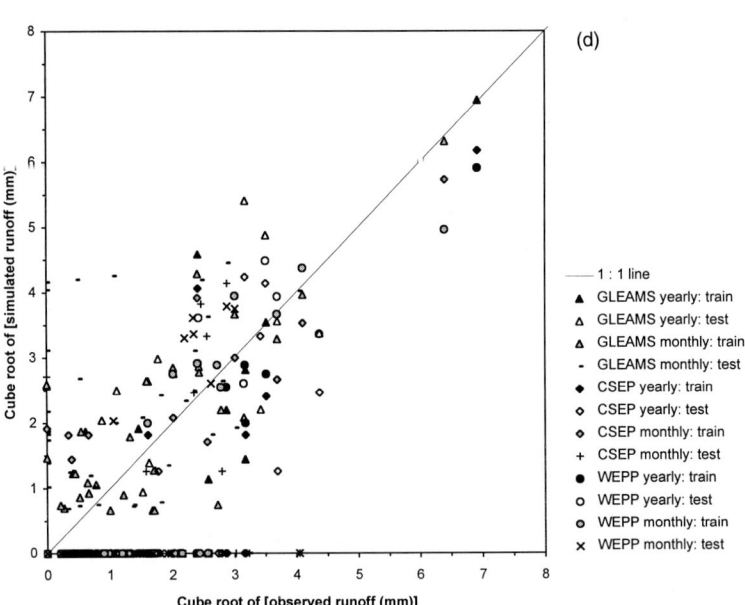

Validation of field-scale soil erosion models using common datasets

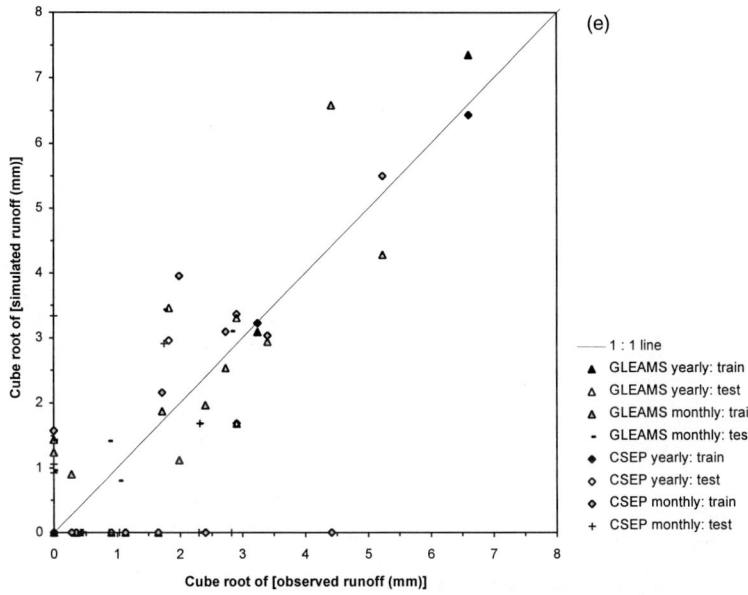

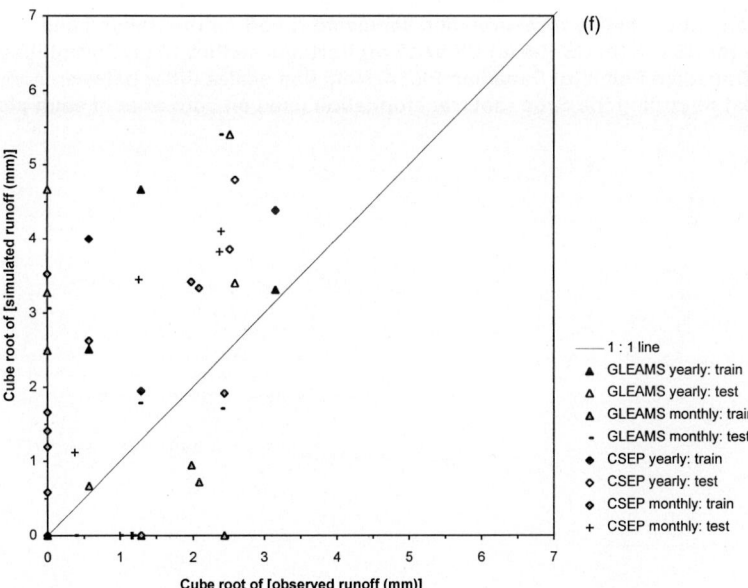

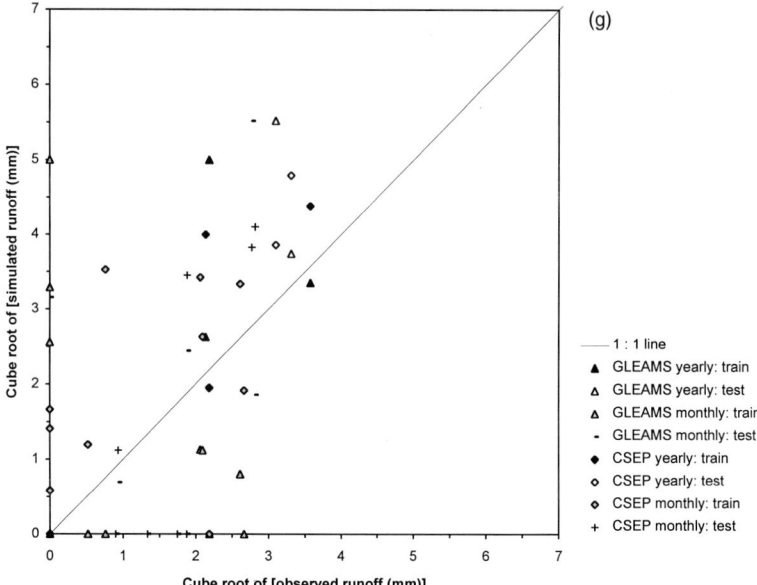

Figure 1. Yearly and monthly measured and simulated runoff for the training and testing sets: (a) US C-5 (b) US R-5 (c) US W-12 (d) Portuguese Plot 10 (e) Portuguese Plot 13A (f) Canadian Plot 3 (g) Canadian Plot 4. Note that scales differ between some plots. See text regarding the cube root transformation used on both axes of each plot

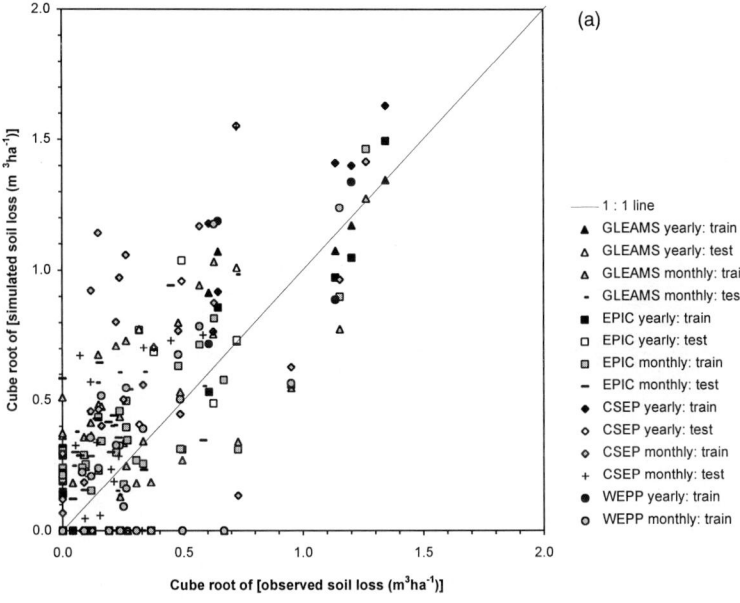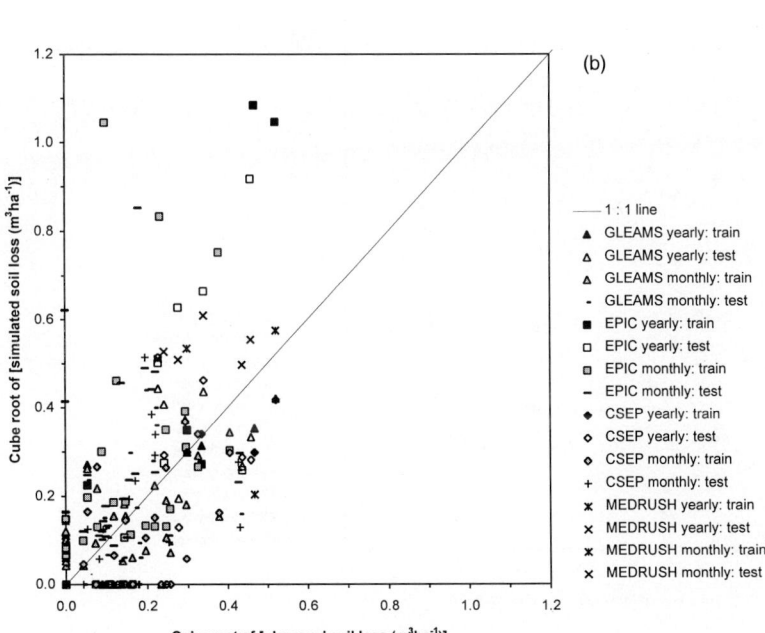

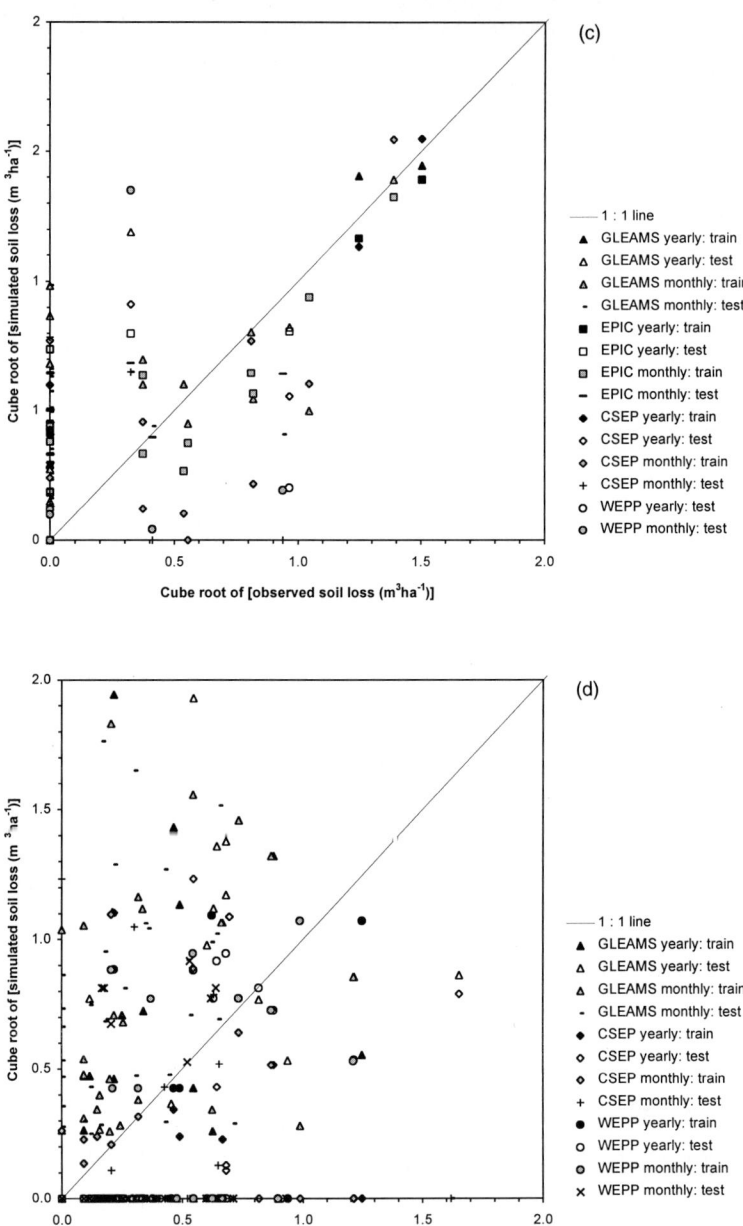

Validation of field-scale soil erosion models using common datasets   101

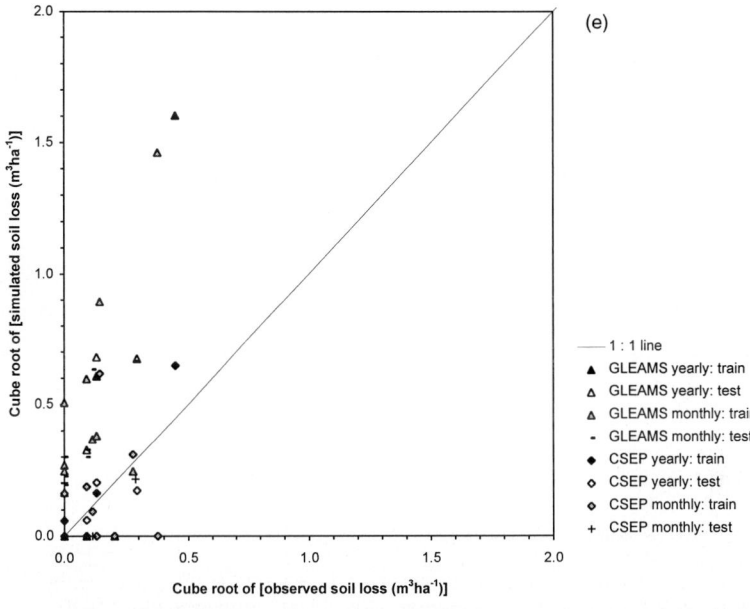

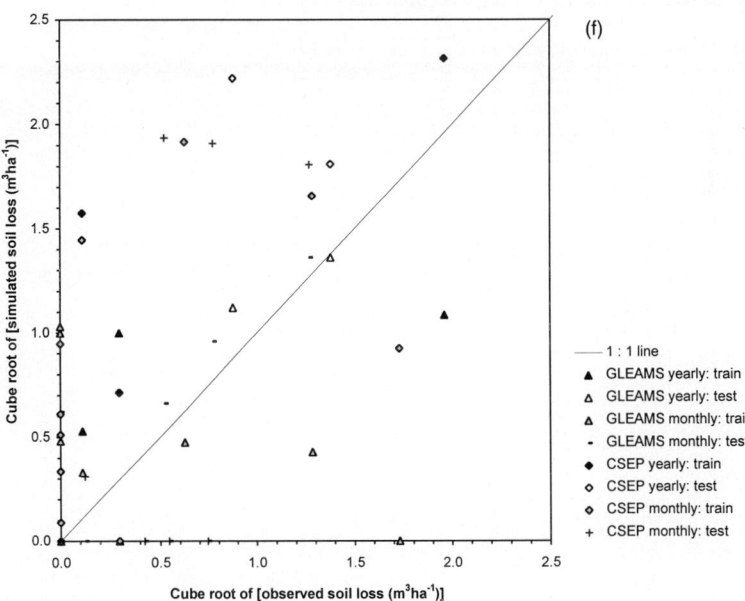

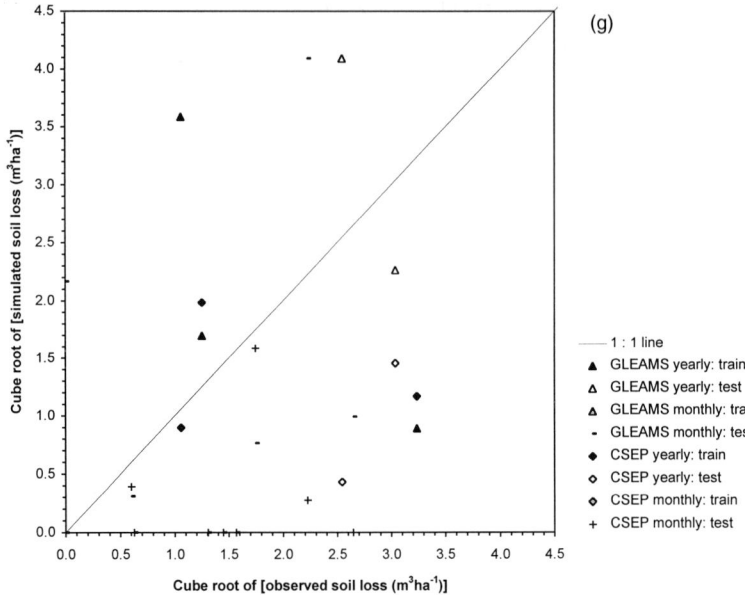

Figure 2. Yearly and monthly measured and simulated soil loss for the training and testing sets: (a) US C-5 (b) US R-5 (c) US W-12 (d) Portuguese Plot 10 (e) Portuguese Plot 13A (f) Canadian Plot 3 (g) Canadian Plot 4. Note that scales differ between some plots. See text regarding the cube root transformation used on both axes of each plot

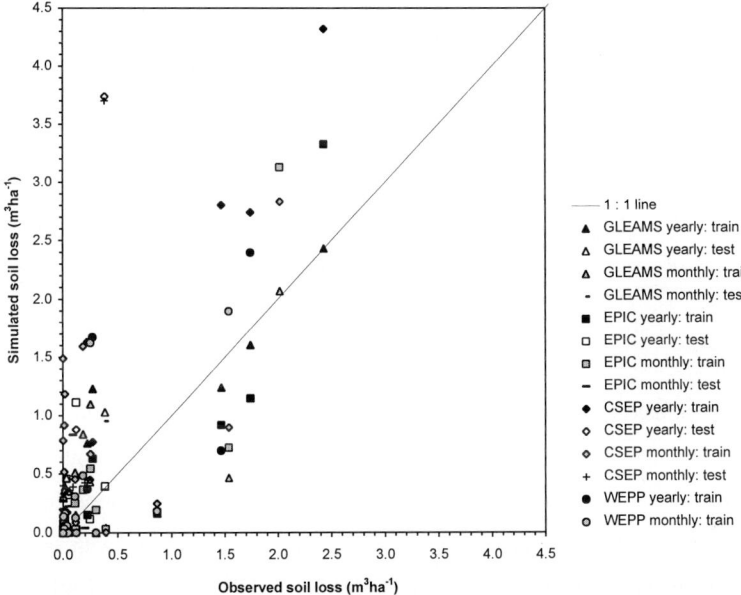

Figure 3. Yearly and monthly measured and simulated soil loss for the US C-5 site, plotted with unscaled axes. Compare Figure 2a

Individual annual values

Absolute results

Results for individual years from the training and testing sets are plotted in Figure 1 (runoff) and Figure 2 (soil loss). Since all distributions are highly skewed, plots using a linear scale show considerable clustering around the origin (Figure 3). Figures 1 and 2 have been plotted using a cube root transformation on both axes to 'unpack' this clustering (compare Figures 2a and 3).

Predictably, there is a wider scatter in every case for individual annual values compared with their averages. As with the average annual values, results for soil loss are in general worse than those for runoff. There are marked outliers for soil loss on the US R-5 and both Portuguese plots (note that all such outliers are overestimates). Results for both Canadian plots are poor for both runoff and soil loss. Again, a particularly good fit to measured data for one annual value in the GLEAMS training set results for US C-5 soil loss is probably due to calibration.

Table 9. Correlation coefficients between measured and simulated annual runoff and soil loss for training (upper) and testing (lower) sets

Model	GLEAMS		EPIC		CSEP		MEDRUSH		WEPP	
	Runoff	Soil loss	Runoff	Soil loss	Runoff	Soil loss	Runoff	Soil loss	Runoff	Soil loss
US C-5	0.74	0.88	0.81	0.86	0.67	0.95	-	-	-0.28	0.44
US R-5	0.97	0.96	0.99	0.93	0.90	-0.37	0.67	0.02	-	-
Port. Plot 10	0.96	-0.11	-	-	0.95	0.12	-	-	0.96	0.45
US C-5	-0.15	0.82	-0.43	-0.25	-0.38	0.85	-	-	-	-
US R-5	0.58	-0.45	0.64	0.56	0.49	0.81	0.62	0.10	-	-
Port. Plot 10	0.46	-0.08	-	-	0.45	-0.07	-	-	0.71	0.21

Relative results

Correlations between measured and simulated annual values for the training and testing sets are given in Table 9 (note that these have been calculated only for those datasets with four or more years data available for both training and testing sets). There are high correlations for GLEAMS, EPIC and CSEP on the data for the training set at the US C-5 site, with soil loss correlations being rather better than those for runoff. However, while soil loss correlations for GLEAMS and CSEP remain high for the testing set, runoff correlations for all models are poorer; EPIC does badly for both runoff and soil loss. WEPP correlations (testing set only) are moderate for soil loss and poor for runoff. Time series for this site (not shown) indicate that GLEAMS' estimates for annual runoff tend to err on the 'insensitive' side, while by contrast those from EPIC are somewhat 'oversensitive' (particularly for the testing set). CSEP is also noticeably oversensitive with respect to annual erosion at this site.

At the US R-5 site, runoff for the testing set correlates extremely well for GLEAMS, EPIC and CSEP; correlations for testing set soil loss are also excellent for GLEAMS and EPIC. Testing set results for runoff are poorer for all models. For erosion, testing set correlations decrease both for EPIC and GLEAMS. Curiously, those for CSEP and MEDRUSH are both better than the training set equivalents. Time series plots (not shown) show that EPIC is much too sensitive with regard to annual runoff, while MEDRUSH and GLEAMS are rather too insensitive for annual soil loss.

GLEAMS, CSEP and WEPP all achieve impressive correlations for runoff in the testing set for Portuguese Plot 10, but do less well for soil loss. Training set correlations for all models decrease for both runoff and soil loss. Interestingly, although all models both overpredict and underpredict annual runoff from year to year, there is some measure of inter-model agreement regarding which years are underestimated and which are overestimated (not shown). However, some entirely spurious peaks for annual soil loss were also produced, particularly by GLEAMS and CSEP.

Time series results for US W-12 (not shown) are broadly similar for all models, and show no great departure from measured values. On the Portuguese Plot 13A, GLEAMS greatly overestimates the initial value for annual soil loss (but not runoff). Both GLEAMS and CSEP are much too sensitive in estimating runoff on the two Canadian sites; however while GLEAMS is a great deal less sensitive than CSEP in estimating soil loss on Canadian Plot 3, both models' responses are reversed on Canadian Plot 4, where CSEP is excessively responsive.

Table 10. Correlation coefficients between measured and simulated monthly runoff and soil loss for training (upper) and testing (lower) sets

Model	GLEAMS		EPIC		CSEP		MEDRUSH		WEPP	
	Runoff	Soil loss	Runoff	Soil loss	Runoff	Soil loss	Runoff	Soil loss	Runoff	Soil loss
US C-5	0.78	0.73	0.75	0.86	0.72	0.65	-	-	0.69	0.75
US R-5	0.74	0.66	0.96	0.19	0.82	0.60	0.70	0.48	-	-
US W-12	0.87	0.85	0.94	0.98	0.83	0.91	-	-	-	-
Port. Plot 10	0.95	0.11	-	-	0.95	0.02	-	-	0.89	0.47
Port. Plot 13A	0.69	0.90	-	-	0.80	0.02	-	-	-	-
Can. Plot 3	-0.07	-0.06	-	-	0.49	0.26	-	-	-	-
Can. Plot 4	-0.08	-0.05	-	-	0.53	0.22	-	-	-	-
US C-5	0.79	0.90	0.49	0.48	0.74	0.92	-	-	-	-
US R-5	0.73	0.21	0.65	0.06	0.67	0.18	0.48	0.31	-	-
US W-12	0.66	0.00	0.80	0.49	0.52	-0.04	-	-	0.96	0.01
Port. Plot 10	0.47	0.03	-	-	0.34	0.00	-	-	0.49	0.25
Port. Plot 13A	0.62	0.78	-	-	0.07	0.75	-	-	-	-
Can. Plot 3	0.68	0.96	-	-	0.93	0.61	-	-	-	-
Can. Plot 4	0.65	0.44	-	-	0.93	0.17	-	-	-	-

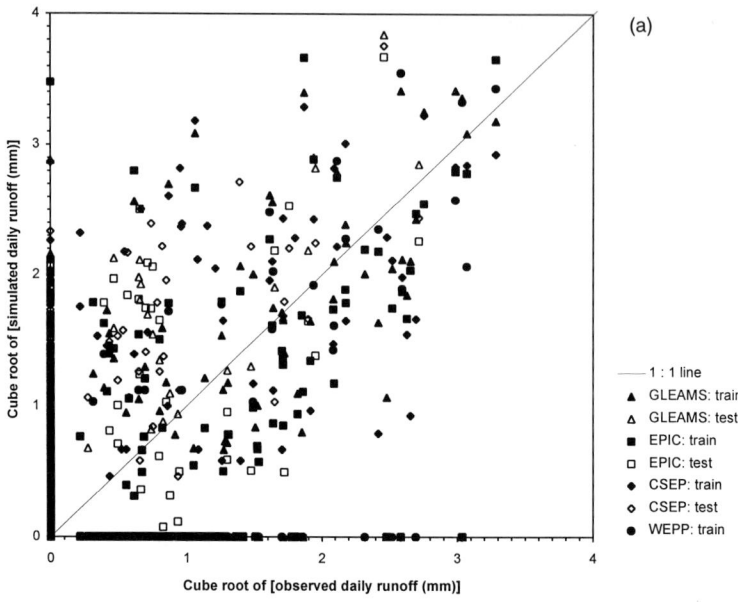

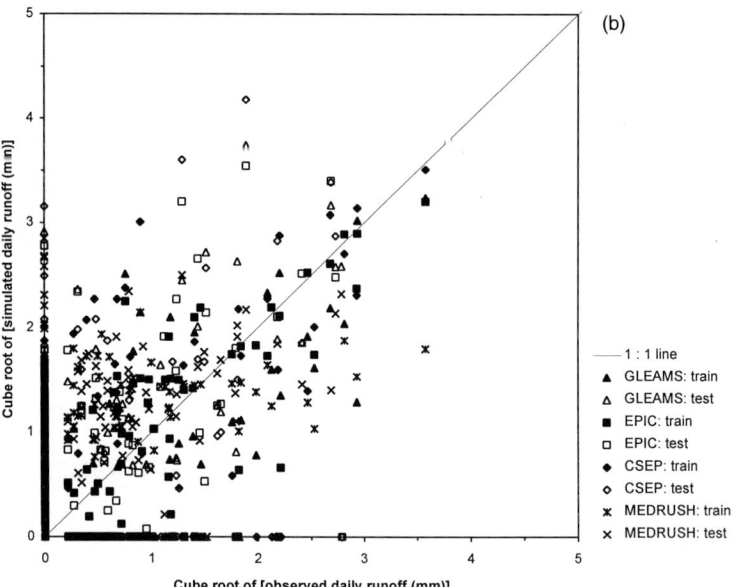

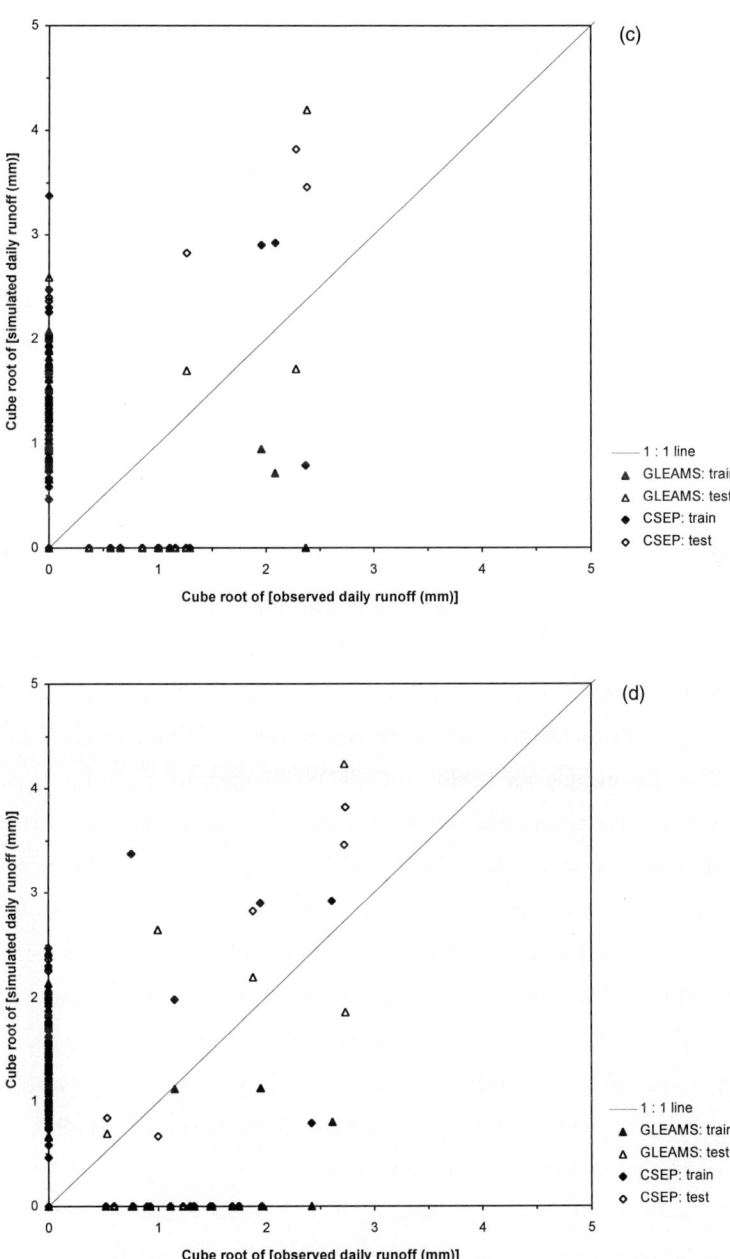

Figure 4. Daily measured and simulated runoff for (a) US C-5 (b) US R-5 (c) Canadian Plot 3 (d) Canadian Plot 4. Note that scales differ between plots, and also in some cases from those in Figure 1

Monthly values

Absolute results

Figures 1 and 2 also show simulated and measured monthly runoff and soil loss. Results for soil loss generally show greater scatter than those for runoff; this is particularly noticeable for US R-5 and both Portuguese plots, where there are several overestimated outliers. Note that for the Portuguese data, uncertainty regarding the exact days on which runoff and erosion occurred during multi-day events could mean that in those events which straddled a monthly boundary, an unknown proportion of the values for runoff and erosion could be allocated to the incorrect month. Monthly results for both runoff and soil loss are poor for both Canadian plots. In general, monthly results are not as good as annual results.

Relative results

Table 10 gives correlation coefficients between simulated and observed monthly values. Overall, monthly correlations are less good than yearly, though more consistent. All models achieve good results for both runoff and erosion for the training set at the US C-5 site; while EPIC's results worsen with the testing set, those of GLEAMS and CSEP improve. At US R-5, simulated runoff values are all good or excellent for the training set, decreasing somewhat with the testing set. Simulated values of erosion are more variable: all models except EPIC do moderately well on the training set (CSEP and MEDRUSH improving greatly on their monthly correlations), while correlations worsen for all models on the testing set. Results are also good for all models for training-set runoff on US W-12, decreasing in quality with the testing set; values for soil loss during the training set are also all very good, but are much worse for GLEAMS and CSEP with the testing set. On both Portuguese plots there is a similar pattern, with runoff faring better than erosion, and training set results better than those from the testing set (except on Plot 13A, where CSEP achieves a good result for testing-set soil loss). On the Canadian plots, GLEAMS curiously does very poorly for both runoff and erosion in the training set, but well or adequately for both in the testing set. CSEP is more consistent, though it also does better on the testing set results.

Monthly time series (not shown) again show the models both over- and underestimating runoff and soil loss for different months during the simulations, but with some degree of inter-model consistency regarding which months are overestimated and which underestimated.

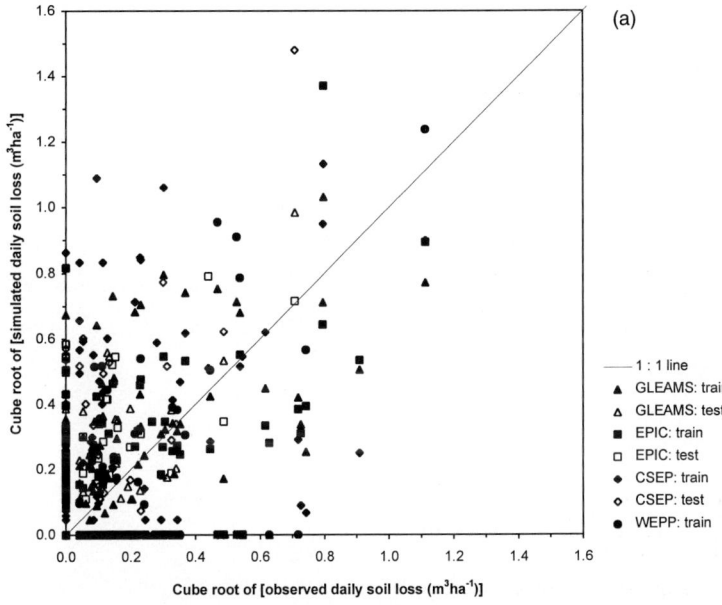

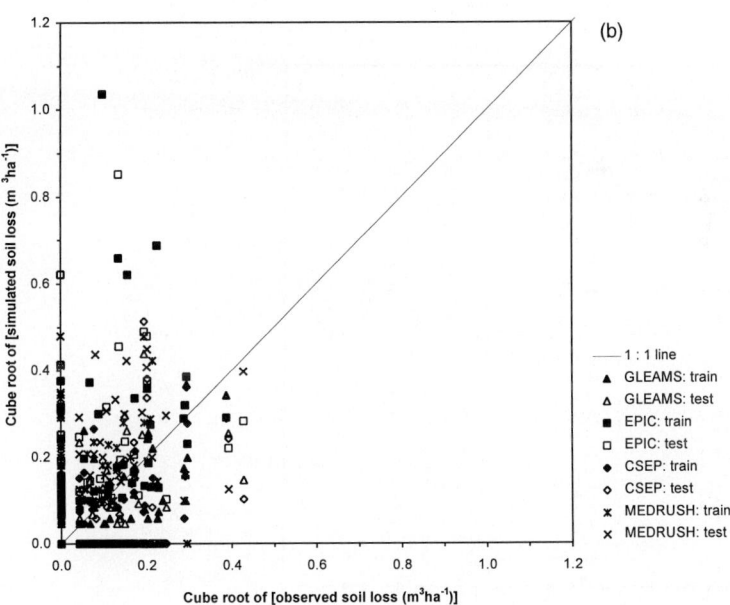

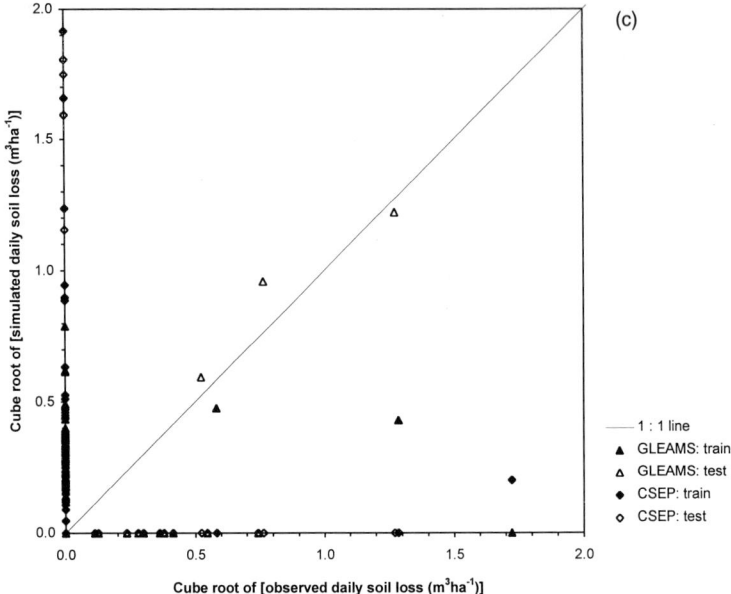
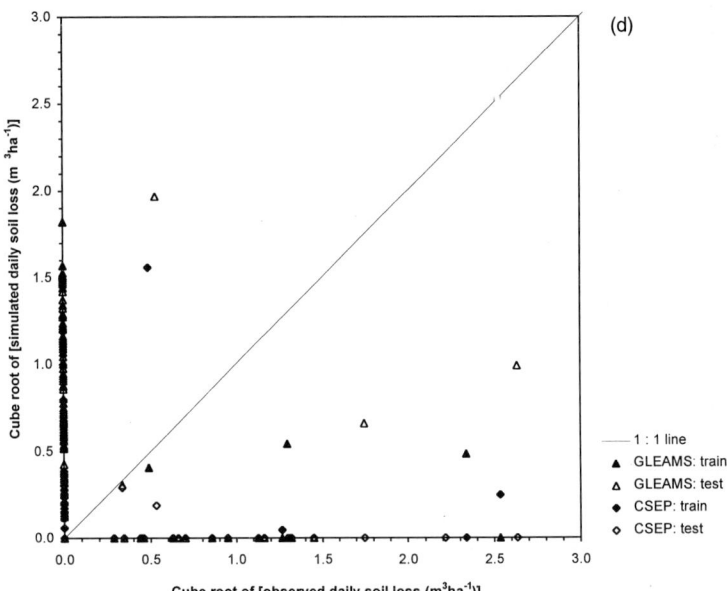

Figure 5. Daily measured and simulated soil loss for (a) US C-5 (b) US R-5 (c) Canadian Plot 3 (d) Canadian Plot 4. Note that scales differ between plots, and also in some cases from those in Figure 2

Daily values

Absolute results

Simulated and measured daily values of runoff and soil loss are compared in Figures 4 and 5 respectively. (Note that results from US W-12 are omitted, since daily values for measured runoff and soil loss are unavailable; also both Portuguese sites are omitted due to the uncertainty regarding measured daily runoff and soil loss in multi-day events. The 'banding' of the lowest values of the R-5 results for soil loss is an artefact, resulting from rounding of the very small numbers involved.) Again there is some tendency toward wider scatter compared with monthly results. At all sites there is a poorer fit for soil loss compared to runoff. Daily soil loss results are very poor at both Canadian sites.

Table 11. Correlation coefficients between measured and simulated daily runoff and soil loss for training (upper) and testing (lower) sets

Model	GLEAMS		EPIC		CSEP		MEDRUSH		WEPP	
	Runoff	Soil loss	Runoff	Soil loss	Runoff	Soil loss	Runoff	Soil loss	Runoff	Soil loss
US C-5	0.76	0.52	0.58	0.50	0.54	0.41	-	-	0.76	0.84
US R-5	0.79	0.73	0.89	0.11	0.79	0.66	0.42	0.24	-	-
Can. Plot 3	0.01	0.05	-	-	0.41	0.00	-	-	-	-
Can. Plot 4	0.01	0.00	-	-	0.44	0.00	-	-	-	-
US C-5	0.80	0.93	0.63	0.62	0.70	0.93	-	-	-	-
US R-5	0.61	0.23	0.55	0.09	0.48	0.18	0.36	0.32	-	-
Can. Plot 3	0.73	0.94	-	-	0.90	0.00	-	-	-	-
Can. Plot 4	0.69	0.48	-	-	0.91	0.00	-	-	-	-

Relative results

Table 11 gives correlation coefficients between measured and simulated daily values (again US W-12 and the Portuguese sites are omitted). A similar pattern to the monthly results can be seen, with values generally lower than their longer-timescale (in this case monthly) equivalents, and poorer for soil loss than runoff. Again though there are exceptions, with some very high values for GLEAMS (Canadian Plot 3, testing set erosion) and CSEP (training-set runoff at both Canadian sites). WEPP does well for daily runoff and (notably so) for daily soil loss.

Results for different timescales

The effects of timescale on model performance are further examined in the following sections. Results from the training and testing sets are combined.

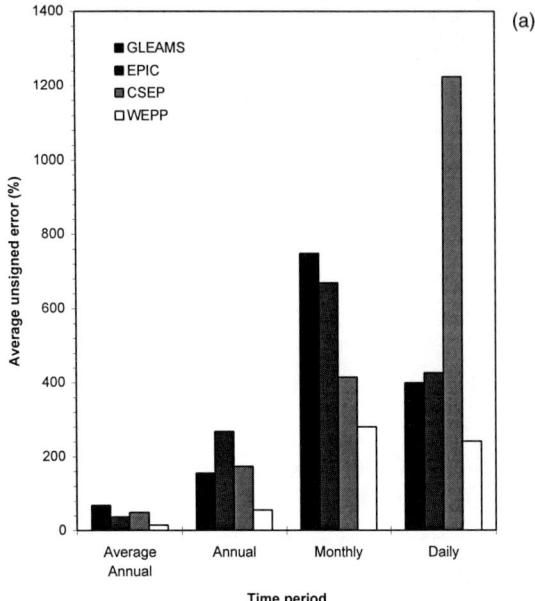

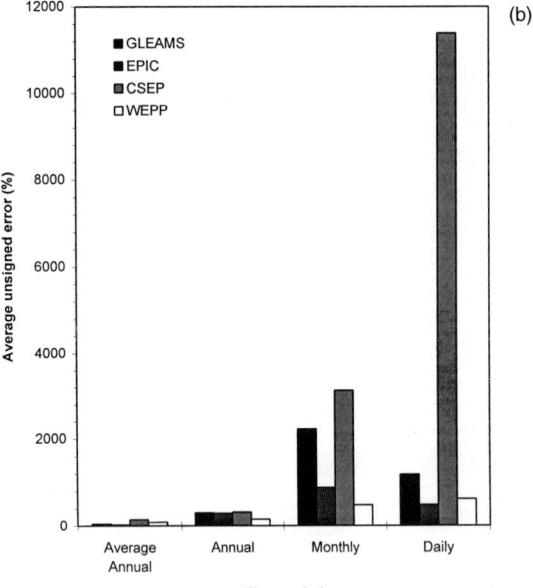

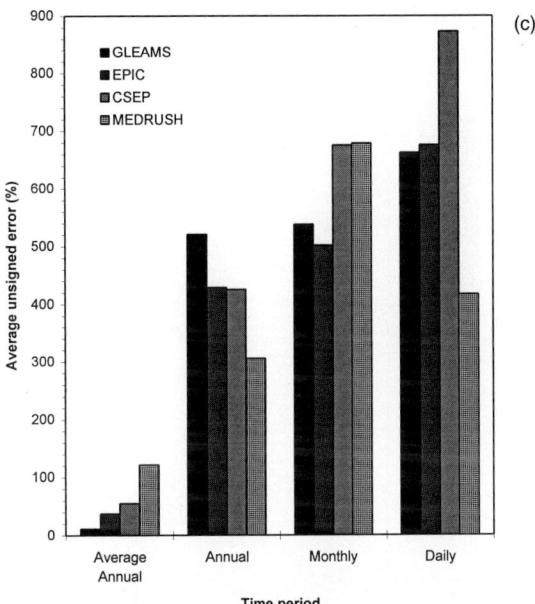
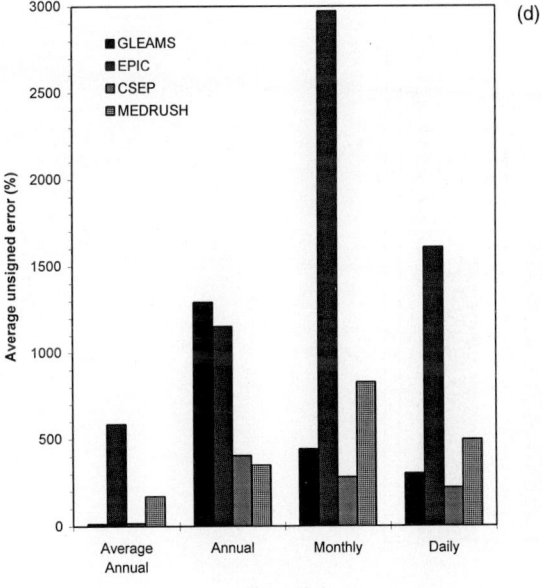

Figure 6. Absolute difference (%) between simulated and measured average annual, annual, monthly and daily values (whole dataset) for (a) US C-5 runoff (b) US C-5 soil loss (c) US R-5 runoff (d) US R-5 soil loss

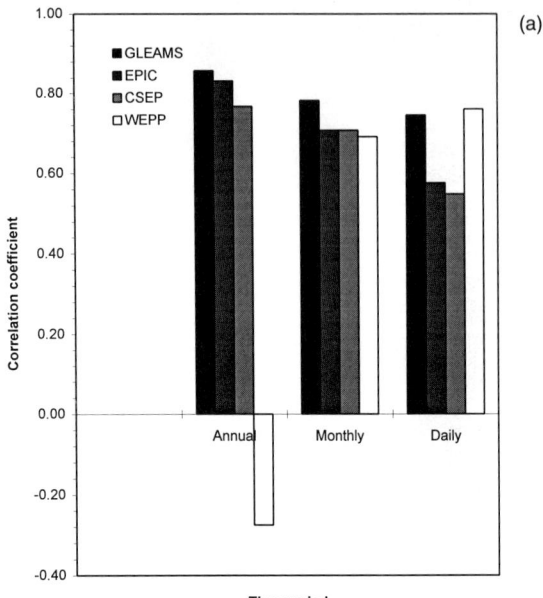

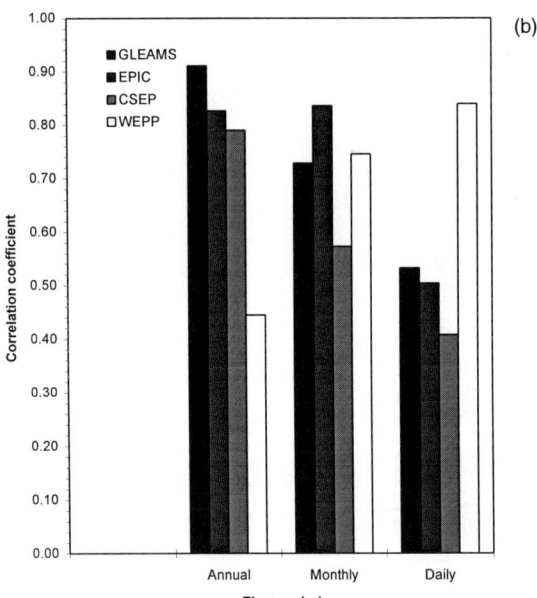

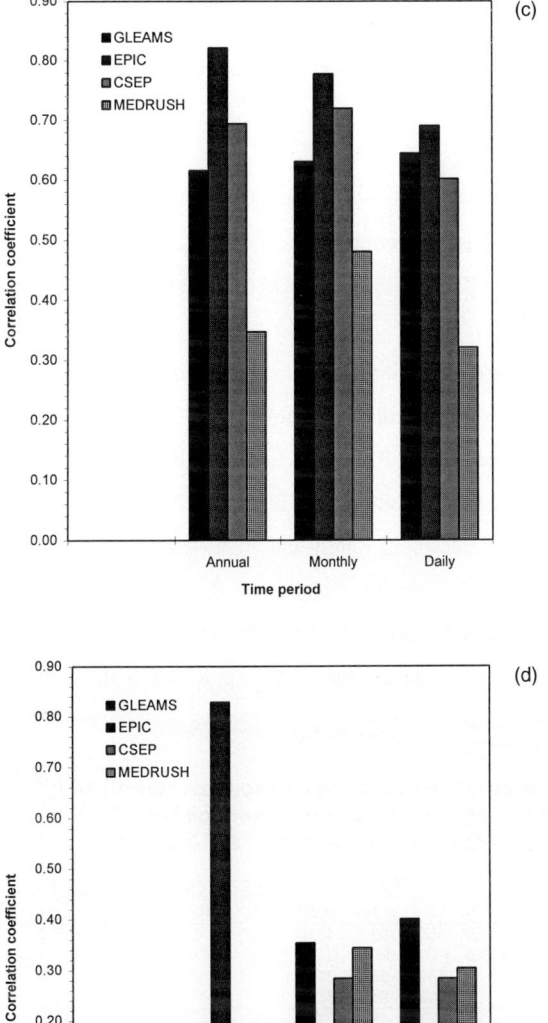

Figure 7. Correlation coefficients between simulated and measured average annual, annual, monthly and daily values (whole dataset) for (a) US C-5 runoff (b) US C-5 soil loss (c) US R-5 runoff (d) US R-5 soil loss

Effects on absolute results

As a measure of absolute error at any timescale, 'average unsigned error' was defined to be the average of the unsigned error $|e|$ (in per cent) for all results at that timescale, where:

$$|e| = 100* \frac{|sim - obs|}{obs}$$

If the observed value is zero and the simulated is non-zero, then $|e|$ is arbitrarily assumed to be 100 per cent. Figure 6 shows average unsigned error values for average annual, individual annual, monthly and daily values of runoff and erosion at the US C-5 and R-5 sites.

In general, average unsigned error increases with shortening timescale. At C-5, average unsigned error is highest at the monthly timescale for runoff from GLEAMS, EPIC and WEPP, but highest at the daily timescale for CSEP. For soil loss, monthly average unsigned error is again highest for GLEAMS and EPIC, while for CSEP and WEPP, the daily value is highest. CSEP's error at the daily timescale (both runoff and erosion) is the highest of all models. At R-5, average unsigned error for runoff is highest at the daily timescale from GLEAMS, EPIC and CSEP, with the monthly value from MEDRUSH being highest. For soil loss at R-5, GLEAMS and CSEP have their highest values at the annual timescale, with EPIC and MEDRUSH highest for monthly values.

Table 12. Percentage of simulated daily runoff (upper) and soil loss (lower) which occurred on correct day irrespective of magnitude of runoff or soil loss. T = total number of daily values; NR = no runoff; R = runoff; NSL = no soil loss; SL = soil loss

Model	GLEAMS			EPIC			CSEP			MEDRUSH			WEPP		
	T	NR	R	T	NR	R	T	NR	R	T	NR	R	T	NR	R
US C-5	3287	98.6	49.7	3653	96.6	52.9	3652	98.6	45.9	-	-	-	1461	100.0	27.9
US R-5	4017	99.5	39.2	4017	98.0	60.5	4017	99.2	34.7	2902	77.7	89.4	-	-	-
Can Plt 3	1826	95.6	31.3	-	-	-	1826	98.2	37.5	-	-	-	-	-	-
Can Plt 4	1826	95.6	30.8	-	-	-	1826	98.4	38.5	-	-	-	-	-	-
	T	NSL	SL	T	NSL	SL	T	NSL	SL	T	NSL	SL	T	NSL	SL
US C-5	3287	98.5	55.2	3653	98.6	39.2	3652	98.5	45.8	-	-	-	1461	100.0	29.6
US R-5	4017	99.1	45.5	4017	99.3	44.6	4017	99.6	30.6	2902	96.2	54.9	-	-	-
Can Plt 3	1826	95.6	31.3	-	-	-	1826	98.2	6.3	-	-	-	-	-	-
Can Plt 4	1826	95.6	30.8	-	-	-	1826	98.7	19.2	-	-	-	-	-	-

Effects on relative results

Figure 7 plots correlation coefficients at the same sites for annual, monthly and daily timescales. At C-5, error in general increases (i.e. correlations decrease) with shortening timescale for both runoff and soil loss from GLEAMS, EPIC and CSEP. WEPP, however, shows the reverse trend, with much better correlations at shorter timescales (compare Tables 9 to 11). At R-5, correlations for runoff either vary little with timescale (GLEAMS), decrease with shorter timescale (EPIC), or peak at the monthly timescale (CSEP and MEDRUSH). For erosion, error generally decreases with shortening timescale (noticeably so for MEDRUSH), although EPIC shows the reverse trend with better correlations at longer timescales.

Further analysis at daily timescale

Occurrence of runoff and erosion

Table 12 shows the percentage of daily results for all models (both training and testing sets) which correctly forecast the occurrence of runoff or soil loss on that day (not calculated for US W-12 or the Portuguese sites due to lack of reliable daily data). The magnitude of forecasts is ignored.

While almost all models are very successful at predicting days on which runoff or soil loss did not occur, they are much worse at forecasting those days on which runoff or erosion did take place. This is due to the comparative rarity of such days (Matthews, 1996). Generally the success rate is less than 50 per cent; results are worst for soil loss on the Canadian plots.

Frequency distributions

Frequency distributions of non-zero daily runoff and daily soil loss are plotted in Figures 8 and 9 respectively (again not calculated for US W-12 or the Portuguese sites). The frequency distributions are expressed as cumulative probabilities.

At C-5, EPIC most closely matches the measured distributions of daily runoff and soil loss, except for the highest daily amounts. However it overestimates the number of days on which runoff occurs; all other models underestimate it (note though that the WEPP results cover a much shorter period of record than the measured values). All models except EPIC tend to overpredict both runoff and soil loss, WEPP notably so for runoff. At R-5 EPIC again best reproduces the observed distribution of runoff, though it strongly overestimates the probability of high values of daily soil loss. GLEAMS and CSEP overestimate the occurrence of daily runoff, while MEDRUSH underpredicts it (but correspondingly overpredicts the number of days on which it occurs). All models apart from EPIC reproduce the observed soil loss distribution adequately.

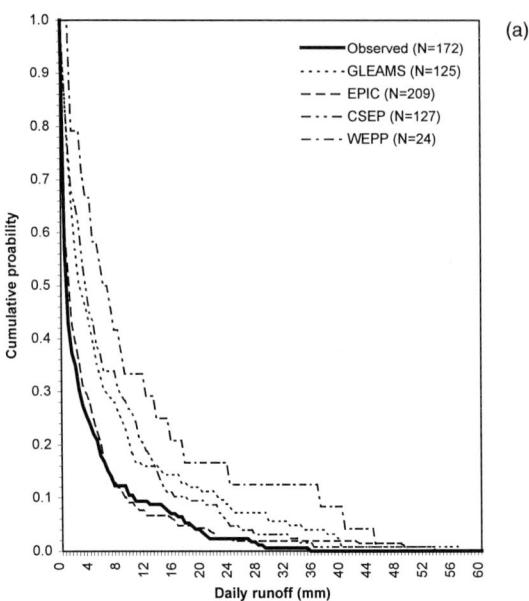

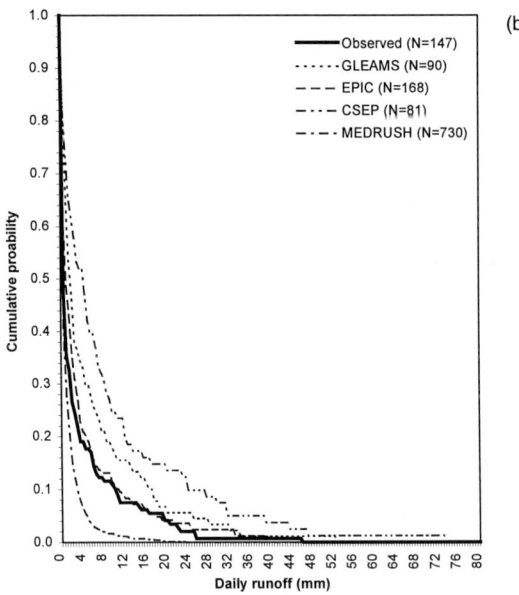

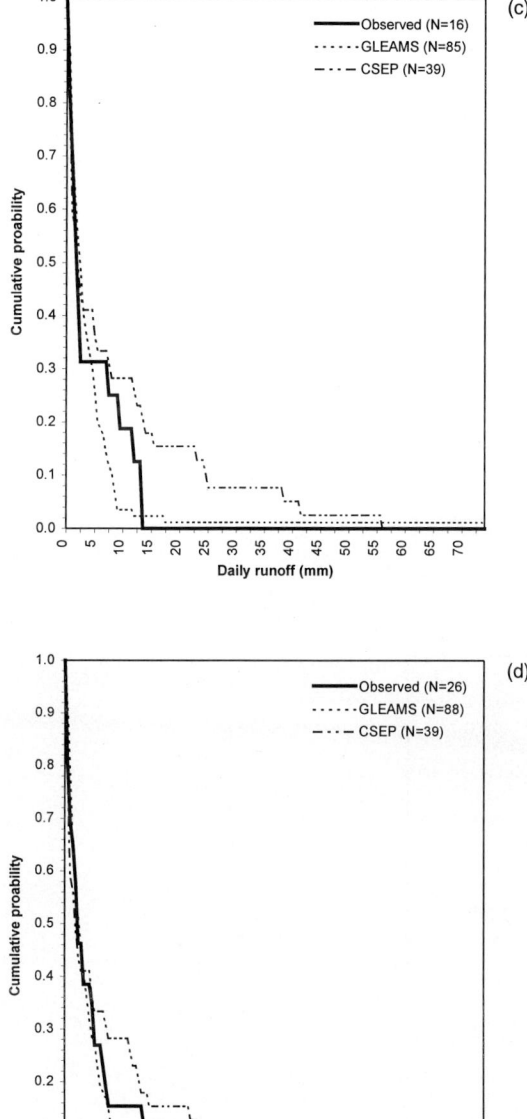

Figure 8. Frequency distributions of non-zero daily runoff expressed as cumulative probabilities for (a) US C-5 (b) US R-5 (c) Canadian Plot 3 (d) Canadian Plot 4

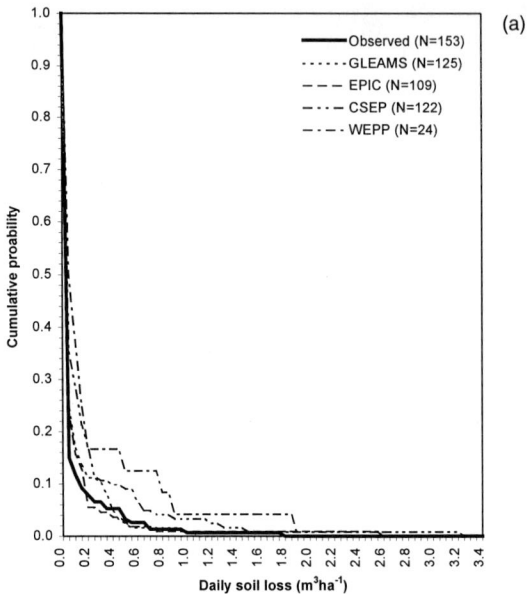

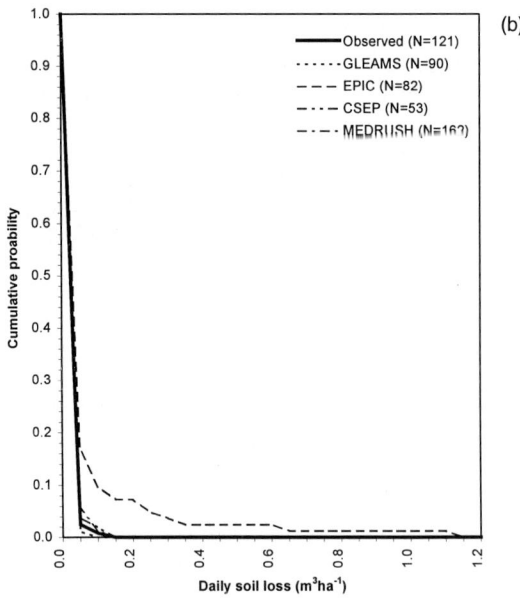

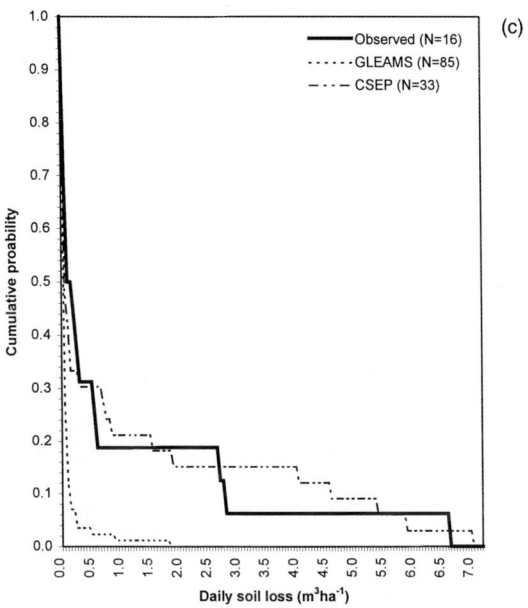

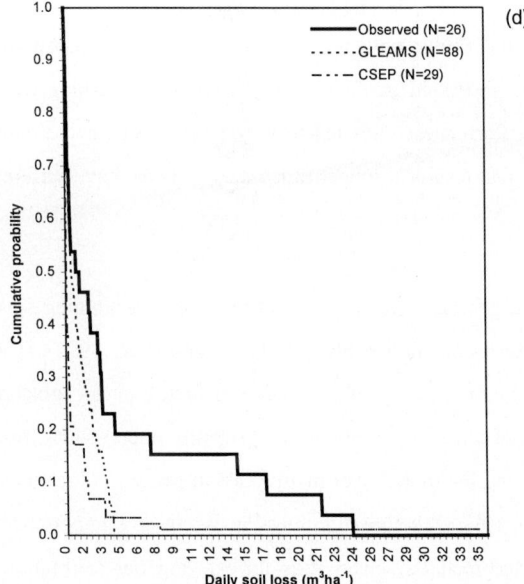

Figure 9. Frequency distributions of non-zero daily soil loss expressed as cumulative probabilities for (a) US C-5 (b) US R-5 (c) Canadian Plot 3 (d) Canadian Plot 4

On the Canadian sites, behaviour is more consistent. On both plots, GLEAMS underpredicts the occurrence of moderate amounts of daily runoff (but overestimates high values), while CSEP provides consistent overestimates. For soil loss, CSEP reproduces the observed distribution on Plot 3 reasonably well, but underestimates the occurrence of moderate values on Plot 4 while overestimating large values. On both plots, GLEAMS underestimates the probability of daily erosion events in each size class.

Discussion

Data requirements and calibration

A first finding is that (notwithstanding practical constraints such as lack of time) not all models were able to run with the supplied datasets. While the variables specified in the GCTE minimum dataset (Ingram, 1994) are appropriate for the majority of erosion models which took part in this evaluation, other models (e.g. GUEST) may have differing data requirements and may as a result be debarred from future evaluations. Further, just how much of a 'minimum' is the variable list? Rainfall data with high temporal resolution, for example, appears to be more essential for some models (e.g. WEPP) than others.

Data quality certainly has an impact on results, both for runoff (e.g. Xu and Vandewiele, 1994) and soil loss (e.g. Jetten *et al.*, 1996). In particular, the uncertainty regarding daily values during multi-day events at the Portuguese sites led to very poor results at the daily timescale (not shown). It seems likely that results at longer timescales will also have suffered (cf. Figure 1d and Table 10).

Since results for testing set data were generally poorer than results from the training sets, calibration appears to have improved results noticeably (cf. Favis-Mortlock, 1994). The preferred strategy for calibration in all cases was to adjust for runoff first, then for erosion. However this is clearly a subjective procedure: for example, the generally good results from GLEAMS are 'due to experience in using the model over many years at many US locations' (Arlin Nicks, personal communication, 1996). The question must be asked: how important is prior knowledge of the site (and/or the model's sensitivities) in carrying out calibration? (Compare the findings of Botterweg, 1995.)

It is however encouraging that nowhere were testing-set results for erosion consistently (i.e. at all timescales) better than results for runoff. This implies that all models are operating — at least notionally — in a 'process-based' manner, in the sense that good prediction of runoff is a prerequisite for good prediction of erosion: in other words, testing-set erosion estimates are not merely the 'right answer for the wrong reason' (Favis-Mortlock *et al.*, 1996. See also Kirkby, 1998a). Since future conditions are necessarily unknown, calibration is particularly undesirable for global change studies (Favis-Mortlock *et al.*, 1996). In reviewing results from the WEPP model for the two US sites, it must be borne in mind that the model was not calibrated for these runs.

The effects of timescale
Not unexpectedly, models in general do better at estimating long term — i.e. multi-year — averages. This seems to be true whether absolute or relative results are considered. However, when the estimation of results for a particular time period (e.g. a year, month or day) is considered, the findings are less clear cut. In some runs, daily results are best: presumably this occurs when the accumulation of same-signed errors over longer timespans becomes significant. In others, annual results are best: this is presumably due to the cancelling of opposite-signed errors.

There is however some evidence that particular models are better when longer or shorter time periods are considered. For example, CSEP's absolute error increases steeply with decreasing timespan at the C-5 site (Figures 6a and 6b) and for runoff at R-5 (Figure 6c); WEPP's absolute and relative error increases only marginally with shortening timespan (Figures 6a and 6b) while relative error actually decreases with shortening timespan (Figures 7a and 7b). Further work is needed here.

Relative vs. absolute results
As discussed above, there appears to be little consistency in the ability of models to better predict absolute or relative results at varying timescales. However, some very high correlations are possible, particularly when models are calibrated (cf. Tables 9 to 11): thus there seems little reason to doubt accepted wisdom (e.g. Barfield *et al.*, 1991) that in many circumstances, relative results from models are more reliable than absolute results. In some cases (e.g. WEPP) relative results somewhat counter-intuitively improve with shortening

timespan, so that daily correlations are better than yearly. This does not necessarily apply for all events in the simulation, however. Overall, there were more overestimates than underestimates, particularly of larger events (cf. Figures 8 and 9). At the monthly timescale, there appears to be some inter-model consistency regarding which values are under- or overestimated. This may result from some deficiency in process representation which operates on all models for particular events.

It is also noteworthy that models find it much more difficult to predict days on which runoff and erosion will occur than days on which it will not occur (Table 12). While in practical terms this may not be a major failing (if only minimal runoff or erosion occurs on the wrongly-predicted days), it is something which requires attention in any model with pretensions to realistic process representation.

Runoff vs. erosion.
In general, runoff is better predicted than erosion. As discussed above, this is to be expected in any 'process-based' model. However, on both Canadian sites (Plot 4 in particular), runoff is consistently over-predicted (Figures 1f, 1g, 4c and 4d) while soil loss, though it has a wider scatter, has no such consistent bias (Figures 2f, 2g, 5c and 5d). This may be due to the effects of snowmelt on these plots (Ramesh Rudra, personal communication, 1996). If so, the results would represent at least a partial case of 'the right answer for the wrong reason'.

Conclusions
- The differing data requirements of erosion models may disadvantage or debar them from participating in model evaluations. Data quality is an important consideration.
- Calibration appears to improve model results, strongly so in some cases. However calibration is undesirable for global change studies, thus models which do not require calibration have some advantage.
- While all models appear to be best at estimating long term (i.e. annual average) values, results for individual time periods can be better or worse for longer or shorter time periods. Nonetheless there are some indications that particular models are best suited to longer or shorter timespans.
- Particularly when calibrated, erosion models can produce excellent correlations for relative results. However more events (particularly the larger ones) were overestimated. There

appears to be some inter-model agreement regarding which events are over- and underpredicted, which may imply that all models evaluated share some common deficiency in process representation. Models also find it harder to correctly estimate the number of days on which runoff or erosion occurs than days on which it does not.
• Runoff is better predicted than erosion.

Acknowledgements

My thanks to all who provided data — the late Arlin Nicks (USDA), Anton Imeson and Mike Kirkby (MEDALUS), Pedro Tomás (CEHIDRO), Trevor Dickinson and Ramesh Rudra (Guelph) — and to all who did the model runs. I am also grateful to Mike Kirkby and Arlin Nicks for their comments on an earlier draft. This paper is a contribution to the Soil Erosion Network of the GCTE, which is a Core Research Project of the International Geosphere-Biosphere Programme.

References

Barfield, B.J., Haan, C.T. and Storm, D.E. (1991). Why model? In, Beasley, D.B., Knisel, W.G. and Rice, A.P. (eds), *Proceedings of the CREAMS/GLEAMS Symposium,* Agricultural Engineering Dept, University of Georgia, Athens, Georgia, USA. pp. 3-8.

Boardman, J. (1994). Property damage by runoff from agricultural land. *Town and Country Planning* **63**(9), 249-251.

Boardman, J. (1995). Damage to property by runoff from agricultural land, South Downs, southern England, 1976-93. *Geographical Journal* **161**(2), 177-191.

Boardman, J., Evans, R., Favis-Mortlock, D.T. and Harris, T.M. (1990). Climate change and soil erosion on agricultural land in England and Wales. *Land Degradation and Rehabilitation* **2**(2), 95-106.

Boardman, J., Ligneau, L., de Roo, A. and Vandaele, K. (1994). Flooding of property by runoff from agricultural land in northwestern Europe. *Geomorphology* **10**, 183-196.

Boardman, J., Burt, T.P., Evans, R., Slattery, M.C. and Shuttleworth, H. (1995). Soil erosion and flooding as a result of a summer thunderstorm in Oxfordshire and Berkshire, May 1993. *Applied Geography* **16**(1), 21-34.

Boardman, J. and Favis-Mortlock, D.T. (1998). Modelling soil erosion by water. In, Boardman, J. and Favis-Mortlock, D.T. (eds). *Modelling Soil Erosion by Water.* Springer-Verlag NATO-ASI series, Heidelberg, Germany.

Botterweg, P. (1995). The user's influence on model calibration results: an example of the model SOIL, independently calibrated by two users. *Ecological Modelling* **81**, 71-81.

Evans, R. (1990). Water erosion in British farmers' fields — some causes, impacts, predictions. *Progress in Physical Geography* **14**(2), 199-219.

Favis-Mortlock, D.T. (1994). *Use and Abuse of Soil Erosion Models in Southern England,* Unpublished PhD Thesis, University of Brighton, Brighton, UK. 310 pp.

Favis-Mortlock, D.T. (1995). *GCTE evaluation of field-scale water erosion models: common data sets.* Unpublished document.

Favis-Mortlock, D.T., Evans, R., Boardman, J. and Harris, T.M. (1991). Climate change, winter wheat yield and soil erosion on the English South Downs. *Agricultural Systems* **37**(4), 415-433.

Favis-Mortlock, D.T. and Boardman, J. (1995). Nonlinear responses of soil erosion to climate change: a modelling study on the UK South Downs. *Catena* **25**(1-4), 365-387.

Favis-Mortlock, D.T. and Savabi, M.R. (1996). Shifts in rates and spatial distributions of soil erosion and deposition under climate change. In, Anderson, M.G. and Brooks, S.M. (eds), *Advances in Hillslope Processes,* Wiley, Chichester, UK. pp. 529-560.

Favis-Mortlock, D.T., Quinton, J.N. and Dickinson, W.T. (1996). The GCTE validation of soil erosion models for global change studies. *Journal of Soil and Water Conservation* **51**(5), 397-403.

Favis-Mortlock, D.T., Boardman, J. and Bell, M. (1997). Modelling long-term anthropogenic erosion of a loess cover: South Downs, UK. *The Holocene* **7**(1), 79-89.

Hunt, L.A., Jones, J.W., Hoogenboom, G., Godwin, D.C., Singh, U., Pickering, N., Thornton, P.K., Boote, K.J. and Ritchie, J.T. (1994). General input and output file structures for crop simulation models. In, Uhlir, P.F. and Carter, G.C. (eds). *Crop Modelling and Related Environmental Data*, CODATA, Paris, France.

IBSNAT (1989). *Technical Report 5: Documentation for the IBSNAT Crop Model Input and Output Files, Version 1.1.* Department of Agronomy and Soil Science, College of Tropical Agriculture and Human Resources, University of Hawaii, USA.

Ingram, J.S.I. (ed.) (1994). *Report of the GCTE Workshop 'Soil Erosion under Global Change'*, Paris, France, 29-31 March 1994, GCTE Focus 3 Associate Office, Oxford, UK. 22 pp.

Jetten, V., Boiffin, J. and De Roo, A. (1996). Defining monitoring strategies for runoff and erosion studies in agricultural catchments: a simulation approach. *European Journal of Soil Science* **47**(4), 579-592.

King, D., Fox, D.M., Le Bissonnais, Y. and Danneels, V. (1998). Scale issues and a scale transfer method for erosion modelling. In, Boardman, J. and Favis-Mortlock, D.T. (eds). *Modelling Soil Erosion by Water.* Springer-Verlag NATO-ASI series, Heidelberg, Germany.

Kirkby, M.J. (1998a). Evaluation of plot runoff and erosion forecasts using the CSEP and MEDRUSH models. In, Boardman, J. and Favis-Mortlock, D.T. (eds). *Modelling Soil Erosion by Water.* Springer-Verlag NATO-ASI series, Heidelberg, Germany.

Kirkby, M.J. (1998b). Modelling across scales: the MEDALUS family of models. In, Boardman, J. and Favis-Mortlock, D.T. (eds). *Modelling Soil Erosion by Water.* Springer-Verlag NATO-ASI series, Heidelberg, Germany.

Lee, J.J. (1998). Cross-scale aspects of EPA erosion studies. In, Boardman, J. and Favis-Mortlock, D.T. (eds). *Modelling Soil Erosion by Water.* Springer-Verlag NATO-ASI series, Heidelberg, Germany.

Matthews, R. (1996). Base-rate errors and rain forecasts. *Nature* **382**, 766.

Nearing, M.A. and Nicks, A.D. (1998). Evaluation of the Water Erosion Prediction Project (WEPP) model for hillslopes. In, Boardman, J. and Favis-Mortlock, D.T. (eds). *Modelling Soil Erosion by Water.* Springer-Verlag NATO-ASI series, Heidelberg, Germany.

Nicks, A.D. (1998). GLEAMS model evaluation — hydrology and erosion components. In, Boardman, J. and Favis-Mortlock, D.T. (eds). *Modelling Soil Erosion by Water.* Springer-Verlag NATO-ASI series, Heidelberg, Germany.

Quinton, J.N. and Morgan, R.P.C. (1998). EUROSEM: an evaluation with single event data from the C5 watershed, Oklahoma, USA. In, Boardman, J. and Favis-Mortlock, D.T. (eds).

Modelling Soil Erosion by Water. Springer-Verlag NATO-ASI series, Heidelberg, Germany.

Ramanarayanan, T.S., Padmanabhan, M.V., Gajanan, G.N., and Williams, J.R. (1998). Comparison of simulated and observed runoff and soil loss on three small United States watersheds. In, Boardman, J. and Favis-Mortlock, D.T. (eds). *Modelling Soil Erosion by Water.* Springer-Verlag NATO-ASI series, Heidelberg, Germany.

Rudra, R.P., Dickinson, W. T. and Wall, G. J. (1998). Problems regarding the use of soil erosion models. In, Boardman, J. and Favis-Mortlock, D.T. (eds). *Modelling Soil Erosion by Water.* Springer-Verlag NATO-ASI series, Heidelberg, Germany.

Tomas, P.P. and Coutinho, M.A. (1995). *Data compendium from the Vale Formoso Experimental Erosion Center 1963/64 to 1992/93. Version 95.1.* Report 1/95, IST/CEHIDRO, Technical University of Lisbon, Portugal.

Wilcox, B.P. and Simanton, J. R. (1998). Predicting runoff in semiarid woodlands: evaluation of the WEPP model. In, Boardman, J. and Favis-Mortlock, D.T. (eds). *Modelling Soil Erosion by Water.* Springer-Verlag NATO-ASI series, Heidelberg, Germany.

Xu, C.-Y. and Vandewiele, G.L. (1994). Sensitivity of monthly rainfall-runoff models to input errors and data length. *Hydrological Sciences Journal* **39**(2), 157-176.

SECTION 3. MODEL EVALUATION WITH USER-SUPPLIED DATASETS

10. PREDICTING RUNOFF IN SEMIARID WOODLANDS: EVALUATION OF THE WEPP MODEL

Bradford P. Wilcox[1] and J. R. Simanton[2]

[1] Environmental Science Group, MS J495
Los Alamos National Laboratory
Los Alamos
NM 87544
USA

[2] USDA - ARS Southwest Watershed Research Center
2000 E. Allen Road
Tucson
AZ 85719
USA

Abstract

Dramatic environmental changes, either from changing climate or land use, or both, will strongly affect erosion processes in semiarid environments. Forecasting the extent and magnitude of these effects will require models in which we have some degree of confidence — or, at the very least, the limitations of which we fully understand. Validating these models, therefore, is critical. In this study, we evaluate the runoff prediction capabilities of the WEPP model, applied to a semiarid pinyon-juniper woodland of the southwestern United States. The model's parameters were drawn from detailed site characterisation data, including data from rainfall simulation studies (the rainfall line studies involved four 30m² plots, two of which were undisturbed; later, the four plots and a nearby 2000 m² hillslope were instrumented to gather data on naturally occurring runoff). WEPP was quite successful at simulating naturally occurring runoff from the disturbed plots, but underpredicted runoff from the undisturbed plots by nearly a factor of three. This was a surprising result, given the quantity of detailed information from the plots used as input for the model. One possibility is that the hydraulic conductivity of the soil — which can vary over time depending on surface soil conditions and, perhaps, rainfall characteristics — was different during the studies of naturally occurring runoff than during the rainfall simulation experiments. With respect to runoff from the hillslope, WEPP's estimates were high; actual measured runoff was considerably lower than that from the plots, presumably because of the greater opportunity for storage and subsequent infiltration on the hillslope — a scale effect that the model's design does not take into account. These results point to the need for caution in applying models to scales significantly different from those for which their parameters were developed.

Introduction

Semiarid landscapes may be among the most sensitive to environmental change (Schlesinger *et al.*, 1990). This is especially true for 'ecotonal' zones, in which a small change in soil moisture can have a large impact on vegetation characteristics (Gosz and Sharpe, 1989) — and changes in vegetation, in turn, can have a large impact on soil erosion. Predicting the magnitude of this impact, however, is much like trying to use a crystal ball: mathematical models, the modern version of the crystal ball, are being and will continue to be used to predict how environmental changes influence soil erosion, but our understanding of the processes and feedbacks involved is incomplete and there are many uncertainties. Model validation studies are necessary to identify inherent model weaknesses and areas of uncertainty.

A new soil erosion model, WEPP (Water Erosion Prediction Project: Flanagan and Nearing, 1995) holds considerable promise as a tool for predicting the impact of environmental change on soil erosion. A process-oriented model based on fundamentals of hydrology and on soil erosion mechanics, WEPP was developed to allow extrapolation to a broad range of conditions. It is, therefore, ideally suited for predicting how climate change affects soil erosion. In this study, we evaluate one component of the WEPP model, the runoff component, for a semiarid woodland in the southwestern United States. Specifically, we evaluate runoff estimates by WEPP for a pinyon-juniper woodland at two scales: (1) the plot scale (30 m²) and (2) the hillslope scale (2000 m²).

Study Area and Methods

Pinyon-juniper woodlands represent an important vegetation type in the southwestern United States, where they cover some 24 million hectares. The range of these woodlands has been in almost continuous flux over the last 12,000 years, expanding and contracting with climatic change; however, their expansion during the last century has been unprecedented and is almost certainly related to land-use patterns (and, possibly, climatic changes as well). What has already happened in the pinyon-juniper woodlands is perhaps a good analogue for future climate- and land-use-induced changes (both are components of global change) in semiarid landscapes (Miller and Wigand, 1994).

The data used for this model validation study were collected from a pinyon-juniper site on the Pajarito Plateau in northern New Mexico. It comprises four 30 m² plots and a 2000 m² hillslope having a well-established herbaceous component and showing little sign of accelerated erosion. This site is one of several in the western United States that were selected to provide data for developing baseline parameters for WEPP for US rangelands (Simanton *et al.*, 1991).

Data collection began in 1987, with detailed rainfall simulation studies on the plots; two of the plots were left undisturbed, whereas the other two were completely denuded of cover, including vegetation, litter, and rocks (the latter we now call the 'disturbed' plots, because there has been regrowth of vegetation since 1987). Rainfall simulation included a 'dry run,' during which water was applied at a rate of about 50 mm/hr for 45 minutes; a 'wet run' 24 hours later, again with rainfall intensities of around 50 mm/hr; and a 'very wet run' 30 minutes after the 'wet run.' For this last run, rainfall was maintained at 50 mm/hr until runoff reached an equilibrium value; rainfall was then increased to 100 mm/hr and again maintained until runoff reached an equilibrium value; at that point, rainfall was returned to 50 mm/hr. This procedure was used for both the disturbed and the undisturbed plots (except, towards the end of the very wet run, additional water was applied to the disturbed plots only, to simulate overland flow). In 1991, the four plots were instrumented for collection of naturally occurring runoff. At first, runoff was simply routed into steel tanks at the downslope end of the plots. In 1995, the collection system was upgraded with the addition of small fibreglass flumes that permit continuous collection of runoff.

The hillslope experiment began in 1994, when we instrumented the 2000 m² pinyon-juniper hillslope for runoff collection. Because there are no permanent channel features on the hillslope, runoff is captured by a 12 m-long gutter installed perpendicular to the hillslope, which routes it through a fibreglass flume. Besides scale, the major difference between the hillslope and the plots is the vegetation cover. The plots are in an intercanopy area, whereas the hillslope encompasses both canopy and intercanopy areas. The canopy areas — pinyon and juniper — cover about 50 per cent of the hillslope; in the intercanopy areas, understory vegetation (mostly sod-forming grasses and semi-shrubs) is well established.

The hillslope version of WEPP (95.7) was used to estimate runoff from the plots and from the hillslope. For the plots, we modelled both the rainfall simulation events and a natural event; for the hillslope, we modelled a natural event. The key parameters used are listed in Table 1. The hydraulic conductivity parameter selected, which was derived from the rainfall simulation experiments conducted on the plots, was one-half the final infiltration rate of the very wet run.

Table 1. Key parameter values by location

Parameter	Undisturbed Plot	Disturbed Plot		Hillslope
		Natural event	Simulated event	
K_e* (mm/hr)	6.4	0.7	0.7	6.4
Antec. soil moisture (%)	60	60	60	60
Interrill basal cover (%)	22	13	0	53
Rill basal cover (%)	48	40	0	19
Canopy cover (%)	27	15	0	60
Random roughness (m)	0.01	0.007	0.005	0.01
Width (m)	3	3	3	12
Length (m)	10	10	10	40
Slope (%)	6	6	6	4
Sand (%)	50	50	50	50
Clay (%)	8	8	8	8
Organic matter (%)	1.4	1.4	1.4	1.4
CEC (meq/100 g)	7.2	7.2	7.2	7.2

* K_e = hydraulic conductivity

Results

Phase I: rainfall simulation

Figure 1 shows the results of the WEPP runoff simulations for one of the undisturbed plots and one of the disturbed plots. The model was able to accurately mimic both the pattern of runoff and the volumes generated during the rainfall simulation experiments, with the exception of peak flow from the disturbed plot, which it overpredicted.

Phase II: natural rainfall at the plot scale

In a second phase, we evaluated WEPP's ability to estimate naturally occurring runoff. The event selected was a storm that occurred on September 8, 1995, and produced 18 mm of rain in about 20 minutes. Antecedent soil moisture was high as a result of a smaller, more gentle rain (11 mm) the previous day. The parameters used for WEPP were the same as those used in the rainfall simulation model runs, except the vegetation cover parameter for the disturbed plots was modified to reflect the regrowth of vegetation that had taken place since the clearing of the plots 8 years earlier (Table 1).

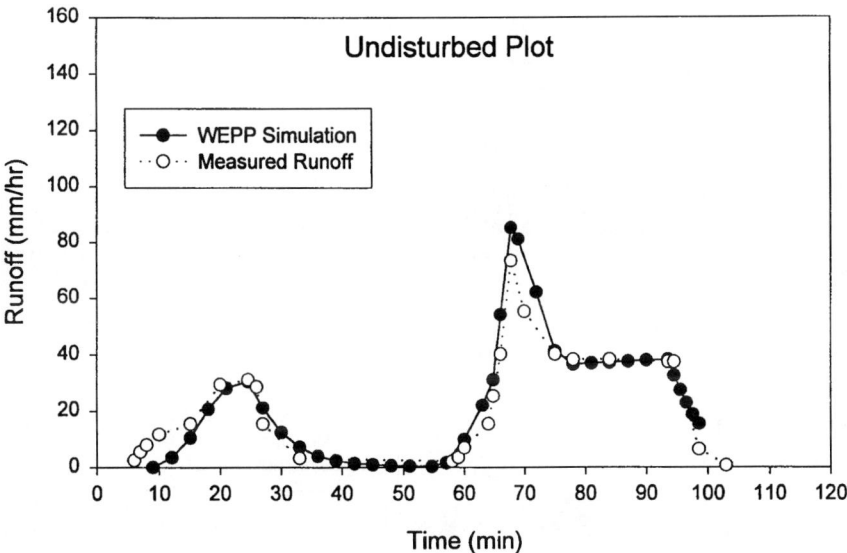

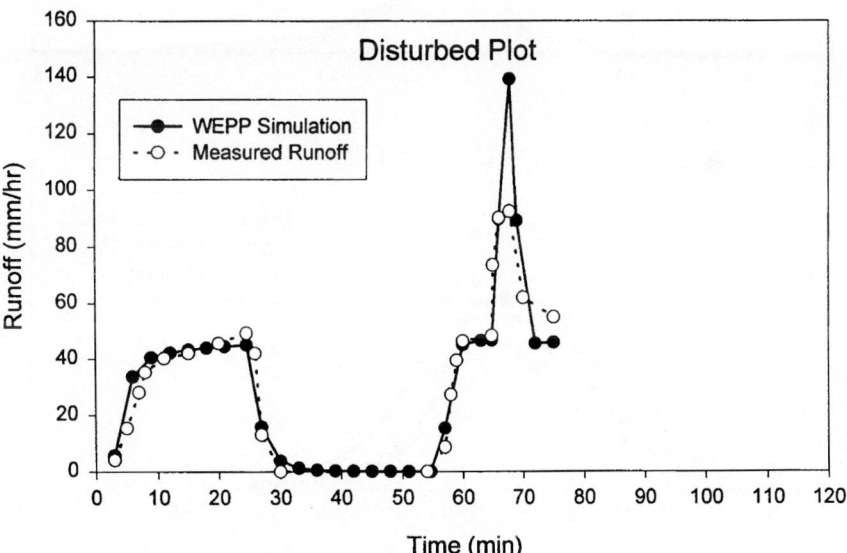

Figure 1. Measured vs. modelled runoff for rainfall simulation experiments conducted on the plots in 1987

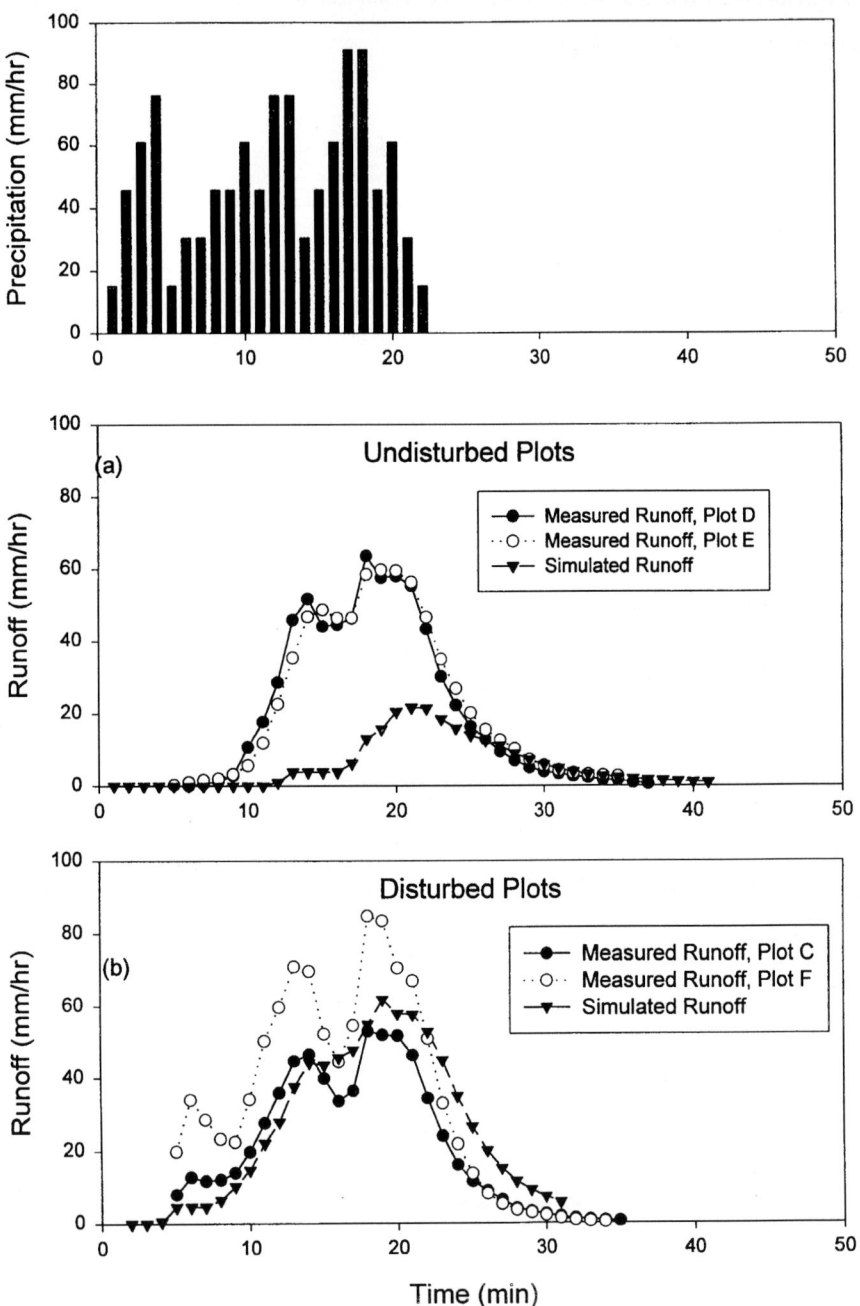

Figure 2. Measured vs. modelled runoff from the undistrubed and disturbed plots for the storm on September 8, 1995

The measured runoff for this event, for each of the four plots, is compared with the WEPP simulation in Figure 2. Runoff efficiency from all of the plots was quite high, ranging from 60 to 90 per cent; this is clearly because of the high antecedent soil moisture, as we have observed with other storms (Wilcox, 1994).

The results for the undisturbed plots were surprising: measured runoff volumes for the two plots were almost identical, but in both cases WEPP greatly underestimated the volumes (Figure 2a). In contrast, the two disturbed plots produced rather different volumes of runoff (we measured about 17 mm for one plot and about 11 mm for the other); in this case, the WEPP simulations were quite good, lying between the 17 mm and 11 mm hydrographs (Figure 2b).

Phase III: natural rainfall at the hillslope scale
In the third phase of our evaluation, we consider runoff from the same storm at a larger scale: that of the 2000 m² hillslope adjacent to the plots. We retained the hillslope version of the model for this phase, rather than use the watershed version, because no permanent channel features are evident on the hillslope. Parameters for the model were based on site characteristics and data from the rainfall simulation experiments (Table 1): hydraulic conductivity was set at 6.4 mm/hr — the same as for the undisturbed plot — but vegetation-cover parameters were set higher, to reflect the presence of tree canopies over about half the hillslope.

Runoff from the hillslope, on a unit-area basis, was markedly lower than that from the plots. We estimate that less than 2 mm of runoff was produced from the hillslope, in contrast to the 11 - 17 mm from the plots. In other words, as scale increased from 30 to 2000 m², runoff efficiency decreased from 60 per cent to less than 10 per cent. Two factors come into play as scale increases from plot to hillslope: (1) the presence of tree canopy areas, from which runoff is lower, reduces overall runoff volumes; and (2) as slope length increases, opportunities for storage and subsequent infiltration of water increase. At the same time, the overall dynamics of runoff from the hillslope were similar to those of the plots, even showing similar bimodal peaks (reflecting rainfall patterns) in the hydrograph. As would be expected, runoff from the hillslope was more prolonged than that from the plots.

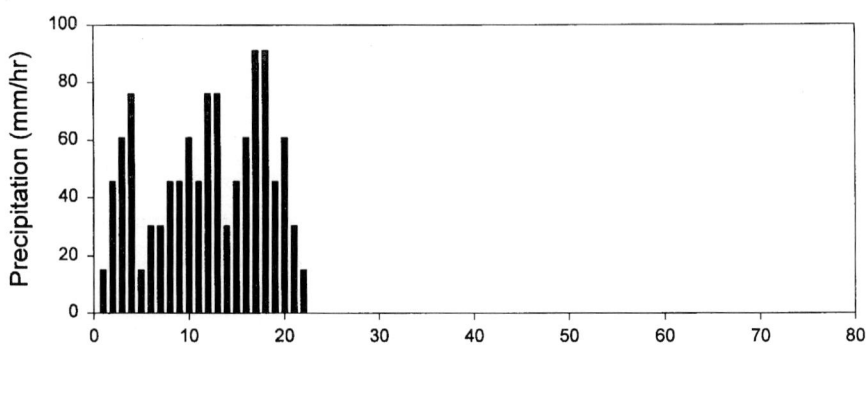

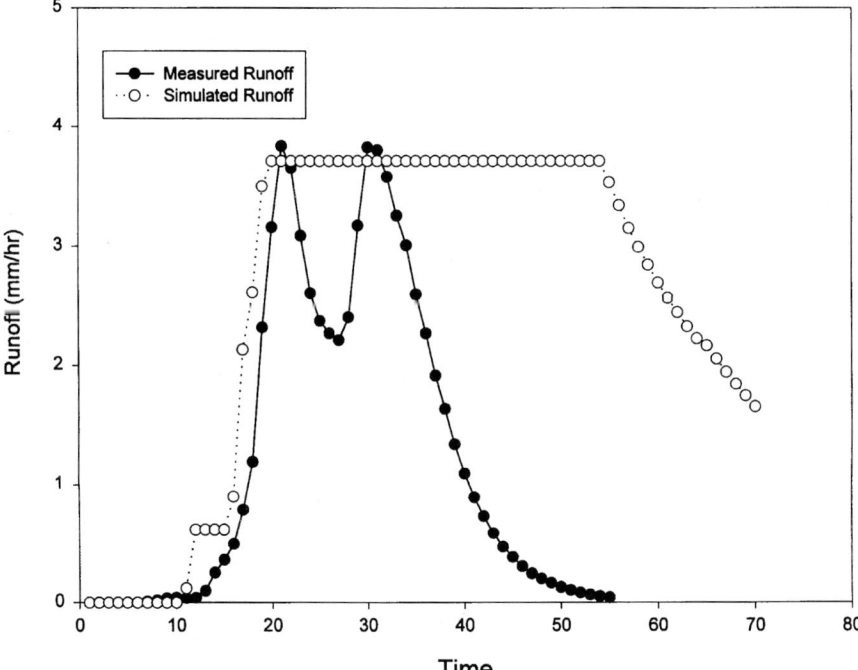

Figure 3. Measured vs. modelled runoff from the hillslope for the storm on September 8, 1995

In this case, the hydrograph produced by WEPP was very close to measured data only for peak flow. For both timing and volume of runoff, the match was poor; for runoff volume, the model's estimation was similar to that for the undisturbed plots (Figure 3). These poor results suggest that even for this relatively small hillslope, channel routing procedures — which are incorporated into the watershed version of WEPP — are needed to adequately allow for water movement on the hillslope.

Discussion and conclusions

This validation study, although limited in scope, has been enlightening. The most surprising result was the model's inability to accurately estimate runoff for the undisturbed plots. Given the detailed characterisation of the hydraulic characteristics of the soil using rainfall simulation, and the fact that these same plots provided base data for the development of the WEPP default parameters for pinyon-juniper woodlands, one would expect model simulations to have been better.

A possible explanation for the discrepancy between the results for the undisturbed plots and those for the disturbed plots is that hydraulic conductivity may have changed in the case of the former. Hydraulic conductivity is the key parameter for runoff prediction by WEPP, and our findings suggest that the hydraulic conductivity of the undisturbed plot was lower during the natural event than during the simulated event. The surface soils at the site are alternately loosened with the freezing/thawing cycles of spring and then recompacted by summer rains. It is possible that at the time of the rainfall simulation experiments, in the early summer, the soil was more permeable than during the natural runoff event, at the end of the summer. Another possibility is that soil hydraulic conductivity is affected by rainfall characteristics. This explanation is not completely satisfying, however, given that rainfall intensities during the natural event were well within the range of intensities applied during the rainfall simulation.

A second important result was the dramatic difference in per-unit-area runoff with scale: much less runoff was generated from the hillslope than from the plots, presumably because of increased opportunity for surface storage and subsequent infiltration. Because the surface conditions on the hillslope are similar to those on the undisturbed plots, the same hydraulic conductivity value was used in the model for both sites. The volume of runoff predicted for the hillslope, therefore, was about the same as that predicted for the undisturbed plot — which in this case constituted an overprediction. Perhaps it is because scale relationships are really poorly understood, particularly for semiarid environments, that the hillslope version of WEPP does not explicitly consider the effect of scale. This result, then, highlights the need for caution in applying a model to scales other than the scale to which it was calibrated, and the importance of understanding the general scale relationships in the environment to be modelled.

The WEPP model, which incorporates significant advances over more mature, less flexible modelling technologies, represents the state-of-the-art in modelling the effects of changes in surface characteristics on soil erosion. Used with caution and a little scepticism, it can be a powerful tool — but as our study has shown, verification through field observations is critical.

Acknowledgements

This work was supported by the Environmental Restoration Project, Los Alamos National Laboratory. Technical editing was provided by Vivi Hriscu. This work contributes to the Global Change and Terrestrial Ecosystem (GCTE) Soil Erosion Network, which is a Core Project of the International Geosphere-Biosphere Program (IGBP).

References

Flanagan, D. C., and Nearing, M. A. (1995). USDA — Water Erosion Prediction Project Hillslope Profile and Watershed Model Documentation. NSERL Report No. 10, National Soil Erosion Research Laboratory.

Gosz, J. R., and Sharpe, P. J. H. (1989). Broad-scale concepts for interactions of climate, topography, and biota at biome transitions. *Landscape Ecology* **3**, 229-243.

Miller, R. F., and Wigand, P. E. (1994). Holocene changes in semiarid pinyon-juniper woodlands. *BioScience* **44**(7), 465-474.

Schlesinger, W. H., Reynolds, J. F., Cunningham, G. L., Huenneke, L. F., Jarrell, W. M., Virginia, R. A., and Whitford, W. G. (1990). Biological feedbacks in global desertification. *Science* **247**, 1043-1048.

Simanton, J. R., Weltz, M. A., and Larsen, H. D. (1991). Rangeland experiments to parameterize the water erosion prediction project model: vegetation canopy cover effects. *Journal of Range Management* **44**, 276-282.

Wilcox, B. P. (1994). Runoff and erosion in intercanopy zones of pinyon-juniper woodlands. *Journal of Range Management* **47**(4), 285-295.

11. EVALUATION OF FIELD-SCALE EROSION MODELS ON THE UK SOUTH DOWNS

David Favis-Mortlock

University of Oxford
Environmental Change Unit
5 South Parks Road
Oxford OX1 3UB
UK

Abstract
This study evaluates three field-scale erosion models — GLEAMS, EPIC and WEPP — against high-quality measured erosion data for a hillslope site in the UK South Downs, collected during the period 1982-88. Calibrated and uncalibrated runs were carried out; however the values used for calibration were constrained so that they remained within an 'acceptable' range, and were consistent between models.

Despite the relatively undemanding nature of this model evaluation, calibration is seen to be essential for all models used. Model results exhibit a wide inter-model scatter. This appears to be in part due to the constraints placed upon the values of calibrated variables: these appear to have prevented models from reproducing measured values with any exactitude. For this dataset at least, there may well be limitations in each model's process descriptions (e.g. regarding crusting for GLEAMS, and regarding the hydraulic implications of soil stoniness for WEPP) which necessitate 'excessive' compensatory calibration. In addition, it appears that identical input parameters can take on subtly different process 'meanings' in different models, and thus may require rather different values for each model.

Introduction

Three field-scale erosion models — GLEAMS, EPIC and WEPP (Table 1) — are evaluated in this study. The models are all used in continuous simulation mode. Model output is compared with measured erosion data for a hillslope field on the UK South Downs: this is within an area of the Downs monitored by Boardman from 1982-91 (e.g. Boardman, 1990; 1993). The study's main aim — as with the other 'user-supplied data' model evaluation in this volume (Wilcox and Simanton, 1998) — is to provide a second opinion on model performance as evinced from the GCTE common-dataset model evaluation (Favis-Mortlock, 1998). Clearly, if a model does well or poorly in one validation exercise, it might be expected to perform similarly in another.

However, the two own-dataset evaluations are not merely replicates, since they are able to supply other, complementary, information on model behaviour (Favis-Mortlock et al., 1996).

- Since the data is supplied by the model users, they will be more familiar with it than is usually the case for common-dataset model runs. Values abstracted from measured data and supplied to the models are therefore likely to be near optimum for that dataset (cf. Quinton and Morgan, 1998). In this respect, the 'user-supplied data' approach to model evaluation tends toward best-case conditions.
- Model users in this instance are not model developers, but more nearly represent the 'typical' user. Intimate knowledge of a model's strengths and weaknesses — in particular, of the sensitivity and acceptable range of input parameters for calibration purposes (below) — may give a considerable advantage to the model developer (cf. Botterweg, 1995). Thus, the common-dataset evaluations are in this sense best-case.
- A degree of inter-model consistency in the choice of parameter values ('constrained calibration' — see below) is possible if one user runs several models.
- The extra datasets extend the range of conditions covered.

Table 1. Model versions and dates

Model	Version	Date	Reference
GLEAMS	2.03	1993	Leonard et al., 1987
EPIC	3090	1993	Williams, 1985
WEPP	95.7	1995	Nearing et al., 1989a

Calibration

One finding of the common-dataset model evaluation (Favis-Mortlock, 1998) is that calibration is still a necessity for many erosion models. Nonetheless, in some cases WEPP appears to be able to produce acceptable results without calibration. Accordingly, the GLEAMS and EPIC runs in this study all made use of calibration, while both calibrated and uncalibrated runs were carried out with WEPP (Table 2).

Whenever calibration is carried out, there is a danger of 'getting the right answer for the wrong reasons', i.e. of merely forcing the model to reproduce some observed value of soil loss (Favis-Mortlock, 1994; Favis-Mortlock et al., 1996). While this may (or may not) be of great practical significance when the calibrated model is used to forecast erosion rates in closely similar situations, use of such a model to predict the erosion likely to result from future climate and land-use will probably be highly dubious. Thus special care must be taken to ensure that calibration does not substitute for adequate process representation. One procedure which can help to guard against this is to compare one or more 'intermediate' results from erosion models, such as runoff amount and timing, against measured data. If — as in this study — such measured data is unavailable, 'intermediate' results may still be usefully compared as a check on inter-model consistency.

Table 2. Simulations carried out in this study

Model	Slope morphology	Weather data	Calibrated?
GLEAMS	concave/convex	• measured 1975-88	yes
		• 30 years simulated by CLIGEN 4.1	yes
EPIC	plane	• measured 1975-88	yes
		• 30 years simulated by EPIC	yes
WEPP	concave/convex	• measured 1975-88	no
		• measured 1975-88	yes
		• 30 years simulated by CLIGEN 4.1	yes

Another, related, aspect of calibration is of particular importance in inter-model evaluations such as this. While all models will of course use the same values for quantities such as daily rainfall amount, other model parameters have a less direct physical expression, e.g. Manning's n or SCS Curve Number. As a result, exact values for such quantities are less easily specified. This is one reason why these parameters are frequently the modeller's first choice for adjustment when calibration is required. A second reason is that parameters describing infiltration are often among the most sensitive in erosion models (e.g. Nearing *et al.*, 1989b; Favis-Mortlock and Smith, 1990). In evaluations where several modelling groups use common datasets, any calibration must necessarily be 'independent' in the sense that individual modellers may independently adjust values of selected input variables. This is likely to result in different values of the same variable being used for different models. The range within which these parameter values may be adjusted is of course constrained by what is considered realistic; but different modellers are free to choose any value within this range.

This study employs an alternative approach, termed 'constrained calibration' (Favis-Mortlock, 1994; Favis-Mortlock *et al.*, 1996). Here, the values of those input variables which are adjusted for calibration purposes are constrained so that the same values are used for all models. Constrained calibration thus has the advantage of ensuring consistency between models. However, it is only practically possible when the same person is carrying out all simulations.

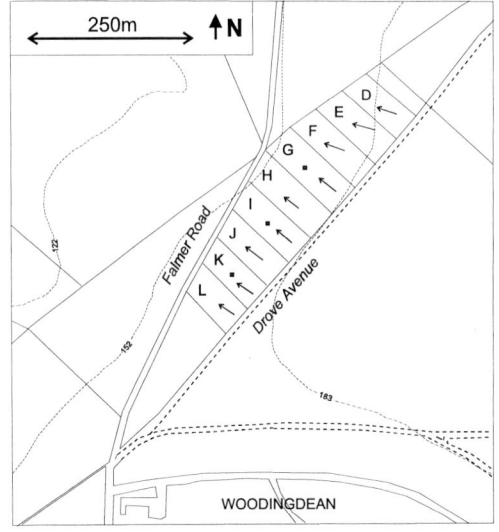

Figure 1. The hillslope field at Woodingdean, East Sussex. (From Favis-Mortlock and Savabi, 1996)

The simulations

The soil loss data are from a 7.7 ha hillslope site at Drove Road, Woodingdean (Figure 1): this is in the UK South Downs, about 6 km south-west of Lewes. Several simulations of erosion have been carried out for this site (e.g. Boardman *et al.*, 1990; Favis-Mortlock *et al.*, 1991; Boardman and Favis-Mortlock, 1993; Favis-Mortlock, 1994; Favis-Mortlock and Boardman, 1995; Favis-Mortlock and Savabi, 1996). Slope angles within the field range from 12 to 20 per cent, with a convexity toward the centre. Rates of soil loss for the whole field for the period 1982-88 were determined by measurement of rill dimensions. The mean annual erosion rate during this period is approximately 2.8 m^3ha^{-1}. This is quite high; a more typical value for the South Downs is around 1 $m^3ha^{-1}yr^{-1}$ (Boardman, 1990). For 1985-86 only, the site was conceptually divided into nine subareas, approximately down the line of greatest slope. Rates for each of the subareas were then measured (Table 3). No measurements of runoff are available.

Table 3. Measured rates of soil loss (m^3ha^{-1}) on subareas D-L, 1985-86

D	E	F	G	H	I	J	K	L	Mean
3.4	7.8	13.7	17.5	21.4	9.6	11.6	11.2	8.1	11.6

Slope profiles for each of the subareas were input to GLEAMS and WEPP. However EPIC cannot consider complex slopes, thus a plane slope was assumed (Table 2).

Mean annual rainfall in the Downs is between 750 and 1000 mm with an autumn peak; mean annual temperature is 9.8°C, with a January mean of 3.9°C and a July mean of 16.3°C (Potts and Browne, 1983). Daily meteorological data are available from a station situated approximately 4 km west-north-west of the site. Weather data for 1975-88 (i.e. 14 years) were used with the models directly. Distributional characteristics extracted from the same dataset were also input to EPIC's weather generator (Richardson and Nicks, 1990) and CLIGEN (Nicks and Lane, 1989) in order to create 30 year time series of synthetic weather (Table 2).

The soil in the Woodingdean field is a shallow (around 20 cm to chalk) and stony silty rendzina of the Andover series (Jarvis et al., 1984). This was parameterised using the values in Table 4.

Table 4. Andover soil details. Not all parameters are used by all models

	Layer				
	1*	2	3	4	5
Depth to bottom of layer (m)	0.01	0.15	0.20	0.30	1.00
Bulk density (t m^{-3})	1.35	1.35	1.39	1.54	1.54
Wilting point (m m^{-1})	0.23	0.23	0.23	0.18	0.05
Field capacity (m m^{-1})	0.45	0.45	0.45	0.31	0.31
Sand (%)	18.9	18.9	18.9	25.0	25.0
Silt (%)	77.6	77.6	77.6	51.0	51.0
Organic N (g t^{-1})	3600	3600	2800	1300	700
pH	7.5	7.5	7.6	8.2	8.2
Sum of bases (cmol kg^{-1})	100.0	100.0	100.0	100.0	100.0
Organic carbon (%)	4.09	4.09	2.79	1.30	0.70
Calcium carbonate (%)	60.0	60.0	60.0	80.0	80.0
Cation exchange capacity (cmol kg^{-1})	45.0	45.0	39.0	30.0	14.0
Coarse fragments (% vol)	38.1	38.1	50.0	90.0	90.0
Nitrate concentration (g t^{-1})	15	15	15	10	5
Labile P concentration (g t^{-1})	15	15	12	8	3
Crop residue (t ha^{-1})	8	8	8	0	0

*This layer used only with EPIC

GLEAMS and EPIC partition rainfall into runoff and infiltration by means of the SCS Curve Number approach (USDA - Agriculture Soil Conservation Service, 1972; Rallison, 1980; Nearing et al., 1996). In keeping with the concept of constrained calibration, this sensitive parameter was set to the same base value for both models (Table 5). Values for Manning's n were similarly set for both models.

Table 5. Hydrological and erosional parameters

	GLEAMS	EPIC	WEPP uncalibrated	WEPP calibrated
SCS Curve Number	78	78	n/a	n/a
Manning's n for overland flow	0.010-0.023 (varies with cropstage)	0.010	n/a	n/a
Manning's n for channel flow	0.030	0.030	n/a	n/a
Effective hydraulic conductivity of top soil layer (mm hr^{-1})	1.8	1.8*	2.1[b]	3.0[b]
Erodibility		0.31[a]	0.35[a]*	5502700 (K_i) 0.0871 (K_r) 3.5 (τ_c)	2000000 (K_i) 0.0050 (K_r) 7.0 (τ_c)

* estimated internally by the model
[a] USLE erodibility (Wischmeier and Smith, 1978)
[b] 'baseline' effective hydraulic conductivity for WEPP (K_b)

Values for hydraulic conductivity however could not be maintained across models, in part due to each model's differing conceptualisation of the effects of this parameter. For both GLEAMS and WEPP, this was one of the parameters used for calibration. In GLEAMS, the value used for top-layer hydraulic conductivity was adjusted downward, with the aim of representing the effects of crusting on infiltration on these silty soils (e.g. Boardman, 1993). GLEAMS does not explicitly simulate crusting; nor does EPIC. WEPP, by contrast, has a well developed crusting submodel which assumes an exponential decrease of hydraulic conductivity with total rainfall since tillage (Alberts et al., 1995). Values for the uncalibrated runs were calculated using the empirical relationships given in the WEPP documentation (Alberts et al., 1995, p 7.1 and Alberts et al., 1995, p 7.9). However, values for WEPP's parameter for effective hydraulic conductivity, together with values for its three erodibility parameters, had to be adjusted in the calibrated runs to compensate for the effects of soil stoniness (Tables 4 and 5). Stone content may affect infiltrability and (more particularly) erodibility, which can be increased or decreased (e.g. Poesen, 1992). The presence of a stone cover decreases apparent erodibility (Evans, 1996), due to the stones' shielding effect (Gerard Govers, personal communication, 1995). On the South Downs, Boardman (1992) noted a long-term decrease in erodibility of soils as a result of increasing stoniness. In a sensitivity analysis of WEPP, Nearing et al. (1989) found hydraulic conductivity to be the most sensitive parameter tested. The parameter for rill erodibility (K_r) was also very sensitive: Mark Nearing (personal communication, 1995) recommends that K_r should be the first choice for adjustment if calibration of WEPP is to be undertaken, since its value is usually the primary source of uncertainty.

As described by Favis-Mortlock and Savabi (1996), values for WEPP's three erodibility parameters and effective hydraulic conductivity parameter were therefore subjectively adjusted (Table 5). The parameters for interrill and rill erodibility (K_i and K_r) needed to be reduced, while the critical shear stress parameter τ_c was increased. The value for the baseline effective hydraulic conductivity parameter K_b was also increased. As in all calibrations, adjusted values were constrained to remain within the range of recommended values: however the WEPP documentation recommends a maximum value of 6 for cropland τ_c.

Table 6. Tillage operations for Woodingdean

Operation	Date	Model(s)
chisel plough	20 Aug	all
harrow	15 Sep	all
drill winter wheat	28 Sep	all
roll	1 Oct	all
start of USLE establishment cropstage* (< 50% cover)	1 Jan	GLEAMS only
start of USLE development cropstage (< 75% cover)	1 Mar	GLEAMS only
start of USLE mature:90 cropstage (< 90% cover)	1 Apr	GLEAMS only
harvest	29 Jul	all

* Wischmeier and Smith, 1978

Continuous winter wheat was simulated (Table 6). Management details for all simulations are based on practices commonly adopted by South Downs farmers (Favis-Mortlock, 1994).

Table 7. Simulated soil loss (t ha^{-1} yr^{-1}) using 1975-88 measured weather

	D	E	F	G	H	I	J	K	L	Mean
GLEAMS	2.3	2.4	2.5	2.6	2.5	2.4	3.0	2.5	2.6	2.5
EPIC	6.1	4.5	7.7	10.4	14.4	12.6	12.1	10.5	10.0	9.8
WEPP uncalib.	99.1	85.1	146.5	186.7	228.6	220.1	211.9	167.5	161.5	167.5
WEPP calib.	8.8	8.4	22.5	33.8	43.3	41.1	35.0	21.1	18.4	25.8

Table 8. Simulated soil loss (t ha^{-1} yr^{-1}) using 30 years of simulated weather

	D	E	F	G	H	I	J	K	L	Mean
GLEAMS	2.3	2.8	2.4	2.8	2.6	3.2	2.7	2.4	2.4	2.6
EPIC	3.7	2.8	4.7	6.3	8.9	7.7	7.4	6.4	6.1	6.0
WEPP	6.5	6.0	17.3	26.4	33.5	31.6	26.6	16.1	13.8	19.8

Results

Absolute values

Values for average annual soil loss from the runs which used 14 years of real weather data are presented in Table 7; results for the runs which used 30 years of simulated weather are in Table 8. For GLEAMS, these results are the sum of the erosion attributed to overland flow and rill flow. Results for GLEAMS and EPIC are for the whole of each subarea: WEPP results are for that part of each subarea which experienced a net soil loss in each event.

Measured average annual soil loss for the whole field during 1982-88 was 3.8 t ha^{-1}. Table 9 compares each model's estimate of soil loss for the whole field with this value. In general, GLEAMS has underestimated annual average soil loss, while EPIC and WEPP have overestimated it. The uncalibrated WEPP runs show a considerable overestimation.

Table 9. Measured and simulated soil loss (all strips)

	Measured weather		Simulated weather	
	Mean soil loss (t ha^{-1}yr^{-1})	% of 1982-88 measured soil loss	Mean soil loss (t ha^{-1}yr^{-1})	% of 1982-88 measured soil loss
GLEAMS	2.5	74	2.6	77
EPIC	9.8	289	6.0	177
WEPP calib.	25.8	763	19.8	585

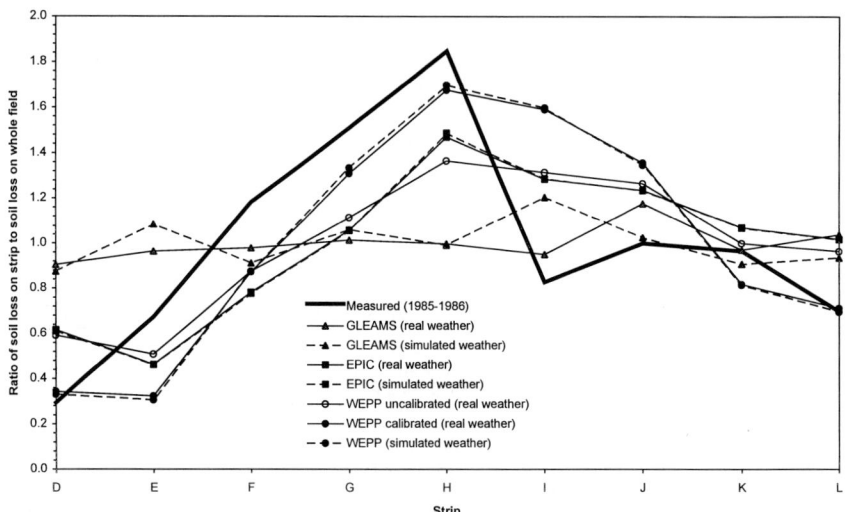

Figure 2. Strip-to-strip variability of erosion rates

Relative results

A somewhat tentative conclusion of the common-dataset model evaluation (Favis-Mortlock, 1998) is that current erosion models can give more reliable estimates of relative, rather than absolute, rates of erosion. This idea is not new (e.g. Barfield *et al.*, 1991). Figure 2 shows the ratios of average annual soil loss on each strip to the average for the whole field, for each model as well as the 1985-86 measurements.

Results for individual years

A second conclusion from the common-dataset evaluation is that erosion rates are better simulated over longer periods than shorter periods. Each model's estimate of annual erosion rate for the whole field is plotted in Figures 3 and 4. GLEAMS' results for the measured-weather data (Figure 3a) show a surprisingly low rate of erosion for 1987, a year with heavy autumn rainfall (e.g. Boardman, 1988). The EPIC results (Figure 3b) are better; the very high rates in 1976 are somewhat unexpected although not unrealistic given the wetness of the year. WEPP's results (Figure 3c) satisfactorily capture the high rates observed in 1987, as well as a secondary peak for the wet autumn of 1982 (Boardman and Robinson, 1985).

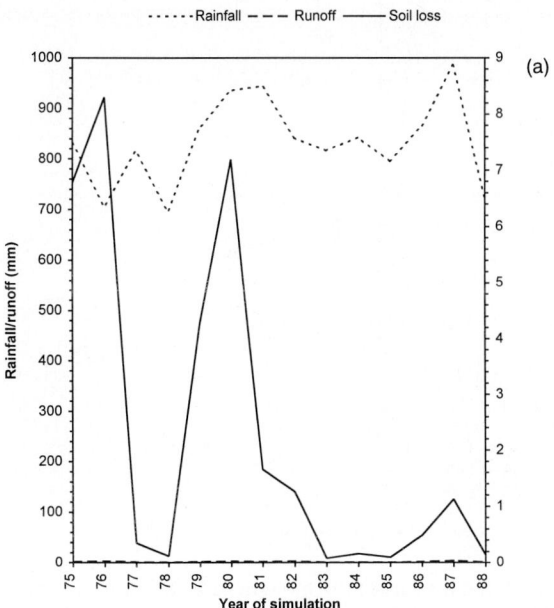

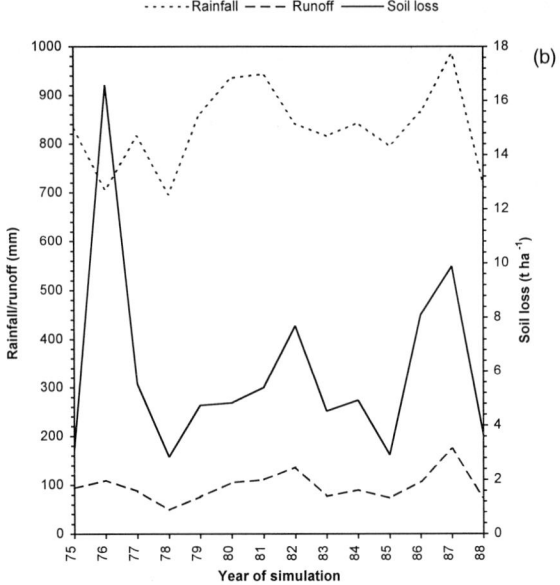
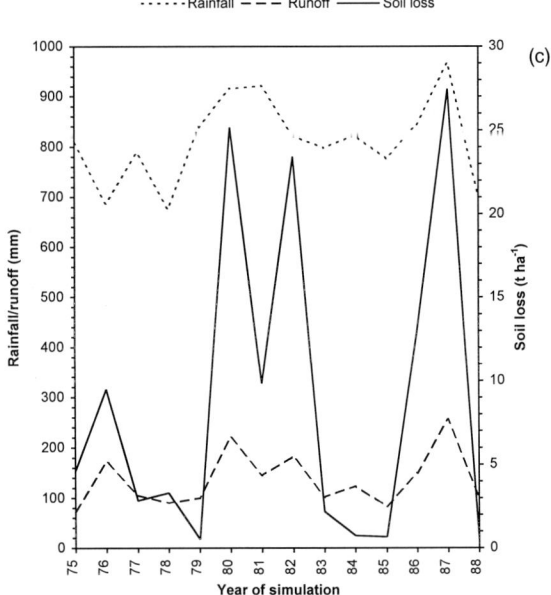

Figure 3. Annual rainfall, runoff and soil loss from the measured-weather runs: (a) GLEAMS (b) EPIC (c) WEPP (calibrated run only). Note the different scales of the soil loss axes

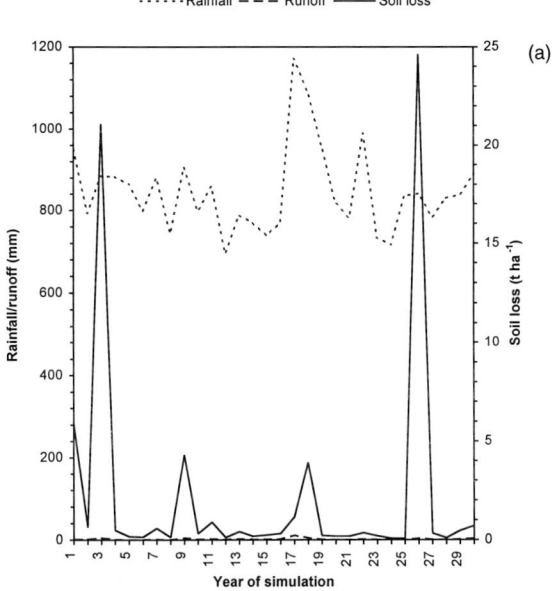

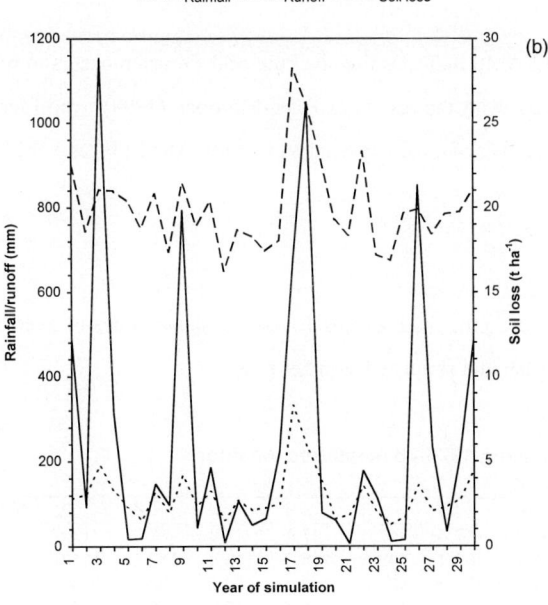

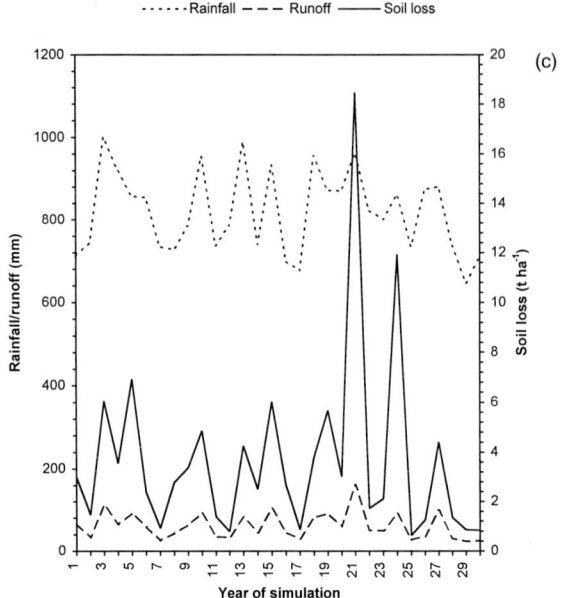

Figure 4. Annual rainfall, runoff and soil loss from the simulated-weather runs: (a) GLEAMS (b) EPIC (c) WEPP. Note the different scales of the soil loss axes, and that the time series of rainfall are not the same for each model

For the simulated-weather runs, GLEAMS indicates that the bulk of the erosion occurs in just two of the 30 years: this is consistent with the results of Favis-Mortlock (1994), who found that GLEAMS underpredicted the number of erosion events on this site. The EPIC and WEPP results are more reasonable.

Intermediate results

Finally, some 'intermediate' results are presented. Tables 10 and 11 show values of average annual runoff for the measured-weather and simulated-weather runs.

Table 10. Simulated runoff (mm) using 1975-88 measured weather

	D	E	F	G	H	I	J	K	L	Mean
GLEAMS	17.4	17.4	17.4	17.4	17.4	17.4	17.4	17.4	17.4	17.4
EPIC	97.6	89.3	101.7	109.1	119.1	115.6	115.4	112.1	111.9	108.0
WEPP uncalib.	177.9	174.5	177.7	178.5	180.1	182.5	184.0	182.0	184.3	180.2
WEPP calib.	136.4	132.6	135.8	136.5	138.1	141.4	142.8	140.9	143.6	138.7

Table 11. Simulated runoff (mm) using 30 years of simulated weather

	D	E	F	G	H	I	J	K	L	Mean
GLEAMS	20.9	20.9	20.9	20.9	20.9	20.9	20.9	20.9	20.9	20.9
EPIC	61.6	55.9	64.5	69.8	77.3	74.7	74.6	72.0	71.9	69.1
WEPP	119.1	116.2	118.2	118.1	119.1	121.2	122.6	121.9	123.9	120.0

To a large extent, the results for runoff mirror those for soil loss. Note though that the range between the results for runoff from the uncalibrated and calibrated WEPP runs is not as large as the range between uncalibrated and calibrated soil loss. Table 12 expresses runoff as a percentage of rainfall. In an experimental study on the South Downs, Robinson and Naghizadeh (1992) determined that 1.8 to 12.9 per cent of rainfall ran off on unwheeled areas, and 8.4 to 19.9 per cent on areas with wheelings. While all model results are within this range, the results for GLEAMS are the least plausible.

Table 12. Fraction of annual rainfall running off (all strips)

	Measured weather		Simulated weather	
	Average annual rainfall (mm)	% running off	Average annual rainfall (mm)	% running off
GLEAMS	832.8	2.1	846.9	2.5
EPIC	832.8	13.0	818.5	8.4
WEPP calib.	812.1[a]	17.1	801.2	15.0

[a]Note that the use of CLIGEN to add time-to-peak values etc. to the measured daily rainfall for WEPP has resulted in a 2.5% reduction in daily rainfall amounts. The reason for this is unknown

Figures 3 and 4 also show year-to-year runoff. Again, GLEAMS predicts very low values.

Discussion

In this model evaluation, the 'correct' value of soil loss was known, and calibration — albeit constrained by the need to keep parameter values within realistic limits and the requirements of inter-model consistency — was permitted. The dataset is of a high standard, with (unlike the common-dataset evaluation) no missing or uncertain values. Why then did the three models still produce estimates of annual average soil loss which range over an order of magnitude?

One part of the answer to this question may be the nature of the soil at the Woodingdean site. The Andover silt loams of the South Downs are both stony and prone to crusting. Since the GLEAMS results suggest that infiltration was too high, results would have been improved if

— as in the GLEAMS runs of Favis-Mortlock (1994) — a lower value had been used for hydraulic conductivity, and a higher value for SCS Curve Number. But since the values used appear satisfactory for EPIC, this would have then resulted in inter-model inconsistency. One conclusion may be, therefore, that constrained calibration is not possible when models make subtly different assumptions regarding the 'meaning' of apparently identical parameters. To determine whether this is the case demands a very thorough knowledge of each model.

For WEPP, it appears that further calibration of the hydraulic conductivity parameter and the three erodibility parameters is needed, even though such calibration would certainly result in values which are outside the range which is considered to be reasonable in the model's documentation. At best, though, such calibration is merely substituting for an acceptable process-representation of the effects of soil stoniness on infiltration and erodibility. Further work is needed here (Favis-Mortlock and Savabi, 1996).

Certainly, the subjective calibration employed here is unsatisfactory. But since calibration appears to be a prerequisite for all three models on this South Downs site (and presumably on other sites), other approaches to parameter fitting should be investigated (cf. Poethke et al., 1994). Recent work on the estimation of WEPP's hydraulic parameters (e.g. Rawls and Brakensiek, 1995; Zhang et al., 1995a,b; Nearing et al., 1996) is surely an implicit recognition of the difficulty of using 'off-the-shelf' values for such sensitive and inherently variable quantities.

Regarding the other conclusions from the common-dataset evaluation, the narrower spread of runoff estimates compared with soil loss estimates is consistent with the notion that runoff is better simulated than soil loss. The notable differences in model estimates for individual years is likewise consistent with the supposition that long-term values are better simulated. Do these results support the hypothesis that models perform better in relative terms? Certainly the models show a broadly similar strip-to-strip variability to that measured in 1985-86: there are higher rates toward the centre of the field, where slopes are longest and steepest, and there is a convexity. But this is hardly surprising! In addition, the spatial variability of erosion is temporally dynamic (e.g. Wendt et al., 1986): thus the spatial distribution of erosion measured in a single year may well differ from longer-term spatial variability, so that (for example) the lower measured rate on strip I is not a feature of the long-term pattern. Thus while the response to the question is somewhat inconclusive, Figure 2 does bring out some interesting trends: note that GLEAMS is in general the least responsive in estimating strip-to-strip change, EPIC intermediate, and WEPP (once calibrated) the most responsive. Strip-to-strip sensitivity is here ranked in the same sequence as the three models' absolute values for soil loss on the whole field.

Conclusions

Even in this relatively undemanding model evaluation, there is still a wide scatter in model results. For this dataset, calibration is essential for all models. Constraining calibrated values so that they remain within the range of 'acceptable' values, and are consistent between models, has here prevented models from reproducing measured values with any exactitude. Two implications follow from this: one is that, for this dataset at least, there may well be limitations in each model's process descriptions which necessitate 'excessive' compensatory calibration. The other is that identical parameters can take on subtly different process 'meanings' in different models, and thus may require rather different values for each model.

Acknowledgements

I wish to thank Mark Nearing and Gerard Govers for discussion regarding WEPP and soil stoniness, and the former National Rivers Authority for climate data. This paper is a contribution to the Soil Erosion Network of GCTE, a Core Research Project of the IGBP.

References

Alberts, E.E., Nearing, M.A., Weltz, M.A., Risse, L.M., Pierson, F.B., Xhang, X.C., Laflen, J.M. and Simanton, J.R. (1995). Soil component. In, Flanagan, D.C. and Nearing, M.A. (eds), *USDA - Water Erosion Prediction Project Hillslope Profile and Watershed Model Documentation,* NSERL Report No. 10, USDA-ARS, West Lafayette, Indiana, USA. pp. 7.1-7.20.

Barfield, B.J., Haan, C.T. and Storm, D.E. (1991). Why model? In, Beasley, D.B., Knisel, W.G. and Rice, A.P. (eds), *Proceedings of the CREAMS/GLEAMS Symposium,* Agricultural Engineering Dept, University of Georgia, Athens, Georgia, USA. pp. 3-8.

Boardman, J. (1988). Severe erosion on agricultural land in East Sussex, UK, October 1987. *Soil Technology* 1, 333-348.

Boardman, J. (1990). Soil erosion on the South Downs: a review. In, Boardman, J., Foster, I.D.L. and Dearing, J.A. (eds), *Soil Erosion on Agricultural Land,* Wiley, Chichester. pp. 87-105.

Boardman, J. (1992). Current erosion on the South Downs: implications for the past. In, Bell, M. and Boardman, J. (eds), *Past and Present Soil Erosion,* Oxbow Monograph 22, Oxbow Books, Oxford, UK. pp. 9-19.

Boardman, J. (1993). The sensitivity of Downland arable land to erosion by water. In, Thomas, D.S.G. and Allison, R.J. (eds), *Landscape Sensitivity,* Wiley, Chichester. pp. 211-228.

Boardman, J. and Robinson, D.A. (1985). Soil erosion, climatic vagary and agricultural change on the Downs around Lewes and Brighton, autumn 1982. *Applied Geography* **5**, 243-258.

Boardman, J., Evans, R., Favis-Mortlock, D.T. and Harris, T.M. (1990). Climate change and soil erosion on agricultural land in England and Wales. *Land Degradation and Rehabilitation* **2**(2), 95-106.

Boardman, J. and Favis-Mortlock, D.T. (1993). Climate change and soil erosion in Britain. *Geographical Journal* **159**(2), 179-183.

Botterweg, P. (1995). The user's influence on model calibration results: an example of the model SOIL, independently calibrated by two users. *Ecological Modelling* **81**, 71-81.

Evans, R. (1996). Some soil factors influencing accelerated water erosion on arable land. *Progress in Physical Geography* **20**(2), 205-215.

Favis-Mortlock, D.T. (1994). *Use and Abuse of Soil Erosion Models in Southern England*, Unpublished PhD Thesis, University of Brighton. 310 pp.

Favis-Mortlock, D.T. (1998). Validation of field-scale soil erosion models using common datasets. In, Boardman, J. and Favis-Mortlock, D.T. (eds), *Modelling Soil Erosion by Water*. Springer-Verlag NATO-ASI Series.

Favis-Mortlock, D.T. and Smith, R.F. (1990). A sensitivity analysis of EPIC. In, Sharpley, A.N. and Williams, J.R. (eds), *EPIC (Erosion/Productivity Impact Calculator). 1. Model Documentation*, USDA-ARS Technical Bulletin 1768, 178-190.

Favis-Mortlock, D.T., Evans, R., Boardman, J. and Harris, T.M. (1991). Climate change, winter wheat yield and soil erosion on the English South Downs. *Agricultural Systems* **37**(4), 415-433.

Favis-Mortlock, D.T. and Boardman, J. (1995). Nonlinear responses of soil erosion to climate change: a modelling study on the UK South Downs. *Catena* **25**(1-4), 365-387.

Favis-Mortlock, D.T., Quinton, J.N. and Dickinson, W.T. (1996). The GCTE validation of soil erosion models for global change studies. *Journal of Soil and Water Conservation* **51**(5), 397-403.

Favis-Mortlock, D.T. and Savabi, M.R. (1996). Shifts in rates and spatial distributions of soil erosion and deposition under climate change. In, Anderson, M.G. and Brooks, S.M. (eds), *Advances in Hillslope Processes*, Wiley, Chichester, UK. pp. 529-560.

Jarvis, M.G., Allen, R.H., Fordham, S.J., Hazelden, J., Moffat, A.J. and Sturdy, R.G. (1984). *Soils and their Use in South-East England*, Soil Survey of England and Wales Bulletin 15, Harpenden, Herts, UK.

Leonard, R.A., Knisel, W.G. and Still, D.A. (1987). GLEAMS: Groundwater Loading Effects of Agricultural Management Systems. *Transactions of the American Society of Agricultural Engineers* **30**(5), 1403-1418.

Nearing, M.A., Foster, G.R., Lane, L.J. and Finckner, S.C. (1989a). A process-based soil erosion model for USDA - Water Erosion Prediction Project Technology. *Transactions of the American Society of Agricultural Engineers* **32**, 1587-1593.

Nearing, M.A., Ascough, L.D. and Chaves, H.M.L. (1989b). WEPP model sensitivity analysis. In, Lane, L.J. and Nearing, M.A. (eds), *USDA- Water Erosion Prediction Project: Hillslope Profile Version*, USDA-ARS National Soil Erosion Research Laboratory Research Report No. 2, West Lafayette, Indiana, USA. pp. 14.1-14.33.

Nearing, M.A., Liu, B.Y., Risse, L.M. and Zhang, X. (1996). Curve numbers and Green-Ampt effective hydraulic conductivities. *American Water Resources Association Bulletin* **32**(1),

Nicks, A.D. and Lane, L.J. (1989). Weather generator. In, Lane, L.J. and Nearing, M.A. (eds), *USDA - Water Erosion Prediction Project: Hillslope Profile Model,* USDA-ARS National Soil Erosion Research Laboratory Report No. 2, West Lafayette, Indiana, USA. pp. 2.1-2.19.

Poesen, J.W.A. (1992). Mechanisms of overland-flow generation and sediment production on loamy and sandy soils with and without rock fragments. In, Parsons, A.J. and Abrahams, A.D. (eds), *Overland Flow,* UCL Press, London. pp. 275-305.

Poethke, H.J., Oertel, D. and Seitz, A. (1994). Parameter sensitivity and the quality of model predictions. In, Grasman, J. and van Staten, G. (eds), *Predictability and Nonlinear Modelling in Natural Sciences and Economics,* Kluwer, Dordrecht, Holland. pp. 389-397.

Potts, A.S. and Browne, T.E. (1983). The climate of Sussex. In, Geographical Editorial Committee (eds), *Sussex: Environment, Landscape and Society*, Alan Sutton, Gloucester. pp. 88-108.

Quinton, J.N. and Morgan, R.P.C. (1998). EUROSEM: an evaluation with single event data from the C5 watershed, Oklahoma, USA. In, Boardman, J. and Favis-Mortlock, D.T. (eds), *Modelling Soil Erosion by Water*. Springer-Verlag NATO-ASI Series.

Rallison, R.E. (1980). Origin and evolution of the SCS runoff equation. In, *Symposium on Watershed Management, Volume 2,* American Society of Civil Engineers,

Rawls, W.J. and Brakensiek, D.L. (1995). Using fractal principles for predicting soil hydraulic properties. *Journal of Soil and Water Conservation* **50**(5), 463-465.

Richardson, C.W. and Nicks, A.D. (1990). Weather generator description. In, Sharpley, A.N. and Williams, J.R. (eds), *EPIC - Erosion/Productivity Impact Calculator. 1. Model Documentation,* U.S. Department of Agriculture Technical Bulletin 1768, 93-104.

Robinson, D.A. and Naghizadeh, R. (1992). The impact of cultivation practice and wheelings on runoff generation and soil erosion on the South Downs: some experimental results using simulated rainfall. *Soil Use and Management* **8**(4), 151-156.

USDA - Agriculture Soil Conservation Service (1972). *National Engineering Handbook. Section 4, Hydrology,* 548 pp.

Wendt, R.C., Alberts, E.E. and Hjemfelt Jr, A.T. (1986). Variability of runoff and soil loss from fallow experimental plots. *Soil Science Society of America Journal* **50**(3), 730-736.

Wilcox, B.P. and Simanton, J.R. (1998). Predicting runoff in semiarid woodlands: evaluation of the WEPP model. In, Boardman, J. and Favis-Mortlock, D.T. (eds), *Modelling Soil Erosion by Water*. Springer-Verlag NATO-ASI Series.

Williams, J.R. (1985). The physical components of the EPIC model. In, El-Swaify, S.A., Moldenhauer, W.C. and Lo, A. (eds), *Soil Erosion and Conservation*, Soil Conservation Society of America, Ankeny, Iowa. pp. 272-284.

Wischmeier, W.H. and Smith, D.D. (1978). *Predicting Rainfall Erosion Losses,* US Department of Agriculture, Agricultural Research Service Handbook 537, 58 pp.

Zhang, X.C., Nearing, M.A. and Risse, L.M. (1995). Estimation of Green-Ampt conductivity parameters: Part I. Row crops. *Transactions of the American Society of Agricultural Engineers* **38**(4), 1069-1077.

Zhang, X.C., Nearing, M.A. and Risse, L.M. (1995). Estimation of Green-Ampt conductivity parameters: Part II. Perennial crops. *Transactions of the American Society of Agricultural Engineers* **38**(4), 1079-1087.

SECTION 4. MODELLING ISSUES

12. MODELLING ACROSS SCALES:
THE MEDALUS FAMILY OF MODELS

Mike Kirkby

School of Geography
University of Leeds
Leeds LS2 9TJ
UK

Abstract
Soil erosion forecasting is relevant at a wide range of time and space scales, from the field scale up to national or global scales. Changes of time and space scales generally go together, because finer time resolution requires better understanding of hydrological and sediment transport processes, and generally implies a more detailed spatial scale. At different scales, different groups of processes tend to become dominant, so that the effective focus of the models also changes. At the scale of the single erosion plot up to the hillslope catena, the timing and volume of overland flow hydrographs is critical, together with its distribution across rill and inter-rill areas. At the scale of the catchment, topography, soil and vegetation patterns become more important, and it is essential to consider periods from single storm events up to several decades, over which these patterns may change. At the coarser national to global scales, climate and lithology become the critical variables, with associated time spans from a few years up to the hundreds or thousands of years over which significant climatic change naturally occurs.

Space and time scales in erosion models

Changes of scales in time and space generally go together, because finer time resolution requires better understanding of hydrological and sediment transport processes, which, in turn, generally implies a more detailed spatial scale. At scales from the single erosion plot to the hillslope catena, for example, the timing and volume of overland flow hydrographs is critical, together with its distribution across rill and inter-rill areas. At the regional or catchment scale, topography, soil and vegetation patterns become more important, and it is essential to consider periods from single storm events up to several decades, over which these patterns may change. At still coarser national to global scales, climate and lithology become the critical variables, with associated time spans from a few years up to the hundreds or thousands of years over which significant climatic change naturally occurs.

At each upward change in scale, there is a choice between computation for a larger area or longer period, with the attendant problems of computer size and data provision; and the use of a more lumped model with longer time steps or spatial cells. As an applicable research and management tool, the latter is generally to be preferred, although it is eventually necessary to reconcile models at each successive scale, which usually involves an integration, implicit or explicit, over relevant distributions in time and/or space. In this process of up-scaling, some of the distributed variables must be replaced by average values, although not necessarily arithmetic averages, if the response is not linear in the variable. It may be more appropriate to use the Root Mean Square value (e.g. for turbulent velocity fluctuations) or the logarithmic mean (e.g. for grainsize), or other physically based weightings. In some cases, however, the mean (first moment) of the distribution is not the only relevant value, and it is important to use higher moments, primarily the standard deviation (second moment), but occasionally higher moments (e.g. skewness). Examples where higher moments are important include hydrograph distributions, where the mean flow is of little relevance, and surface micro-topography (discussed below) where the roughness, which may be expressed as the standard deviation of elevation, becomes a crucial variable.

At coarser scales, additional processes are also generally observed for a variety of reasons, many of them associated with non-linearities in the relationships. These coarse scale processes can rarely be predicted at the finer scale, but the onset of critical conditions may sometimes be analysed. An important example of the response to this kind of scale change is the development of discrete channels and valleys, which cannot be modelled simply by scaling up from small plot experiments.

Plot-scale models may represent physical and other processes at the limits of current understanding. Inevitably there is a strong concentration on the spatial distribution of flow and sediment movement at the scale of the microtopography and small rills or gullies. These models are usually only for small areas and over periods of one to a few storms. Their value lies in extending process understanding knowledge through detailed field and laboratory experimentation, but they are generally too demanding in data requirements, and often in computing time, to be widely applicable as a management tool. Examples of such models are WEPP (Foster and Lane, 1987; Lane and Nearing, 1989), EUROSEM (Morgan *et al.*, 1990) and MEDALUS (Kirkby *et al.*, 1993; in press).

Catchment scale models generally integrate over the details of micro-topography, and simplify the detailed structure of the hydrograph to make use of available data at the larger scale. This generally consists of topographic data from maps or DEMs, together with data on soil and vegetation types, based on thematic maps or remote sensing. This regional scale is important for decision makers, but has proved difficult to address from a physical basis. The Universal Soil Loss Equation is essentially at this scale, although its technical approach has largely been superseded. Work is in progress at this scale within the framework of the MEDALUS project (MEDRUSH model, Abrahart *et al.*, 1994), and in association with other programmes related to global change, including TIGER and LOIS in Britain. Modelling at catchment scales is inevitably a compromise between detailed process understanding and reasonable computation times. In the MEDRUSH model, large catchments are divided into 200-500 sub-catchments, each represented by a family of flow strips, of which a single example is modelled in detail. This strategy provides comparability with the MEDALUS catena model and detailed forecasts of areas liable to degradation, while at the same time allowing integration of water and sediment yields up to the regional scale.

For national or global scale models, the crucial variables become climate and lithology. It is necessary to ignore much topographic and soil detail, retaining only broad regional differences associated with lithology and climate or vegetation. Such models attempt to simulate hydrology, vegetation, runoff and erosion from regional or global data on climate, land cover, lithology and topography, and are generally based on a one dimensional model of surface properties and hydrology.

Hydrology in catena and catchment models

The slope catena model represents four interacting sub-models for the atmosphere, vegetation, soil and surface systems (Figure 1). These four sub-systems are defined at each of a series of points down a hillslope catena or flow strip, which are connected by kinematic routing of overland flow, where appropriate subsurface flow, and grain-size selective sediment transport. Simulation of vegetation growth is through eco-physiologically based shrub and/ or grass models, which also gradually modifies soil organic matter. The MEDALUS model is developed for a single flow strip over a period of up to 10 years, whereas in MEDRUSH up to 200 simplified flow strips are included within a catchment of up to 2000 km^2, and model is intended to run for up to 100 years.

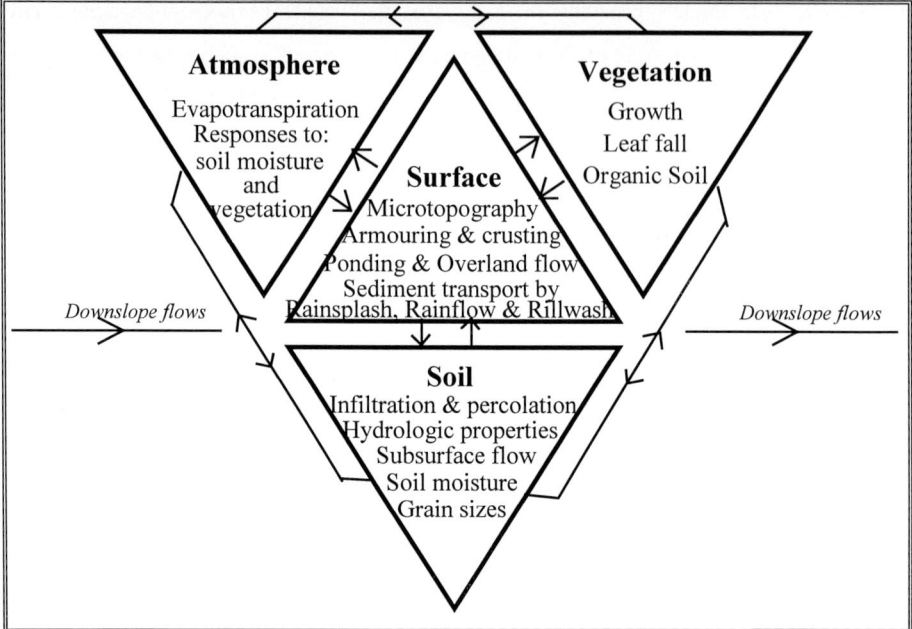

Figure 1. Relationships between components of MEDALUS and MEDRUSH models along a catena

The hydrology, for example, is therefore handled somewhat differently in the MEDALUS and MEDRUSH models. For MEDALUS, evapotranspiration is estimated from a modified Shuttleworth-Wallace (1985) sparse vegetation model. Soil water movement is obtained by solving the Richards equation for up to 10 layers, using a moisture-tension curve based on the van Genuchten equations. Flow on the surface is explicitly related to microtopography, which controls the spatial distribution of overland flow, infiltration and sediment transport. The model uses 5-15 minute interval data from automatic weather stations to simulate conditions using a variable time step.

In MEDRUSH model, the hydrological routines are inevitably somewhat simpler. Evapotranspiration is estimated from net radiation using the a Priestley-Taylor relationship (1972). Soil moisture is represented by one unsaturated and one saturated store, the latter based on TOPMODEL (Beven and Kirkby, 1979). The model runs in basic time steps of one hour, using a fractal distribution of rainfall intensity within the hour to take account of short periods bursts of intense rainfall, which have a significant effect on overland flow runoff. Similarly, the flow is statistically distributed across a rough topography. This allows explicit estimation of ponded areas, and of exfiltrating return flow along depressions. The conceptual basis of the model is outlined in Figure 2.

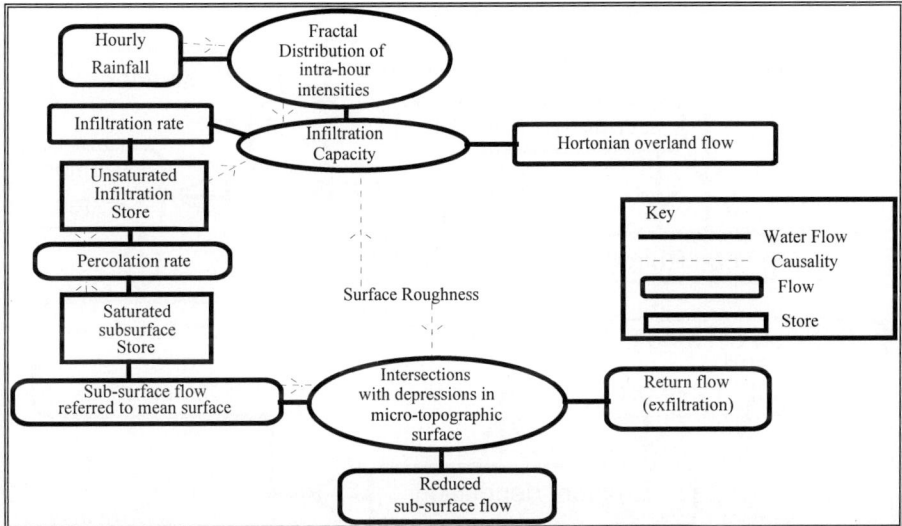

Figure 2. Relationships between overland and subsurface flow components in MEDRUSH model

Long term interactions

In order that the models can run effectively for periods of a decade or more, it is critical that they address the relevant interactions for this time scale. The long-term behaviour of the model simulations is determined mainly by interactions between the four sub-models (Figure 3). By developing a dynamic vegetation sub-model, the plants grow and their morphology changes in response to the actual sequence of climatic and soil conditions, and will, in principle, continue to do so if these conditions change substantially over a period of years. Their growth and ageing respond to photosynthetically active radiation and to current soil moisture status, and changes in conditions, seasonally and from year to year, are able to make substantial changes in the partition of nutrients between roots, stems, leaves and fruits. Leaf litter, dead stems and dead root material all contribute to the organic matter in the soil, which decomposes in response to soil temperature and, to a less extent, moisture conditions. These inputs may be substantially modified in response to cultivation, grazing and fire, although these impacts are only partly incorporated into the model. Over a period of 5-10 years, the concentration of soil organic matter may change substantially, and in turn influences the soil hydraulic properties. The main effect of increased organic matter is to increase the amount of effective soil water storage, which has an important effect on subsequent plant growth except under extremely wet or dry conditions.

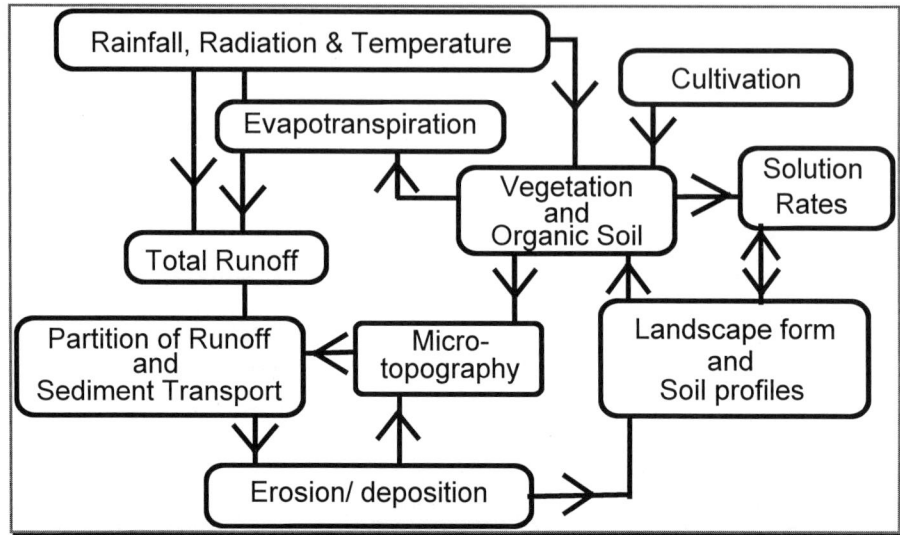

Figure 3. Climate, vegetation and landform relationships in the MEDALUS and MEDRUSH models

Changes in climate and/or vegetation cover tend to changes in the amount of overland flow runoff generated. Increases in overland flow lead to increased sediment transport, producing greater erosion upslope and greater deposition at some suitable downslope sites. Erosion, as well as being important in its own right, has a strong influence on subsequent hillslope response in at least two ways; through grainsize and through microtopography.

First, sediment transport selects for grainsize, leading to the production of an armour layer, particularly in eroded areas, over periods of a few years. Erosion also gradually strips the topsoil, over periods of decades, usually bringing coarser subsoil material to the surface. Armouring generally limits subsequent runoff and erosion, although the exact relationship between stones and the surface may be critical in small plots (Poesen *et al.*, 1994). The critical role of the armour layer is sometimes to enhance infiltration around the margins of the stones, where they provide some protection against surface sealing. Stones also provide some physical protection. Stones within the soil also have a beneficial effect up to concentrations of 30-50%. The soil organic matter is only mixed into the non-stony (fine earth) fraction, so that its concentration increases with stoniness. This increases the total availability of water to plants, so that modest increases in stoniness partly counteract the effects of erosion by increasing vegetation cover.

Second, overland flow and sediment transport are concentrated in depressions on the surface, particularly depressions running up- and down-slope. Thus erosion and deposition are most effective in changing the depth of the deepest depressions. Under high rates of erosion, the microtopography therefore becomes rougher, and under deposition smoother. Roughness is described by the standard deviation of elevation relative to the local mean surface, and measurements suggest that this is commonly normally distributed, although with a scale dependence which appears to be approximately fractal. For example small positive features are associated with soil grains, aggregates, clods and stones and larger negative features with linear depressions, rills and areas of flow convergence. Even where there is not an established or recurring pattern of rills and/ or channels, it is not unusual for roughness to increase systematically downslope on eroded areas, and to decrease downslope in areas of deposition. In the MEDALUS model changes in grain size and roughness are explicitly modelled, while in MEDRUSH they are again treated statistically. The roughness model, for example, is based on rates of rainsplash which reduces roughness, vegetation growth which increases it, and rillwash which may increase or decrease roughness according to whether the site is eroding or depositing respectively.

These long-term interactions have little effect in the short run, although the impact of individual intense storms may be seen on armouring and surface roughness. Over a period, however, they control the positive and negative feedbacks which promote or constrain erosional degradation, and may lead to irreversible desertification of soils and landscapes. The most important feedback loop is through soil stoniness. For a while, increased erosion of stony soils creates negative feedbacks through increased water availability in the soil organic matter, and irreversible degradation only occurs when stoniness becomes very high, as described above. Nevertheless erosion may be continuing if driven by poor landuse practices, so that the buffering effect of stones may provide a false sense of security. This buffering effect is, of course, completely absent in deep fine-grained soils, allowing badland development in extreme cases. Armouring provides a third loop, of less importance, but also providing a degree of buffering and protection against erosion.

Whereas the MEDALUS model is for a single slope catena, the MEDRUSH model treats each catena as a representative for a sub-catchment within a larger catchment. The model is being applied to the Agri (Basilicata, S. Italy) and Guadalentin (S.E. Spain) catchments, each of 1,000-2,000 km^2 area. Each catchment is subdivided into 100-500 sub-catchments, using a topographic analysis based on DEMs. From an analysis of every pixel in each sub-catchment, a representative flow-strip or catena is selected for modelling, and net changes in vegetation, erosion or surface properties are simulated for the flow strip and then re-distributed to its whole sub-catchment. In this way simplified catena models, which are cross-validated against the more detailed MEDALUS model, can be used to forecast global change across areas which are large enough to have significance for regional planners.

Large area models

The third group of models which have been developed are based on the hydrological balance at a point, and are related to one-dimensional SVATs, but concentrating on vegetation growth and erosion time spans. The hydrology is used to simulate the climate and vegetation input to topographically based slope evolution models. This allows examination of the basis for the widely accepted relationship, first noted by Langbein and Schumm (1958), between rainfall and sediment yield, mediated by vegetation. The CSEP point model (De Ploey et al, 1991; Kirkby and Neale, 1987; Kirkby and Cox, 1995) is a model of this type, which simulates hydrology, vegetation and runoff distribution derived from IIASA climatic data (Leeman and Cramer, 1991) to provide a physically based erosivity model, and further work is ongoing to relate it with global DEM and lithological data. This approach has applications to planning at national and larger scales, and for comparison between remotely sensed actual land cover and the climate-derived potential cover.

The CSEP is essentially a climatic index which integrates the effect of daily storm rainfalls by fitting their distribution to the sum of two exponential distributions. It assumes a constant threshold storage before overland flow runoff occurs, a linear accumulation of discharge downslope, and a power law relating sediment transport capacity to overland flow discharge. Here the power exponent is set at 2.0, which lies within the range of empirical values used and, along with other integer values, allows an explicit integration of the equations below. With this value, the CSEP may be summarised as follows.

In a single storm of rainfall r, the overland flow production:

$$j = (r - h) \tag{1}$$

where h is the storage threshold capacity of the soil.

For a single exponential distribution, the density of days with rainfall r is:

$$N(r) = N_0 \exp(-r/r_0) \tag{2}$$

where N_0, r_0 are empirical parameters fitted to the distribution in the neighbourhood of a one year recurrence interval. Summing over this distribution, we have the total overland flow production:

$$J = \int_h^\infty (r-h) N(r) dr = N_0 r_0 \exp(-h/r_0) \tag{3}$$

Similarly for sediment yield, assumed proportional to discharge squared, the yield from a rainfall event of r is:

$$t \propto (r-h)^2 \tag{4}$$

Similarly summing over the frequency distribution, the total sediment yield is:

$$T \propto \int_h^\infty (r-h)^2 N_0 \exp(-r/r_0) dr = 2 N_0 r_0^2 \exp(-h/r_0) \tag{5}$$

This final expression is defined as the Cumulative Erosion Potential.

$$CSEP = 2 N_0 r_0^2 \exp(-h/r_0) \tag{6}$$

Thus it simply and rationally combines the effects of the daily rainfall distribution, through N_0 and r_0, and a significant soil hydraulic parameter, h_0, to give a measure of the climatic component in soil erosion transport. This calculation is made over the sum of two exponential distributions, which give a better fit to the extremes of the rainfall distribution, and the summation is made over each month separately.

The values of the storage capacity, h, in equations (1) to (6), is explicitly linked to the vegetation, and to a less extent, the soil. For a fixed perennial vegetation cover, or lack of it, a fixed storage term may be used in equation (1) above. Alternatively different fixed values may be adopted for each month of the year. This approach seems appropriate for simple cultivated areas with a known cover, although an explicit crop model may be preferable.

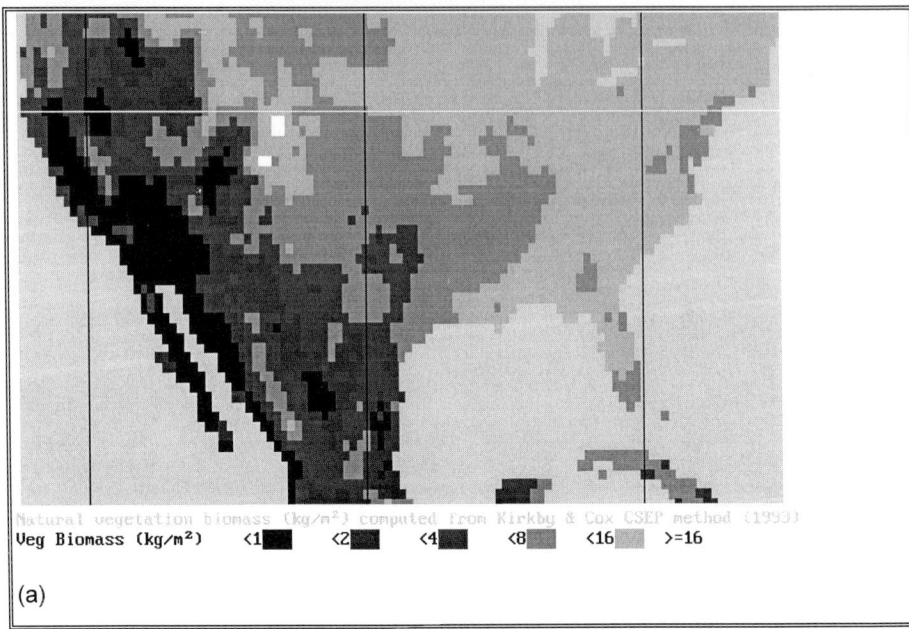

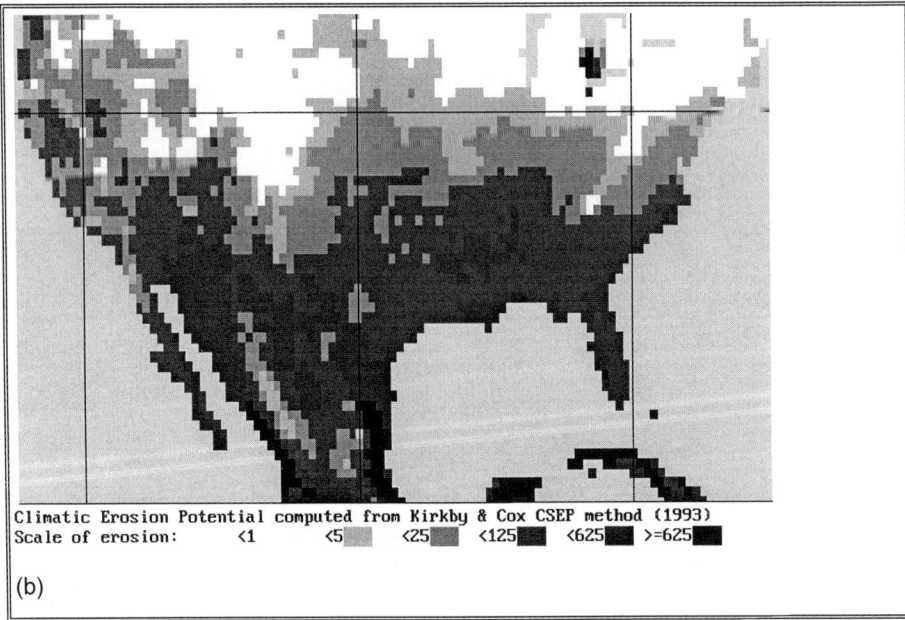

Figure 4. Maps of erosion related forecasts for southern North America, based on the IIASA 0.5°× 0.5° latitude-longitude climatic database. (a) Vegetation biomass (kg m^{-2}) (b) Climatic Seasonal Erosion Potential (CSEP)

For uncultivated vegetation, an explicit growth model is used, which is normally allowed to run for long enough to reach an annual repeating cycle, although it may also be used in a transient mode. In this form the CSEP is intended to provide a measure of erosion potential with uncultivated, semi-natural vegetation cover (Figure 4). The growth of living vegetation is simulated using a gross primary productivity, which is directly proportional to actual evapotranspiration. Losses are related to respiration and leaf fall. Leaf fall is added to the organic soil, and material is lost through decomposition. Rates are derived from monthly mean climate data for rainfall, potential evapotranspiration and temperature.

Actual evapotranspiration is estimated from potential on the basis of available water. Gross primary productivity is estimated as proportional to actual evapotranspiration, using a water use efficiency approach. GPP is added to the biomass, with losses due to respiration and leaf fall, which is added to the soil organic matter, and decomposed. For non-crop land uses such as grazing or fuel wood collection, part of the leaf fall or standing crop may be removed from the system, usually as a proportion of the total available.

This type of model is able to simulate some of the main transient responses to changes in climate or landuse from year to year through the vegetation, and is therefore well suited to address global change issues. Their simplicity allows them to be used for national to global scales of generalisation. Their major shortcoming is in simulating interactions at the large scale, where they need to be coupled to GCMs for the atmosphere, and to hydrological routing schemes for sediment transport in large rivers. Ideally, such models should be able to extract the essential one-dimensional processes from the catchment scale, again taking account of relevant average values for landscape and soil variables, and significant descriptors for their distributions.

Conclusion

In comparing simulation models at a range of time and space scales, it is essential to retain a clear view of the scope and limitations of each. Choices of either time or space scales tend to constrain the reasonable, and in some cases the possible scope for the other. Nevertheless there is some range in choice, and some models are now being constructed with a time perspective which is related to global change issues, and lies between the traditional scales of detailed process studies and long term geomorphological evolution. At the same time, there has been a heightened awareness of the possibility of obtaining direct field evidence of environmental change over periods of a few centuries. There is thus a convergence of models and evidence in an important middle ground, reinforcing the need to ensure that new models are, and can be shown to be, consistent with both the process and evolutionary time scales.

Acknowledgement

This work has been supported in part by EC funding through the MEDALUS project (Contract EPOCC-CT90-0014(SMA) co-ordinated by J.B. Thornes).

References

Abrahart, R.J., Kirkby, M.J. and McMahon, M.L. (1994). MEDRUSH - a combined Geographical Information System and large scale distributed process model, *Proceedings 2nd National Conference on GIS Research.* Leicester, 67-76.

Beven, K.J. and Kirkby, M.J. (1979). Towards a simple, physically based, variable contributing area model of catchment hydrology. *International Association of Hydrological Sciences Bulletin* **24**, 43-69.

Foster, G.F. and Lane, L.J. (1987). *User requirements. USDA — Water Erosion Prediction Project (WEPP).* NSERL Report No **1**, National Soil Erosion Laboratory, USDA Agricultural Research Station, West Lafayette, Indiana 47907.

Kirkby, M.J., Baird, A.J., Lockwood, J.G., McMahon, M.D., Mitchell, P.J., Shao, J., Sheehy, J.E., Thornes, J.B. and Woodward, F.I. (1993). *MEDALUS Project A1: Physically Based Process Models: Final report.* (Part of MEDALUS I final report, edited by J.B. Thornes)

Kirkby, M.J., Abrahart, R., McMahon, M.D., Shao, J. and Thornes, J.B. (in press). *MEDALUS Soil Erosion Models for Global Change.* Paper presented at 3rd International Geomorphology Conference, Hamilton Ontario (for publication in an edited volume).

Kirkby, M.J. and Cox, N.J., (1995). A climatic index for soil erosion potential (CSEP), including seasonal factors. *Catena* **25**, 333-352.

Kirkby, M.J. and Neale, R.H., (1987). A soil erosion model incorporating seasonal factors. In, Gardiner, V. (ed.), *International Geomorphology1986, Part II,* Wiley, Chichester, 189-210.

Lane, L.J. and Nearing, M.A. (1989). *USDA — Water Erosion Prediction Project: Hillslope Model Documentation.* NSERL Report No **2**, National Soil Erosion Laboratory, USDA Agricultural Research Station, West Lafayette, Indiana 47907.

Langbein, W.B. and Schumm, S.A. (1958). Yield of sediment in relation to mean annual precipitation. *Transactions, American Geophysical Union* **39**, 1076-84.

Leeman, R. and Cramer, W.P. (1991). *IAASA Data Base for Mean Monthly Values of Temperature, Precipitation and Cloudiness on a Global Terrestrial Grid.* Report **RR-91-18**. IAASA, Laxenburg, Austria, 62 pp.

Morgan, R.P.C., Quinton, J.N. and Rickson, R.J., (1990). Structure of the soil erosion prediction model for the EC. In, *Proceedings of InternationaL Symposium on water erosion, sedimentation and reservoir construction,* Central Soil and Water Conservation Research and Training Institute, Dehradun, India, 49-59.

De Ploey, J., Kirkby, M.J. and Ahnert, F. (1991). Hillslope erosion by rainstorms - a magnitude-frequency analysis. *Earth Surface Processes and Landforms* **16**, 399-409.

Poesen, J.W., Torri, D. and Bunte, K. (1994). Effects of rock fragments on soil erosion by water at different spatial scales: a review. *Catena* **23**, 141-66.

Priestley, C.H.B. and Taylor, R.J. (1972). On the assessment off surface heat flux and evaporation using large scale parameters. *Monthly Weather Review* **100**(2), 81-92.

Shuttleworth, J.W. and Wallace, J.S., (1985). Evaporation from sparse crops - an energy combination. *Quarterly Journal of the Royal. Meteorological Society* **111**, 839-55.

13. PROBLEMS REGARDING THE USE OF SOIL EROSION MODELS

R. P. Rudra[1], W. T. Dickinson[1] and G. J. Wall[2]

[1]School of Engineering
University of Guelph
Guelph
Ontario
Canada N1G 2W1

[2]Agriculture and Agro-Food Canada
Guelph
Ontario
Canada

Abstract

This paper focuses on problems regarding the use of erosion models. Though substantial progress has been made regarding the development and use of erosion models, still significant improvements are needed to describe many fundamental processes and their spatial and temporal variations. The important weak links identified include the role of precipitation inputs, infiltration, soil surface characteristics, soil erodibility, and evapotranspiration and soil water distribution. These weak links cannot be resolved until they are accepted and confronted with collective honesty and openness.

Introduction

Soil erosion has been recognised to be a serious environmental and soil degradation problem: it can reduce soil productivity, and it can also increase sediment and other pollution loads in receiving waters. In attempts during the last two decades to better manage soil and water resources, computer modelling has received increasing attention, including the development of numerous soil erosion models. These erosion models have been used in investigative, evaluative, predictive and learning modes. In the investigative mode, models have been used as research tools to help improve our understanding of erosion processes. Models have also been used as an aid to compare the possible effects of land use changes on soil and water quality, including impacts of agricultural practices, urbanisation and deforestation. In the predictive mode, erosion models have been used for data generation and risk assessment; and in the classroom, they have become important learning tools.

Available soil erosion models cover a wide range of temporal and spatial scales. Regarding time, an erosion model can be classified as:

- a long term average annual model, such as the USLE (Wischmeier and Smith, 1978) and RUSLE (Renard *et al.*, 1991);
- a seasonal model, such as GAMES (Rudra *et al.*, 1986) and GAMESP (Rousseau *et al.*, 1988);
- an event model, such as WEPP (Laflen at al., 1991), EPIC (Jones *et al.*, 1991) and ANSWERS (Beasley *et al.*, 1980); or
- a continuous simulation model, such as WEPP (Laflen at al., 1991), EPIC (Jones, *et al.*, 1991), CREAMS (Knisel, 1980) and GLEAMS (Leonard *et al.*, 1987). Most of the continuous models can also be used for an event. Note that in this paper, only the hill-slope version of WEPP is considered; field and grid versions are not discussed.

On the spatial scale, an erosion model can be classified as:

- a hill-slope model, such as WEPP (Laflen *et al.*, 1991);
- a field-scale model which aims to consider both hill-slope and valley bottom, such as CREAMS (Knisel, 1980), GLEAMS (Leonard *et al.*, 1987) and EPIC (Jones *et al.*, 1991); or
- a watershed or grid model, such as GAMES (Rudra *et al.*, 1986), GAMESP (Rousseau *et al.*, 1988), ANSWERS (Beasley *et al.*, 1980) and AGNPS (Young *et al.*, 1985).

In the past, most soil erosion models were based primarily on empirical relationships. More recently, there has been a shift towards more physically-based models. Whether empirical or physically-based, many of the models, such as WEPP, CREAMS and EPIC, have been configured in a piggyback fashion. That is, algorithms regarding hydrologic processes form the fundamental components in such models, onto which algorithms associated with soil erosion and sediment transport have been 'piggybacked'. If the model includes other components, involving nutrients, pesticides, etc., they are 'piggybacked' onto the erosion and/or runoff components. Although these piggyback configurations are both convenient and rational, the quality of simulation with such models becomes quite hierarchical. Errors in soil erosion estimates may be due to errors in runoff estimates and/or errors in soil erosion components; and errors in estimates involving nutrients, pesticides, etc., may be due to errors in estimates of runoff, soil erosion and/or nutrients.

There is little question that the development, calibration and validation of soil erosion models have helped to improve our understanding about soil erosion processes, including the identification of various sub-processes involved and the effect of changes in land use on these sub-processes. Some achievements include the development of improved data bases, the identification of critical processes, and confirmation of temporal and spatial trends. However, there are still many dilemmas in soil erosion modelling which have not received adequate attention. In this paper the authors focus attention on some of the 'weak links' identified during their personal experience with erosion modelling.

General observations

In the literature and in our experience, it is clear that predictions (in the form of soil erosion model outputs) do not always match observed values. Further, the variance of the discrepancies between predicted and observed values has tended to vary inversely with the length of the time unit used for the model and for the comparison, and with the size of the area under consideration. For example, the variance of discrepancies between hourly values (say of runoff or soil loss) is inevitably larger than the variance of differences between predicted and observed daily values, which is again larger than the variance associated with weekly values. The situation has been quite similar as we have gone from small to larger spatial units. In other words, predictions have definitely been best for lumped time scales (e.g. monthly or seasonal) and for lumped spatial scales (e.g. medium to large basins).

Further, when possible, the modeller has preferred to use measured values to estimate input parameters. However, when model predictions and field observations have not shown a reasonable match, inappropriate values of the input parameters have usually been considered to be the main culprit. In such situations, optimum values of sensitive parameters have been sought by means of calibration. That is, predicted and observed values of response have been matched by adjusting the values of one or more parameters. In many of these situations, proper attention has not always been given to an acceptable range for each parameter. Calibration of the parameters has thus becomes a 'twiddling of the knobs'! — with little or no attention given to the processes being modelled. Subsequent use of a model so calibrated, for management purposes, is prone to yield questionable results. These problems have not been addressed to any extent for soil erosion models.

Some weak links

- *Precision of precipitation data: calibrated model parameters are unfortunately dependent upon the time step selected for precipitation input data*

Precipitation has been recognised to be a major driving force in every hydrologic and soil erosion model; and precipitation data, discretised into specified time intervals, are prime model inputs. Ideally, parameter values associated with the various runoff and erosion algorithms included in a model should not be linked to or dependent on the duration of the precipitation time step selected. Further, it would be desirable if this time step could be selected in accord with the time scale expected to be appropriate for the modelling exercise being undertaken. However, the basic working time interval for the model is often chosen to be relatively small (e.g. 1 hour, 30 minutes, or even less), and the time interval selected for the precipitation data may be in the order of 12 hours or 1 day, the latter interval usually being determined by the temporal precision associated with the precipitation data base. Almost no attention has been given to the role which this time interval selected for the precipitation data plays *vis-á-vis* calibrated parameter values. Yet, as the following two examples reveal, soil erosion modelling is very sensitive to the precision of precipitation data.

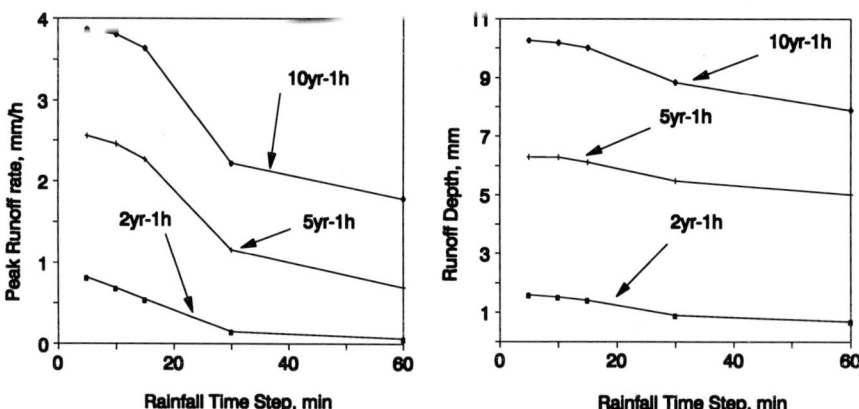

Figure 1. Variations in peak runoff rate and runoff depth with rainfall time step in modelling soil erosion processes at the watershed scale

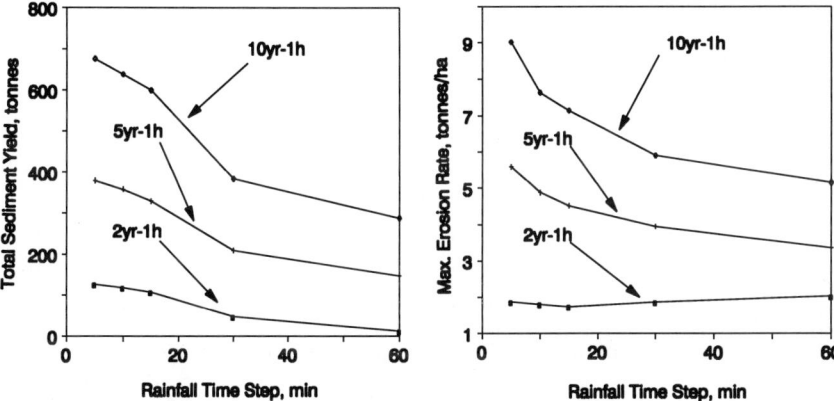

Figure 2. Variations in total sediment yield and maximum erosion rate with rainfall time step in modelling of soil erosion at the watershed scale

Some results regarding application of the ANSWERS model in Southern Ontario conditions are presented in Figures 1 and 2 (Rudra et al., 1991). In this study, the model was calibrated for runoff and sediment yield on the basis of hourly rainfall data. Then, since more detailed data were in fact available, the model was run using a variety of smaller time intervals for the precipitation input. Clearly, runoff amount, peak runoff rate, maximum erosion rate and total sediment yield were highly dependent/sensitive to the time interval selected for the rainfall data. In almost every case, the magnitude of the output variable was inversely related to the rainfall time step, and the sensitivity was greatest for the larger rainfall events on discretised time step. At least for these Southern Ontario conditions, the ANSWERS model appears to be quite sensitive to the rainfall time step selected, and the nature of the sensitivity is a function of the output variable and the magnitude of the rainfall event.

Similar problems regarding the temporal precision of precipitation data have been observed in field-scale erosion modelling. Rudra et al. (1991) evaluated the effect of rainfall time steps in the CREAMS model on the magnitude of hydrologic and soil erosion parameters calibrated on two plots having a similar land use but different land management practices. These results, presented in Figure 3, indicate that calibrated values of RC (an index of saturated hydraulic conductivity) revealed a significant inverse relationship with the size of the rainfall time step. That is, the larger rainfall time steps resulted in lower values of calibrated RC. On the other hand, calibrated values of the I/I_o ratio (an index of the relative contribution of rainfall and runoff to soil erosion) had a direct relationship to the size of the rainfall time step. Larger values of the time step resulted in an apparent higher contribution to total erosion measured at the bottom of the slope from runoff (rill erosion).

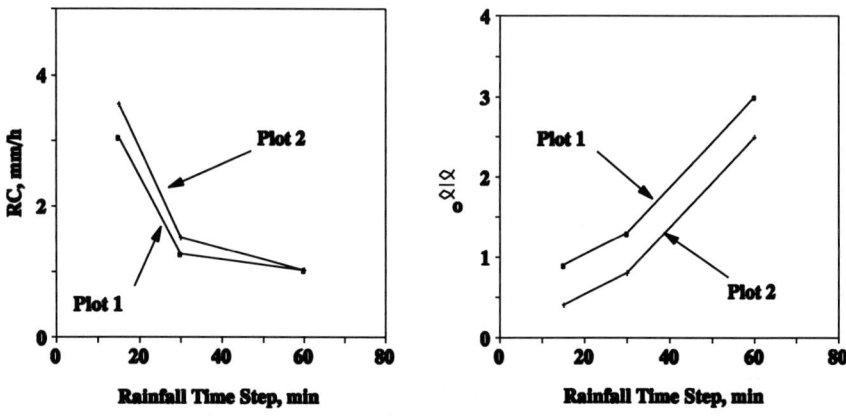

Figure 3. Effect of rainfall time step on hydraulic and erosion characteristics of soil for modelling of soil erosion on the field scale

These latter results raise fundamental questions about acceptable values of RC and I/I_o. Which of these calibrated values should be used to index 'true' hydraulic conductivity and the contribution of rainfall and runoff to soil loss? Do these values really represent saturated hydraulic conductivity and I/I_o, respectively, or are they just some fitted parameters? It is possible that the calibrated RC value includes effects of surface storage and surface sealing. Since hydraulic conductivity, rainfall and runoff all play important roles in erosion modelling, the quality of predictions will depend upon the confidence of the modeller in estimates of the associated parameter values.

- *Spatial and temporal variations in infiltration: parameters associated with infiltration of water into soil are in reality substantially more variable in time and space than we are able to model; and techniques used to measure infiltration parameters in the field yield widely different values*

Infiltration is an important hydrologic process. In modelling it controls the split between the amount of water that moves as surface runoff and the water that enters the soil water regime and ground water. The quality of soil erosion predictions made from physically-based models is highly dependant upon our ability to model infiltration and runoff, since the soil erosion algorithms are closely linked to runoff amount and runoff peak, which in turn depend upon infiltration. A primary problem in infiltration modelling is an inadequate representation of infiltration and hydraulic characteristics of soils, including their variability in time and space. Commonly-used infiltration modelling approaches do not adequately describe apparent changes in infiltrability/hydraulic conductivity in time and space, with surface crusting and with method of determination.

Gupta *et al.* (1993 and 1994) observed variations in hydraulic conductivity with time and method of determination, and some of their results are presented in Figures 4 and 5. These data show that different determination techniques yield quite different values of hydraulic conductivity. The techniques involving a relatively shallow depth of the soil profile (i.e. Guelph Infiltrometer and Rainfall Simulator) yield higher values of saturated hydraulic conductivity in all seasons than the techniques involving a relatively deeper depth of the soil profile (Guelph Permeameter and Ring Infiltrometer). The surface layer usually possesses more crop residue, cracks and macro-pores than deeper soil layers. Further, the methods involving a smaller sample size generally have a higher coefficient of variation (60.7 per cent for Guelph Permeameter and 88.3 per cent for Guelph Infiltrometer) than the methods involving a larger sample size (49.7 per cent for Rainfall simulator and 43.6 per cent for Ring Infiltrometer). An implication of these differences is that substantially more test runs are required when using techniques which cover a small sampling area, to obtain a mean value of hydraulic conductivity in which one has a similar level of confidence. Such points have not normally been addressed when infiltration parameter values have been determined for physically-based soil erosion and hydrologic models from field measurements.

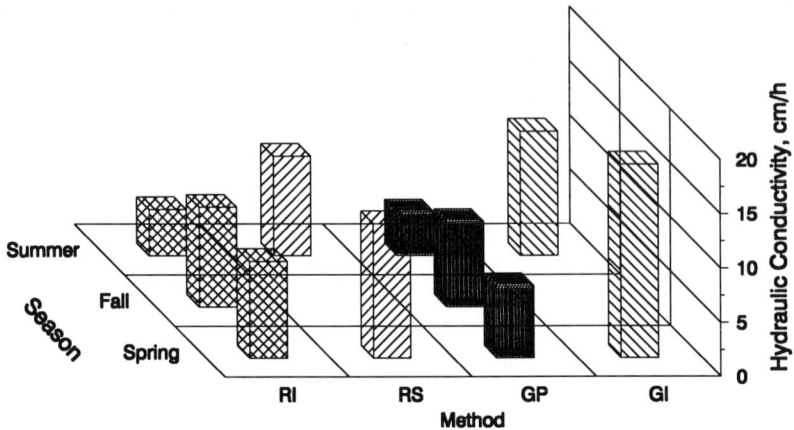

Figure 4. Variations in hydraulic conductivity with season and measurement techniques

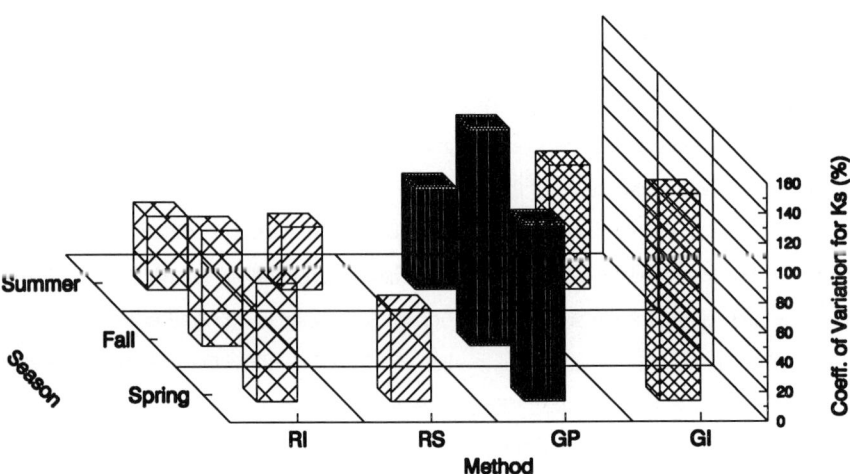

Figure 5. Coefficient of variation for measurement techniques used for determination of hydraulic conductivity

The data of Gupta *et al.* (1994) also indicate that saturated hydraulic conductivity can exhibit substantial seasonal variations. In a temperate and humid climate like that in Southern Ontario, where more than 75 per cent of stream flows occur during late winter and early spring periods, proper description of seasonal variations in soil hydraulic properties is very important for adequate modelling of soil erosion. Gupta *et al.* (1994) have shown that the mean value of saturated hydraulic conductivity for fall and spring conditions can be one and a half to two times greater than summer season values. These differences can be attributed to changes in soil surface and hydraulic conditions from season to season. Tillage operations during the fall and spring can create highly disturbed and non-uniform conditions in the soil profile and at the soil surface. These operations too can result in a highly variable spatial structure, with cracks of various sizes over the soil surface affecting total porosity, bulk density and pore size distribution, all responsible for variations in hydraulic conductivity. Climatic characteristics such as freeze-thaw cycles also create spatial and temporal variations in surface soil structures affecting hydraulic and physical characteristics of soils. In some empirical infiltration/runoff models (such as those involving the SCS Curve Number Method), the effect of land management and land use on infiltration and runoff has been qualitatively described; but many physically-based infiltration equations (such as the Green-Ampt equation) either ignore or offer something of a 'cosmetic touch' to these effects.

- *Soil surface characteristics: the roles of surface depressions, surface roughness and surface crusting/sealing are quite inadequately dealt with in soil erosion models*

Characterisation of the soil surface is perhaps one of the weakest links in hydrologic and soil erosion models. In many models the topography has been represented by an average slope length and slope gradient; and in some models (such as EPIC), the slope has been represented by a set of planes. In reality, the land surface is likely to have numerous temporary and permanent depressions, either connected to each other or isolated. Runoff and sediment loads measured at the bottom of a hill-slope depend upon the spatial distribution of surface depressions. Isolated depressions act as sinks, and do not contribute runoff for small to medium storms. Surface roughness, such as that created by tillage, results in temporary depressional storage. Though some encouraging efforts have been made to include such effects in hydrologic modelling, much more work is needed to understand and quantify the effects of depression storage on runoff, overland flow and soil deposition.

Surface crusting, present on many if not most soils, is another surface phenomenon which has significant effects on soil detachment and runoff rates on slopes. These effects too have not been adequately quantified in soil erosion models. The changes in hydraulic conductivity which were noted with rainfall time step may in fact be partially due to surface effects, including effects of micro depressions, surface crusting and/or crop cover. Rudra *et al.* (1985), while applying the

CREAMS model, observed different values of calibrated hydraulic conductivity on replicated plots, where soil, land use and rainfall characteristics were similar. Such differences could well be due to the presence of and differences in surface characteristics.

- *Variations in soil erodibility: the erodibility of soils can vary much more widely from season to season than has been acknowledged to date in soil erosion models; and laboratory and field determinations of soil erodibility, considered to be independent of the temporal pattern of simulated rainfall, are indeed likely to be highly dependent on that pattern*

Although soil erodibility has always been recognised to be a very important parameter in soil erosion models, it has been dealt with incompletely and simplistically. For example, the soil loss component of many of the models has been based on the Universal Soil Loss Equation, employing a constant or near constant value of the K factor for a particular soil with time. Yet there is growing evidence that at least in temperate and humid regions, soil erodibility can vary considerably from season to season. (Mutchler and Carter, 1983; Coote *et al.*, 1988; Rudra *et al.*, 1989; Wall *et al.*, 1988). Rudra *et al.* (1988) have observed higher values of soil erodibility factor during late winter and early spring than in summer periods in the temperate and humid region of Southern Ontario (Figure 6). During late winter and spring periods the soils often still have a frost layer at shallow depth, and the surface layer is very unstable due to the thawing processes involved. The water status of such soil conditions is very high, being saturated or even super-saturated; and the bulk density and surface shear strength are quite low. Any disturbance at the surface, either by rain or by runoff, detaches and moves soil down slope very readily. These effects are not well represented in erosion models.

Mackenzie (1995) has observed not only temporal changes in erosion rates during simulated rainfall events, but also dramatic differences in erosion rate/erodibility from tests using discontinuous versus continuous applications of rainfall, as shown in Figures 7 and 8. (Please note difference in ordinate scale in Figures 7 and 8). These figures reveal that, for a continuous and constant simulated rainfall intensity, the erosion rate was initially low or zero until runoff developed on the 1m x 1m plots. Once runoff began, the erosion rate increased rapidly to a relatively high value, then decreased to a more or less steady state value. The general temporal pattern of erosion rate was essentially the same for the discontinuous rainfall runs, involving equal, brief on-and-off intervals of rainfall, with a constant rainfall intensity during the 'on' periods that was double the intensity used for the continuous rainfall runs. Note that the average rainfall rates (and total rainfalls) were equal for comparable continuous and discontinuous rainfall test runs. However, it is quite clear that the erosion rates and the total accumulated soil loss (i.e. the integral of the erosion rate curve) were substantially greater for the discontinuous rainfall runs than for the continuous runs. The direct implication is that erodibility values determined with equipment involving discontinuous applications of simulated rainfall can be expected to be orders of magnitude larger than erodibility values determined from continuous applications of rainfall of the same average intensity. Obviously, if two soil erosion modellers use different rainfall simulators to estimate inter-rill erodibility, other calibrated parameters will inevitably be quite different, to accommodate the large differences in erodibility. The dilemma of erodibility estimates being dependent on the simulation equipment used has not yet received any attention in the soil erosion modelling literature.

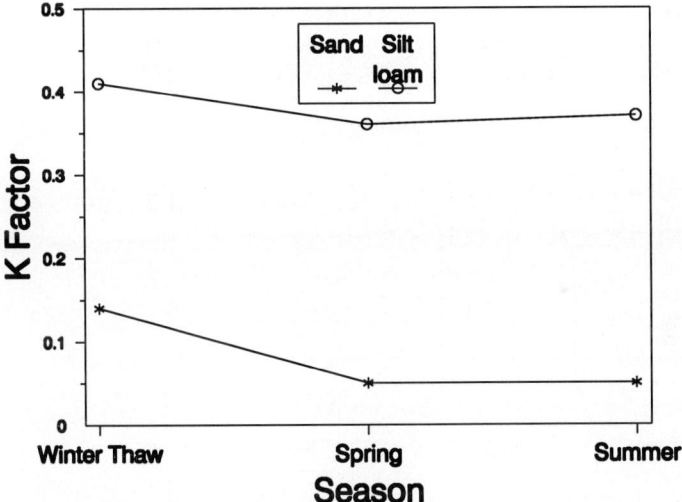

Figure 6. Variations in soil erodibility with season

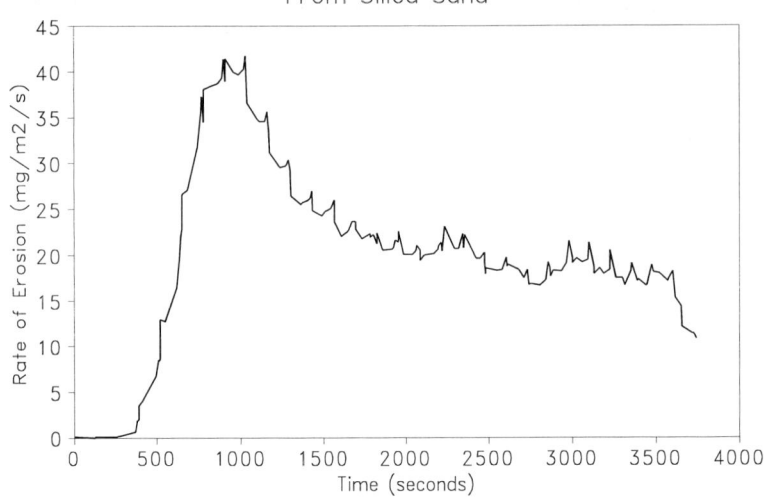

Figure 7. Change in soil erosion during a constant intensity rain storm

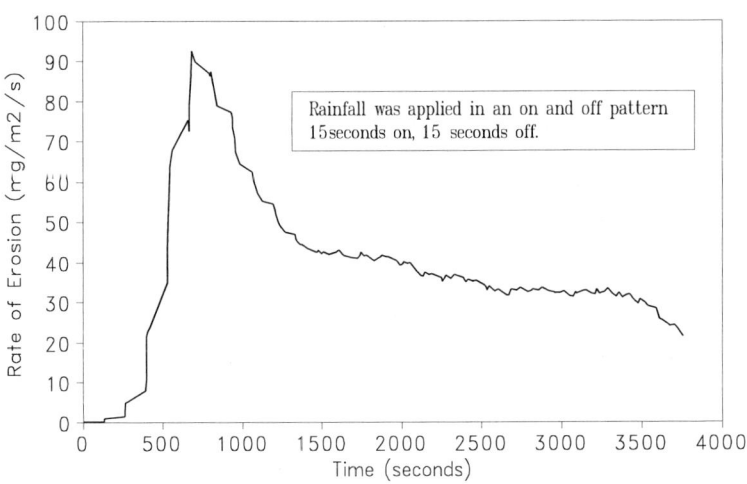

Figure 8. Change in erosion rate during an intermittent intensity rain storm

- *The role of evapotranspiration and soil water distribution: evapotranspiration and percolation are other processes which have not received serious attention by soil erosion modellers*

In many hydrologically-based erosion models, the focus of the water budget has been runoff and infiltration. Evapotranspiration, the soil moisture distribution in the soil profile and percolation have been treated in a relatively simple and crude fashion, even though evapotranspiration and percolation are major hydrologic processes and have a major influence on the moisture distribution in the soil profile. These processes dominate the intervals between rainfall events. In many erosion models, evapotranspiration has been estimated using a temperature-based equation, and percolation has been based on a field capacity approach. It is also obvious from the work of McKenney and Rosenberg (1993) that estimates of potential evapotranspiration are sensitive to climatic change. Recently, while applying a continuous drainage model, DRAINMOD, Thooko et al. (1990) observed a strong dependence between the relative rating of parameters to which model outputs were sensitive (i.e. the output was most sensitive to the parameter rated first, and so on), the quality of evapotranspiration estimates and the length of the simulation period (Table 1). These data indicate that, on a yearly basis, a rating of the sensitive parameters depended on the method used to estimate evapotranspiration. Over long periods of record, due to a lumping effect, the rating of parameters became relatively stable. For shorter periods, discrepancies between the ratings were attributable to errors in the estimates of evapotranspiration in the water budget. Commonly, for the determination of water budgets, runoff and infiltration are estimated on the basis of a reasonably detailed equation, evapotranspiration is estimated with relatively simple methods based on easily available climatic parameters, and the remainder is allocated to percolation.

Further, a major focus regarding water budgets in soil erosion models is the matching of predicted and observed runoff. The comparison of other components, such as evapotranspiration and percolation, is generally ignored. Such a practice inevitably results in poor estimation of initial conditions for subsequent rain events. As evapotranspiration and percolation estimates diverge from correct values, the errors are transferred in the modelling exercise to the redistribution of soil moisture. On the occasion of the next rainfall/runoff event, the initial soil moisture condition can be grossly in error. More attention needs to be paid to getting the evapotranspiration component right!

Table 1. The change in sensitivity rating of input parameters with the method of determination of evapotranspiration

Sensitivity Rating	Observed PET			PET estimated by temperature based method		
	1988	1989	1988/89	1988	1989	1988/89
1	RD	DP	DP	KS	DV	DP
2	KS	DV	RD	RD	DP	KS
3	DP	KS	KS	DP	RD	RD
4	UF	RD	UF	WP	UF	UF
5	WP	UF	DV	UF	KS	WP
6	DV	WP	WP	DP	WP	DV

RD - Rooting depth
DP - Depth of Impermeable layer
WP - Permanent wilting point
KS - Saturated hydraulic conductivity
UP - Upward flux
DV - Drained volume

Concluding remarks

Substantial progress has been made regarding the development of soil erosion models. However, for improvements to be realised, weak links in the models must be identified and addressed. Some of the present modelling dilemmas appear to include improper descriptions of fundamental processes, poor or inappropriate methods of estimating some critical model parameters, and a lack of ways to take differences in scale into account. Problems with fundamental processes include an inability to represent variations in infiltration, soil erodibility and surface storage in time and space; and a number of key parameters are too dependent on the nature/precision of inputs and measurement techniques. The scale of model elements tends to be rather large (e.g. fields and watersheds); the scale at which many input parameters are measured is much smaller (e.g. point sites or small plots); and relationships or scaling factors between the measurements and appropriate model parameter values have not yet been developed. Better model results cannot be expected until these dilemmas are resolved. Such resolution cannot occur until the problems are openly acknowledged and collectively addressed! That is our challenge in the coming years.

References

Beasley, D.B., Huggins, L.F. and Monke E.J. (1980). ANSWERS: A model for watershed planning. *Transactions of the American Society of Agricultural Engineers*, **23**(4), 938-.944.

Coote, D.R., Malcolm-Mcgroven, C.A., Wall G.J., Dickinson, W.T. and Rudra, R. P. (1988). Seasonal variations in erodibility indices based on shear strength and aggregate stability in some Ontario soils. *Canadian Journal of Soil Science* **68**, 405-416.

Gupta, R.K, Rudra, R.P., Dickinson, W.T. and Wall, G.J. (1994). Spatial and temporal variations in hydraulic conductivity in relation to four determination techniques. *Canadian Water Resources Journal* **19**(2), 1-11.

Gupta, R.K., Rudra, R.P., Dickinson, W.T., Patni, N.K. and Wall, G.J. (1993). Comparison of hydraulic conductivity measured by various field methods. *Transactions of the American Society of Agricultural Engineers* **36**(1), 51-55.

Jones, C.A., Dyke, P.T., Williams, J.R., Kiniry, J.R., Benson, V.M. and Griggs, R.H. (1991). An operational model for evaluation of agricultural systems. *Agricultural Systems* **37**, 341-350.

Knisel, W.D. (1980) (ed.), *CREAMS: A Field Scale Model for Chemicals, Runoff and Erosion from Agricultural Management Systems.* USDA Conservation Research Report No. 26, 643 pp.

Laflen, J.M., Lane, L.J. and Foster, G.R. (1991). WEPP: A new generation of erosion prediction technology. *Journal of Soil and Water Conservation* **46**, 34-38.

Leonard, R.A., Knisel, W.G. and Still, D.A. (1987). GLEAMS: Groundwater Loading Effects of Agricultural Management Systems. *Transactions of the American Society of Agricultural Engineers* **30**, 1403-1418.

Mackenzie, K.M. (1995). *Investigations of the Effect of Temporally Distributed Rainfall on Inter-Rill Soil Detachment Processes.* Unpublished M.Sc. Thesis, University of Guelph, Guelph, Ontario, Canada.

McKenney, M.S and Rosenberg, N.J. (1993). Sensitivity of some potential evapotranspiration methods to climate change. *Agricultural and Forest Meteorology* **64**, 81-110.

Mutchler, C.R. and Carter, C.E. (1983). Soil erodibility variation during the year. *Transactions of the American Society of Agricultural Engineers* **26**, 1102-1104, 1108.

Renard, K.G, Foster, G.R., Weesies, G.A. and Porter, J.P. (1991). RUSLE: Revised universal soil loss equation. *Journal of Soil and Water Conservation* **46**, 30-33.

Rousseau, A., Dickinson, W.T, Rudra, R.P. and Wall, G.J. (1988). A phosphorus transport model for small agricultural watersheds. *Canadian Agricultural Engineering* **30**(2), 213-220.

Rudra, R.P., Dickinson, W.T. and von Euw, E.L. (1993). The importance of precise rainfall inputs in nonpoint source modelling. *Transactions of the American Society of Agricultural Engineers* **36**(2), 445-450.

Rudra, R. P., Dickinson, W.T. and Wall, G.J. (1989). The role of hydrometeorological and soil conditions in soil erosion and fluvial sedimentation. *Canadian Agricultural Engineering* **31**(2), 107-115.

Rudra, R.P., Dickinson, W.T, Clark, D.J. and Wall G.J. (1986). GAMES - A screening model of soil erosion and fluvial sedimentation in agricultural watersheds. *Canadian Water Resources Journal* **11**(4), 58-71.

Thooko, L. W., Rudra, R.P., Patni, N.K. and Dickinson, W.T. (1990). *DRAINMOD in Ontario conditions.* American Society of Agricultural Engineers Paper No. 90- 2064, American Society of Agricultural Engineers, St. Joseph, MI.

Wall, G. J., Dickinson, W.T., Rudra, R.P. and Coote, D.R. (1988). Seasonal Soil Erodibility Variations in Southwestern Ontario. *Canadian Journal of Soil Science* **68**, 417-424.

Young, R. A., Onstad, C.A., Bosh, D.B. and Anderson, W.P. (1985). *AGNPS, Agricultural Non-Point-Source Pollution Model.* U.S. Department of Agriculture Research Report 35. Agriculture Research Service, Washington, D.C.

14. CROSS-SCALE ASPECTS OF EPA EROSION STUDIES

Jeffrey J. Lee[a]

US *Environmental Protection Agency*
National Health and Environmental Effects Laboratory
Western Ecology Division
200 SW 35th Street
Corvallis, OR 97333
USA

Abstract
Studies at US EPA's NHEERL/WED facility have focused on two broad aspects of soil erosion within the context of global change: (1) the effects of soil erosion on carbon budgets, especially on the release or sequestration of CO_2 by soils; and (2) the potential effects of global change on soil erosion as an environmental issue. These issues generally are addressed on regional and decadal scales. The main approach has been to apply a site scale, daily timestep model (the Erosion/Productivity Impact Calculator, EPIC) to selected sites for decades to centuries, and to aggregate the outputs to estimate regional responses to changes in climate and management. Two methods of aggregation are used. In the first, sites are selected or constructed to represent categories of land use, land management, and/or climate. The definition of 'representative' is usually at least somewhat qualitative, and might be based primarily on availability of data. In contrast, the second method employs a statistical approach. The target population, the selection criteria, and the probability of selection are rigorously defined. Means, standard deviations, and other descriptors of the population are obtained through the statistical structure of the sampling scheme. In this paper, results from a sensitivity analysis of wind and water erosion to climate change are used to illustrate the statistical approach.

Introduction

The atmospheric concentrations of CO_2 and other greenhouse gases (e.g. CH_4, N_2O) are increasing. Results from climate models indicate that the changes in radiative forcing from these increased concentrations could alter atmospheric temperature, precipitation patterns and wind speeds (IPCC, 1990). The projected changes in climate and increases in CO_2 could have significant direct and indirect impacts on agricultural systems, including effects on crop growth and yield (e.g. Rosenzweig and Parry, 1994; Crosson, 1993), soil organic carbon (Jenkinson *et al.*, 1991) and soil erosion (Boardman *et al.*, 1990; Phillips *et al.*, 1993; Boardman and Favis-Mortlock, 1993).

---
[a] Deceased

The Erosion-Productivity Impact Calculator (EPIC) model (Sharpley and Williams, 1990) was used to simulate a corn site in eastern Nebraska, which is the extreme western (i.e. driest) extent of the US cornbelt. Annual precipitation is 637 mm, while mean monthly wind speeds range from 4.6 to 6.7 m s^{-1}. Wind erosion rates, as simulated by EPIC, are extremely high at this site (82 t ha^{-1} yr^{-1}). Simulated corn yields are low (4.1 t ha^{-1}) and highly variable (coefficient of variation of 58 per cent) because of variability of precipitation. Simulated soil organic C (SOC) in the top 15 cm remains fairly constant for about 60 years under current climate (Figure 1), until soil water holding capacity becomes limiting for yield. SOC then begins to drop sharply, decreasing by 50 per cent in the next 40 years. A century after the accelerated decline begins, SOC is only about 1/7 of its original value. An additional EPIC run (not shown) suggests that, if wind speeds increased by 20 per cent, increased wind erosion would accelerate loss of SOC, with 90 per cent depletion within 50 years of the start of simulation.

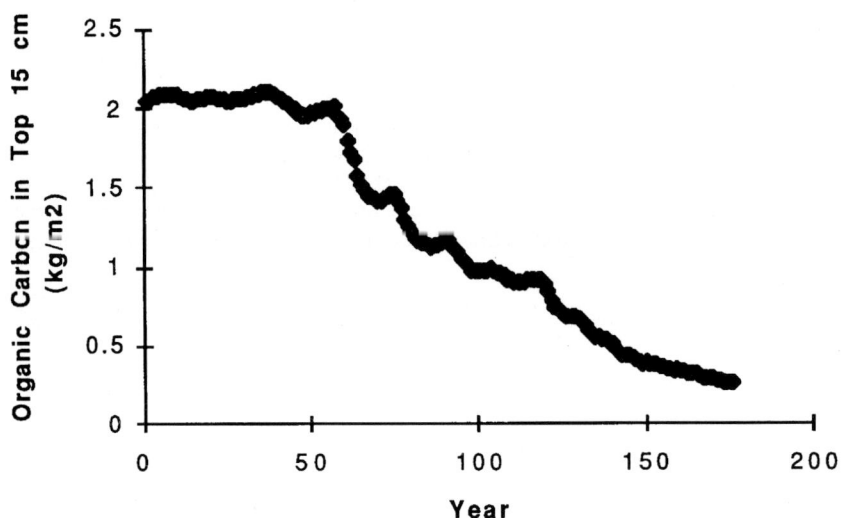

Figure 1. Long-term effect of soil erosion on soil organic carbon as simulated by EPIC for a vulnerable site in the extreme western US cornbelt

Erosion as an environmental issue

An important question within the environmental issue is: What are the potential impacts of soil erosion on productivity and sustainability of agricultural and grassland ecosystems? The EPIC results discussed above (Figure 1) illustrate that the effects of climate change on yield and erosion can be a tightly coupled positive feedback loop, capable of causing rapid change after a prolonged period of slow change. Thus, it is critical to consider these effects together.

Soil erosion can also cause secondary environmental effects, such as increased dust in the air or increased transport of sediments to rivers and lakes. Increased dust ('particulate matter') can have adverse effects on human health. The atmospheric concentration of dust particles which are smaller that 10 µm in diameter (PM10) is regulated as an air pollutant in the US (Stetler *et al.*, 1994). Increased sediment transport to rivers, lakes and reservoirs can cause many adverse effects such as reducing fisheries, increasing flooding, hampering transportation by boat, reducing water storage in reservoirs, and decreasing hydroelectric power (Robinson, 1979; Tagwira, 1992; Phillips *et al.*, 1993).

Cross-scale issues

The questions and concerns raised above apply to regions (i.e. 1-100 x 106 ha) or larger areas. Sites (i.e. 1-1000 ha) might be considered, but still in the context of regional impacts. For example, it would be important to know the density of especially vulnerable sites (e.g. Figure 1) within a region, by not necessarily the location of individual sites. Similarly, the temporal scale is decades to centuries. While the frequency of extreme events might be important, the actual timing of specific events is generally not of concern. In contrast, soil erosion processes occur on micro (i.e. 1-10 mm) to site spatial scales, and on minute to event temporal scales.

Three approaches to resolving these cross-scale issues can be identified:

1. Scale the input: in this approach, the driving variables (e.g. soil, climate) are averaged spatially and temporally and used as input to a high-resolution erosion model. The resulting output is then taken to represent the average erosion rate for the region and time period being considered. This approach is appropriate only to the extent that the relationship between the driving variables and the erosion rate can be treated as linear. Thresholds and other non-linear elements of erosion processes restrict this approach to extremely localised analyses. It is not, in general, appropriate for regional or long-term analyses.

2. Scale of model: in this approach, a model is developed explicity for scale of analysis desired. The input data and the predictions of the model are consistent with this scale. Site

models such as USLE (Wischmeier and Smith, 1978), WEQ (Woodruff and Siddoway, 1965), and EPIC (Sharpley and Williams, 1990) use this approach. For example, a 'roughness coefficient' might be used to represent the average effects of inhomogenous microtopography. Aggregated parameters such as 'mean wind speed at 10 m' would be used instead of detailed descriptions of variability across a field. Model predictions can be compared to measurements at the same (e.g. field) scale. Thus, the entire procedure from data acquisition through validation is conducted at a consistent scale. Application of this approach has been limited to watershed or smaller areas.

3. Scale the output: in this approach, a site model is used in conjunction with data at the same scale. Thus, the non-linear relationships between input and prediction are preserved. Model outputs are then scaled to a region using either of two approaches. In the 'Representative Approach' the individual sites are considered representative of categories of sites. Theses categories might, for example, consist of combinations of land use, cropping systems, climate, soil, or other relevant attributes. In some cases, the sites might be synthesised as ideal expressions of the categories, and might not exist as physical sites. Model outputs are extrapolated to the region according to the estimated area of each category. An example of this approach is the use of EPIC for analyses required by the US Resources Conservation Act (Williams and Renard, 1985).

In the 'Statistical Approach' sites are randomly selected within the region. Data from each site are used in a sited model, and the model outputs are extrapolated to the region through the statistical framework (Lee and Lammers, 1990). This approach avoids the somewhat subjective step of developing categories. It is, however, more data intensive than the representative approach. USLE, WEQ (Kellogg *et al.*, 1994), and EPIC (Lee *et al.*, 1993) have been used within the statistical approach.

Both output-scaling approaches assume that the entities being scaled (categories, sites) are independent. This assumption would not be reasonable for entities with strong interactions, such as microsites within a single field.

Statistical scaling: an example from the US cornbelt

The primary approach to resolving cross-scale erosion issues at NHEERL/WED has been the statistical approach to scaling model outputs (Figure 2). The steps in this approach are illustrated by recently completed analysis of the potential effects of climate change on soil erosion in the US cornbelt. Details of the implementation of this analysis for current climate (i.e. steps 1-6 below) have been presented by Lee *et al.* (1993).

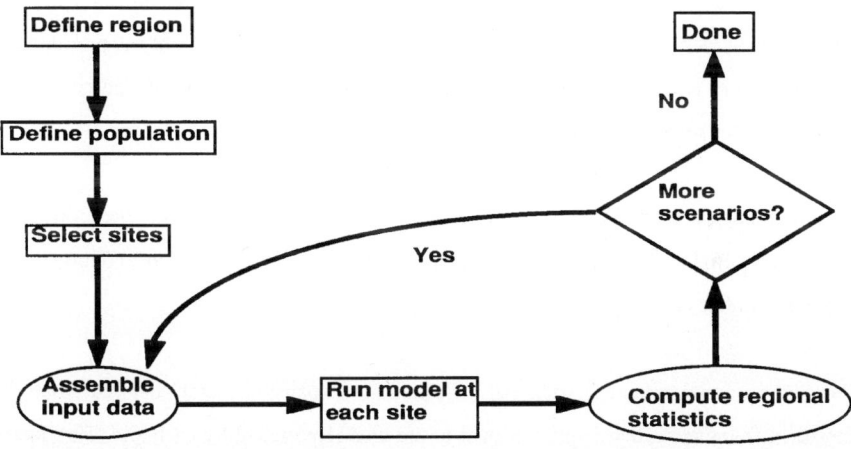

Figure 2. Procedure for statistical approach used to scale results from sites to regions

Implementation of the approach

1. Define the region: the approach requires a rigorous definition of the geographical region of interest. In the case, the 'US cornbelt' was defined as a set of Major Land Resource Areas (SCS, 1981). In general, a geo-referenced representation of the region is required.

2. Define the population: not all locations within a region will necessarily be included in the analysis. Thus, it is essential to develop criteria by which a site can unambiguously be defined as included or excluded from the analysis. In this case, only sites which were used exclusively for corn or corn-soybean rotation during 1984 to 1987 were included. This would exclude, for example, sites that were set-aside status for any of that period.

3. Select the sites: sites must be selected according to a rigorous statistical procedure. In this case, 100 sites were randomly selected from more than 11,000 National Resources Inventory (NRI; SCS, 1989) sites that met the selection criteria. The NRI is itself a statistically based recurring inventory of non-federal lands within the US

4. Assemble the input data: the NRI provides descriptive data for each site, including approximate location, cropping history, tillage and erosion control practices, and soil series and phase. The site can be linked to climate and soil databases through the location and soil series, respectively. Strictly speaking, the climate and soil data are only approximately correct for the site. This is not necessarily a limitation as the ultimate purpose is to characterise the region, not individual sites.

5. Run the model at each site: EPIC was run for 100 years at each site. This was consistent with characterising long-term response, while including the effects of common and extreme events.

6. Compute regional statistics: each NRI site has an 'expansion factor' derived from the statistical framework, that is proportional to the area represented by each site. This factor provides an appropriate weighting factor to estimate means, standard deviations, and other statistical parameters for the region

7. Repeat for alternative scenarios: in this case the objective was to compare erosion under several assumed climate regimes. This was accomplished by altering the climate variables in the EPIC input datasets, including mean temperatures (current, +2°C), precipitation (current, ±10%, ±20%), CO_2 (350, 625 ppmv), and wind speed (current, ±10%, ±20%).

Example results
A full presentation of the results of the cornbelt climate change analysis is beyond the scope of this paper. Two examples (Figure 3, Table 1) illustrate some of the uses of the results.

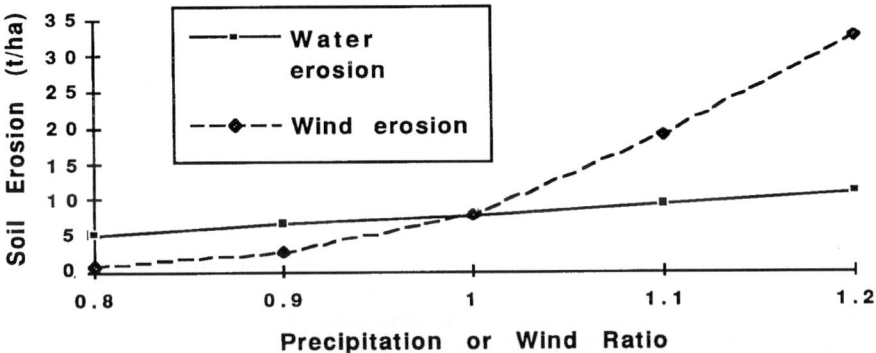

Figure 3. Regional sensitivity of erosion by wind and water to changes in mean wind speed or precipitation volume, respectively, as simulated by EPIC. Each point represents the 100-year weighted average of 100 randomly selected sites in the US cornbelt

One way to summarise the results is to aggregate the erosion rates spatially and temporally for the entire region and period of interest. Viewed in this way, wind erosion is much more sensitive to changes in wind speed than water erosion is to changes in precipitation volume (Figure 3). This suggests that, at least for this region and this issue, it is critical to understand the thresholds and mechanisms for wind erosion. For other regions or cropping systems, water erosion might be more sensitive to climate change.

A somewhat less aggregated approach is to consider the frequency with which erosion events occur. For example, under current climate, mean wind erosion for the cornbelt is 8.0 t ha^{-1} yr^{-1}, with a coefficient of variation (CV) of 70 per cent (Table1). A decrease of 20 per cent in mean wind speed would cause wind erosion to decrease ten-fold, while becoming relatively more variable (100% CV). In contrast, a 20 per cent increase in wind speed would cause a four-fold increase in mean wind erosion and a relative decrease in variability (59% CV). For current climate, mean wind erosion rate was as large as 32 t yr^{-1} for only 1 of the 100 simulated years. In contrast, with wind speed increased by 20 per cent, annual values greater than 27.3 t ha^{-1} yr^{-1} occurred for 50 of 100 simulated years. This means that an event which was unlikely to be experienced during the course of an average human lifetime for current climate would be experienced very frequently with a 20 per cent increase in wind speed. This suggest the need for a finer-scale analysis within the region to determine the frequency and locality of wind erosion events with implications for human health.

Mean Wind Speed (%)	Mean Erosion Rate	Standard Deviation	Number of Years with Lower or Equal Annual Wind Erosion Rates		
			10 yr	50 yr	100 yr
80	0.8	0.8	0.2	0.5	6.1
90	2.8	2.5	0.8	2.0	15.2
100	8.0	5.6	2.8	6.3	32.0
110	19.0	11.7	6.9	16.2	62.8
120	32.7	19.2	14.2	27.3	94.9

Table 1. Effect of mean wind speed (as per cent of current value) on mean and annual wind erosion rates (in t ha^{-1} yr^{-1}) averaged across the US cornbelt, as simulated by EPIC. Example: for 20 per cent greater mean wind speed, wind erosion was less than or equal to 14.2 t ha^{-1} yr^{-1} for 10 of 100 simulated years

References

Boardman, J. and Favis-Mortlock, D.T. (1993). Climate change and soil erosion in Britain. *Geographical Journal* **159**, 179-183

Boardman, J., Evans, R., Favis-Mortlock, D.T. and Harris, T.M. (1990). Climate change and soil erosion on agricultural land in England and Wales. *Land Degradation and Rehabilitation* **2**, 95-106.

Crosson, P. (1993). Impacts of climate change on agriculture and economy of the Missouri, Iowa, Nebraska, and Kansas (MINK) Region. In, Kaiser, H.M. and Drennen, T.E. (eds), *Agricultural Dimensions of Global Climate Change*. St. Lucie Press, Delray Beach, FL. pp. 117-135.

Follett, R.F. and Stewart, B.A. (eds) (1985). *Soil Erosion and Productivity*. American Society for Agronomy, Crop Science Society of America, Soil Science Society of America, Madison, WI.

Intergovernmental Panel on Climate Change (IPCC). (1990). *Climate Change: The IPCC Assessment*, eds Houghton, J.T., Jenkins, G.J. and Ephraums, J.J. Cambridge University Press, Cambridge, UK.

Jenkinson, D.S., Adams, D.E. and Wild, A. (1991). Model estimates of CO_2 emissions from soil in response to global warming. *Nature* **351**, 304-306.

Kellogg, R.L., TeSelle, G.W. and Goebel, J.J. (1994). Highlights from the 1992 National Resources Inventory. *Journal of Soil and Water Conservation* **49**, 521-527.

Kern, J.S., and Johnson, M.G. (1993). Conservation tillage impacts on national soil and atmospheric carbon levels. *Soil Science Society of America Journal* **57**, 200-210.

Lee, J.J. and Lammers, D.A. (1990). An approach to the regional evaluation of the responses of soils to global climate change. In, Bouwman, A.F. (ed.), *Soils and the Greenhouse Effect*. Wiley, Chichester, UK. pp. 395-399.

Lee. J.J., Phillips, D.L. and Liu, R. (1993). The effect of trends in tillage practices on erosion and carbon content of soils in the US corn belt. *Water, Air, and Soil Pollution* **70**, 389-401.

Phillips, D.L., White, D. and Johnson, C.B. (1993). Implications of climate change scenarios for soil erosion potential in the USA. *Land Degradation and Rehabilitation* **4**, 61-72.

Robinson, A.R. (1979). Sediment yield as a function of upstream erosion. In, Peterson, A.E. and Swan, J.B. (eds), *Universal Soil Loss Equation: Past, Present and Future*. Soil Science Society of America, Madison, WI, pp. 7-16.

Rosenzweig, C. and Parry, M.L. (1994). Potential impact of climate change on world food supply. *Nature* **367**, 133-138.

Sharpley, A.N. and Williams, J.R. (eds) (1990). EPIC - *Erosion/Productivity Impact Calculator*, US Department of Agriculture Technical Bulletin No. 1768, Washington, DC.

Soil Conservation Service (SCS) (1981). *Land Resource Regions and Major Land Resource Areas of the United States*, US Department of Agriculture Agricultural Handbook 296, Washington, DC.

Soil Conservation Service (SCS) (1989*). Summary Report, 1987 National Resources Inventory*, Iowa State University Statistical Laboratory, Statistical Bulletin Number 790, Ames, IA.

Stetler, L.D., Saxton, K.E. and Fryrear, D.W. (1994). *Wind Erosion and PM_{10} Measurements from Agricultural Fields in Texas and Washington.* Paper 94-FA145.02, Annual Meeting, Cincinnati, June 19-24, Air and Waste Management Association, Pittsburgh

Stocking, M.A. (1986). *The Cost of Soil Erosion in Zimbabwe in Terms of the Loss of Three Major Nutrients.* Consultant's working paper No. 3, FAO, Rome, Italy.

Tagwira, F. (1992). Soil erosion and conservation techniques for sustainable crop production in Zimbabwe. *Journal of Soil and Water Conservation* **47**, 370-374.

Williams, J.R., and Renard, K.G. (1985). Assessments of soil erosion and crop productivity with process models (EPIC). In, Follett, R.F. and Stewart, B.A. (eds) (1985). *Soil Erosion and Productivity*, American Society for Agronomy, Crop Science Society of America, Soil Science Society of America, Madison, WI. pp 67-103.

Wischmeier, W.H. and Smith, D.D. (1978*). Prediction of Rainfall Erosion Losses: a Guide to Conservation Planning*, US Department of Agriculture Agricultural Handbook No. 537, Washington, DC.

Woodruff, N.P. and Siddoway, F.H. (1965). A wind erosion equation. *Soil Science Society of America Proceedings* **29**, 602-608.

World Resources Institute (WRI) (1992). *World Resources*. Oxford University Press, Oxford, UK.

15. SCALE ISSUES AND A SCALE TRANSFER METHOD FOR EROSION MODELLING

D. King, D.M. Fox, Y. Le Bissonnais and **V. Danneels**

Institut National de la Recherche Agronomique (INRA)
Unité de Science du Sol
Service d'Etudes des Sols et de la Carte Pédologique de France
Centre de Recherches d'Orléans
45160 Ardon
France

Abstract

Most soil databases contain a geometrical dataset, which describes the spatial objects; and a semantic dataset, which manages the non-spatial attributes of these objects. Scale issues for soil databases are related to the precision of the data in each of the two datasets. Scale concepts are reviewed briefly and a scale transfer method is proposed to link results from a small catchment scale to a larger regional scale. A scale transfer method is demonstrated by upscaling a rill erosion model. Four stages are examined when extending results from small reference areas to a region: (1) the verification of the accuracy of the predictive method, (2) the impact of the change in spatial resolution on erosion prediction, (3) the impact of the change in semantic precision on erosion prediction, and (4) the production of an erosion estimation map at the regional scale. According to a sensitivity test of the method, the semantic precision was sufficient. However, the spatial resolution was too low to provide a good prediction of rill erosion, so a statistical relationship was used as an alternative to produce a rill erosion hazard map. In this case, the scale transfer method was used for upscaling rill erosion, but the method can be applied to other processes.

Introduction

Runoff and erosion models require several parameters, particularly soil parameters. Soils are generally highly variable and this entails difficulties in providing precise temporal and spatial data for modelling. Furthermore, soil databases are often elaborated without considering their possible use. In consequence, such databases often do not have the soil parameters required for modelling (Msanya *et al.*, 1987). This is especially true for large areas where data collection is expensive and time-consuming.

Geographic Information Systems (GIS) facilitate the management of geographical information. Their current uses include digitisation, spatial analysis, and producing maps (Burrough, 1986). GIS tools provide technical solutions to some of the problems described above by inferring, combining and scaling data. However, the use of these tools without adequate consideration for the processes occurring in the field can lead to irrelevant results.

Soil databases are developed from soil mapping at various scales. GIS tools are helpful in defining scale concepts and in developing new techniques for quantifying a scale transfer method (Sivapalan and Kalma, 1995). In this paper, a scale transfer method is defined as a method of generalising knowledge from a small catchment scale to a larger regional scale. The structure of current soil databases will first be presented. Some scale issues in soil mapping will then be discussed. Finally, a scale transfer method for a simple empirical erosion model will be described.

Soil databases
Soil data can be split into two sets. The first is the geometrical dataset which describes the spatial objects (e.g. points, lines, and areas). The second is the 'semantic' dataset which manages the non-spatial attributes of these objects (e.g. soil depth, hydraulic conductivity, etc.) (Brabant, 1992; Jamagne *et al.*, 1994). These two datasets are linked by relational tables in a database management system.

Geometrical data
Most soil databases contain digitised soil maps. Data for the maps originate from conventional soil surveys using expert knowledge. Soil boundaries are estimated empirically and are scale-dependent due to their representation on a map. Scale is defined here as the ratio of the distance between two points on a map to its distance on the ground. GIS tools are used in the final stage to digitise soil boundaries. Once in a vector format, the use of scale as defined above is inapplicable. The concept of scale is replaced by one of 'spatial resolution' or 'spatial tolerance.' This defines the geometrical precision of the elementary points.

An important concept in GIS is topology: this describes the spatial relationships between objects. For example, arcs of a polygon have a spatial relationship with one another, just as do neighbouring polygons. A decrease in the spatial resolution leads to a loss of topology: two distinct points in reality may be indistinguishable on a map and be represented as a single point.

Soil databases also contain local information stored as point data. These are usually called soil profile databases. These databases provide the basic soil information for modelling and deserve extensive development (Madsen, 1991). However, the stored data profiles are often limited to representative profiles, and their representivity often remains an intuitive estimation.

Finally, databases contain information related to elementary features of the surface. These are comprised primarily of relief parameters obtained from Digital Elevation Models (Bell *et al.*, 1994; Gessler *et al.*, 1995) and of landuse data obtained from satellite images (Girard and Girard, 1994). The covariation of these easily observed variables with soil parameters can be represented as a soil-landform conceptual model (Hewitt, 1993). The automated GIS output maps use a grid system similar to Digital Elevation Models (DEM) and images. Scale in such maps is replaced by the spatial resolution corresponding to the pixel size.

Semantic data

Data stored in a soil profile database are measured directly with instruments, but the range of error of the measurement is rarely indicated. Moreover, soil parameters needed by hydrological models are often absent from the databases. The missing soil parameters are frequently estimated using available surrogate (covariate) data and pedotransfer functions (Bouma and Van Lanen, 1987). The final uncertainty in the estimated data is therefore a combination of the unexplained variance in the relationship between the estimated parameter and the surrogate data and the error involved in measuring the surrogate data itself (Van Ranst *et al.*, 1995).

In general, measurements cannot be performed for a spatial area. With the exception of remotely sensed data, soil data are measured at a point. The variables in a Soil Mapping Unit (SMU) are estimated using statistical methods from point data within the SMU (Webster and Oliver, 1990) or with empirical methods from soil surveyor expertise (King *et al.*, 1994). Since a single value does not describe the spatial variability within the SMU, a distribution function is better (standard deviation, Kurtosis coefficient, skewness, etc.). In conventional mapping, pedologists use the concept of soil association. Several Soil Typological Units (STU) are described within a SMU and values of soil parameters are estimated for each STU. The soil association method provides a description of the spatial structure of the soil, and this is important in defining the relationships between neighbouring soil bodies (King et al., 1994). The result, however, is a description of the STU that does not include its geometrical and topological information.

Scale issues

The noun 'scale'

The use of soil databases necessitates a definition of 'scale' which is different from the one provided above, since the words 'scale' and 'scaling' are used in a broader sense than a simple ratio. More generally, the word 'scale' indicates (1) the spatial extent of a study or (2) the resolution of the data. We have seen above that this concept of resolution can be defined as either the minimum distance between two points or the integration volume (surface) of the data (for example, 1 dm^3 for a soil sample, 900 m^2 for a Landsat pixel, several km^2 for a catchment).

Blöschl and Sivapalan (1995) distinguish three concepts: process scale, observation scale, and representation scale (scale is used for both space and time). The *process scale* corresponds to the minimum area required for the process of interest to occur. For instance, peaks, ridges and breaks in slope have a natural variation in a landscape which is measured using a spatial correlation. The *observation scale* is associated with the data collection, each measurement having a geometrical resolution and a semantic precision. In building a DEM, for example, data are collected with a known precision and resolution which depend on the instrument used. As with soil databases, many DEMs do not contain the original measurements but only derived products which are easier to manage. As a consequence, there is no precise information about the observation scale, which is the most important when using databases. The *representation scale* is the precision with which we describe and model reality. In this case, values are estimated on a grid with a pixel size chosen according to the objectives and computing limitations.

The verb 'to scale'

'To scale' means to zoom in and out (Blöschl and Sivapalan, 1995). For example, upscaling refers to transferring information from a given scale to a larger one. Downscaling is the opposite process. This definition is extended to scale transfer where knowledge from a small area is generalised to a larger area.

In scale transfer, two approaches related to the geometrical and semantic datasets can be taken. The first attempts to keep the same resolution between the small and large areas, so the same density of observations is maintained. Due to the cost of instrumentation, it is usually necessary to use cheaper but less precise equipment. Measuring substitute covariate data is another option for reducing costs when trying to obtain high spatial resolution data for large areas. With either method, the precision decreases. Scale transfer must therefore include a sensitivity test for evaluating the effect of the loss in precision on the output of runoff and erosion models.

The second approach has implications for the semantic dataset. It is possible to preserve a high precision in the measurements by reducing the spatial resolution. In this approach, a limited number of samples are measured with a high precision. However, the variability over large areas is often more poorly described by fewer samples. In general, the larger the area, the more heterogeneous are the soil properties, and the less relevant is the mean value of a parameter for this area. A decrease in the number of samples or an increase in the size of an area entails an increase in the standard deviation of the measured parameter and the loss of some of the topology. One of the aims of a scale transfer method is to show the relevance of a model when applied at different scales of soil pattern heterogeneity (Wood et al., 1988).

A scale transfer method for erosion modelling

The aim of the study was to produce a map of the winter rill erosion hazard for the Nord-Pas-de-Calais administrative region of France, an area of about 10,000 km^2. Due to the difficulties in obtaining precise information for such a large area, only easily available data were used. The elementary spatial unit at this scale is an administrative unit called the 'commune.' It generally has a surface area of no more than about 10 km^2. Mean values of soil variables are available for each 'commune' from data stored in a regional soil associations database. Similarly, landuse is available in a regional statistical office. Soil type and crop areas are stored in the regional database as percentages of the total surface area of each 'commune.' The data are estimated since they combine several sources of elementary data and are derived from empirical estimation and statistical methods.

In order to test the feasibility of modelling at a regional scale, two representative reference areas were investigated with a high precision. Each representative area had a surface area of about 50 km². Combined, they included 219 elementary catchments and 18 'communes.' Soil mapping at a scale of 1:25,000 was carried out in the reference areas and the crop type was noted for each agricultural field.

In the scale transfer method used, four stages can be identified. The first defines the erosion prediction method. The second examines the effect of the change in spatial resolution on the predictive power of the model. 'Communes' are administrative units that do not correspond with the limits of the catchments which are environmental units. The third stage considers the effect of the decrease in the semantic precision when passing from the reference areas to the 'communes.' The final stage is the production of a soil erosion hazard map and its validation with other sources of estimated erosion risk.

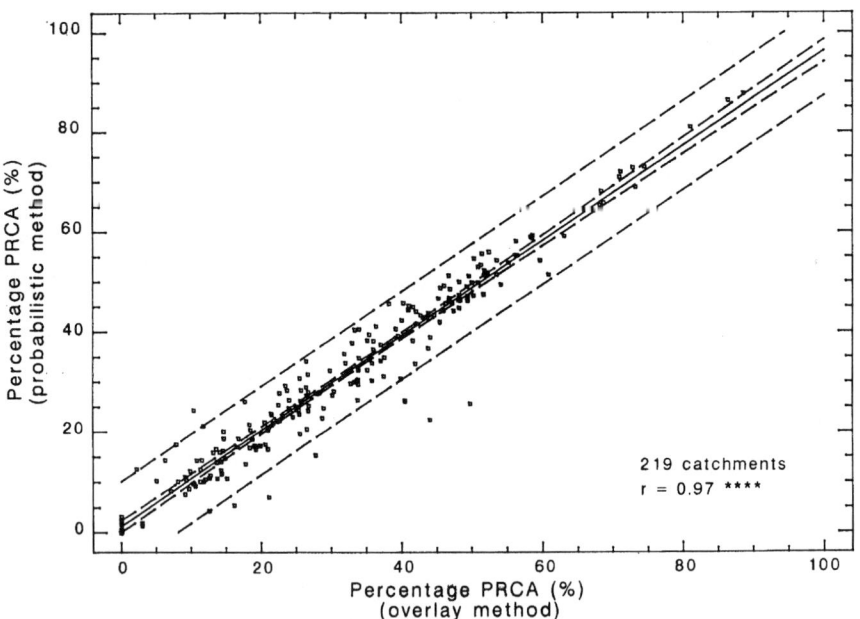

Figure 1. Relationship of the percentage PRCA predicted by the probabilistic method to the percentage PRCA predicted by the overlay method

Stage 1: erosion prediction

Predicting erosion at the catchment scale has been done successfully using a simple empirical function, linking winter rill erosion to the pedological and landuse characteristics of cultivated catchments (Auzet *et al.*, 1993). The percentage Potential Runoff Contributing Area (PRCA) per catchment was computed by overlaying a map of the soil prone to crusting with a map of the landuse susceptible to producing runoff. Rill erosion was measured in catchments and correlated with the percentage PRCA. The method was used on the two reference areas of the Nord-Pas-de-Calais with satisfactory results (Chery *et al.*, 1992).

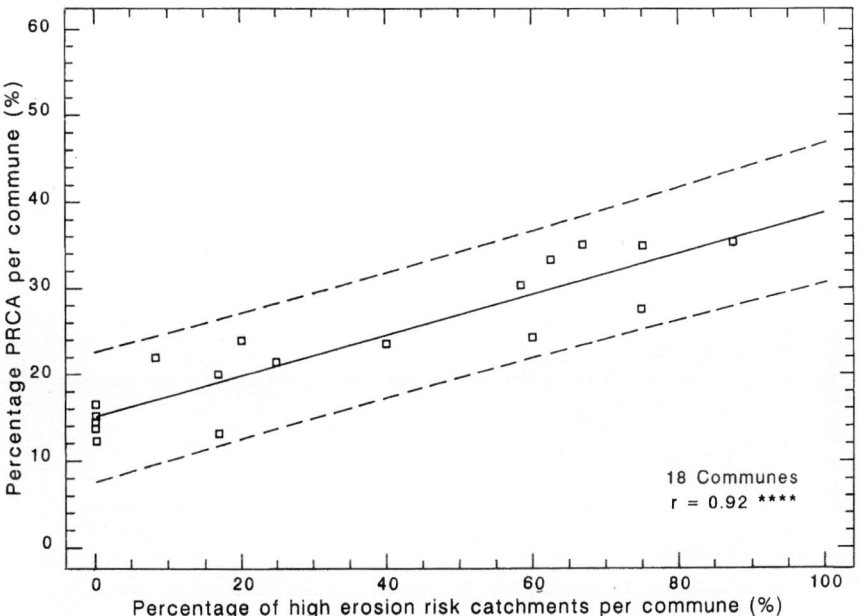

Figure 2. Relationship between the percentage PRCA per 'commune' and the percentage of catchments per commune with a high risk of rill erosion (PRCA per catchment >30%)

The overlay method, however, could not be used at the 'commune' scale since the data were expressed as percentages within the 'communes' and were not geometrically defined. It was therefore necessary to develop a predictive relationship that could be used with percentage data. The runoff area was estimated using a probabilistic method which assumed that soil type and landuse were spatially independent. The percentage of PRCA was then equal to the percentage of the soil prone to crusting multiplied by the percentage of the landuse capable of generating runoff.

Data from the reference areas was used to compare the predicted potential PRCAs for the 219 catchments using the two methods (Figure 1). The strong correlation ($R^2=0.94$) indicates that the probabilistic method gave similar results to the overlay method at the catchment scale. Furthermore, the slope of the regression was approximately unity, indicating that there was no bias. It was therefore concluded that the probabilistic method could be applied to the 'commune' level, where data limitations outside the reference areas did not permit the use of the overlay method.

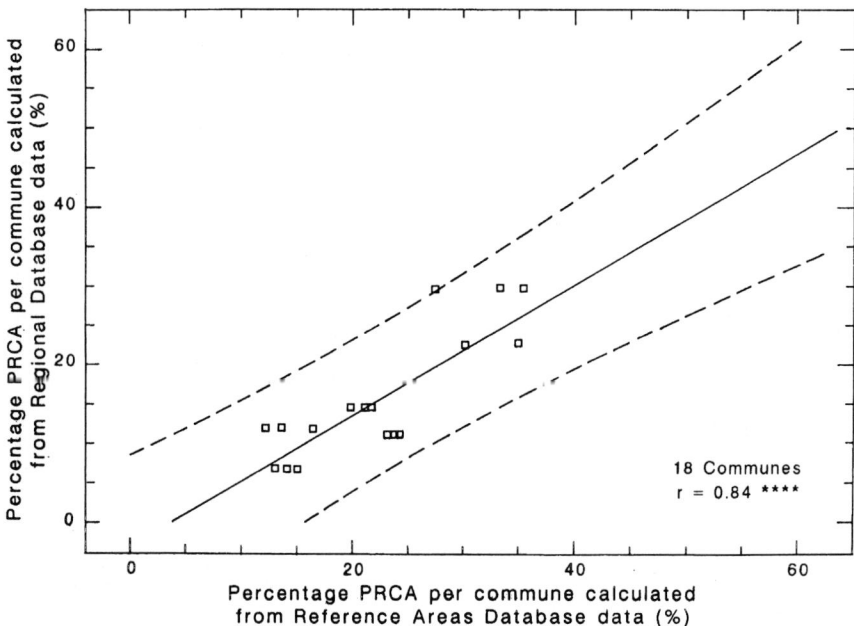

Figure 3. Comparison between the percentage PRCA estimated using data from the regional database and the percentage PRCA estimated from data in the reference areas database

Stage 2: spatial resolution, from the catchment to the 'commune'

Since the objective was to identify and map high risk erosion areas, 'communes' were classified according to the number of catchments with a high risk of rill erosion. Catchments with a percentage of PRCA greater than 30 were considered high risk. The 30 per cent threshold corresponds to a rate of eroded soil of about 1 m^3 ha^{-1} in this region (Auzet et al., 1993). Figure 2 shows the relationship between the percentage of PRCA area within a 'commune' and the percentage of high risk catchments within a 'commune.' The predictive power of this relationship appears quite good under the study conditions.

Stage 3: semantic precision, from the Reference Area 'commune' to the region 'commune'

As mentioned above, data available for the 'communes' in the regional database were based on empirical relationships and were less precise than the information gathered for the reference areas. The purpose of this stage was to examine the effect of the loss of precision in the soil type and landuse percentages on the predicted PRCA. The PRCAs for the eighteen 'communes' in the reference areas were estimated using data in the regional database. These results were compared to the potential PRCAs estimated using the more precise data in the reference areas database. The relationship is fairly good (Figure 3), but there exists a bias. Predicted values using the regional data underestimate the PRCA within the 'communes.' This is probably due to errors in the estimated non-agricultural area in the regional database.

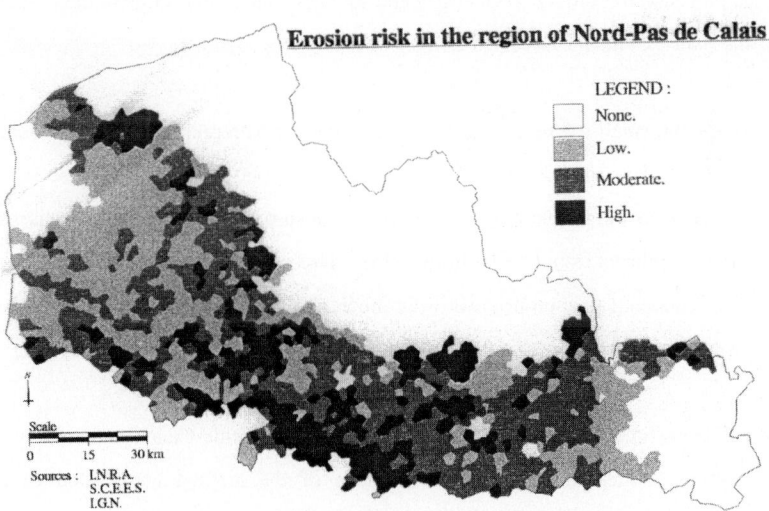

Figure 4. Map of the rill erosion risk per commune of the Nord-Pas-de-Calais region

Stage 4: a rill erosion risk map for the Nord-Pas-de-Calais region

The percentage PRCA for each 'commune' within the region was calculated using data in the regional database, and the 'communes' were classified into four erosion risk categories (Figure 4). The effects of the change in geometrical and semantic resolution were taken into account, and the thresholds were established arbitrarily based on the percentage of high risk basins in the 'commune.' A 'commune' considered high risk had more than 90 percent of its catchments with a percentage PRCA greater than 30. A low risk 'commune' had less than 10 percent, and moderate risk ranged from 10 to 90 percent. The area classified as no risk had topographic and landuse characteristics that made the occurrence of rill erosion extremely unlikely. A quantitative evaluation of the map was not possible, but regional experts confirmed that it corresponded well with their observations.

Conclusion

Most soil databases correspond to either profile databases or digital map databases. The development of GIS has enabled the evolution of new concepts related to scale. Spatial resolution is associated with the geometrical precision. Uncertainties in non-spatial attributes or pedotransfer functions are related to the semantic precision. Studies of large areas entail a decrease in the precision of both the geometrical and semantic data. This is due to the cost and time consuming nature of the measurements.

Before upscaling data, small reference areas should be chosen as representative of the larger territory under study. The smaller area of these reference areas allows an increase in the number and precision of the measurements. The scale transfer process is then a comparison between the results produced from observations in the reference area and predictions based on regional datasets. Choosing an appropriate erosion prediction method and verifying the effects of the loss in precision of the geometrical and semantic data are essential steps in upscaling.

The example of the estimation of rill erosion risk in the Nord-Pas-de-Calais region shows a possible scale transfer method. The main limitations of the method are related to the irrelevance of the administrative unit for the erosion process. Moreover, the erosion model used was elementary, requiring only two input parameters and ignoring temporal processes. More complex models could be used where better data are available. Despite the limitations, the scale transfer method presented here was well-suited to the data and objectives of the study and could be applied under similar conditions elsewhere.

References

Auzet, V., Boiffin, J., Papy, F., Ludwig, B. and Maucorps, J. (1993). Rill erosion as a function of the characteristics of cultivated catchments in the North of France. *Catena* **20**(1/2), 41-62.

Bell, J.C., Thompson, J.A., Butler, C.A. and McSweeny, K. (1994). Modeling soil genesis from a landscape perspective. *15th International Congress of Soil Science*, Acapulco 10-16/07/1994. Vol **6a**, 179-195.

Blöschl, G. and Sivapalan, M. (1995). Scale issues in hydrological modelling: a review. *Hydrological Processes* **9**, 251-290.

Bouma, J. and Van Lanen, H.A.J. (1987). Transfer functions and threshold values: from soil characteristics to land qualities. In, Beek, K.J., Burrough, P.A. and McCormack, D.E. (eds), *Quantified Land Evaluation Procedures*. ITC Publication No. 6, pp 106-110.

Brabant, P. (1992). Pédologie et système d'information géographique. *Cahiers Orstom, série Pédologique* **27**, 315-345.

Burrough, P.A. (1986). Principles of geographical information systems for land resources assessment. *Monographs on Soil and Resources Survey* **12**. Oxford Science Publications, 193 pp.

Chery, P., Le Bissonnais, Y., King, D. and Daroussin J. (1992). Définition et délimitation des Unités Spatiales de Fonctionnement (USF) du ruissellement et de l'érosion (Région Nord-Pas-de-Calais). In, Florac. P., Buche, D., King, D. and Lardon, S. (eds), *Gestion de l'Espace Rural et SIG*., INRA Editions. pp 133-148.

Gessler, P.E., Moore, I.D., McKenzie, N.J. and Ryan, P.J. (1995). Soil-landscape modelling and spatial prediction of soil attributes. *International Journal of Geographical Information Systems* **9**(4), 421-432.

Girard, C.M. and Girard, M.C. (1994). Applications de la télédétection à la connaissance et gestion de l'environnement: exemple du programme CORINE. *Cahiers des Ingénieurs Agronomes* **435**, 6-9.

Hewitt, A.E. (1993). Predictive modelling in soil survey. *Soil and Fertilizers* **56**(3), 305-314.

Jamagne, M., King, D., Le Bas, C., Daroussin, J., Burrill, A. and Vossen, P. (1994). Creation and use of a European Soil Geographic Database. *15th International Congress of Soil Science*, Acapulco 10-16/07/1994. Vol **6a**., 728-742.

King, D., Jamagne, M., Chretien, J. and Hardy, R. (1994). Soil-space organization model and soil functioning units in Geographic Information Systems. *15th International Congress of Soil Science*, Acapulco, 10-16/07/1994. Vol **6a**, 743-757.

Madsen, H.B. (1991). The principles for construction of an EC-Soil database system. In, Hodgson, J.M. (ed.), *Soil Survey, a Basis for European Soil Protection*. CEC. 173-180.

Moore, I.D., Grayson, R.B. and Ladson, A.R. (1991). Digital terrain modelling: a review of hydrological, geomorphological, and biological applications. *Hydrological processes* **5**, 3-30.

Msanya, B.M., Langomr, R. and Lopulisa, C. (1987). Testing and improvement of a questionnaire to users of soil maps. *Soil Survey and Land Evaluation* **7**, 33-42.

Sivapalan, M. and Kalma, J.D. (1995). Scale problems in hydrologiy: contributions of the Robertson workshop. *Hydrological Processes* **9**, 243-250.

Van Ranst, E., Vanmechelen, L., Thomasson, A.J., Daroussin, J., Hollis, J.M., Jones, R.J.A., Jamagne, M. and King, D. (1995). Elaboration of an exended knowledge database to

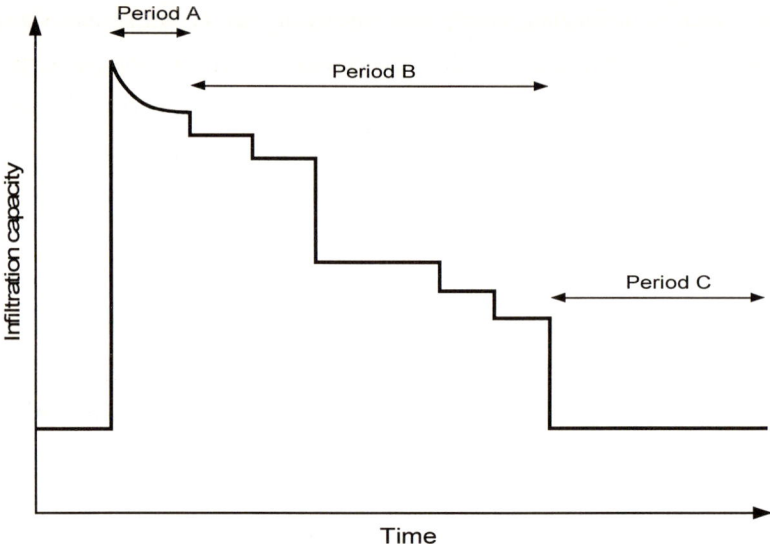

Figure 1. Evolution in soil surface condition (using infiltration capacity as an index) from tillage to harvest (Adapted from Imeson and Kwaad, 1990, and Boardman, 1992)

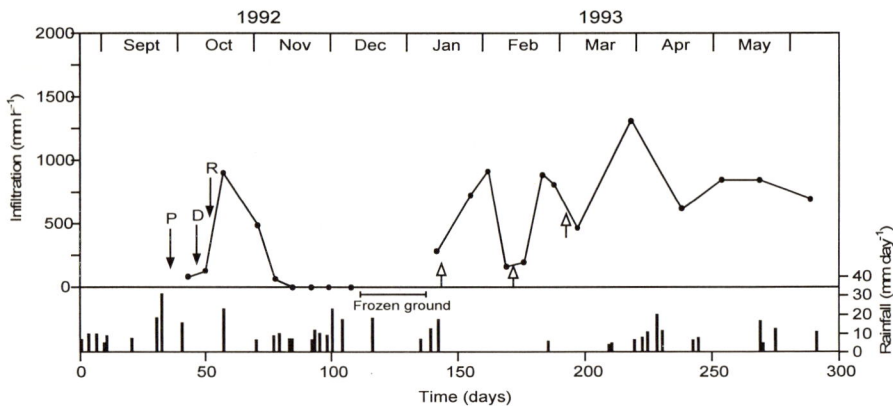

Figure 2. Measured changes in infiltration capacity for a field in the Stour Source catchment, Cotswold Hills, England, over a 9-month period (From Slattery, 1994. Reproduced by kind permission of the author)

Macropore flow

A number of studies in the 1980s emphasised the hydrological importance of macropore flow, a topic which had previously been largely neglected (Beven and Germann, 1981, 1982; Reid and Parkinson, 1984; Kneale and White, 1984). Macroporosity comprises pores and voids in the soil which are larger than capillary size. Where macropores are open to the surface they may catch any free water available and conduct it rapidly down through the soil profile, by-passing some or all of the soil matrix. Flow in macropores is assumed to be fast and unrestricted by the size of the hole (Rawls *et al.*, 1992). By-pass flow is most likely to be generated during intense rain or when antecedent soil moisture content is high (Coles and Trudgill, 1985; Germann, 1986), conditions when an unstructured soil might be expected to produce infiltration-excess overland flow. Kneale (1986) estimated the saturated hydraulic conductivity of soil aggregates in a cracked clay to be 1.4 mm hr^{-1}; by contrast, the bulk hydraulic conductivity of the topsoil (dominated by macropores) was 1800 mm hr^{-1}. Indeed, one practical approach to modelling infiltration in macroporous soil is to adjust the effective hydraulic conductivity (Rawls *et al.*, 1992).

The disruption of crusts by cracks has already been mentioned in relation to Figure 2. For both crusting and macropore flow, inclusion of cracking in erosion models must be extended to the seasonal timescale in order to forecast differences between events. For within-event modelling, the challenge will be to define the changing impact of cracks and worm-holes as these become progressively blocked by splashed soil and inwashing of fines. These and related issues are further explored in Le Bissonnais *et al.* (1998).

Rainfall intensity and infiltration capacity

A number of researchers have noted that the infiltration capacity measured using a rainfall simulator is not independent of the applied rainfall intensity. There may be several reasons for this. One important factor is disturbance of the soil surface by large raindrops at high intensities; this keeps splashed particles in suspension and prevents depositional crusts from forming. A further explanation may be that a greater depth of ponding at higher rainfall intensity means that there is a larger surface area for ponded water to infiltrate (e.g. Romkens *et al.*, 1990). Figure 3 shows infiltration in a loess soil measured using a drip-type simulator (Bowyer-Bower and Burt, 1989). As part of the experimental design, rainfall intensity was doubled part-way through the experiment. Although runoff rates increased greatly as a result, so too did infiltration, though to a lesser extent (Baade, Boardman and Burt, unpublished data). Such results emphasise the need for models to include sophisticated routines to deal with within-event changes in infiltration capacity. Moreover, such models should also include some measure of soil surface microtopography in order to be able to simulate details of surface detention storage. Distributed event-based models may well need to include some spatial variability in crust characteristics in order to incorporate features such as row and inter-row microtopography. Continuous simulators should, in addition, be able to cope with seasonal evolution of soil surface microtopography and condition.

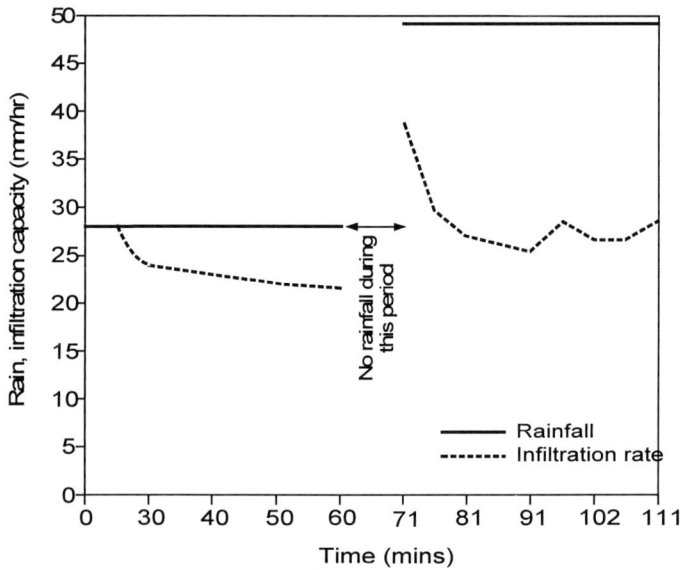

Figure 3. Changes in infiltration and runoff during a rainfall simulation experiment on a loess soil near Jena, Germany (Baade, Boardman and Burt, unpublished data)

Spatial patterns of infiltration loss and runoff generation

Infiltration into layered soils

Classical infiltration models have assumed a semi-infinite soil where storage effects have no importance on the penetration of the wetting front; overland flow occurs only when and where rainfall intensity exceeds infiltration capacity. Such an assumption is clearly inappropriate in many cases: where there is a less permeable soil horizon at depth or where more permeable soil overlies less permeable bedrock. Infiltration into soils of limited moisture storage has been considered by Knapp (1978), Zaslavsky and Sinai (1981) and Burt (1986). Saturation-excess overland flow can occur even though rainfall intensity is well below infiltration capacity, once the upper soil layer becomes saturated. Rainfall simulator experiments on loess soils show that infiltrating water frequently ponds above the plough pan (c. 25 cm depth); during prolonged natural rainfall, the effect on a long slope would be to encourage saturation of the entire plough layer, especially downslope (Baade, Boardman and Burt, unpublished data; see also Kwaad, 1998).

The spatial distribution of soil moisture is controlled mainly by topography. Soil saturation is encouraged at the foot of any slope where the upslope drainage area is at a maximum, and in topographic hollows, where convergence of flow lines favours the accumulation of soil water (Anderson and Burt, 1978). In areas of reduced storage, the transmissivity of the soil profile is lessened and moisture levels can be expected to rise. *Saturation-excess overland flow* may therefore be generated by two mechanisms: *direct runoff*, where rainfall falls onto saturated soil and is unable to infiltrate, and *return flow*, where infiltrated water returns to the land surface having flowed for some distance within the soil profile.

Two points follow from this analysis which are relevant to the generation of overland flow. Firstly, the source areas for infiltration-excess overland flow and saturation-excess overland flow are very different. The Partial Area model of Betson (1964) remains the best guide to the location of source areas for the former type of flow i.e. those areas in the catchment where rainfall intensity exceeds infiltration capacity; this topic is discussed further below. Production of saturation-excess overland flow is best described by the Variable Source Area model of Hewlett (1961), with surface runoff being produced in areas of high soil moisture content as noted above. Secondly, these two mechanisms are dependent in very different ways on hydraulic conductivity and rainfall intensity. Soils of high hydraulic conductivity are likely to produce only saturation-excess overland flow, and this may happen at rainfall intensities well below infiltration capacity, since it depends on soil moisture content and not on soil surface condition. Soils of low infiltration capacity are likely to be dominated by infiltration-excess overland flow. These two runoff domains are shown schematically in Figure 4. Church and Woo (1990) point out that soils tend to vary much more widely than climate, concluding that soil properties may therefore be viewed as *the* dominant control of surface runoff production and erosion.

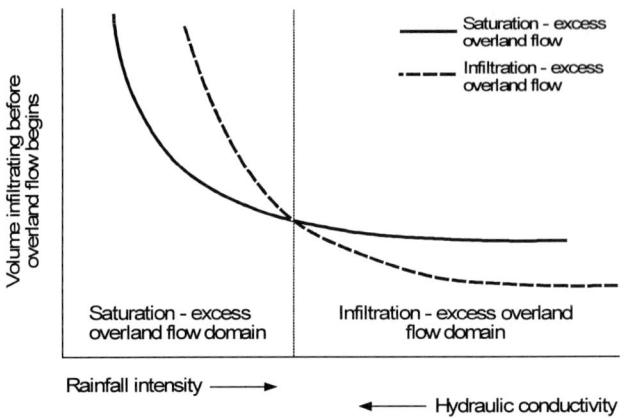

Figure 4. The relationship between the volume of rainfall which infiltrates before overland flow commences and rainfall intensity. Soils with low hydraulic conductivity will be dominated by infiltration-excess overland flow; those with high hydraulic conductivity by saturation-excess overland flow (Adapted from Burt, 1989)

Models of surface runoff and erosion which ignore saturation-excess overland flow may well fail to predict important erosional features within a catchment, most especially ephemeral gullies formed along the talweg of some dry valleys. These features have become common on tilled land in north west Europe in recent years (e.g. Slattery et al., 1994). Following Meyer (1986), it can be inferred that ephemeral gullies will form, *ceteris paribus*, where upslope drainage area exceeds a certain threshold. Auzet et al. (1995) show that larger contributing areas produce more infiltration-excess overland flow and, consequently, more rill erosion. What is not clear, however, is the extra influence of saturation-excess overland flow. If this can generate additional volumes of overland flow, or indeed cause surface runoff at times and in places where infiltration-excess overland flow has not occurred, then erosional models based solely on infiltration will fail to predict such features completely. More field observation is needed to confirm the importance of saturation-excess overland flow in ephemeral rill formation. Kirkby (1978) has modelled the effect of both runoff mechanisms on erosion and sediment transport. Thorne et al. (1987) have shown how topography controls the location of gullies on the High Plains of the USA; saturation-excess overland flow was identified as the dominant flow process causing erosion.

One other source of sediment not included in plot-scale erosion models is the stream channel. Distributed catchment models must be able to cope with channel as well as slope erosion. As authors like Foster et al. (1990) and Slattery (1994) have shown, hillslope and channel systems may be so poorly coupled that sediment delivery at the catchment outlet is dominated by instream sources.

Modelling disjunct source areas

The Partial Area model of Betson (1964) refined Horton's model for the generation of infiltration-excess overland flow. Betson recognised that only parts of a basin will produce runoff at any one time; Horton had assumed that the whole basin would generate runoff simultaneously. The difficulty though, is to predict which partial areas, not contiguous to the channel, can introduce runoff and sediment to the stream. Jones (1979) has labelled such distant sources 'disjunct contributing areas'. As attempts are made to scale up soil erosion models from single hillslope elements to small catchments (<10 km^2), this routing problem becomes a significant one. A distributed model must therefore be capable, not only of predicting which fields will generate infiltration-excess runoff during a given storm, but also of routing that runoff out of the field, down the hillslope, and into the stream channel, accounting for sediment losses (or possibly gains) along the way. The use of digital terrain models may help a good deal in this regard. Incorporation of linear elements such as roads and tracks into such schemes is of paramount importance since these can be significant runoff sources in their own right and may also convey large quantities of sediment-laden water from distant fields to the stream. The earlier discussion of ephemeral rills demonstrates the importance of scaling up erosion models from the plot scale, if important erosion features are not to be missed. Slattery et al. (1994) provide a relatively straightforward example of runoff routing from an ephemeral rill to a stream over a sequence of storms. Concentration of infiltration-excess overland flow from adjacent hillslopes along the floor of a dry valley caused a large rill to be eroded. Nevertheless, despite the size of the feature, very little of the eroded sediment reached the stream channel, most eroded sediment being deposited en route behind field boundaries.

Conclusions

It is clear that the soil surface is not an unchanging uniform porous medium into which water infiltrates only via capillary-sized pores. Soil erosion models must cope with a soil surface where macropores may dominate infiltration losses, at least during some parts of the year, and where the formation of a crust may greatly lower the infiltration capacity at other times. Such time-dependent changes in soil surface condition are poorly handled at present, especially by continuous models where the changing nature of the soil surface must itself be predicted as part of the simulation. Even within an individual event, clogging of macropores by splashed particles, development of depositional seals and changing depths of ponding may all influence infiltration capacity, detention storage and runoff rates. There is still much to learn about these topics in the field and yet it is clear that, even now, there is a need to provide relevant quantitative information so that a start may be made in modelling these subtle but important changes in the soil surface.

To characterise and model the relationship between soil surface condition and infiltration capacity is not, however, enough. As demands increase for modelling of off-site impacts as well as on-site erosion, it becomes ever more important to be able to scale up plot-based models to the small catchment. In so doing, it is vital to be able to route runoff and sediment through the catchment system and to incorporate runoff and erosion processes not covered by traditional models.

References
Ahuja, L.R. (1983). Modelling infiltration into crusted soils using the Green-Ampt approach. *Soil Science Society of America Journal* **47**, 412-418.
Anderson, M.G. and Burt, T.P. (1978). The role of topography in controlling throughflow generation. *Earth Surface Processes* **3**, 331-344.
Auzet, A.V., Boiffin, J. and Ludwig, B. (1995). Concentrated flow erosion in cultivated catchments: influence of soil surface state. *Earth Surface Processes and Landforms* **20**, 759-768.
Betson, R.P. (1964). What is watershed runoff? *Journal of Geophysical Research* **69**, 1541-1552.
Beven, K. and Germann, P.F. (1981). Water flow in macropores. II: A combined flow model. *Journal of Soil Science* **32**, 15-29.
Beven, K. and Germann, P.F. (1982) Macropores and water flow in soils. *Water Resources Research* **18**, 1311-25.
Boardman, J. (1992). Agriculture and erosion in Britain. *Geography Review* **6**(1), 15-19.
Boiffin, J. (1984). *La Degradation Structurale des Couches Superficielles des Sols sous l'Action des Pluies*. These Docteur Ingenieur, Paris. INA-P.G., 320 pp + annexes.
Bowyer-Bower, T.A.S. and Burt, T.P. (1989). Rainfall simulators for investigating soil response to rainfall. *Soil Technology* **2**, 1-16.
Burt, T.P. (1986). Runoff processes and solutional denudation on humid temperate hillslopes, in S.T. Trudgill (ed.), *Solute Processes*, John Wiley and Sons, Chichester, 193-250.

Burt, T.P. (1989). Storm runoff generation in small catchments in relation to the flood response of large basins. In, Beven, K.J. and Carling, P. (eds), *Floods*, Wiley, Chichester, 11-35.

Chorley, R.J. (1978). The hillslope hydrological cycle. In, Kirkby, M.J. (ed.), *Hillslope Hydrology*, Wiley, Chichester, 1-42.

Church, M. and Woo, M-K. (1990). Geography of surface runoff: some lessons for research. In, Anderson, M.G. and Burt, T.P. (eds), *Process Studies in Hillslope Hydrology*, Wiley, Chichester, 299-325.

Coles, N. and Trudgill, S.T. (1985). The movement of nitrate fertilizer from the soil surface to drainage waters by preferential flow in weakly structured soils, Slapton, south Devon. *Agriculture, Ecosystems and Environment* **13**, 241-59.

Foster, I.D.L., Grew, R. and Dearing, J.A. (1990). Magnitude and frequency of sediment transport in agricultural catchments: a paired lake-catchment study in Midland England. In, Boardman, J., Foster, I.D.L. and Dearing, J.A. (eds.). *Soil Erosion on Agricultural Land*, Wiley, Chichester, 153-172.

Germann, P.F. (1986). Rapid drainage response to precipitation. *Hydrological Processes* **1**, 3-14.

Germann, P.F. (1990). Macropores and hydrologic hillslope processes. In Anderson, M.G. and Burt, T.P. (eds.). *Process Studies in Hillslope Hydrology*, Wiley, Chichester, 325-363.

Green, W.H. and Ampt, G.A. (1911). Studies on soil physics: 1. Flow of air and water through soils. *Journal of Agricultural Science* **4**, 1-24.

Hewlett, J.D. (1961). Watershed management. In, *Report for 1961 Southeastern Forest Experiment Station*, US Forest Service, Asheville, North Carolina, pp 62-66.

Horton, R.E. (1933). The role of infiltration in the hydrological cycle. *Transactions of the American Geophysical Union* **14**, 446-60.

Imeson, A.C. and Kwaad, F.J.P.M. (1990). The response of tilled soils to wetting by rainfall and the dynamic character of soil erodibility. In, Boardman, J., Foster, I.D.L. and Dearing, J.A. (eds), *Soil Erosion on Agricultural Land*, Wiley, Chichester, 3-14.

Jones, J.A.A. (1979). Extending the Hewlett model of stream runoff generation. *Area* **11**, 110-114.

Kirkby, M.J. (1978). Implications for sediment delivery. In, Kirkby, M.J. (ed.), *Hillslope Hydrology*, Wiley, 325-364.

Knapp, B.J. (1978). Infiltration and storage of water. In, Kirkby, M.J. (ed.), *Hillslope Hydrology*, Wiley, 43-72.

Kneale, W.R. (1986). The hydrology of a sloping, structured clay soil at Wytham, near Oxford, England. *Journal of Hydrology* **85**, 1-14.

Kneale, W.R. and White, R.E. (1984). The movement of water through cores of a dry (cracked) clay-loam grassland topsoil. *Journal of Hydrology* **64**, 361-365.

Kwaad, F.J.P.M (1998). Saturation overland flow on loess soils in the Netherlands. In, Boardman, J. and Favis-Mortlock, D.T. (eds), *Modelling Soil Erosion by Water*, Springer-Verlag NATO-ASI Global Change Series, Heidelberg.

Le Bissonnais, Y., Fox, D. and Bresson, L.-M. (1998). Incorporating crusting processes in erosion models. In, Boardman, J. and Favis-Mortlock, D.T. (eds), *Modelling Soil Erosion by Water*, Springer-Verlag NATO-ASI Global Change Series, Heidelberg.

Meyer, L.D. (1986). Erosion processes and sediment properties for agricultural cropland. In, Abrahams, A.D. (ed.), *Hillslope Processes*, Allen and Unwin, 55-76.

Monnier, G. and Boiffin, J. (1986). The effect of the agricultural land use of soils on water erosion: the case of cropping systems in western Europe. In, Chisci, G. and Morgan, R.P.C. (eds), *Soil Erosion in the European Community*, Balkema, 17-32.

Philip, J.R. (1969). Theory of infiltration. *Advances in Hydroscience* **5**, 215-305.

Rawls, W.J., Ahuja, L.R., Brakensiek, D.L. and Shirmohammadi, A. (1992). Infiltration and soil water movement. In, Maidment, D.R. (ed.), *Handbook of Hydrology*, McGraw-Hill, 5.1-5.51.

Reid, I. and Parkinson, R.J. (1984). The wetting and drying of a grazed and ungrazed clay soil. *Journal of Soil Science* **35**, 607-14.

Römkens, M.J.M., Prasad, S.N. and Whisler, F.D. (1990). Surface sealing and infiltration. In, Anderson, M.G. and Burt, T.P. (eds), *Process Studies in Hillslope Hydrology*, Wiley, Chichester, 127-172.

Slattery, M.C. (1994). *Contemporary Sediment Dynamics and Sediment Delivery in a Small Agricultural Catchment, North Oxfordshire, UK.* Unpublished D.Phil. thesis, University of Oxford, U.K.

Slattery, M.C., Burt, T.P. and Boardman, J. (1994). Rill erosion along the thalweg of a hillslope hollow: a case study from the Cotswold Hills, central England. *Earth Surface Processes and Landforms* **19**, 377-385.

Thorne, C.R., Zevenbergen, L.W., Burt, T.P. and Butcher, D.P. (1987). Terrain analysis for quantitative description of zero-order drainage basins. *International Association of Hydrological Sciences Publication* **165**, 121-30.

Zaslavsky, D. and Sinai, G. (1981). Surface hydrology: 5 parts. *Journal of the Hydraulics Division, Proceedings of the American Society of Civil Engineers* **107**, 1-93.

17. SATURATION OVERLAND FLOW ON LOESS SOILS IN THE NETHERLANDS

F.J.P.M. Kwaad

Laboratory of Physical Geography and Soil Science
University of Amsterdam
Nieuwe Prinsengracht 130
1018 VZ Amsterdam
The Netherlands

Abstract

Evidence of saturation overland flow in Dutch South Limburg is discussed. Besides Horton overland flow saturation overland flow may be a cause of rainfall-induced accelerated erosion, which has not been given due consideration in the area so far. Implications for soil erosion modelling and soil conservation are: (1) that a soil erosion model should have a saturation overland flow module, and (2) that soil conservation measures that only protect the soil surface will have little or no effect on soil loss caused by saturation overland flow.

Introduction

In Dutch South-Limbourg surface slaking and crusting of structurally unstable loess soils, causing a lowering of the infiltration capacity of the soil and giving rise to Horton overland flow, is thought to be the prime cause of soil erosion on 40,000 ha of sloping crop land (Bouten et al., 1985; Kwaad and Mücher, 1994). In autumn 1985 a field study was started in the area to evaluate the effectiveness of several soil conservation measures and to develop a soil erosion model for the region. Measurements were carried out at several spatial scales ranging from experimental plots to nested drainage basins (Kwaad, 1991, 1993, 1994a, 1994b; Kwaad and Van Mulligen, 1991; De Roo et al., 1994; Van Dijk and Kwaad, 1995; Van Dijk et al., 1995). In the course of the work overland flow was observed under conditions of low intensity rainfall. This raised the question of how to identify the cause of overland flow and erosion. Very often a certain mechanism of overland flow generation is assumed as being most plausible, not corroborated by field evidence. In this paper evidence of saturation overland flow is discussed for the area of study. This is an important issue when dealing with soil erosion in humid-temperate western Europe (Chisci and Morgan, 1986; Schwertmann et al., 1988; De Ploey, 1989; Boardman et al., 1990; Wicherek, 1993; Rickson, 1994). In parts of this region, rainfall induced soil erosion occurs predominantly either in winter, e.g. England (Evans, 1990), north of France (Auzet et al., 1990), or in summer, e.g. west and central Germany (Baade et al., 1993), or both in winter and summer, e.g. Belgium (De Ploey, 1986), the Netherlands (Kwaad, 1993). Presumably, different mechanisms of overland flow generation are involved in summer and winter.

From the point of view of soil conservation it must be stressed that in the Horton model the state of the soil surface is the key soil characteristic, whereas in the saturation model the moisture storage capacity of the soil above an impeding layer is the key soil condition. Soil conservation measures that only protect the soil surface may have little or no effect on soil loss caused by saturation overland flow.

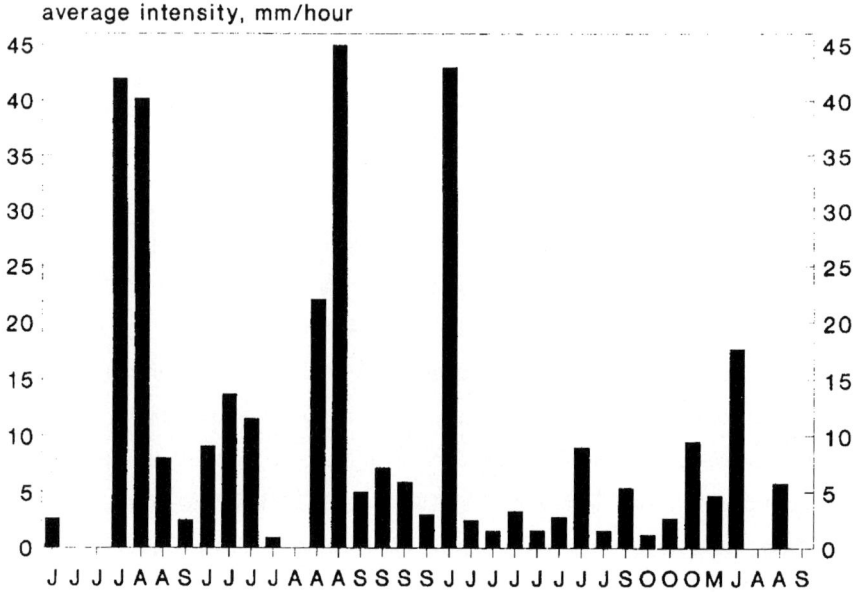

Figure 1. Average intensity of runoff-generating summer rains, Wijnandsrade, 1986-1989

The research area

Dutch South Limburg has a hilly relief dominated by numerous dry valleys. Land surface elevations range from 40 to 321 m a.s.l. A large part of the area is covered with a 2-20 m thick layer of Pleistocene loess (Mücher, 1986; Van den Broek, 1966), overlying coarse-grained Quaternary river sediments, Tertiary sands and Cretaceous chalk. In the loess, Holocene soils with a clay illuviation or Bt horizon have developed. South Limburg is part of the European loess belt that extends across SE England, NW France, Belgium, parts of Germany and further into Poland and Russia. South Limburg has a temperate oceanic climate with rainfall in all seasons. Average annual precipitation is 750 mm. High intensity rainfall is restricted to the period April-October (Levert, 1954). Land use has been agricultural for many centuries (Renes, 1988). Erosion risk is highest in April-June, when the surface coverage by crops is small and high intensity rainfall may occur. Prolonged wet weather and occasional rapid snowmelt may cause overland flow and erosion in winter.

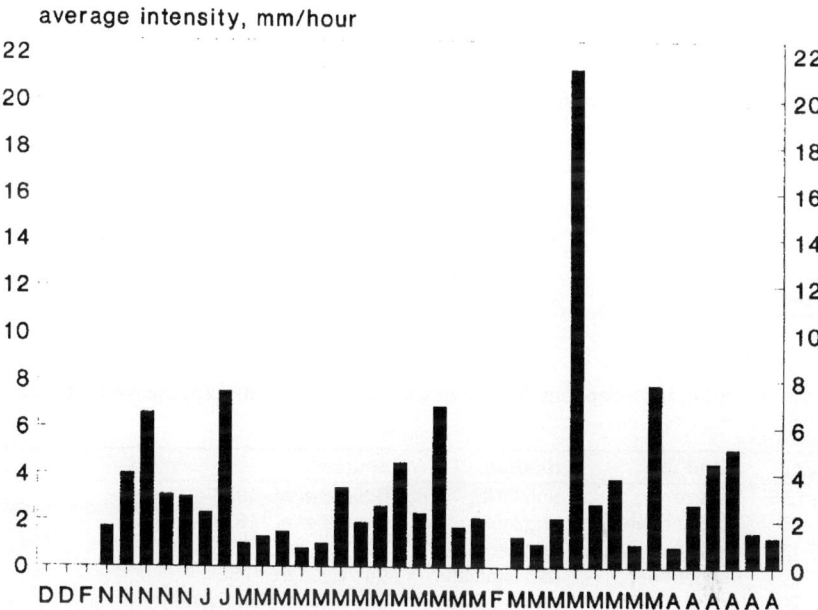

Figure 2. Average intensity of runoff-generating winter rains, Wijnandsrade, 1986-1989

Methods

Various types of field observations and measurements have been employed to determine or infer the occurrence of saturation overland flow in the area of study. Most data were collected on Wischmeier plots. Details of the applied measurement techniques are given by Kwaad (1991, 1993, 1994a), De Roo et al. (1994) and Van Dijk and Kwaad (1995).

Results and discussion

Rainfall intensity and final infiltration capacity of the soil

In 1986-1989 73 runoff events were recorded on a permanently bare Wischmeier plot, of which 35 (48%) occurred from May to October and 38 (52%) from November to April (Kwaad, 1993). Values of the average and maximum 5 minute intensity of the individual runoff generating storms are shown in Figures 1 to 4. In Table 1 average values for all runoff generating summer and winter storms are given. By comparison with Table 2 it appears that many overland flow events were caused by rain that fell at intensities that were lower than the final infiltration capacity of the soil.

Table 1. Average values of characteristics of runoff-generating rains, Wijnandsrade, 1986-1989

	amount (mm)	duration (min.)	intensity (mm/hour)	max. 5 min. intensity (mm/hour)
all rains	7.6 ± 5.9	187.1 ± 232.3	7.0 ± 10.2	20.0 ± 17.1
summer rains	7.9 ± 5.9	118.9 ± 117.7	11.0 ± 13.3	27.1 ± 19.9
winter rains	7.4 ± 5.8	247.3 ± 285.7	3.4 ± 3.7	13.7 ± 10.8

Table 2. Final infiltration capacity (I_f) of loess soils, South-Limbourg, the Netherlands

I_f (mm/h)	land use	method	source
1.8 -12.6	maize	ring infiltr.	Bouma *et al.*, 1978
21 - 60	sugar beet	ring infiltr.	Bouma *et al.*, 1978
3 - 18	maize	rain simulator	Van Eysden, 1986
7.2 - 27	sugar beet	rain simulator	Van Eysden, 1986
37 - 70	sugar beet	rain simulator	Sapoera, in Van Eysden
10 - 21	bare soil	rain simulator	Sapoera, in Van Eysden
7	arable	not mentioned	Anonymous, 1987
15 - 55.2	arable	rain simulator	Hollemans and van Dijk, 1988
24.6	maize	rain simulator	Hollemans and van Dijk, 1988
29.4	potatoes	rain simulator	Hollemans and van Dijk, 1988
27.0	sugar beet	rain simulator	Hollemans and van Dijk, 1988
25.2	fallow	rain simulator	Hollemans and van Dijk, 1988
33.3- 40.5	maize	rain simulator	Van Mulligen, 1990
61.3- 91.6	maize	rain simulator	Wansink, 1991
29.2	maize, direct	rain simulator	Van Dijk *et al.*, 1995
27.1	maize, mulch	rain simulator	Van Dijk *et al.*, 1995
22.6	maize, bare	rain simulator	Van Dijk *et al.*, 1995
44.6	maize, straw	rain simulator	Van Dijk *et al.*, 1995
31.2	maize, conv.	rain simulator	Van Dijk *et al.*, 1995

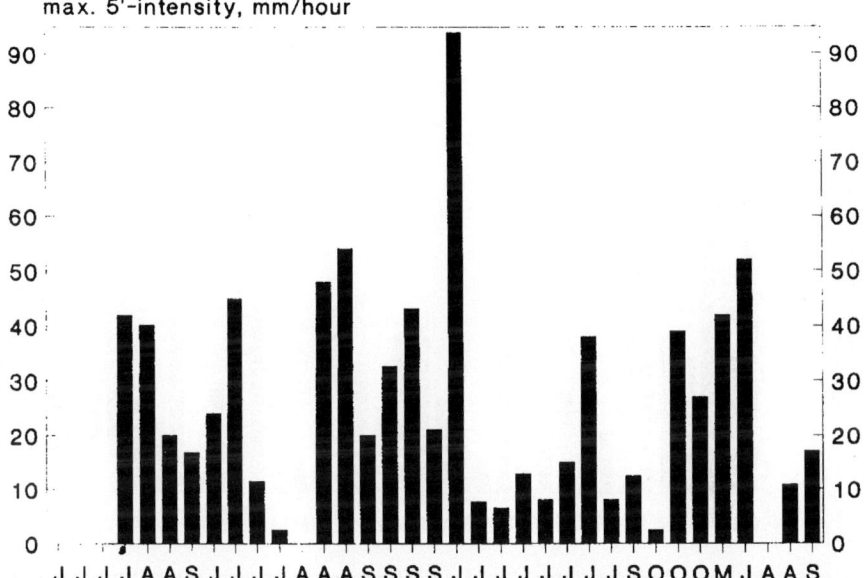

Figure 3. Maximum 5 minute intensity of runoff-generating summer rains, Wijnandsrade, 1986-1989

Overland flow after cessation of rainfall

On one occasion in March 1988 water was observed flowing in 0.20 m deep erosion rills near the field site with experimental plots until at least 38 hours after cessation of the last rainfall. The rills had formed on a long gentle slope leading down from a broad flat-topped ridge. Shallow pits, dug in the plough layer, immediately filled with water at that time, while the underlying Bt-horizon was relatively dry. Obviously, water was draining from the waterlogged plateau soils as shallow subsurface flow which emerged in the rills.

Shallow groundwater

Shallow groundwater was observed on a number of occasions in the soil pits in which the storage tanks of the Wischmeier plots were installed. Monthly measurements in 2 m long groundwater tubes revealed that groundwater could rise to within 10 cm of the soil surface in winter, spring and even early summer and drop to below 2 m in summer and autumn (Figure 5). Within a small area like the plot site (100 x 22 m) large spatial variations in groundwater depth can occur (Figure 6). The plots were aligned on a contour.

Figure 4. Maximum 5 minute intensity of runoff-generating winter rains, Wijnandsrade, 1986-1989

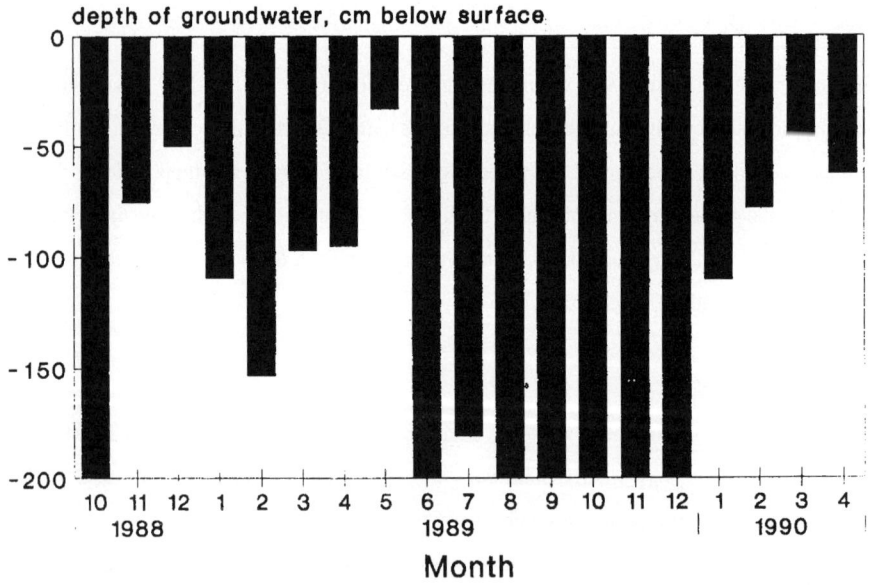

Figure 5. Temporal variation of groundwater depth, average of 24 tubes, experimental plots, Wijnandsrade

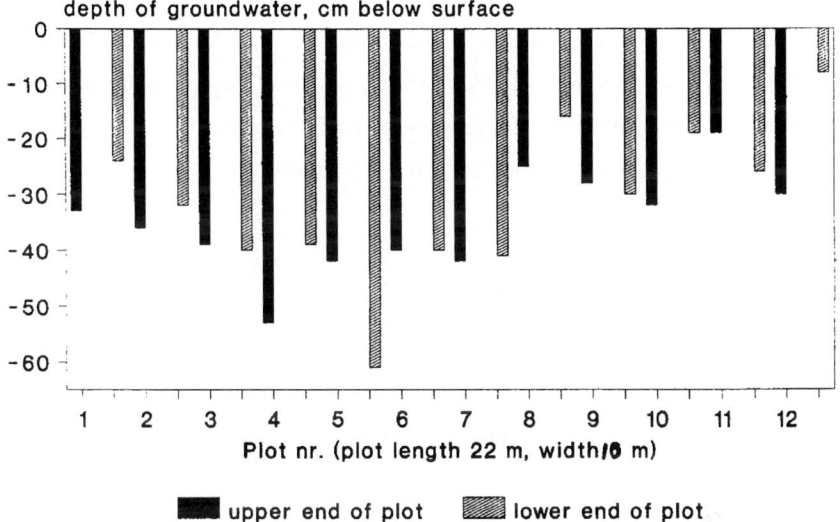

Figure 6. Spatial variation of groundwater depth on 27 April 1989, experimental plots, Wijnandsrade

Hydromorphic soils

At the site of the experimental plots near Wijnandsrade soils showed hydromorphic mottling directly below the plough layer. Soils were classified as truncated gleyic luvisols in loess with locally overlying colluvium which also showed shallow hydromorphic mottling. This is a sure indication of high groundwater during part of the year.

Winter as overland flow season

From experimental plot data it appears that both summer and winter conditions can be conducive to runoff and erosion on fallow loess soils in South-Limbourg (Kwaad, 1991). However, runoff and erosion regimes are very different in summer and winter. Summer is characterised by relatively high soil losses and low runoff volumes and winter by relatively low soil losses and high runoff volumes. Summer runoff volume does not increase with increasing monthly rainfall amount but significantly increases with increasing maximum I_{30}. Winter runoff volume, on the other hand, does strongly increase with increasing rainfall amount, and monthly runoff volumes are much higher in winter than in summer.

Runoff percentage

Monthly runoff percentages of permanently bare soil (average of three Wischmeier plots) ranged from less than 5% in summer to about 90% in winter. On one of the three plots values up to 107% were reached in winter '87/'88. The high winter runoff percentages on a monthly basis can only be explained by assuming saturation of the soil and possibly even return flow in the plots originating from water that has infiltrated upslope.

On a catchment scale it appeared, that in the years 1992-1994 runoff percentages were low in summer in all four catchments in which runoff has occurred and much higher in winter in two of the four catchments in which runoff was observed (Van Dijk and Kwaad, 1995). Local subsoil conditions influenced the occurrence of soil saturation in winter.

Soil moisture tension

Measurements of soil moisture tension at various depths in the soil have been carried out with tensiometers by the Staring Centre (De Roo, 1994). An example result is that the top 5 cm of soil remained saturated until at least two hours after a rainfall event of 6.2 mm. This led to the generation of overland flow only two minutes after the start of a second rainfall event, which began three hours after the 6.2 mm event. Antecedent rainfall had been 10 mm on the day before.

Effect of autumn tillage on winter runoff

By autumn tillage of the soil in the form of ploughing and harrowing or rough chiselling, winter runoff was generally strongly reduced on experimental plots (Kwaad, 1994). However, in some very wet months the autumn ploughed and harrowed plots did give rise to important runoff volumes. This is ascribed to an increased storage capacity of the autumn tilled layer, which was exceeded by rainfall amount in very wet months in the case of autumn ploughing and harrowing. Apparently, coarse chiselling of the soil gave a still higher pore volume to the soil.

Conclusions

In soil erosion studies we must be more aware of the mode of runoff generation. Too often a certain mode is tacitly assumed. Evidence is presented in this paper by which saturation overland flow can be recognised. From this evidence it can be inferred, that overland flow generation can shift from surface controlled Hortonian mode to storage controlled saturation mode, i.e. there is temporal and spatial variability in the mode of overland flow generation:

a. larger areas comprise parts which are prone to Hortonian overland flow and at the same time other parts which are prone to saturation overland flow;

b. the same area can give rise to Hortonian overland flow at one moment and to saturation overland flow at another moment; this shift can even take place within one and the same rainfall event.

These variations in space and time have to be accounted for by soil erosion models which aim at predicting soil loss over longer periods of time and larger areas. This requires detailed knowledge of the temporal and spatial variability of the relevant soil properties.

Acknowledgements

The experimental plot research was carried out in close co-operation with the Experimental Farm Wijnandsrade. Part of the work was financed by the Government of the Province of Limburg. Dr. Mücher helped with classifying and mapping the soils at the plot site.

References

Anonymous (1987). *Onderzoek wateroverlast en bodemerosie nabij Catsop (Gem. Stein, Limburg)*. Uitg. Landinrichtingsdienst Roermond, 104 pp. + bijlagen.

Auzet, A.V., Boiffin, J, Papy, F., Maucorps, J. and Ouvry, J.F. (1990). An approach to the assessment of erosion forms and erosion risk on agricultural land in the northern Paris Basin, France. In, Boardman, J., Foster, I.D.L. and Dearing, J.A. (eds), *Soil Erosion on Agricultural Land*, Wiley, Chichester, pp. 383-400.

Baade, J., Barsch, D., Mausbacher, R. and Schukraft, G. (1993). Field experiments on the reduction of sediment yield from arable land to receiving water courses (N. Kraichgau, SW. Germany). In, Wicherek, S. (ed.), *Farm Land Erosion in Temperate Plains Environment and Hills*. Elsevier, Amsterdam, pp. 471-480.

Boardman, J., Foster, I.D.L. and Dearing, J.A. (eds) (1990). *Soil Erosion on Agricultural Land*. Wiley, Chichester, 687 pp.

Bouma, J, Stoffelsen, G.H., Teunissen van Manen, T.C., Dekkers, J.M.J. en Poelman, J.N.B. (1978). Onderzoek naar de hydrologische situatie in de onverzadigde zone op het Plateau van Margraten. *Rapport Stiboka* no. **1403**.

Bouten, W., Van Eijsden, G., Imeson, A.C., Kwaad, F.J.P.M., Mücher, H.J., and Tiktak, A. (1985). Loessial soils in Dutch South-Limbourg: their origin and removal by accelerated erosion (paper in Dutch with original title 'Ontstaan en erosie van de lössleemgronden in Zuid-Limburg'), *K.N.A.G. Geografisch Tijdschrift* **19**(3), 192-208.

Chisci, G. and Morgan, R.P.C. (eds) (1986). *Soil Erosion in the European Community*. Balkema, Rotterdam.

De Ploey, J. (1986). *Bodemerosie in de lage landen, een Europees milieuprobleem*. Acco, Leuven/Amersfoort, 108 pp.

De Ploey, J. (1989). *Soil Erosion Map of Western Europe.* Catena Verlag.

De Roo, A.P.J., Van Dijk, P.M., Ritsema, C.J., Cremers, N.H.D.T., Stolte, J., Offermans, R.J.E., Kwaad, F.J.P.M., and Verzandvoort, M.A. (1994). *Soil Erosion Normalisation Project South Limburg* (report in Dutch). University of Utrecht, University of Amsterdam and Winand Staring Centre, 184 pp.

Evans, R. (1990). Water erosion in British farmers' fields - some causes, impacts, predictions. *Progress in Physical Geography* **14**, 199-219.

Hollemans, W.A. and Van Dijk, P.M. (1988). *Faktoren van invloed op Infiltratie en Bodemerosie. Regionale vergelijking tussen drie Gebieden in Zuid-Limburg*, M.Sc. thesis. University of Amsterdam, 77 pp.

Kwaad, F.J.P.M. (1991). Summer and winter regimes of runoff generation and soil erosion on cultivated loess soils (The Netherlands), *Earth Surface Processes and Landforms* **16**, 653-662.

Kwaad, F.J.P.M. (1993). Characteristics of runoff generating rains on bare loess soil in South-Limbourg (The Netherlands), In, Wicherek, S. (ed.), *Farm Land Erosion in Temperate Plains Environment and Hills*, Elsevier, 71-86.

Kwaad, F.J.P.M. (1994a). Cropping systems of fodder maize to reduce erosion of cultivated loess soils. In, Rickson, R.J. (ed.), *Conserving Soil Resources, European Perspectives*, CAB International, pp. 354-368.

Kwaad, F.J.P.M. (1994b). A splash delivery ratio to characterize soil erosion events. In, Rickson, R.J. (ed.), *Conserving Soil Resources, European Perspectives*, CAB International, pp. 264-272.

Kwaad, F.J.P.M. and Mücher, H.J. (1994). Degradation of soil structure by welding - a micromorphological study, *Catena* **23**, 253-268.

Kwaad, F.J.P.M. and Van Mulligen, E.J. (1991). Cropping system effects of maize on infiltration, runoff and erosion on loess soils in South-Limbourg (The Netherlands): a comparison of two rainfall events, *Soil Technology* **4**, 281-295.

Levert, C. (1954). Regens. Een statistische studie. *Mededelingen en Verhandelingen KNMI* **102-62**, Staatsdrukkerij, Den Haag, 246 pp.

Mücher, H.J. (1986). Aspects of loess and loess derived slope deposits: an experimental and micromorphological approach, *Netherlands Geographical Studies* **23**, 267 pp.

Renes, J. (1988). *De geschiedenis van het Zuidlimburgse cultuurlandschap.* Van Gorcum, Assen, 265 pp. + maps.

Rickson, R.J. (ed.) (1994). *Conserving Soil Resources, European Perspectives.* CAB International, Wallingford, 425 pp.

Schwertmann, Rickson, R.J. and Auerswald, K. (eds) (1988). Soil Erosion Protection Measures in Europe. Catena Verlag, *Soil Technology Series* No. **1**, 216 pp.

Van den Broek, J.M.M. (1966). *De bodem van Limburg.* Stichting voor Bodemkartering, Wageningen, 147 pp.

Van Dijk, P.M. and Kwaad, F.J.P.M. (1995). Runoff generation and soil erosion in small agricultural catchments with loess derived soils. *Hydrologic Processes*, in press.

Van Dijk, P.M., Van der Zijp, M., and Kwaad, F.J.P.M. (1995). Soil erodibility parameters under various cropping systems of maize, *Hydrologic Processes*, in press.

Van Eijsden, G.G. (1986). Bodemerosie op landbouwpercelen in een loessgebied. *Rapport Fysisch-geografisch en bodemkundig laboratorium, Universiteit van Amsterdam*, 61 pp.

Van Mulligen, E., (1990). Erosieonderzoek voor mais-, suikerbieten- en aardappelteelt in 1990 te Wijnandsrade. *Rapport ROC Wijnandsrade-FGBL Univ. van Amsterdam, 92 pp. + bijlagen.*

Wansink, A.G., (1991). Onderzoek naar erosiebestrijdende teeltsystemen voor mais en suikerbieten in 1991 te Wijnandsrade. *Rapport Vakgroep Fysische Geografie en Bodemkunde, Universiteit van Amsterdam, Stichting Proefboerderij Wijnandsrade, 44 pp. + bijlagen.*

Wicherek, S. (ed.) (1993). *Farm Land Erosion in Temperate Plains Environment and Hills.* Elsevier, Amsterdam, 587 pp.

18. INCORPORATING CRUSTING PROCESSES IN EROSION MODELS

Y. Le Bissonnais[1], D. Fox[1], L.-M. Bresson[2]

[1] Institut National de la Recherche Agronomique
Science du Sol
Centre de recherche d'Orléans
45160 Olivet
France

[2] Institut National Agronomique-Paris-Grignon
Département Agronomie-Environnement
78850 Thiverval-Grignon
France

Abstract

In most soils, surface crusting strongly influences the infiltration and erosion processes, so there is a need to include crusting in infiltration and erosion modelling. The processes involved in surface crusting, however, are extremely dynamic and crust characteristics are often difficult to measure. Significant progress has been made in describing crust formation. Aggregate breakdown and the displacement and reorganisation of detached soil units give rise to two general categories of surface crusts: structural and depositional. Modelling infiltration into these crusts has led to the development of equations of varying complexity, ranging from simple empirical equations to numerical solutions of the Richards equation. Obtaining the parameters for the more mechanistic approaches remains a challenge, and the equations need to be evaluated under field conditions where crust characteristics have a high spatial variability. For erosion modelling, surface crusting and erosion have many common processes, so crusting is implicitly present in some models. Further research should consider the potential for estimating the quantitative parameters needed for infiltration and erosion modelling from the descriptive studies of crusting processes and crust micromorphology.

Introduction

Soil surface crusting is an incipient process of water erosion, influencing both the runoff rate and soil erodibility. It is therefore of considerable interest to include crusting in process-based erosion models. The processes influencing surface crusting (a distinction between a crust and a seal will not be maintained here) have been well-studied, but their dynamic nature make quantitative modelling extremely complex. Crusting reduces infiltration significantly, thereby increasing runoff risks. It also alters soil characteristics, such as shear strength and microtopography, which influence sediment detachment and transport processes.

In this paper, the basic crusting processes and crust morphologies will be described, and the first attempts to model crusting will be discussed. This will be followed by a review of the current state in modelling the influence of crusting on infiltration and of the interactions between crusting and erosion.

Crusting processes and modelling: crust formation

It is possible to distinguish between two groups of processes in crust formation: (i) the breakdown processes, producing microaggregates or primary soil particles from initial clods, and (ii) the displacement and reorganisation of these breakdown products into new, denser and more continuous layers.

Breakdown processes

Aggregate breakdown by water may result from various physico-chemical and physical mechanisms and may involve different levels of soil structure. Four main mechanisms, producing different sizes of fragments or particles, can be identified from the various reviews already available (Le Bissonnais, 1990; Loch, 1994; Hairsine and Hook, 1995): (i) slaking, i.e. disaggregation by compression of entrapped air during wetting, (ii) microcracking by differential swelling, (iii) shearing by raindrop impact, (iv) physico-chemical dispersion. The main soil characteristics influencing aggregate stability and erosion are soil texture, clay mineralogy, organic matter content, type and concentration of cations, iron and aluminium oxides, pH and $CaCO_3$ content (Wischmeier and Mannering, 1969). The initial water content and roughness of the soil surface layer strongly affects disaggregation, crusting dynamics and subsequent runoff and erosion (Le Bissonnais and Singer, 1992; Bradford and Huang, 1992).

Displacement and reorganisation

Vertical translocation of detached units by gravity, or with infiltrating water, may lead to the clogging of pores within the upper few millimetres of soil. Splash induced by raindrop impact is the main process of particle displacement under rainfall before runoff begins (Nearing and Bradford, 1985; Farres, 1987). Compaction of the surface layer by the rearrangement and coalescence of fragments upon wetting (slumping) and under the impact of raindrops is of major importance in reducing porosity and hydraulic conductivity.

Runoff and turbulence in overland flow and puddles can detach and transport fragments. Deposition of fragments in microdepressions occurs after an initial period of water excess and structural crust development. It contributes to the reduction in infiltration by building a more continuous layer. This depositional crust is often formed by several microbeds resulting from the sorting and differential deposition of fragments in puddles. Further compaction may also occur with changes in matric potential upon drying.

Crust morphology

The main morphological types of crusts differ in the combination of the subprocesses involved in their formation (Bresson and Valentin, 1994). The structural or disruptional crusts are formed by an in-situ reorganisation with very limited particle displacement and without sorting or sedimentation. They result from the gradual packing and coalescence of soil units produced through breakdown. The depositional or sedimentary crusts result from particle displacement and sorting in puddling conditions. Bresson and Boiffin (1990) showed that structural and depositional crusts correspond to two successive stages in a general pattern of crust development (Valentin, 1986; Boiffin, 1986): sealing of the surface by a structural crust, and then formation of a depositional crust. The change from the first to the second stage depends on the hydrodynamic behaviour of the soil surface, which is partly controlled by the structural crust development.

Crust evolution: changes in crust characteristics from its initial formation to harvest/tillage

During a cultivation period, a crust can be affected by sheet erosion or incised by concentrated flow. Crust development interacts with vegetation emergence and growth: crusting can reduce crop emergence. Conversely, vegetation cover can protect the surface from further crusting. Several biological processes may also affect crust structure and properties: termites, ants and earthworms may be very active at the soil surface. They generally increase infiltration. Cracking of the crust during dry periods enhances infiltration (Levy *et al.*, 1986); however, this effect is generally of short duration and the cracks close quickly when the rain starts again (Le Bissonnais and Singer, 1992). This evolution in crust characteristics concerns cultivated areas primarily.

Modelling crusting

Boiffin (1986) proposed an empirical model which takes the crust development rate into account. It was demonstrated that the size of the smallest clods not included in the crust (Dmin) is related to the cumulative kinetic energy of rainfall and is a good index of crust development. Therefore, crusting intensity and the associated decrease in infiltrability are predicted as a function of the cumulative rainfall kinetic energy for intensities above a threshold. This threshold depends on the soil characteristics (Boiffin and Monnier, 1986). Despite its simplicity, the model has shown good results for the soils tested. However, soil moisture conditions and tillage practices are not taken into account. Furthermore, the model requires an extensive period of calibration and is site specific.

Modelling infiltration into crusted soils

Duley (1939) was among the first to cite the deleterious effect of crusting on infiltration, and modelling infiltration in crusted soils has been the subject of intense investigation since the late 1960's. Crust infiltration models can be divided into three general categories: Horton-type regression equations (Morin and Benyamini, 1977), adaptations of the Green-Ampt equation (Hillel and Gardner, 1970; Ahuja and Ross, 1983; Brakensiek and Rawls, 1983), and the Richards equation for layered soils (Aboujaoudé *et al.*, 1991; Bristow *et al.*, 1995). The Horton-type regression equations are the simplest to use, requiring only measured infiltration rate curves to calibrate the regression constants. They are, however, sensitive to initial and boundary conditions and therefore of limited application for predictive purposes. The Green-Ampt and Richards equations become increasingly more mechanistic in approach and rely on parameters derived from soil physics theory. The Richards approach is computationally more time-consuming and requires more data to calibrate; however, it is more flexible when attempting to model spatial variability or multi-layered systems. Several adaptations of the Green-Ampt equation have been developed: they have a much more solid basis in soil physics theory than the Horton-type equations and are generally easier to parameterise and computationally less time-consuming than the Richards equation. The suitability of any particular equation depends on the specific objectives, site heterogeneity, and the resources and data available for parameterisation.

For the physically-based equations, the hydraulic properties are difficult to estimate accurately. The seal is generally in the order of a few mm's to a few cm's thick, and its thickness is generally fixed for practical reasons related to the functioning of the model. Data which can be gained from micromorphology are generally not taken into consideration. Mualem *et al.* (1990a) reviewed much of the crusting literature and suggested that compaction was the dominant process reducing porosity in most of the cases studied. They also concluded that the crust generally shows no sharp boundary with the undisturbed soil beneath.

Measuring or calculating changes in hydraulic conductivity with cumulative kinetic energy for thin seals is extremely time-consuming and complex. In some models, the hydraulic conductivity is assumed to change as a decaying exponential function of cumulative kinetic energy (Moore, 1981; Brakensiek and Rawls, 1983; Römkens *et al.*, 1986). Others have simplified the problem by combining the seal thickness and hydraulic conductivity into a single term of seal impedance or its reciprocal, hydraulic conductance (Römkens, 1990). Most of the equations presented in the literature have performed well within the limitations of their experimental conditions. The limitations of these conditions become more restrictive when attempting to pass from the laboratory into the field.

Initial models described infiltration within the context of a two-layer system with constant hydraulic conductivity and thickness (Hillel and Gardner, 1969). This evolved into a two-layer system with changing seal hydraulic characteristics (Ahuja and Ross, 1983). More recent developments include more than two layers (Brakensiek and Rawls, 1983) or treat the seal-subseal system as a single discontinuous layer where hydraulic conductivity decreases non-linearly near the surface (Mualem *et al.*, 1990b). Apart from the difficulty in obtaining the necessary parameters, the greatest limitation of the models published to date is their lack of consideration for the high spatial variability observed in the field.

Field and laboratory observations of crusts suggest that they can vary in thickness and hydraulic conductivity by an order of magnitude or more within a spatial scale of a few cm's (Casenave and Valentin, 1989). Two recent models have made first attempts at modelling infiltration under such varied conditions. Aboujaoudé *et al.* (1991) modelled infiltration into a surface depression with varying seal thickness: the maximum thickness was in the centre of the depression and it thinned toward the crest. Although the model predicted that infiltration does not vary greatly from that into a uniform seal thickness over long times, it did not test for interactions with rainfall intensity or runoff depth, conditions which would most likely generate significant differences. Bristow *et al.* (1995) simulated infiltration into a surface capped with a crust only in the depressions and not on adjoining mounds. They found that significant differences in infiltration could arise depending on if water stored in the depression was allowed to run off or not. Neither of these models has yet been validated under field conditions, but they point the direction in crust infiltration modelling.

Micromorphologists and modellers must combine their efforts to characterise crust variability and express it in terms of the physical parameters required to model infiltration. Ideally, these parameters could be derived from a crusting model that is sensitive to rainfall kinetic energy, aggregate stability, initial water content and surface roughness. In other words, is it possible to predict infiltrability from a morphological model such as the one presented in section 1? This is the challenge that morphologists and physicists must tackle together.

Modelling the interactions between crusting and erosion

Surface crusting can be expected to have major consequences for both interrill and rill erosion. Reduced infiltrability and surface roughness increase runoff, potentially leading to more rill erosion downslope (De Ploey and Poesen, 1985). However, because the crust has a greater shear strength than the uncrusted soil, it may impede raindrop detachment (interrill erosion) or headcutting by concentrated overland flow (rill erosion inducement). The effects of crusting on detachment and infiltration greatly depend on soil and climatic conditions (Le Bissonnais and Singer, 1992).

Change in microtopography
The decrease in microrelief is a function of the volume of displaced soil, and it generally results in the irreversible filling of microdepressions with products from larger clods and micromounds. It is therefore closely related to breakdown, splash and deposition processes. The effects of surface roughness on interrill erosion processes are not well documented, mainly

because of limitations in instrumentation and analytical techniques (Nearing *et al.*, 1990). However, new technologies (laser scanning, stereophotography) for getting detailed digital elevation models (DEM) of the soil surface, including clod, furrow or residue roughness, allow (1) the assessment of changes in microtopography with time, as structural and depositional crusts develop, (2) the measurement of related changes in depressional storage, and (3) the quantification of erosion and deposition processes (Moran and Vézina, 1993). These measurements could be used in order to validate empirical erosion models (Boiffin, 1986).

Increase in shear strength
Crusting modifies the shear strength of the soil surface, mainly through a change in microstructure and sometimes through a change in particle size distribution. Mechanical properties of crusts have long since been found to be related to particle size distribution (Lemos and Lutz, 1957). Crust texture may differ from the texture of the uncrusted soil. Coarse-textured surface layers are related to fine particle depletion by splash (Moss, 1991), by overland flow (Norton, 1987), or by raindrop winnowing and filtration (Valentin, 1986). Fine-textured surface layers are formed by the deposition of suspended particles when rainfall stops, or they result from the erosion of the coarse-textured upper layers of sieving crusts ('erosion crust', Valentin, 1986).

The size distribution of the breakdown fragments (Le Bissonnais *et al.*, 1989) which form the crust is likely to control its mechanical properties. The effects of microstructure on shear strength are poorly documented because few mechanical studies of crusting involve micromorphological characterisation (Bresson, 1995). In this respect, the genetically based classification suggested by Valentin and Bresson (1992) could provide useful guidelines. The related microstructural diagnostic features include solid continuity, porosity and microfabric (coarse/fine related distribution fabric) (Bresson and Valentin, 1994). Other properties are expected to help in predicting shear strength. Bulk density should be considered because it allows the detection of ultramicrostructural changes resulting from raindrop compaction which are not visible under microscopic observation. Bulk density appears to be an important descriptor of fall-cone shear strength (Sharma *et al.*, 1991), so appropriate measurement techniques should be set up for the development of crust bulk density versus rainfall kinetic energy functions.

Common processes
Crusting and erosion both involve particle detachment and transport processes. Although most erosion models do not take crusting into account for adjusting interrill erosion, new concepts have been introduced which involve the explicit delineation between detachment and transport processes (Nearing *et al.*, 1990). These models are likely to apply to crust formation, so it is of interest to compare their basic concepts with the morphological crusting model presented

above. In shallow flow which would otherwise be incapable of transporting sediments, raindrop impacts cause fragments to be repeatedly ejected into the flow and transported downstream. Introduced by Walker et al. (1977), the concept of raindrop induced flow transport was further developed by Kinnell (1988), and it was improved by considering the protective effect of the layer formed by the deposition of detached fragments (Hairsine and Rose, 1991). Therefore, detachment from the cohesive soil matrix and redetachment from the uncohesive deposited layer are distinguished. The deposited layer model seems to fit well with the morphology of crusts developed in erosional conditions. In a diagram of combined erosion and crusting processes, Hairsine and Hook (1995) introduced the structural crust ('semi-attached degraded layer') but did not show the depositional crust and its relationship with the uncohesive deposited layer. From the morphological description above, a more comprehensive representation of a combined crusting and erosion model can be suggested (Figure 1).

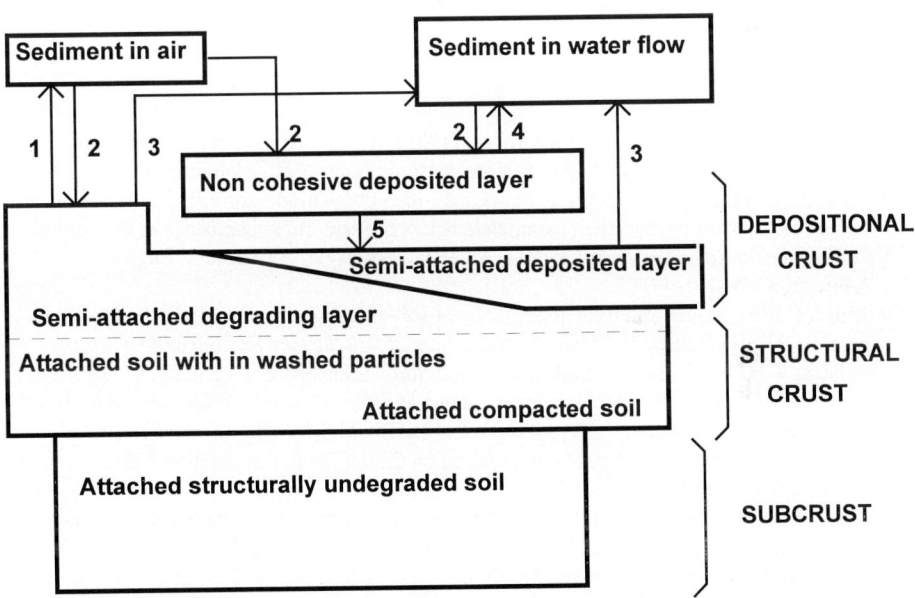

Figure 1. Diagram of the relationships between crusting and erosion processes (after Hairsine and Hook, 1995). 1: splash, 2: deposition, 3: rainwash (detachment and entrainment), 4: rainwash (re-detachment and re-entrainment), 5: deposition and cohesion

Conclusion

In many soils, surface crusting is the dominant factor controlling infiltration and runoff, and it is therefore of critical importance for erosion. Modelling its influence, however, is extremely difficult due to its high temporal and spatial variability. Considerable progress has been made in recent years in modelling crusting, its effect on infiltration, and its relationship to erosion. The following aspects deserve particular attention if further progress is to be made: (i) extend current qualitative descriptions of crusting processes to enable the quantification of parameters required for infiltration and erosion modelling; (ii) include spatial variability in crust characteristics in infiltration and erosion modelling; (iii) extend the time scale of crusting models to a seasonal scale.

Acknowledgements

This paper is a contribution to the Soil Erosion Network of the GCTE, which is a Core Research Project of the IGBP.

References

Aboujaoudé, A., Belleudy, P. and Vauclin, M. (1991). A numerical study of infiltration through crusted soils: flat and other surface configurations, *Soil Technology* **4**, 1-18.

Ahuja, L. R. and Ross, J. D. (1983). A new Green-Ampt type model for infiltration through a surface seal permitting transient parameters below the seal. In, *Proceedings of the National Conference on Advances in Infiltration*, American Society of Agricultural Engineers Publication **11-83**, 147-162.

Boiffin, J. (1986). Stages and time dependency of soil crusting in situ. In, Caillebaud, F., Gabriels, D. and De Boodt, M. (eds), *Assessment of Soil Surface Sealing and Crusting*, Flanders Research Center for Soil Erosion and Soil Conservation, Ghent. pp. 91-98.

Boiffin, J. and Monnier, G. (1986). Infiltration rate as affected by soil surface crusting caused by rainfall. In, Caillebaud, F., Gabriels, D. and De Boodt. M. (eds), *Assessment of Soil Surface Sealing and Crusting*, Flanders Research Center for Soil Erosion and Soil Conservation, Ghent. pp. 210-217.

Bradford, J. M. and Huang, C. (1992). Mechanisms of crust formation: physical components. In, Sumner, M. E. and Stewart, B. A. (eds), *Soil Crusting: Physical and Chemical Processes*. Lewis, Boca Raton. pp. 55-72.

Brakensiek, D. L. and Rawls, W. L. (1983). Agricultural management effects on soil water process. Part II Green-Ampt parameters for crusting soils. *Transactions of the American Society of Agricultural Engineers* **26**, 1753-1757.

Bresson, L.-M., (1995). A review of physical management for crusting control in Australian cropping systems. *Australian Journal of Soil Research* **33**, 195-209.

Bresson, L.-M. and Boiffin, J. (1990). Morphological characterization of soil crust development stages on an experimental field. *Geoderma* **47**, 301-325.

Bresson, L.-M. and Valentin, C. (1994). Soil surface crust formation: contribution of micromorphology. In, Ringrose-Voase A. and Humphreys G. (eds), *Soil Micromorphology*, Elsevier, Amsterdam. pp. 737-762.

Bristow, K. L., Cass A., Smetten, K. R. J., and Ross, P., (1995). Water entry into sealing, crusting, and hardsetting soil: a review and illustrative simulation study. In, So H. B. and al. (eds), *Sealing, Crusting and Hardsetting Soils: Productivity and Conservation.*, Australian Society of Soil Science, Brisbane, pp 183-203.

Casenave, A. and Valentin, C. (1989). Les états de surface de la zone Sahélienne: influence sur l'infiltration. *ORSTOM, Collection 'Didactiques'.* 230 pp.

De Ploey, J. and Poesen, J. (1985). Aggregate stability, runoff generation and interrill erosion. In, Richards, Arnett and Ellis (eds), *Geomorphology and Soils*. Allen and Unwin, pp. 99-120.

Duley, F. I. (1939). Surface factors affecting the rate of intake of water by soils. *Soil Science Society of America Proceedings* **4**, 60-64.

Farres, P. J. (1987). The dynamics of rainsplash erosion and the role of soil aggregate stability. *Catena* **14**, 119-130.

Foster, G. R. (1990). Process-based modelling of soil erosion by water on agricultural land. In, Boardman, J., Foster, I. D. L. and Dearing, J. A. (eds), *Soil erosion on agricultural land.*, Wiley, Chichester. pp. 429-446.

Hairsine, P. B. and Rose, C. W. (1991). Rainfall detachment and deposition, sediment transport in the absence of flow-driven processes. *Soil Science Society of America Journal* **55**, 320-324.

Hairsine, P. B. and Hook, R. A. (1995). Relating soil erosion by water to the nature of the soil surface. In, So H. B. *et al.* (eds), *Sealing, Crusting and Hardsetting Soils: Productivity and Conservation*, Australian Society of Soil Science, Brisbane. pp. 77-91.

Hillel, D. and Gardner, W. R. (1969). Steady infiltration into crust-topped profiles, *Soil Science* **108**, 137-142.

Hillel, D. and Gardner, W. R. (1970). Transient infiltration into crust-topped profiles, *Soil Science* **109**, 149-153.

Kinnell, P. I. A. (1988). The influence of flow discharge on sediment concentrations in raindrop induced flow transport. *Australian Journal of Soil Research* **26**, 575-582.

Le Bissonnais, Y. (1990). Experimental study and modelling of soil surface crusting processes. In, Bryan, R.B. (ed.). *Soil Erosion — Experiments and Models*, Catena supplement **17**, 13-28.

Le Bissonnais, Y., Bruand, A. and Jamagne, M. (1989). Laboratory experimental study of soil crusting: relation between aggregate breakdown and crust structure. *Catena* **16**, 377-392.

Le Bissonnais, Y. and Singer, M. J. (1992). Crusting, runoff and erosion response to soil water content and successive rainfall events. *Soil Science Society of America Journal* **56**, 1898-1903.

Lemos, P. and Lutz, J. F. (1957). Soil crusting and some factors affecting it. *Soil Science Society of America Proceedings* **21**, 485-491.

Levy, G., Shainberg, I. and Morin, J. (1986). Factors affecting the stability of soil crusts in subsequent storms. *Soil Science Society of America Journal* **50**, 196-201.

Loch, R. J. (1994). A method for measuring aggregate water stability with relevance to surface seal development. *Australian Journal of soil Science* **32**, 687-700.

Moore, I. D. (1981). Infiltration equations modified for surface effects. *Journal of the Irrigation Division, Transactions of the American Society of Agricultural Engineers* **107**, 71-79.

Moran, C. J. and Vézina, G. (1993). Visualizing soil surfaces and crop residues. *Computer Graphics* **13-2**, 40-47.

Morin, J. and Benyamini, Y. (1977). Rainfall infiltration into bare soils. *Water Resources Research* **13**, 813-817.

Moss, A. J. (1991). Rain-impact soil crust. 1: Formation on a granite derived soil. *Australian Journal of Soil Research* **29**, 271-290.

Mualem, Y., Assouline S., and Rohdenburg, H. (1990a). Rainfall induced soil seal (A): a critical review of observations and models. *Catena* **17**, 185-203.

Mualem, Y., Assouline S. and Rohdenburg, H. (1990b). Rainfall induced soil seal (B): Application of a new model to saturated soils. *Catena* **17**, 205-218.

Nearing, M. A. and Bradford, J. M. (1985). Single waterdrop splash detachment and mechanical properties of soils. *Soil Science Society of America Journal* **49**, 547-552.

Nearing, M. A., Lane, L. J., Alberts, E. E. and Laflen, J. M. (1990). Prediction technology for soil erosion by water: status and research needs. *Soil Science Society of America Journal* **54**, 1702-1711.

Norton, L. D. (1987). Micromorphological study of surface seals developed under simulated rainfall. *Geoderma* **40**, 127-140.

Römkens, M. J. M., Baumhardt, R. L., Parlange, M. B., Whisler, F. D., Parlange, J.-Y. and Prasad, S. N. (1986). Rain-induced surface seals: their effect on ponding and infiltration. *Annales Geophysicae* **4**, 4417-4424.

Römkens, M. J. M. (1990). Surface sealing and infiltration. In, Anderson, M.G. and Burt, T.P. (eds), *Process Studies in Hillslope Hydrology*, Wiley, Chichester. pp. 127-172.

Sharma, P. P., Gupta, S. C. and Rawls, W. J. (1991). Soil detachment by single raindrops of varying kinetic energy. *Soil Science Society of America Journal* **55**, 301-307.

Valentin, C. (1986). Surface crusting of arid sandy soils. In, Callebaut, F., Gabriels, D. and De Boodt, M. (eds), *Assessment of Soil Surface Sealing and Crusting,* Flanders Research Center for Soil Erosion and Soil Conservation, Ghent. pp. 40-47.

Valentin, C., and Bresson, L.-M. (1992). Soil crust morphology and forming processes in loamy and sandy soils. *Geoderma* **55**, 225-245.

Walker, P. H., Hutka, J., Moss, A. J. and Kinnell, P. I. A. (1977). An experimental facility for dynamic soil erosion studies. *Soil Science Society of America Journal* **41**, 610-612.

Wischmeier, W. H. and Mannering, L. W. (1969). Relation of soil properties to its erodibility. *Soil Science Society of America Proceedings* **33**, 131-137.

19. THE ROLE OF SOIL AGGREGATES IN SOIL EROSION PROCESSES

Dino Torri, Rossano Ciampalini and Pietro Accolti Gil

CNR - Soil Genesis, Classification and Cartography Research Centre
P. le Cascine 15
50144 Florence
Italy

Abstract

Aggregate size and stability represent two soil characteristics that may rapidly change in response to changes in climatic conditions and land-use. Their roles in soil erosion processes have always been recognised but their sizes and stability have rarely been used in formula or algorithms. Several erosion processes are discussed here and possible roles of aggregates are shown using experimental data. Particularly, aggregates interfere with splash detachment and the cushion effect that surface water exerts on drop impact. Overland flow detachment depends on aggregate size. A feedback effect is also shown due to the progressive thinning of stable grains during rainfall detachment and re-detachment.

Introduction

Soil aggregates play an important role in the dynamics of soil erosion. It has been shown that indices of aggregate stability are often good predictors of soil parameters such as erodibility (e.g. Rousseva, 1989). Moreover, aggregates influence erosion both indirectly owing to interaction with processes such as cracking, crusting, and sealing, and directly, owing to their physical characteristics such as bulk density, size and shape, inter- and intra-granular forces. Their stability is also important. Aggregate stability shows a seasonal and spatial variation within the same soil (e.g. Nestroy, 1993). Since organic matter influences aggregates, processes acting on organic matter modify aggregate characteristics e.g. incident radiation, temperature, soil microbiota, etc. (Le Bissonnais *et al.*, 1993).

Aggregate breakdown is a fundamental process in soil erosion (Le Bissonnais *et al.*, 1993) and it may ultimately yield crusting, enhance runoff and reduce roughness. Hence, aggregates play a paramount role in soil erosion and are extremely sensitive to global change.

At present there are several quantitative relationships between aggregate characteristics and soil erosion processes. Nearing *et al.* (1991) showed that soil detachment by shallow flow is related to mean weight diameter of stable grains (where *grain* is a synonym of stable aggregates and single primary particles). Aggregate role in transport and deposition processes has been acknowledged for years and techniques based on laboratory tests (Rhoton *et al.*, 1982) or on algorithms (Foster *et al.*, 1985) have been developed for modelling purposes. An extensive summary is given by Haan *et al.* (1994).

In the following paragraphs some interactions between aggregates and soil erosion processes will be discussed. The relationships presented are not meant for predictive purposes but only as a means for exemplifying the discussion.

Materials and methods

The data used in this paper were collected using different techniques in a period spanning 12 years. Data on rain splash detachment are part of data used by Torri *et al.* (1987), data used by Torri and Borselli (1991), and unpublished data collected more recently. The experiments were conducted as described in the quoted papers, but the rain simulators were completely modified in later experiments (Panini *et al.*, 1993) so that drop size and kinetic energy are not the same all over the final data set. Shear strength was measured using different types of torvane apparatus or Swedish falling cone. All data were transformed into pocket torvane values, using calibration curves (Brunori *et al.*, 1990). Textural analyses and bulk densities of the studied samples were determined using standard methods. Stable grain distributions were obtained by means of wet sieving. The above mentioned data were collected for six different kinds of soil, in four classes (Figure 1).

Data on flow detachment were collected using a methodology described by Ciampalini (1992) and Ciampalini and Torri (in press). Detachment rate was taken as the erosion rate value recorded immediately after runoff first reached a rectangular soil target (with the side parallel to flow as long as 2.5 cm). Soil samples were sieved through a 2mm mesh and saturated for 24h before testing.

Data on aggregate dynamics were collected during rainfall simulation experiments in the field and sediment samples were wet-sieved immediately after collection. Samples collected on the same soils were examined in the laboratory. Here the material (30g) was settled in cylinders with 200 cm^3 of water. The cylinders were then rotated round their height for variable time spans at constant angular velocity (33 r.p.m.). The sample within the cylinder was then moving at an average speed (=average flow velocity) of 0.22 m/s, as the cylinders had 40 cm of inner basal circumference. A rough estimate of the energy applied per unit of surface transverse to the flow direction (Ea) can be calculated using the formula Ea=0.5ρv³ where ρ is the fluid density and v its mean velocity. This gives an applied energy of 9.68 Joule/m²/revolution.

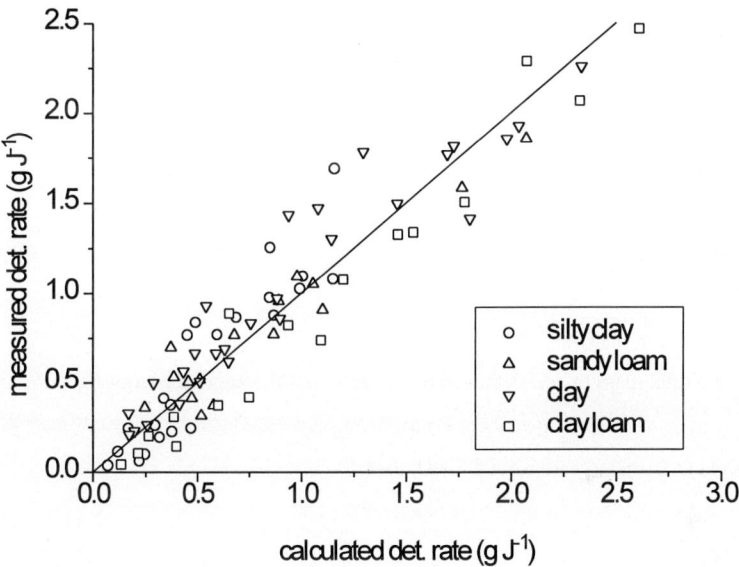

Figure 1. Detachment rate due to raindrop impact: the proposed equation fits the measured data reasonably well

Detachment processes

Detachment can be described as the work W done by the energy E_a input into the system by any detaching agent. If E_p is the part of applied energy that is not used for soil detachment, then it follows:

$$W = E_a - E_p = \frac{E_a - E_p}{E_a} E_a = eE_a \tag{1}$$

where e is the efficiency of the process. The work done can also be described in terms of forces resisting detachment; in particular if w_i is the work done for removing the i-th stable grain, and N is the number of detached grains then it follows:

$$W = \sum_{1}^{N} w_i = N \bar{w} \qquad (2)$$

where $\bar{w}$ is the average work per detached grain.

The average work $\bar{w}$ can be estimated using the resistive pressure T, the cross section A of the representative detached grain, and the average length l over which a grain must be moved to be detached from the soil mass. The same can be done for N, once the volume (V) and the dry bulk density (δ_s) of the representative grain are known. It follows:

$$\left. \begin{array}{l} \bar{w} = TAl \\ N = \dfrac{D}{V\delta_s} \end{array} \right\} \Rightarrow W = \dfrac{DTAl}{V\delta_s} \qquad (3)$$

where D is the total mass of soil detached by the action of the applied energy E_a.

If equations (1), (2) and (3) are solved for D, then:

$$D = \frac{\delta_s}{T} \frac{V}{A} \frac{e}{l} E_a \qquad (4)$$

In the following paragraphs the ratio V/A will be estimated using the median diameter of the distribution by mass of stable grains and supposing grains to be spherical (i.e. $V/A = D_{50}/1.5$). Dry bulk density corresponds to D_{50}.

The ratio e/l depends on several parameters. Certainly the detachment length l must depend on clay content (alternatively on plastic limit or plasticity index), and on the average protrusion of nearby grains over which the detached grain must be transported before being freed from any type of bonding (simple friction enclosed). Since the efficiency also changes following the degree of inelasticity of the soil, clay content should be a good indicator. This ratio e/l is certainly influenced by many other parameters. Particularly, initial soil water content, soil/water chemistry, mineralogy, organic matter, degree of soil hydrophobicity, etc. should be included in the discussion but we will limit ourselves to the characteristics relevant to the database that will be used: e.g., soil samples were always saturated by capillary rise and organic matter was always close to 1-2%. Only one kind of soil has a significantly larger amount of expandable clay minerals. Hence, the set of possible soil parameters to be included in this analysis are limited to primary particle distribution, stable grain distribution and a few others, such as Atterberg's consistency limits. The form of the mathematical relationships describing the ratio *e/l* will possibly be defined by non linear best fitting techniques.

Splash detachment

Let us now apply equation (4) for describing drop impact. The applied energy Ea is the kinetic energy of rain (KE) which is expressed in Joules. Now we need something to sort out how the efficiency factor varies with several other parameters, such as slope and overland flow depth. It is well known that detachment rate is positively influenced by slope, as pointed out by many authors (e.g. De Ploey and Savat, 1968) and explained by Torri and Poesen (1992). In this case an exponential effect is expected because of the type of trends observed experimentally.

Moreover, splash decreases exponentially with surface water depth (Palmer, 1963; Torri *et al.*, 1987). Applying the above considerations, equation (4) becomes:

$$\frac{D}{KE} = a \frac{\delta_S D_{50}}{1.5T} \exp(-qh + b\tan\alpha) f(s) \qquad (5)$$

where *a* and *b* are best fitting coefficients, *h* is runoff depth (mm), *f(s)* stands for *e/l* and s indicates any set of soil characteristics among those previously mentioned. In these paragraphs *D* is expressed in kg m^{-2}, grain sizes in micrometres, densities in kg m^{-3} and shear strength in Pascal.

The term *q* expresses how adequately the soil surface is represented by a plane surface. In other terms, while one fraction of the soil surface is well submerged by a water layer deeper than *h*, the other is submerged by a layer thinner than *h*. As the amount of detached material decreases exponentially with h, it follows that the contributions of the two fractions are not equivalent to the detachment from a surface uniformly submerged under the average depth h. Hence q must depend on roughness. If roughness is due to soil grains alone, then q is a function of how well sorted the grain size distribution is. This may be written as follows:

$$q = g(\sigma) = g(g^1(\frac{d_i}{g_i})) \qquad (6)$$

where σ is the geometric standard deviation, g and g^1 two unknown functions, d_i and d_j two percentiles to be used for estimating σ.

The following relationship was found:

$$\frac{D}{KE} = 0.036 \frac{\delta_s D_{50}}{1.5T} \exp(-qh + 1.31\tan\alpha) f(c) \qquad (7)$$

where:
$$f(c) = \exp(0.067c)$$
$$q = -0.36\ln(6\frac{d_{95}}{d_{50}})$$

and r=0.921, n=82.

Detachment by flow

If equation (4) is rewritten for flow detachment, the applied energy can be estimated using flow stream power (S_p, calculated as unit discharge times slope) and comparing it to detachment rate (Dr).

$$Dr = \frac{\delta D_{50} e S_p}{1.5Tl} \qquad (8)$$

Nearing et al. (1991) and Parker et al. (1995) demonstrated empirically that slope affects flow detachment to a greater extent than described by stream power. Hence slope should positively influence the efficiency of the process.

Our data, once elaborated using the same type of approach as discussed for splash detachment, lead to a reasonable degree of fit when equation (6) is modified as follows:

$$Dr = 0.00057 \frac{\delta_s D_{50}}{1.5T} S_p \exp\left[6.4 \frac{c}{D_{50}} + 7.1\tan\alpha - 1.3\left(\frac{\delta - \delta_w}{\delta_w}\right)\right] \qquad (9)$$

where Dr is expressed in kg m^{-2} s^{-1}, δ is the wet bulk density of the soil grains (kg m^{-3}) and δ_w is the water density (r = 0.94, n=66).

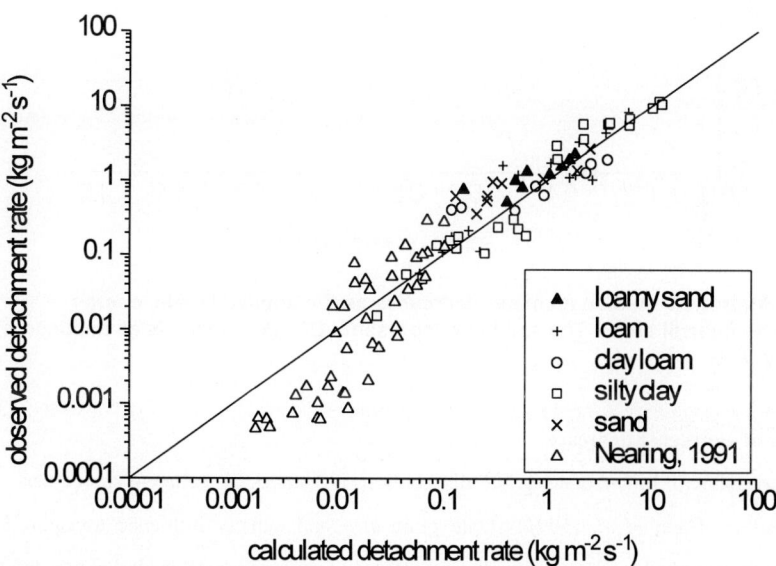

Figure 2. Calculated Vs. measured detachment rate due to flow. The data collected by Nearing et al. (1991) were multiplied by 10 before plotting (see text for explanation)

Predicted vs. observed values are shown in Figure 2 using a logarithmic scale. Despite a good fit, great uncertainty still remains as data are scattered. The data collected by Nearing et al.

(1991) are also included. They are as shown in the figure if the predicted values (equation 9) are multiplied by 0.1. The extra coefficient is due to the fact that initial detachment values measured following Ciampalini and Torri's (in press) methodology are larger than the corresponding values measured following Nearing et al.'s (1991) methodology. The fact that Nearing et al.'s data follow a similar trend confirms the general form of equation (9). The S-shaped form of the Nearing et al.'s data is mainly due to slope, indicating that this parameter is not yet well described in equation (9).

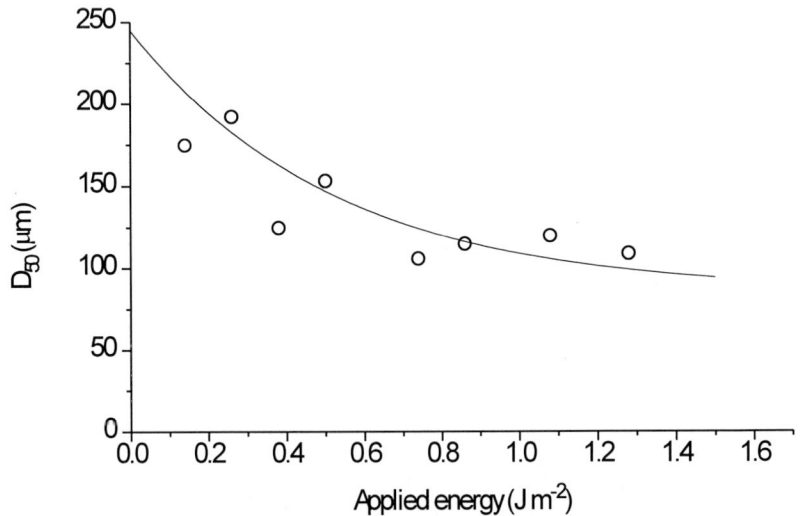

Figure 3. Aggregate median diameter decreases as the applied kinetic energy increases in interrill areas. The solid line represents D50 decay trends as deduced in laboratory tests

Dynamics of aggregate decrease

It has been shown how stable grain distribution influences detachment processes. As summarized by Haan et al. (1994) stable grain size and density influence transport and deposition of detached grains. Hence the whole soil erosion process is dominated by the dynamics of stable grains. The problem now is how to define what the correct stable grain distribution is to be used. There is a noticeable difference between aggregate distributions obtained with minimal energy (i.e., aggregates sieved after wetting by slow capillary rise) and primary particle distributions. Chisci et al. (1993) described this clearly. They also suggested using a distribution resistant to intense shaking in distilled water. Unfortunately the situation is not so simple because grain-size distribution changes even during single rainfalls. An example is given in Figure 3 with data collected during a rain simulation in a potato field, where erosion was entirely due to interrill processes. Hence, the progressive decrease of the stable grain distribution was due to mechanisms of detachment and re-detachment of grains by raindrop impact, as pointed out by Rose *et al.* (1990).

Similar trends can be obtained in the laboratory, using the rotating cylinder apparatus previously described. The laboratory measurements of aggregate size decay (represented by the progressive decrease of the median diameter, measured on samples collected near the rainfall simulation plot) are represented in Figure 3 by a continuous line.

Conclusions

The processes discussed above clearly show that soil erosion is partly defined by stable grain characteristics and dynamics. As soil aggregates influence other mechanisms such as runoff generation, soil surface sealing, etc., it may be concluded that stable grains play a paramount role in soil erosion.

Aggregates are the product of rapidly evolving processes of physical, chemical and biological nature and their characteristics are strongly influenced by human activities. It has been shown that their characteristics can change even during single rainstorms. Hence, they are the soil characteristics that more promptly react to global changes.

This implies that models describing soil erosion during single events (such as EUROSEM, Morgan *et al.*, 1998) should incorporate feedbacks similar to the ones envisaged in the discussed equations. Middle and long term models should introduce routines accounting for changes in grain stability and, consequently in erosion rates. These are necessary conditions if models want to deal with the dynamics of global change within their own codes, otherwise they will simply compare steady states and long term soil responses to different climatic and land use situations.

Acknowledgements

This research was partly supported by an European programme CEC-STEP-CT90-0026. The authors acknowledge M. Del Sette, M.P. Salvador and T. Panini for their contributions in conducting the experiments.

References

Brunori, F., Penzo, M.C. and Torri, D. (1989). Soil shear strength: its measurement and soil detachability. *Catena* **16**, 59-71.

Chisci, G., and Martinez, V. (1993). Environmental impact of soil erosion under different cover and management systems. *Soil Technology* **6**, 239-249.

Ciampalini, R. (1992). Erosione del suolo: distacco e trasporto di sedimenti ad opera delle acque di deflusso superficiale. Unpublished Thesis, Department of Earth Sciences, University of Florence, Florence, Italy.

Ciampalini, R. and Torri, D. (in press). Soil particles detachment by shallow flow: sampling methodology and observations. *Catena*.

De Ploey, J. and Savat, J. (1968). Contribution à l'étude de l'érosion par le splash. *Zeitschrift für Geomorphologie* **12**, 174-193.

Foster, G.R.., Young, R.A. and Neibling, W.H. (1985). Sediment composition for nonpoint source pollution analyses. *Transactions of the American Society of Agricultural Engineers* **28**(1), 133-146.

Haan, C.T., Barfield, B.J. and Hayes, J.C. (1994). *Design Hydrology and Sedimentology for Small Catchments*. Academic Press, San Diego. 588 pp.

Le Bissonnais, Y., Singer, M.J., and Bradford J.M. (1993). Assessment of soil erodibility: the relationship between soil properties, erosion processes and susceptibility to erosion. In, Wicherek, S. (ed.). *Farm Land Erosion in Temperate Plains, Environments and Hills*, Elsevier, pp. 87-96

Morgan, R.P.C., Quinton, J.N., Smith, R.E., Govers, G., Poesen, J.W.A., Chisci, G. and Torri, D. (1998). The EUROSEM Model. In, Boardman, J. and Favis-Mortlock, D.T. (eds), *Modelling Soil Erosion by Water*, Springer-Verlag NATO-ASI Global Change Series, Heidelberg

Nearing, M.A., Bradford, J.M. and Parker, D.B. (1991) Soil detachment by shallow flow at low slopes. *Soil Science Society of America Journal* **55**, 339-344.

Nestroy, O. (1993). On the question of aggregate stability of soils. In, Wicherek, S. (ed.). *Farm Land Erosion in Temperate Plains, Environments and Hills*. Elsevier, pp. 107-109.

Palmer, R. (1963). *The Influence of a Thin Water Layer on Waterdrop Impact Forces*. IAHS Publication 65, pp. 141-148.

Parker, D.B., Michel, T.G. and Smith, J.L. (1995). Compaction and water velocity effects on soil erosion in shallow flow. *Journal of Irrigation and Drainage Engineering* **121**(2), 170-178.

Panini, T., Salvador Sanchis, M.P. and Torri, D. (1993). A portable rain simulator for rough and smooth morphologies. *Quaderni di scienza del del suolo* **V**, Firenze, 47-58.

Rhoton, F.E., Meyer, L.D. and Whisler, F.D. (1982). A laboratory method for predicting the size distribution of sediment eroded from surface soils. *Soil Science Society of America Journal* **46**, 1259-1263.

Rose, C.W., Hairsine, P.B., Proffitt, A.P.B. and Misra, R.K. (1990). Interpreting the role of soil strength in erosion processes. *Catena Supplement* **17**, 153-165.

Rousseva, S. (1989). A laboratory index for soil erodibility assessment. *Soil Technology* **2**, 287-299.

Savat, J. (1980). Resistance to flow in rough supercritical sheet flow. *Earth Surface Processes* **5**, 103-122.

Torri, D., Sfalanga, M. and Del Sette, M. (1987). Splash detachment: runoff depth and soil cohesion. *Catena* **14**, 149-155.

Torri, D., and Borselli, L. (1991). Overland flow and soil erosion: some processes and their interactions. *Catena Supplement* **19**, 129-137.

Torri, D. and Poesen, J. (1992). The effect of soil surface slope on raindrop detachment. *Catena* **19**, 561-578.

20. PROCESS-BASED APPROACHES TO MODELLING SOIL EROSION

Calvin W. Rose and William L. Hogarth

Faculty of Environmental Sciences
Griffith University
Nathan Campus
Kessels Road
Queensland
Australia 4111

Abstract

The approach to soil erosion by water at the hillslope scale outlined describes erosion and sedimentation as contemporary competing processes, whose net outcome determines sediment concentration. Sedimentation forms a layer of easily-eroded deposited sediment, thus modifying the eroding surface. Rates of erosion and sedimentation are influenced by different soil characteristics, making it desirable to separately describe these two types of competing processes.

Description is limited to erosion processes caused by overland flow and rainfall impact, giving more detail on the latter. The great range in settling velocity of sediment components leads to continuous change in the size composition of the deposited sediment layer. This leads to time variation in concentration and settling velocity characteristics of eroded sediment, even in a constant erosive environment. The ability of dynamic theory to represent and interpret such behaviour measured in a simulated rainfall tilting flume facility is illustrated. Solution of the set of simultaneous partial differential equations describing each size class of sediment presents some numerical challenge, which can be met. Good approximate analytical solutions are also available and the results of one such approximation is illustrated.

Introduction

In order to predict net soil loss, there are a number of different challenges. The first challenge is to predict adequately runoff characteristics from those of rainfall. The second challenge, which requires the first challenge to have been met, is the subsequent dependent prediction of sediment concentration and thus soil loss. There is active debate on which of these two challenges is the more difficult or less adequately dealt with, but this paper deals with the second of these two challenges.

An understanding of the processes involved in soil erosion provides an important key, not only to providing mathematical models of these processes, but also in evaluating the likely effectiveness of the many management methods whose many objectives can include soil conservation. This paper ignores erosion processes other than those due to rainfall impact or flowing water, which dominate in many situations. Seeking an understanding of such processes can involve detailed study, in which controlled-environment experimentation can play an important role. Successful detailed understanding of processes can provide more confidence when making simplifications or approximations in methods or theory which are more suitable for field application. An example of such simplification for field application is given in Rose et al., 1998.

There have been a number of different approaches seeking a more strongly process-based description of soil erosion than that given in the Universal Soil Loss Equation (or USLE), examples being the work of Foster (1982), Govers and Poesen (1988), Nearing et al. (1989), Hairsine and Rose (1991, 1992a, b), and Morgan et al. (1992). The approach illustrated in this paper had its origin in the work of Hairsine and Rose (1991, 1992a, b), which recognises deposition as an ongoing process.

A consequence of the explicit representation of deposition as an ongoing process is that, at any time, some fraction of the soil surface will have upon it sediment previously eroded in the current erosion event which is lodging there, however temporarily. This collection of deposited material will be referred to as the 'deposited layer'. This description is not meant necessarily to imply a continuous blanket of such sediment, though this may be possible, at least temporarily. Certainly the deposited layer must be conceived of as continuously receiving sediment from the water layer, as well as yielding up sediment to the water layer, and is therefore a dynamic process.

It follows from this interpretation that two different kinds of surface are exposed to erosive processes:
(a)   the original soil matrix which, if the soil is cohesive, will have some strength; and also
(b)   sediment deposited during the erosion event, which forms a deposited layer of material covering some fraction of the soil matrix at any time. This deposited material is expected to offer little resistance to removal within the erosion event, which is one reason for distinguishing it from the cohesive soil matrix.

The second reason for distinguishing between the original soil matrix and the layer formed by deposition is that the deposited layer is coarser, and thus the transported sediment richer in fine sediment than would be expected from the size distribution of the original soil. This arises from the much greater settling velocity of larger or denser particles, which therefore have a greater rate of deposition than finer or less dense particles. Though this initially great distortion in size or settling velocity characteristics is found to become less with duration of the erosion event (Proffitt *et al.*, 1991, 1993), it can lead to substantial enrichment of chemicals if these are sorbed preferentially to the finer size fraction, which is not uncommon (Ghadiri and Rose, 1991a, b). There is a need to represent such chemical enrichment for both on-site and off-site water quality reasons (Rose *et al.*, 1990).

Theory introduction

The deterministic approach to describing soil erosion which has been developed at Griffith University will be briefly outlined, and its testing under controlled experimental conditions illustrated. Satisfactory performance of a theory in controlled environmental situations is a necessary, but not sufficient condition for theory's ability to interpret field experimentation adequately. Application of a simplified form of this theory to field experimentation is illustrated by Rose *et al.* (1998).

There are two processes which add to sediment concentration, and one process, namely sedimentation, leading to deposition which withdraws sediment from the water layer on a soil surface.

Deposition

For any size class it follows from the definition of the terms that the rate of deposition of sediment of size class I, d_i, is given by:

$$d_i = v_i c_i \qquad (1)$$

where v_i is the settling velocity and c_i the concentration of sediment of size class i.

Whilst a range of experimental methods can be used to measure or estimate terminal velocities, especially for larger aggregates and for shallow water depths, terminal velocity may not be achieved. We have developed theory which corrects for this if the effect of turbulence on the motion of the larger particles involved can be neglected.

Entrainment and re-entrainment

Even with small cultivated plots of length of order 10 metres, but at significant slopes, these flow-driven processes can be dominant, even without well-developed rills. The rate of working of the mutual shear stress (τ) between water and the soil surface over which it is flowing, is called the stream power (Ω) where:

$$\Omega = \tau V, \tag{2}$$

with V the velocity of the flowing water.

Some threshold stream power Ω_o needs to be exceeded for entrainment to commence, with ($\Omega - \Omega_o$) being called the excess stream power. Some fraction, F, of the excess stream power is effective in erosion, the effective excess stream power thus being $F(\Omega - \Omega_o)$. The remaining fraction, (1 - F), of the excess stream power is dissipated chiefly as heat. The fraction F has been experimentally found to be of order 0.1 - 0.2 (Proffitt *et al.*, 1993).

Denote by H the fraction of the soil surface which is protected from flow-driven erosion by the deposited layer at any time. Since this deposited layer is expected to be mechanically much weaker than a cohesive soil, the maximum sediment concentration would be expected to occur when cohesive soil is completely covered by a deposited layer, so that H = 1. This maximum sediment concentration is identified with the sediment concentration at the transport limit, denoted c_t. Making the assumptions outlined by Rose *et al.* (1998), it has been shown by Equation (2) in that paper that c_t can be related to flow characteristics and other physically measurable parameters. In any particular situation where c_t has a given value, the sediment concentration, c, can fluctuate in time. Such fluctuations in c have c_t as an upper limit and a lower limit which depends on the strength of the soil. If such fluctuations in sediment concentration occur, they are not expected to be describable in detail with deterministic theory. However, the theory of Hairsine and Rose (1992a, b) provides support for dealing with this more complex situation, or with simpler situations, using the approximate theory in program GUEST, yielding a soil erodibility parameter β defined and applied in Rose *et al.* (1998).

The theory of rainfall detachment, and re-detachment of sediment from the deposited layer will now be given in more detail than that of entrainment and re-entrainment, and its testing using data from controlled environmental studies will be illustrated.

Theory of rainfall detachment and re-detachment

As with flow-driven processes, a dynamic interaction between rainfall-driven erosion processes and sediment return to the soil surface in deposition is conceived as resulting, at any time, in some fraction of the soil surface being effectively protected from rainfall erosion by a layer of deposited sediment.

Let m_{dt} = mass/area of this deposited layer, consisting of all sediment size classes present,

and M_d = mass/area of the deposited layer necessary to ensure complete shielding of the soil matrix against erosion by rainfall.

Then the fractional shielding, H, is defined by

$$H(x,t) = m_{dt}/M_d \qquad (3)$$

where x is distance measured in the downslope direction along the surface, and t is time.

Let e denote the rate of detachment of sediment from the soil matrix by the input of raindrops. The magnitude of e has been found to be approximately proportional to the rainfall rate, P (Proffitt et al., 1991). By the definition of H (Equation 3), the fraction of the soil matrix unprotected from rainfall impact is (1 - H). It follows that:

$$e = a (1 - H)P \qquad (4)$$

where a is the soil detachability (Rose, 1985), whose value depends on soil strength and depth of water over the soil surface.

The rate of removal of sediment in the deposited layer by rainfall impact is called the rate of re-detachment, denoted e_d. The rate of re-detachment is proportional both to P and to H so that:

$$e_d = a_d HP \qquad (5)$$

where a_d is the re-detachability of sediment in the deposited layer, whose value depends on soil characteristics and water depth.

Whilst these detachability characteristics, a and a_d, refer to soil as a whole, let us now recognise that soil consists of a wide range of size classes. Let the suffix i denote a typical size class, the total number of such classes being I. As raindrops impact on the soil surface, it is difficult to imagine that the resultant removal of soil from the soil surface can be selective with respect to any particular size class. Even so, raindrops impact the two different types of surfaces described in the Introduction, each consisting of sediment of different size distribution. Taking detachment from the soil matrix to be a non-size-selective process, it follows from Equation (4) that the rate of rainfall detachment of the typical size class i, directed e_i, is given by:

$$e_i = \frac{aP}{I}(1 - H) \tag{6}$$

The size distribution of sediment in the deposited layer is quite different from that of the parent soil because of the very different settling velocity for sediment of different size. An expression for the rate of re-detachment of sediment of any size class i from the deposited layer (denoted e_i) must recognise this strong modification in the size class distribution in that layer, where the mass per unit area of sediment in size class i may be denoted m_{di}. It then follows from the same assumption of non-size selectivity as was made for rainfall detachment, that the rate of re-detachment for size class i, e_{di}, is given by:

$$e_{di} = a_d\, HP\, \frac{m_{di}}{m_{dt}} \tag{7}$$

Since $m_{dt} = \sum_{i=1}^{I} m_{di}$, it follows that the ratio $\frac{m_{di}}{m_{dt}}$ in Equation (7) is the mass fraction of size class i in sediment in the deposited layer.

The theory is continued, and completed, by considering the mass conservation of water and sediment, where sediment is considered in the water flowing over the soil surface, and also in the deposited layer. Let $D(x,t)$ be the depth of water flowing overland. Then mass conservation of water is expressed by the well-known equation:

$$\frac{\partial D}{\partial t} + \frac{\partial q}{\partial x} = R(t) \tag{8}$$

where q is the volumetric water flux per unit width of the planar land surface over which it is assumed the water is flowing, and $R(t)$ is the excess rainfall rate (the difference between rainfall and infiltration rates).

Mass conservation of sediment of size class i on the flowing water requires that:

$$\frac{\partial}{\partial t}(Dc_i) + \frac{\partial}{\partial x}(qc_i) = e_i + e_{di} - d_i \tag{9}$$

where d_i is given by Equation (1), and where i = 1, 2, ... I.

Mass conservation of sediment in the deposited layer recognises that it is augmented by deposition but depleted by re-detachment, so that for sediment of size class i:

$$\frac{\partial}{\partial t}(m_{di}) = d_i - e_{di} \qquad (10)$$

Equations (9) and (10) consist of the number 2I partial differential equations which allow solution of the 2I unknowns given by $c_i(x,t)$ and $m_{di}(x,t)$.

These equations can be solved numerically. Because of the great range of settling velocities of sediment (differing by several orders of magnitude), the partial differential equations (9) and (10) are of the type described as 'stiff'. Such types of equations lead to long computational times with normal upwinding finite-difference numerical techniques, and the 'method of lines' has been found to be far more efficient in their solution.

It should be noted that the Equations (9) and (10) do not explicitly describe the process of structural breakdown under rainfall through time, which can occur. The effect of such structural breakdown can be accommodated by using settling velocity characteristics in the theory which have been determined after the original soil has been exposed to appropriate rainfall.

Comparison of theory with controlled-environment experimentation

The results of numerical solution of the equations presented will be compared with experimental data obtained using the Griffith University Tilting Flume Simulated Rainfall facility described in Misra and Rose (1995). In this facility, soil in the flume bed is subject to rainfall of constant rate and drop size during any experiment, drops falling from a height of some 8 metres. Samples of sediment in overland flow can be taken sequentially at the end of the flume of length 6 metres and width 1 metre.

Though no exact analytical solution is available for the equations described in the theory above, approximations can be made which allow such solutions to be obtained. Justification of such approximations can be obtained both on the grounds of physical plausibility and by comparison of solutions obtained using numerical techniques (described below). An example of an approximate analytical solution has been given by Sander *et al.* (1995), based on the expectation that towards the end of the flume under rainfall-driven erosion, the rate of change of sediment concentration with distance will be much less than its rate of change with time. This physically plausible assumption leads to a simpler set of linear autonomous ordinary differential equations. These equations can be solved using standard methods, the constants of integration being determined by satisfying appropriate boundary conditions.

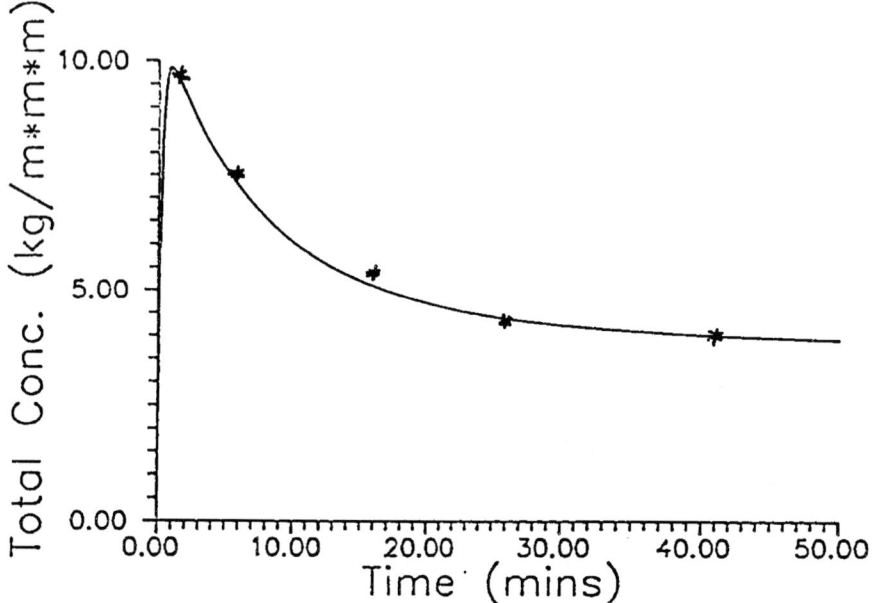

Figure 1. Sediment concentration for a slightly dispersive sandy clay loam (classified as an Aridisol or solanchak) shown as a function of time when measured at the end of the flume. Rainfall rate was 100 mm/h, mean drop size 2.3 mm and depth of water 10 mm. The asterisks are the experimental values from Proffitt *et al.* (1991) and the solid line the approximate analytical solution of Sander *et al.* (1995)

As is illustrated in Figure 1, this approximate analytical solution can be well fitted to the experimental data of Proffitt *et al.* (1991). At long times, this approximate unsteady solution tends to the same equation used to describe the steady state by Hairsine and Rose (1991). Whatever the rainfall characteristics and water depth, the time variation in total sediment concentration measured at the end of the flume has the characteristics illustrated in Figure 2. This figure illustrates the very rapid rise in sediment concentration following the commencement of rainfall, with runoff in these experiments commencing essentially at the same time because of pre-wetting and pre-ponding of the soil bed.

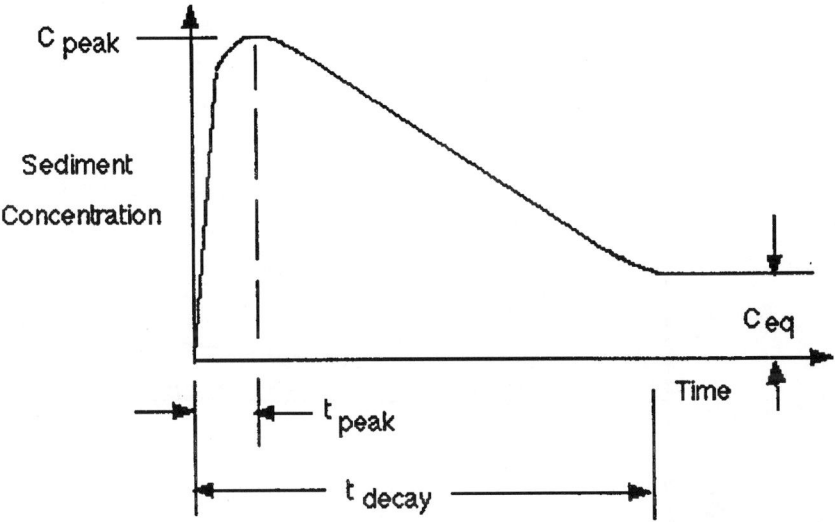

Figure 2. Illustrating salient features of the time variation in sediment concentration measured at the end of a low-slope flume containing pre-wetted soil ponded to some depth. Thus rainfall detachment and re-detachment of the deposited layer formed by deposition are the only erosion processes

Figure 2 shows the rapidly-achieved maximum concentration as c_{peak}, with the steady concentration ultimately achieved in the long term being denoted c_{eq}. Using the symbols given in the theory it has been found using numerical solutions that c_{peak}, is approximately proportional to M_d, and c_{eq} to a_d. The time for the peak in sediment concentration to be achieved (t_{peak}, Figure 2) has been found to depend approximately on a (the detachability), and on M_d. The time for the sediment concentration to fall to values close to c_{eq}, shown as t_{decay} in Figure 2, is dominantly dependent on the hydrologic characteristics of the surface flow.

The results of numerical solution of the equations by Hogarth et al. (1996) are given in Figure 3 at various times (in minutes) from the commencement of rainfall. The solution shows sediment concentration as a function of normalised distance down the flume (X), as well as at various times. This numerical solution, giving results in both space and time, provides support for the assumption of Sander et al. (1995) referred to earlier that the time rate of change of sediment concentration is much greater than its spatial rate of change at the end of the flume (Figure 3). Figure 1 shows the degree of agreement with experimental data at X = 1.

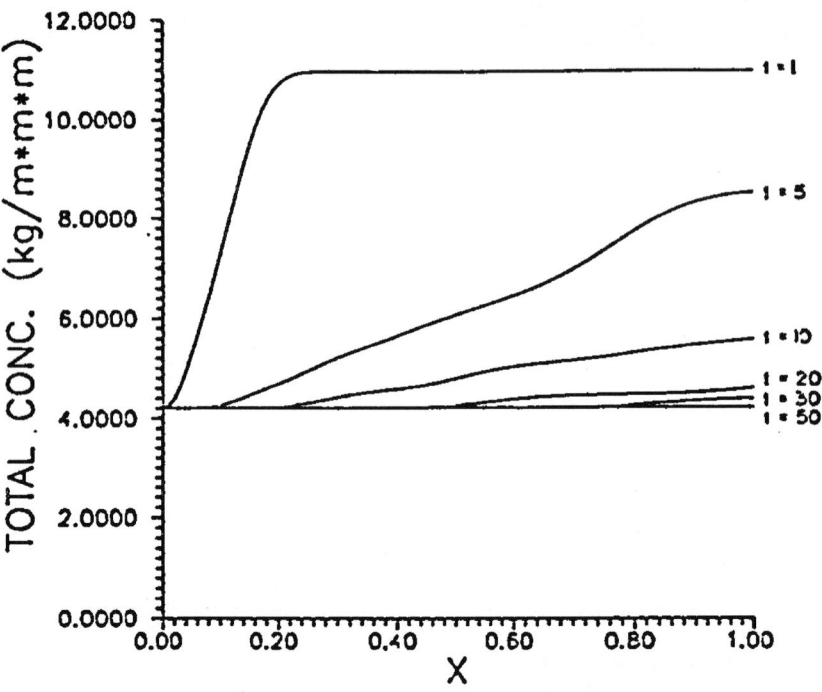

Figure 3. Sediment concentration calculated numerically for the same soil, rainfall and water depth as for Figure 1, showed plotted at various times (t, in minutes) as a function of the non-dimensional ratio X, the fractional ratio of distance down the flume. (X = 1.0 indicates the end of the flume)

Conclusions

The paper has reviewed a particular approach to soil erosion process description, giving more complete attention to the situation where overland flow is limited, so that erosion due to rainfall impact is the dominant process. This approach gives explicit representation of the sedimentation process resulting in deposition of sediment at the soil/water interface. The theory is expressed in partial differential equations which can be solved numerically, or, by making physically plausible assumptions, analytical solutions can be obtained. An example is given, comparing an approximate analytical solution of the theory equations with controlled environmental data. The approximation made which led to the analytical solution is justified, not only by a good fit of theory to the experimental data, but also by results presented for exact numerical analysis.

A major purpose of comparing controlled environmental experimentation with solution of the full equations used to describe the physical processes is to test the veracity of process description, and also to provide greater confidence in using appropriate simplifications of these more fundamental descriptions in field applications.

Acknowledgement

Grant support from the Australian Research Committee for some of the work reported in this paper is gratefully acknowledged.

References

Foster, G.R. (1982). Modelling the erosion process. In, Hann, C.T. (ed.), *Hydrologic Modelling of Small Watersheds*. American Society of Agricultural Engineers Monograph No. 5, St. Joseph, Michigan, USA. pp. 297-399.

Ghadiri, H., and Rose, C.W. (1991a). Sorbed chemical transport in overland flow. I. A nutrient and pesticide enrichment mechanism. *Journal of Environmental Quality* **20**, 628-633.

Ghadiri, H., and Rose, C.W. (1991b). Sorbed chemical transport in overland flow. II. Enrichment ratio variation with erosion processes. *Journal of Environmental Quality* **20**, 634-641.

Govers, G., and Poesen, J. (1988). Assessment of the interrill and rill contributions to total soil loss from an upland field plot. *Geomorphology*. **1**, 343-354.

Hairsine, P.B., and Rose, C.W. (1991). Rainfall detachment and deposition: Sediment transport in the absence of flow-driven processes. *Soil Scence Society of America Journal* **55**, 320-324.

Hairsine, P.B., and Rose, C.W. (1992a). Modelling water erosion due to overland flow using physical principles. I. Sheet flow. *Water Resources Research* **28**, 237-243.

Hairsine, P.B., and Rose, C.W. (1992b). Modelling water erosion due to overland flow using physical principles. II. Rill flow. *Water Resources Research* **28**, 245-250.

Hogarth, W.L., Rose, C.W., Lisle, I., Parlange, J-Y., Sander, G.C., Hairsine, P., and Carey, G. (1996). Water erosion due to rainfall impact. Part II. Numerical solution. *Water Resources Research* (submitted).

Misra, R.K. and Rose, C.W. (1995). An examination of the relationship between erodibility parameters and soil strength. *Australian Journal of Soil Research* **33**, 715-732.

Morgan, R.P.C., Quinton, J.N., and Rickson, R.J. (1992). *EUROSEM: Documentation Manual*. Silsoe College, Silsoe, U.K.

Nearing, M.A., Foster, G.R., Lane, L.J., and Finkner, S.C. (1989). A process based erosion model for USDA water erosion prediction project technology. *Transactions of the American Society of Agricultural Engineers* **32**, 1587-1593.

Proffitt, A.P.B., Hairsine, P.B., and Rose, C.W. (1993). Modelling soil erosion by overland flow: Application over a range of hydraulic conditions. *Transactions of the American Society of Agricultural Engineers* **36**, 1743-1753.

Proffitt, A.P.B., Rose, C.W. and Hairsine, P.B., (1991). Rainfall detachment and deposition: Experiments with low slopes and significant water depths. *Soil Science Society of America Journal* **55**, 325-332.

Rose, C.W. (1985). Developments in soil erosion and deposition modes. *Advances in Soil Science* **2**, 1-63.

Rose, C.W., Dickinson, W.T., Ghadiri, H. and Jorgensen, S.E. (1990). Agricultural non-point source runoff and sediment yield water quality (NPSWQ) models: modeller's perspective. In, *Proceedings of the International Symposium on Water Quality Modelling of Agricultural Non-Point Sources*, Part I, June 19-23, 1988, Utah State University, Logan, Utah, USA pp. 145-169.

Rose, C.W., Coughlan, K.J. and Fentie, B. (1998). Griffith University Erosion System Template (GUEST). In, Boardman, J. and Favis-Mortlock, D.T. (eds), *Modelling Soil Erosion by Water*, Springer-Verlag NATO-ASI Global Change Series, Heidelberg.

Sander, G.C., Hairsine, P.B., Rose, C.W., Cassidy, P., Parlange, J.Y., Hogarth, W.L., and Lisle, I.G. (1995). Unsteady soil erosion model, analytical solutions and comparison with experimental results. *Journal of Hydrology* (in press).

21. SENSITIVITY OF SEDIMENT-TRANSPORT EQUATIONS TO ERRORS IN HYDRAULIC MODELS OF OVERLAND FLOW

John Wainwright[1] and Anthony J. Parsons[2]

[1]Department of Geography
King's College London
Strand
London, WC2R 2LS
UK

[2]Department of Geography
University of Leicester
University Rd
Leicester, LE1 7RH
UK

Abstract

The accuracy of process-based soil-erosion models is restricted not only in terms of the type of equation used, but also in their dependence on the input from models of the hydraulics of overland flow. Ten commonly used equations are applied to model sediment transport to determine their sensitivities to errors in overland flow hydraulics. The most suitable type of sediment-transport equation for a particular modelling setting will depend not only on its ability to fit the data, but also on the model structure and the types of feedback incorporated.

Introduction

Many process-based soil-erosion models use a coupled structure, in which the calculation of sediment-transport rate is dependent on the prior estimation of the hydraulics of overland flow. Therefore the ability of this type of model to predict erosion is ultimately a function of the efficacy of the hydraulic predictions, and any errors in the prediction of the hydraulics will propagate through the estimation process. The ultimate error in predicting soil erosion will thus be a compound of the separate hydraulic and sediment-transport model components. Given that overland-flow models often produce poor results, or provide good results for the wrong reasons (see Grayson *et al.*, 1992a and b; Scoging *et al.*, 1992) this is a significant problem in the prediction of soil-erosion rates.

The literature on sediment-transport rates in overland flow is not as well developed as that for fluvial contexts. However, recent reviews suggest that a number of different equations may be used. These equations depend on a number of hydraulic variables — flow discharge, velocity, shear stress, unit stream power and effective stream power — used either individually or in combination. Given that the hydraulics variables are often raised to a power greater than unity in the sediment-transport equations, the estimation of soil erosion is often highly sensitive to errors in the prediction of hydraulics. The analysis that follows investigates the extent to which this is the case. The question addressed here is how this type of coupled model can be made robust to the errors in its individual components. This is achieved by investigating the different sediment-transport-rate models available and seeing how they would integrate with a particular example of an overland-flow hydraulics model. The use of sensitivity analysis is demonstrated as a means of developing criteria for the robustness of coupled hydraulics-soil erosion models.

Selection of sediment-transport equations

Recent studies suggest that there are a number of equations that could be used to predict the transporting capacity of interrill overland flow. Julien and Simons (1985) demonstrated that the basic equation for interrill sediment-transport rate (q_s in g m^{-1} min^{-1}) is:

$$q_s = \alpha\, q^y\, S^z\, i^c \tag{1a}$$

or, in certain cases:

$$q_s = \alpha\, q^y\, S^z. \tag{1b}$$

where q is overland-flow discharge (cm³ s^{-1}), S is slope (m m^{-1}), i is rainfall intensity (mm h^{-1}), α is a coefficient reflecting soil erodibility and y, z and c are empirical coefficients.

Julien and Simons proceeded to show the equivalence of various other commonly used sediment-transport formulæ to this equation. Their results suggested that the equations of Engelund-Hansen and of Barekyan could be used for sediment transport in all overland flows. The relationships of Shields, Kalinske-Brown and Yalin were also suitable, although the equivalent value of the y exponent was too low in laminar flows. Everaert (1991) demonstrated using laboratory experiments that Bagnold's effective stream power (Ω [g cm$^{-2/3}$ s$^{-4.5}$], defined as $[\rho g q S]^{1.5}/d^{2/3}$) and Yang's unit stream power measurements also provided expressions that were able to predict transport rates well. Govers (1992) compared a wide range of laboratory studies and provided two further relationships that were applicable to interrill conditions. These are the model of Low and secondly a version of Bagnold's approach, using excess effective stream power: The forms of all of these equations as used in the present study are given below. The Greek symbols β - κ have been used in these equations to represent a soil-erodibility coefficient, equivalent to α in equation 1.

Engelund-Hansen	$q_s = \beta \tau^{1.5} \bar{u}^2$	(2)
Barekyan	$q_s = \gamma S q \bar{u}$	(3)
Shields	$q_s = \delta S q (\tau - \tau_c)$	(4)
Kalinske-Brown	$q_s = \varepsilon \tau^{2.5}$	(5)
Yalin	$q_s = \zeta \tau^{0.5} (\tau - \tau_c)^2$.	(6)
Bagnold (1)	$q_s = \eta \Omega^h D_{50}^j$	(7)
Yang	$q_s = \vartheta (S\bar{u})^k D_{50}^l$,	(8)
Low	$q_s = 16.42 ([\rho_s / \rho] - 1)^{-0.5} (Y - Y_c) D S^{0.6} \bar{u} \rho_s$	(9)
Bagnold (2)	$q_s = \kappa (\Omega - \Omega_c)^m$.	(10)

where τ is bed shear stress (g cm^{-1} s^{-2}), $\bar{u}$ mean flow velocity (cm s^{-1}), τ_c critical bed shear stress required for onset of sediment transport (g cm^{-1} s^{-2}), ρ unit density of water (g cm^{-3}), g gravitational acceleration (cm s^{-2}), d flow depth (cm), D_{50} median grain size (μm), ρ_s unit density of the sediment (g cm^{-3}), Y dimensionless shear stress ($\tau/[\{\rho_s - \rho\} g S]$), Y_c the dimensionless shear stress required for entrainment and D is grain size (μm).

These ten equations are tested here in terms of their applicability to field data, and their sensitivity to inclusion in a coupled runoff-erosion model. Finally, the implications of using these equations to estimate the measured erosion are assessed.

Field experiments

The experimental data used to investigate the ten sediment-transport equations described above were obtained from a site within the Walnut Gulch Experimental Watershed, Tombstone, Arizona (31° 43'N, 110° 41'W). The site consists of an experimental plot which is 29 m long by 18 m wide and is located on a grassland hillslope (Parsons *et al.*, 1996a).

Two rainfall simulation experiments were carried out on this plot (see Parsons *et al.*, 1996a). On the plot, three cross sections G1, G2 and G3, located 6 m, 12 m and 20.5 m from the upslope boundary respectively, were established for the determination of the within-plot flow hydraulics and sediment-transport rates. The topography of the plot was surveyed on a 0.50 × 0.50 m grid. For each cell in the grid, the surface characteristics were determined. The following analysis considers data from the first experiment only, because detailed hydraulic modelling of this experiment has been carried out (see Parsons *et al.*, 1996b).

Table 1. χ^2 statistics and parameter estimates for the sediment transport equations 1-10 presented in Figure 1. Critical level of χ^2 statistic for p=0.05 is 1.24

cross-section	χ^2 statistics equation									
	1	2	3	4	5	6	7	8	9	10
G1	0.45	0.72	0.11	0.31	0.41	0.53	0.03	1.67	0.42	0.04
G2	2.68	0.22	0.25	0.43	1.20	1.02	0.66	3.87	2.05	0.89
G3	1.03	0.30	0.03	0.93	1.41	1.58	0.04	5.19	0.55	0.05
parameter values	α	β	γ	δ	ε	ζ	η	θ	ι	κ
G1	1.28×10^{-4}	4.02×10^{-2}	7.50×10^{-3}	8.35×10^{-3}	1.26×10^{-1}	1.51×10^{-1}	3.17×10^{-3}	4.22×10^{-1}	1.20×10^{2}	3.58×10^{-2}
G2	1.53×10^{-4}	7.73×10^{-2}	1.68×10^{-2}	9.56×10^{-3}	1.05×10^{-1}	1.18×10^{-1}	7.25×10^{-3}	1.45×10^{0}	2.18×10^{2}	8.06×10^{-2}
G3	1.24×10^{-5}	5.16×10^{-3}	1.37×10^{-3}	1.20×10^{-3}	2.21×10^{-2}	2.38×10^{-2}	6.32×10^{-4}	7.13×10^{-2}	2.88×10^{1}	6.82×10^{-3}

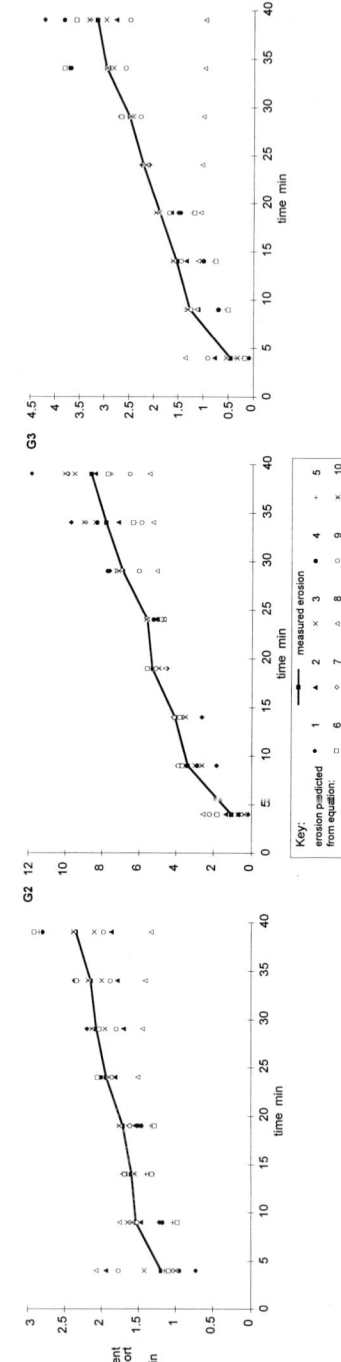

Figure 1. Predicted sediment-transport rates for G1, G2 and G3 using equations 1-10, compared with measured sediment transport rates

Testing of sediment-transport equations and their application to field data

The first step in testing the suitability of the sediment-transport equations 1b to 10 for the field data is to estimate the values of the coefficients α - κ for each of the cross-sections G1, G2 and G3. These estimates have been carried out by cross-section because of the variability of the surface characteristics at each location. However, with only three cross-sections, it is not possible to generalise further by relating the coefficients to the ground surface characteristics. The estimation procedure was carried out by minimising a χ^2 statistic of measured versus predicted sediment-transport rates. The hydraulic variables in equations 1b to 10 are discharge, mean flow velocity, shear stress and effective stream power. At the cross-sections, flow width and depth are measured directly, discharge is obtained using a depth-discharge rating equation (Parsons *et al.*, 1990) and flow velocity is estimated using the continuity equation. Shear stress is estimated (as $\tau = \rho\,g\,d\,S$) and effective stream power is estimated using the relationship given previously. The slope is defined using the topographic survey. The value of D_{50} used in equations 7 to 9 is 176 μm.

Figure 1 shows the predicted sediment-transport rates for the three cross sections (parameter estimates and χ^2 statistics in Table 1). Equation 3 provides the most consistently accurate predictions, followed by equations 7 and 10, the latter two both being dependent on the effective stream power. Equation 8 provides very poor fits, and in the cases of G1 and G3, predicts a declining sediment transport rate through time. With the exception of equations 1 and 9 for cross-section G2, and equation 8 for all three cross-sections, all cases provide significant fits. There is no consistency between cross-sections in terms of the best fit, with equation 7 providing the best fit for G1, equation 2 for G2 and equation 3 for G3. Given that most of the equations presented can produce significant fits to the field data, the question that needs to be addressed is which equation is most suitable for inclusion in a soil-erosion model.

Modelling overland-flow hydraulics and its implications for soil erosion modelling

The criteria that may be used to assess the suitability of the different sediment-transport equations are developed by applying the equations to a model which is capable of predicting flow hydraulics in the environment studied. The model used is the distributed model developed by Scoging (1992) which was capable of predicting the spatial pattern of flow hydraulics on a shrubland experimental plot at Walnut Gulch (Scoging *et al.*, 1992). Parsons *et al.* (1996b) describe the application of the model to the grassland plot, and the modifications required to the routing algorithm so that the pattern of flow hydraulics on this plot can also be simulated.

The basic structure of the model consists of three parts. First, overland flow is generated by a Hortonian excess method, i.e. the difference between the rainfall and the infiltration rate. The latter is predicted using the simplified Green and Ampt equation. Secondly, the kinematic wave simplification is used to distribute the flow using the continuity equation:

$$\frac{\partial q'}{\partial x} + \frac{\partial d}{\partial t} = e_x \qquad (13)$$

where q' is the overland flow discharge per unit width (cm² s⁻¹), x is distance (cm), d is the flow depth (cm) and e_x is the rainfall excess (cm s⁻¹); together with a rating equation:

$$q' = \bar{u}\,d. \qquad (14)$$

The mean flow velocity is calculated using the Darcy-Weisbach friction factor (f):

$$\bar{u} = [(8\,g\,d\,S)\,/\,f]^{0.5}. \qquad (15)$$

Thirdly, flow is routed to one of the four adjacent cells in a finite difference grid using a topographically based algorithm. This algorithm is a function of the greatest difference in altitude of the cells and the microtopography. The model is made spatially distributed by defining the infiltration parameters and friction factor as functions of the measured surface conditions (see Scoging et al., 1992; Parsons et al., 1994; Abrahams et al., 1995; Parsons et al., 1996a, b for further details).

Theoretical Sensitivity to Model Structure

Given the model structure, which has been chosen because it is both simple and capable of satisfactorily reproducing the within-plot hydraulics on the experimental plot, it is possible to define the theoretical sensitivity of the different sediment-transport equations to errors in model hydraulics. Because of the calculation procedure (equations 12 to 15), all errors depend on the calculation of flow depth d, and ultimately on the prediction of infiltration rates, which is notoriously difficult. From equation 15, it can be seen that errors in velocity (Δu) are attenuated in relation to the error in depth (Δd). Errors in shear stress ($\Delta\tau$) will be of the same order of magnitude as errors in depth. In the case of discharge ($\Delta q'$), equation 14 causes the error to be magnified in relation to the error in depth estimation. Finally, the error on effective stream power ($\Delta\Omega$) will be increased further, because of the power term on q. These give:

$$\Delta u \propto \Delta d^{0.5} \tag{16a}$$

$$\Delta \tau \propto \Delta d \tag{17}$$

$$\Delta q' \propto \Delta d^{1.5} \tag{18a}$$

$$\Delta\Omega \propto \Delta d^{1.58}. \tag{19a}$$

If the Chézy coefficient is used to estimate velocity, rather than equation 15, then the exponents in the equations 16a, 18a and 19a remain unchanged. Alternatively, if Manning's n is used, the exponents will increase, with values of 0.67 in equation 16a, 1.67 in equation 18a and 1.83 in equation 19a. Thus, the Darcy-Weisbach or Chézy friction factor is preferable in this type of physically based erosion model, as it will minimise the propagation of errors.

Table 2. Values of the exponent ξ for equation 20: a. assuming f is independent of flow conditions; b. assuming f is a function of Re

equation:	1	2	3	4	5	6	7	8	9	10
a. ξ	3.00	2.50	2.00	2.50	2.50	2.50	1.69	0.76	1.50	1.71
b. ξ	0.88	1.50	0.88	1.94	2.50	2.50	0.80	-0.09	0.94	0.84

Note: this table assumes the following values for coefficients in equations 1 to 10: $y = 2$; $z = 1.66$; $h = 1.07$; $j = 0.47$; $k = 1.52$; $l = 0.56$; and $m = 1.08$.

Using equations 16a to 19a, it is possible to define the compound error on estimating sediment transport (Δq_s) for each of the equations 1 to 10, simply as a function of the error in estimation of the flow depth. In general terms, this can be stated as:

$$\Delta q_s \propto \Delta d^{\xi}, \tag{20}$$

where the calculated exponent ξ for each equation is given in Table 2a. The values indicate that equation 1, the basic equation defined by Julien and Simons (1985), is the most sensitive to errors. Equation 8 is the least sensitive, with a smaller error on erosion relative to that on depth. However, it has been noted above that this equation is particularly poor at representing the experimental sediment-transport rates. The effects of these differing sensitivities are illustrated in Figure 2. Assuming the worst case of equation 1, an error in depth prediction of 50% of the real value will lead to the prediction of sediment transport which is 13% of the measured value. Conversely, an overprediction of 200% of the real depth value would lead to an overprediction of 800% of the measured sediment-transport rate. Clearly, these sensitivities

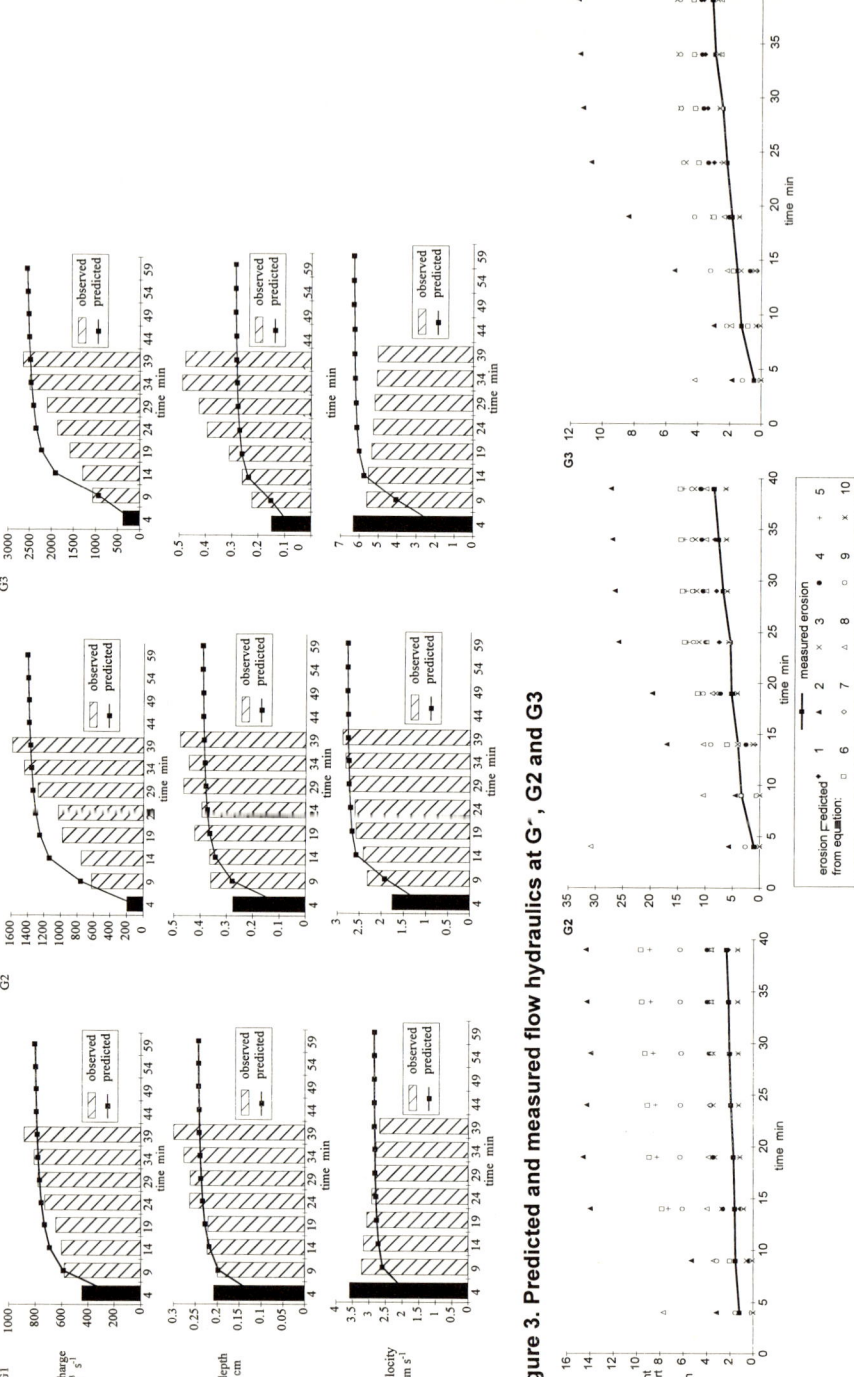

Figure 3. Predicted and measured flow hydraulics at G1, G2 and G3

Figure 4. Predicted and measured sediment-transport rates for selected sediment-transport equations

However, whereas discharge is usually well predicted, especially at or near equilibrium, this is often a function of a cancelling of errors in depth and velocity. Depths tend to be under- and velocities over-predicted. Furthermore, for G1 and G3, the model is unable to reproduce the decreasing trends in measured velocities through time. All of these factors have important implications for predicting sediment-transport rates. Whereas the measured data produce significant fits for most equations at all three cross-sections, the modelled data produce no fits which are significant at the 95% level using the χ^2 statistic (Table 3). Equation 1, which depends on discharge alone, tends to produce the best fits because of the better discharge predictions at the cross sections (Figure 4). Equation 10 also produces reasonable estimates for the same reason and also probably because the calculation of effective stream power causes some cancellation of the errors on discharge and depth measurements. Most other equations produce substantial over-estimates of the sediment-transport rate, which is more extreme in the cases where velocity is the dominant hydraulic variable, as is the case in equations 2 and 3. Finally, equation 8, based on Yang's unit stream power produces decreasing sediment transport rates at all three cross-sections, and thus again proves to be highly unsatisfactory.

These results indicate that even where hydraulics predictions at the intra-plot scale can be thought of as adequate, the resulting estimates of sediment transport can be totally unsuccessful. This implies that very high precision in making hydraulics estimates is required for the use of coupled hydrology-erosion models.

Table 3. χ^2 **statistics for goodness of fit for sediment transport equations 1-10. Critical value of χ^2 statistic for p=0.05 is 1.24**

cross section	χ^2 statistics by equation									
	1	2	3	4	5	6	7	8	9	10
G1	2.76	482.83	8.04	10.36	128.23	157.04	4.16	49.59	59.63	4.36
G2	7.88	280.09	17.34	12.41	32.21	37.49	6.73	882.46	29.20	6.91
G3	3.63	127.54	9.73	3.28	4.24	4.41	2.52	31.94	14.55	2.57

Detachment-limited sediment transport

Only the case where sediment transport is transport limited is discussed above. Frequently, the more complex case will arise where sediment transport is detachment limited, as for example on the shrubland at Walnut Gulch (Parsons and Abrahams, 1992). In this situation, a similar analysis could be undertaken to assess the sensitivity of a detachment-limited sediment-transport model. However, the physical basis of this is much more poorly understood, and empirical work is much more scarce. Taking as an example the experimental work of Torri *et al.* (1987), there is a negative exponential relationship between sediment detachment by raindrops and depth of overland flow. In this case the error in predicted sediment transport will be proportional to $e^{-\Delta d}$, and the effects of under- and overprediction of depth will be highly non-linear. In more realistic simulations, the sensitivity of sediment-transport rates will be a more complex function of the flow conditions occurring through space and time.

Conclusions

The results presented above demonstrate that different sediment-transport equations have different sensitivities to errors in hydraulic predictions. These sensitivities vary in terms of the way the hydraulics model incorporates resistance to flow, and in the way that resistance to flow is a function of flow conditions and thus to surface conditions. One conclusion that can be drawn from this exercise is that the use of the Darcy-Weisbach friction factor or Chézy C to describe hillslope resistance to flow will produce less sensitive results than using Manning's n.

Inasmuch as the results of Everaert (1991), Govers (1992) and Wainwright (1996) indicate that the different sediment-transport equations can produce reasonable estimates of sediment transport, the use of the type of sensitivity analyses presented above can be used to build more robust soil-erosion models. The results presented above suggest that the model of Low (equation 8) is least sensitive where there is no feedback in the model between flow conditions and resistance to flow. In the example presented above where this is not the case, which is developed specifically for the grassland at Walnut Gulch, the two equations that are dependent on Bagnold's effective stream power (equations 7 and 10) are the least sensitive, and would therefore be the best to use in predicting sediment-transport rates at this location. The cases presented above are simplified, and do not take into account the relationship between detachment-limited and transport-limited conditions. In a spatially distributed model, this relationship will be complex, and thus the type of error that occurs in erosion predictions will also be spatially distributed, and this distribution will vary through time.

The structure of the hydraulics model presented above means that the estimation of flow depth is the critical variable. At any location, this will depend on the prediction of infiltration rates, resistance to flow and flow routing. Despite the fact that a great deal of effort has gone into the prediction of infiltration rates, this still remains an area which requires more study. Resistance to flow on hillslopes has only been studied in detail relatively recently, and flow routing is very poorly dealt with in the literature. These deficiencies must be addressed rapidly so that reliable predictions of soil erosion at local, regional and global scales can be made.

Acknowledgements

The field experiments at Walnut Gulch were undertaken with the assistance and co-operation of the USDA-ARS Southwest Watershed Research Unit, Tucson. We thank Leonard Lane for permission to use the facilities at the Tombstone field station and Roger Simanton and the technicians at Tombstone for their generous advice and assistance. We would also like to thank the many friends and colleagues who helped with the experiments. In particular we acknowledge the collaboration with Athol Abrahams. This research was funded by a grant from the Natural Environment Research Council (GR 3/7999).

References

Abrahams, A.D., Parsons, A.J. and Wainwright, J. (1994). Resistance to overland flow on semiarid grassland and shrubland hillslopes, Walnut Gulch, Southern Arizona. *Journal of Hydrology* **156,** 431-446.

Abrahams, A.D., Parsons, A.J. and Wainwright, J. (1995). Effects of vegetation change on interrill runoff and erosion, southern Arizona. *Geomorphology* **13**, 37-48.

Everaert, W. (1991). Empirical relations for the sediment transport capacity of interrill flow. *Earth Surface Processes and Landforms* **16**, 513-32.

Govers, G. (1992). Evaluation of transporting capacity formulae for overland flow. In, Parsons, A.J. and Abrahams, A.D. (eds). *Overland Flow. Hydraulics and Erosion Mechanics*, UCL Press, London. pp. 243-273.

Grayson, R.B., I.D. Moore and T.A. McMahon (1992a). Physically based hydrologic modelling. 1. A terrain-based model for investigative purposes. *Water Resources Research* **28**, 2639-2658.

Grayson, R.B., I.D. Moore and T.A. McMahon (1992b). Physically based hydrologic modelling. 2. Is the concept realistic? *Water Resources Research* **28**, 2659-2666.

Julien, P.Y. and Simons, D.B. (1985). Sediment transport capacity of overland flow. *Transactions of the American Society of Agricultural Engineers* **28**, 755-776.

Parsons, A.J. and Abrahams, A.D. (1992). Field investigations of sediment removal in interrill overland flow. In, Parsons, A.J. and Abrahams, A.D. (eds). *Overland Flow. Hydraulics and Erosion Mechanics*, UCL Press, London. pp. 307-334.

Parsons, A.J., Abrahams, A.D. and Luk, S.-H. (1990). Hydraulics of interrill overland flow on a semi-arid hillslope, southern Arizona. *Journal of Hydrology* **117**, 255-273.

Parsons, A.J., Abrahams, A.D. and Simanton, J.R. (1992). Microtopography and soil-surface materials on semi-arid piedmont hillslopes, southern Arizona. *Journal of Arid Environments* **22**, 107-115.

Parsons, A.J., Abrahams, A.D. and Wainwright, J. (1994). On determining resistance to overland flow. *Water Resources Research* **30**, 3515-3521.

Parsons, A.J., Abrahams, A.D. and Wainwright, J. (1996a). Responses of interrill runoff and erosion rates to vegetation change in Southern Arizona. *Geomorphology* **14**, 311-317.

Parsons, A.J., Wainwright, J., Abrahams, A.D. and Simanton, J.R. (1996b). Distributed dynamic modelling of interrill overland flow. *Hydrological Processes*, in press.

Scoging, H.M. (1992). Modelling overland-flow hydrology for dynamic hydraulics. In, Parsons, A.J. and Abrahams, A.D. (eds). *Overland Flow. Hydraulics and Erosion Mechanics*, UCL Press, London. pp. 89-104.

Scoging, H.M., Parsons, A.J. and Abrahams, A.D. (1992). Application of a dynamic overland-flow hydraulic model to a semi-arid hillslope, Walnut Gulch, Arizona. In, Parsons, A.J. and Abrahams, A.D. (eds). *Overland Flow. Hydraulics and Erosion Mechanics*, UCL Press, London. pp. 105-145.

Torri, D., Sfalanga, M. and Del Sette, M. (1987). Splash detachment: runoff depth and soil cohesion. *Catena* **14**, 149-155.

Turner, A.K., Langford, K.J., Win, M. and Clift, T.R. (1978). Discharge-depth equation for shallow flow. *Journal of the Irrigation Drainage Division, Proceedings of the American Society of Civil Engineers* **104**, 95-110.

Wainwright, J. (1996). Infiltration, runoff and erosion characteristics of agricultural land in extreme storm events, SE France. *Catena*, in press.

22. GULLY EROSION: IMPORTANCE AND MODEL IMPLICATIONS

Jean Poesen[1,2], Karel Vandaele[1] and Bas van Wesemael[1]

[1] Laboratory for Experimental Geomorphology
K.U.Leuven
Redingenstraat 16
3000 Leuven
Belgium

[2] National Fund for Scientific Research

Abstract

This paper deals with gully erosion in agricultural environments and concentrates in particular on types of gullies, importance of gully erosion with respect to sediment production in agricultural environments and modelling needs. Two major gully types can be often observed in agricultural lands: ephemeral gullies and bank gullies. Field measurements combined with aerial photo analysis in three contrasting European environments indicate that ephemeral gully erosion rates represent 44 per cent of total sediment production in central Belgium and up to 80 - 83 per cent of total sediment production in Mediterranean conditions (Portugal and Spain). In contrast, bank gully erosion rates are about one order of magnitude less than ephemeral gully erosion rates in central Belgium. Nevertheless, there is a need to model these erosion features, which are complementary to interrill and rill erosion. More particularly, there is a need for more detailed monitoring, experimenting and modelling of the development and infilling of both ephemeral gullies and bank gullies in a variety of agricultural environments. In other words, there is a need to better predict the location, the total length and the cross-section of gullies. The threshold concept could be a useful tool to help locate ephemeral gullies in the landscape. However, threshold conditions for incipient gullying in a variety of climatological, topographic, pedological and land-use conditions first need to be established. Existing erosion models need to be refined to incorporate the effects of the resistance of various soil horizons to concentrated flow erosion and the effects of other soil detaching mechanisms in gullies such as soil fall, slumping and headcutting. Improved gully models are needed to predict more accurately the effects of environmental change on the intensity of this soil degradation process.

Introduction

When measuring and modelling soil losses due to water erosion in cultivated land and rangeland, most attention has hitherto been given to interrill (sheet) and rill erosion. Gully erosion has received much less attention. Many field observations in a variety of agricultural environments (both temperate humid and Mediterranean environments) have led a number of researchers to conclude that, besides interrill and rill erosion, gully erosion can also be an important sediment source. Therefore, there is a need to predict A) where and when gullies form in the landscape and how their position, frequency of occurrence and erosion intensity is affected by climatic or landuse changes, and B) gully cross-sectional size and shape and soil loss rates due to gully erosion. Therefore, this paper focuses on gully erosion in agricultural environments and concentrates in particular on the following aspects: gully types, importance of gully erosion with respect to sediment production in agricultural environments and gully modelling needs.

Gully types

A gully has been defined as a steep-sided channel, often with a steeply sloping and actively eroding head scarp, caused by erosion due to the intermittent flow of water, usually during and immediately following heavy rains. However, in some cases ephemeral gullies have been observed to have formed by concentrated saturation overland flow or even by snowmelt. These channels are deep enough to interfere with, and not to be obliterated by, normal tillage operations (Bradford and Piest, 1980; Soil Science Society of America, 1984). Because 'normal' tillage operations vary both in space and time, there are no widely agreed dimensions for distinguishing gullies from rills. In this study we use the more rigorous definition proposed by Hauge (1977): i.e. gullies are distinguished from rills by a critical channel cross-sectional area of one square foot (0.10 m^2). This threshold is also perceived by farmers as a critical channel size above which the channels start to interfere with the trafficability of the land (Souchère, 1995).

Based on their location in the landscape, on their morphology as well as on the dominant erosion process leading to their formation, two main gully types have been recognised in European agricultural lands (Poesen, 1989; 1993; 1995; Poesen and Govers, 1990): (ephemeral) gullies and bank gullies.

Ephemeral gullies
Ephemeral gullies form where overland flow concentrates, i.e. either in natural drainage-lines (talwegs of zero order basins or hollows) or along (or in) linear landscape elements such as, for instance, drill lines, dead furrows on parcel borders or at the limit of headlands, tractor ruts or unpaved access roads (Poesen, 1989; 1993; Poesen and Govers, 1990). These features are continuous, temporary channels which are usually erased by tillage operations. By applying the definition of Hauge (1977), the boundary between a rill and an ephemeral gully becomes clear-cut. Ephemeral gullies have been first reported in the USA (e.g. Foster and Lane, 1983;

Thorne et al., 1984) and have later also been described by various authors in Europe (e.g. Boardman (1990) in the UK; Boiffin et al. (1988) and Auzet et al. (1990) in France; Poesen (1989) in Belgium; Baade et al. (1993) in Germany, Vandaele et al. (in press a) in Portugal) and in Australia (e.g. Moore et al., 1988).

Ephemeral gullies seem to result essentially from hydraulic erosion by concentrated overland flow. This implies that sediment detachment and removal is a function of flow intensity (Foster and Lane, 1983; Thorne et al., 1986).

Ephemeral gullies can be further classified on the basis of their topographic location as valley-side, valley-head and valley-bottom gullies (Brice, 1966). However, practical considerations suggest to subdivide ephemeral gullies according to their width-depth ratio (w/d: Poesen and Govers, 1990; Poesen, 1993). Wide ephemeral gullies with a w/d $\gg$ 1 cause important crop damage. In addition, a high percentage of total soil lost through this gully type consists of fertile topsoil with a high organic matter and fertiliser content. These gullies are, however, easily erased by conventional tillage. On the other hand, narrow and deeper ephemeral gullies with a w/d = 1 or even < 1 cause relatively little crop damage and the percentage of total soil loss consisting of fertile topsoil is limited. The narrow gullies, however, are not easily erased by conventional tillage and often heavy equipment is required to reshape the areas where they form. The infilling of these gullies (e.g. by tillage) often creates topographic depressions in which new gullies will develop subsequently.

Bank gullies

Bank gullies form where a wash-line, a rill, a dead furrow or an ephemeral gully crosses an earth bank (e.g. a terrace bank, a lynchet, an exploitation shoulder or a sunken lane bank). These features are discontinuous, permanent channels which usually cannot be obliterated by conventional tillage operations. This gully type was first described in northern Europe (Poesen, 1989; Poesen and Govers, 1990; Farres et al., 1993) and later also in Mediterranean environments (Poesen, 1995).

Bank gully erosion seems to be less controlled by overland flow intensity but more by factors controlling piping and mass movement (i.e. slumping and soil fall). The latter processes are less controlled by the catchment size but depend more on the local site characteristics, such as the presence and density of biopores or cracks and the type of soil material. This implies that prediction of the exact location and the volume eroded by bank gullies is more difficult than that of ephemeral gullies (Poesen, 1989; 1993).

Importance of gully erosion with respect to sediment production in agricultural environments

Ephemeral gullies

In intensively cultivated land, ephemeral gullies occur more frequently than previously thought. One of the problems encountered when studying these erosion phenomena is that these gullies are often rapidly filled in by tillage soon after they have formed. Hence, there is a need for a quick and detailed field monitoring of this erosion process.

In order to assess the contribution of gully erosion to sediment production, soil losses due to interrill, rill and (ephemeral) gully erosion were assessed in three contrasting European environments with different climates: i.e. a temperate humid environment (central Belgium), a Mediterranean environment (south-east Portugal) and an arid environment (south-east Spain). Before discussing the methodology and the results, the studied environments are briefly described. More information regarding the Belgian study sites can be found in Vandaele and Poesen (1995) and in Vandaele *et al.* (in press a).

Description of study sites
Central Belgium

The study area in central Belgium is located between Leuven and Brussels and is part of the north European loess belt. Depth of the loess cover ranges between a few cm to 10 m. The loess sheet covers tertiary sandy deposits. Topsoils have a very high silt content (70-80 per cent) and a moderate clay content (10-20 per cent). The study area is characterised by a dense network of dry valleys. The land in this region has been under cultivation for at least 1000 years and is presently used for agricultural crops, the most important being winter wheat and barley (autumn sown), sugar beet, potatoes, maize and chicory (spring sown). About 4 per cent of the study area has slopes exceeding 10 per cent. Mean annual precipitation ranges between 700 and 850 mm (Brussels and Leuven) and is relatively well distributed over the year

South-east Portugal

The study area in Portugal is located in the Alentejo, 5 km east of Mertola (Vandaele *et al.*, in press a). The typical red schist soils are very shallow (depth ranges between several centimetres and several decimetres: lithosols) due to intense erosion by water as well as by tillage erosion which has occurred since the 1930s (Poesen and Lavee, 1994). Rock fragment content by mass equals about 30 per cent. The area is characterised by a network of dry valleys while most valley bottoms are incised by an intermittent stream network. Around the, 1930s, the matorral (brush and oak vegetation) was cleared on a large scale for autumn-sown winter wheat and barley production (Tomas and Coutinho, 1994). The traditional and most widespread crop rotation is wheat-fallow (De Lima, 1989). Slopes in this study area are gentle with only 8 per cent of the area having slopes exceeding 10 per cent. The climate is typical Mediterranean with a mean annual precipitation of 560 mm (Beja) and a maximum rainfall between October and March. This means that the fields are unprotected during the period with the highest rainfall amounts.

South-east Spain

The study area in south-east Spain is located 25 km east of the town of Almeria at the footslopes of the Sierra de Gata. The typical piedmonts are covered with a coarse weathering mantle on andesitic rock and have, in their basal part, extensive alluvial fan systems of late Pleistocene age (Harvey, 1987). Soil depth ranges between 25 cm in the upper parts to more than 1 m in the lower parts (Poesen *et al.* in prep. a). There is an abundance of angular rock fragments in the upper parts (20 to 50 per cent rock fragment cover). The soils have a sandy loam texture and an organic rich topsoil (mollic epipedon). Before 1983, the study area was cultivated for rainfed cereal production. Since 1983, this land became abandoned and was used as rangeland (regular grazing by sheep and goats). It is sparsely vegetated with annuals and bushes (e.g. *Thymelaea hirsuta*). Since land abandonment, a dense rill and gully network developed which was mapped in 1993. Slopes range between 3 per cent in the lower parts and 20-25 per cent in the upper parts. Thirty per cent of the study area has slopes exceeding 10 per cent. Mean annual precipitation is 180 mm and rains are concentrated in winter and early spring.

Methodology

Different techniques were used to assess soil losses by gully erosion as well as by rill and interrill erosion in the three study sites.

For the study site in central Belgium two techniques were used.

1) Detailed seasonal mapping at a scale of 1: 5000 of all erosion features observed in an intensively cultivated 25 ha catchment (Hammeveld-1 catchment: Vandaele and Poesen, 1995) which occurred in a three year monitoring period (Oct. 1989 - Oct. 1992). At the same time, measurements of the cross-section of erosion channels (rills and gullies) with a minimum depth of one cm and a minimum length of 10 m as well as of their total length were made. Next, eroded volumes were calculated. It should be kept in mind that the calculated soil volumes eroded by ephemeral gullying represent minimum values, since most ephemeral gullies experience repeated cycles of cut and fill between tillage events so that the actual soil loss due to concentrated flow may be several times the volume indicated by the periodic measurement of channel cross-sectional area (Thorne *et al.*, 1986).

2) Detailed measurements of the length of ephemeral gullies in a study area ranging between 272 and 1074 ha using aerial photos at a scale varying between 1:15000 and 1:21000 (Vandaele *et al.*, in press a). The aerial photos were taken in 1963, 1969, 1971, 1981 and 1986. These photos are considered to be a random sample of the time series. Only ephemeral gullies with a width larger than about 1.0 m could be detected on these photos. Since ephemeral gullies with a width less than 1.0 m were not included in these calculations, and since aerial photos were not taken at the optimal period for detecting ephemeral gullies, a percentage of the existing gullies in that year were not visible anymore on the photos (for instance, because they were already filled in by the farmers or because they were overgrown by weeds), Hence, the obtained figures with this method are rather conservative. Based on numerous field measurements of mean ephemeral gully cross-sections in central Belgium, eroded volumes by ephemeral gullies were calculated using a mean minimum and a mean maximum cross-sectional area (depth x width): i.e. 0.20x1.00 m and 0.30x1.50 m (Vandaele *et al.*, in press a). These data were compared with published interrill and rill soil loss data obtained from field plots in the study area.

For the study site in south-east Portugal, detailed measurements of the length of ephemeral gullies in a study area ranging between 270 and 553 ha were made using aerial photos at a scale of 1:15000 (Vandaele et al., in press a). The aerial photos were taken in 1970, 1978 and 1985. Since ephemeral gullies with a width less than 0.5 - 1.0 m were not included in these calculations, the obtained soil loss data for ephemeral gully erosion in south-east Portugal with this method are also underestimations (see previous section). Based on field measurements of mean ephemeral gully cross-sections, eroded volumes by ephemeral gullies were calculated using two sets of cross-sectional data (depth x width): i.e. 0.15x0.50 m (minimum) and 0.25x1.50 m (maximum; Vandaele et al., in press a). These data were compared to reported interrill and rill soil loss data obtained from runoff plots in the study area (Vale Formoso; Tomas, 1992).

In south-east Spain, detailed mapping at a scale of 1:500 of all rills and gullies on a representative 10 ha footslope section (Cerro Pistolas, Sierra de Gata, Almeria) was conducted (Poesen et al., in prep. a). The rills and gullies developed during a period of 10 years (i.e. between 1983, the year of abandonment and 1993, the year of erosion mapping). Transects of 200 m length were established parallel to the contour every 10 m along the slope over a length of 500 m. Along each transect, location, width and depth of all rills and gullies (minimum depth = 1 cm) were recorded. The sum of rill and gully cross sections was calculated for each transect. Total soil loss due to rill and gully erosion was calculated by multiplying the sum of rill and gully cross sections with a slope length of 10 m per transect, and summing these values up over the entire 10 ha field area.

Contribution of ephemeral gully erosion to sediment production in the three study areas
Central Belgium
Table 1 shows a good agreement between ephemeral gully data obtained by detailed field monitoring (mapping) and data extracted from aerial photos. These data indicate that mean soil loss due to ephemeral gully erosion is far from negligible and is, on average, 44 per cent of total soil lost. For another catchment in central Belgium (i.e. the Ganspoel catchment), Vandaele (1995) reported this fraction to amount to 55 per cent of total sediment produced in a three year study period (1989-1992).

South-east Portugal

For south-east Portugal, mean contribution of ephemeral gully erosion to total sediment production amounts to 80 per cent (Table 2), which is more important than the contribution of gully erosion found for central Belgium (Table 1). One of the possible reasons for this observation is the fairly high rock fragment contents in the Portuguese soils (30 per cent by mass) which reduces interrill and rill erosion (Poesen *et al.*, 1994) but not necessarily overland flow production. This overland flow causes ephemeral gully erosion to take place in topographic concavities (hollows) or in linear landscape elements.

Table 1. Sediment production by various water erosion processes in central Belgium

Method	Observational period (y)	Ephemeral gully (m³/ha/y)	Interrill and rill (m³/ha/y)	% gully	Source
field mapping	3	3.4	4.0*	46	(1)
aerial photo	5	2.3-5.1**			(2)
runoff plots	5		5.8		(3)
field mapping	3		4.0*		(4)
mean		3.6	4.6	44	

% gully equals the percentage of total soil loss due to ephemeral gully erosion;
* soil loss due to interrill erosion was assumed to be 10 per cent of total soil lost. This figure is based on data reported in the literature for comparable conditions (Govers and Poesen, 1988).
** eroded volumes by ephemeral gullies were calculated using two sets of cross-sectional data (depth x width): i.e. 0.20x1.00 m and 0.30x1.50 m;
(1) Vandaele and Poesen (1995);
(2) Vandaele *et al.* (in press a);
(3) Bollinne (1982);
(4) Govers (1991): these data only include observations made during the winter period.

Table 2. Sediment production by various water erosion processes in south-east Portugal

Method	Observational period (y)	Ephemeral gully (m³/ha/y)	Interrill and rill (m³/ha/y)	% gully	Source
aerial photo	3	1.1-5.3*			(1)
runoff plots	20		0.2-1.3		(2)
mean		3.2	0.8	80	

% gully equals the percentage of total soil loss due to ephemeral gully erosion;
* eroded volumes by ephemeral gullies were calculated using two sets of cross-sectional data (depth x width): i.e. 0.15x0.50 m and 0.25x1.50 m;
(1) Vandaele *et al.* (in press a);
(2) Tomas (1992): soil loss data are means for 13 plots with different crop rotations (wheat-leguminosa-wheat, wheat-leguminosa, fallow-wheat)

South-east Spain

The contribution of gully erosion to total soil loss by water erosion amounts to 83 per cent for the studied hillslope section in south-east Spain (Table 3). The data in this Table also show that the calculated interrill and rill soil losses for the studied abandoned hillslope are higher than those reported for a similar land-use elsewhere in the Mediterranean: i.e. 0.01- 1.5 m^3/ha/year (Poesen *et al.*, in prep. a). One of the possible reasons is that the vegetation cover on the studied slope was significantly lower than that in other studies.

Table 3. Sediment production by various water erosion processes in south-east Spain

Method	Observational period (y)	Ephemeral gully (m³/ha/y)	Interrill and rill (m³/ha/y)	% gully	Source
field map	10	9.7	2.0*	83	(1)

% gully equals the percentage of total soil loss due to ephemeral gully erosion;
* soil loss due to interrill erosion was assumed to be 10 per cent of total soil lost. This figure is based on data reported in the literature for comparable conditions (Govers and Poesen, 1988).
(1) Poesen *et al.* (in prep. a)

The results from the three studied environments indicate that soil losses due to ephemeral gullying are far from negligible: ephemeral gully erosion rates are almost as important as interrill and rill erosion rates in central Belgium, but are more important in Mediterranean environments, particularly in abandoned cultivated land (rangelands). However, one must not forget that data on ephemeral gullying, extracted from aerial photos result in conservative figures (see above). For abandoned cultivated land, it is concluded that interrill and rill erosion is drastically reduced by the vegetation cover but also by the presence of a well-developed erosion pavement. Hence, sediment concentration in overland flow, generated on the intergully areas will be quite low. Downslope, however, this overland flow can be responsible for rapid gully entrenchment and development, which is partly explained by the 'clear water' effect.

From the data presented in Tables 1, 2 and 3, the following can be concluded.
1) Given that gully erosion contributes significantly to sediment yield in small agricultural catchments, and that in particular environments it is even more important than interrill and rill erosion, more attention should be devoted to the study and the prediction of

(ephemeral) gully erosion and how the intensity of this soil degradation process is affected by environmental change (climatic and landuse changes).

2) Ephemeral gully erosion operates at spatial units which are larger than those of the traditional runoff plot. Therefore, an appropriate (standard) methodology to study ephemeral gully erosion should be developed.

One geomorphological implication of these results is that they indicate whether valley incision or valley aggradation will dominate: interrill and rill erosion mainly occur on upland areas, whereas ephemeral gully erosion dominantly occurs in valley bottoms of zero order catchments (Vandaele and Poesen, 1995). If the contribution of gully erosion to total soil loss exceeds 50 per cent (such as for instance in south-east Portugal and in south-east Spain), one may expect the valley bottoms to deepen if no other erosion process takes place (such as for instance tillage erosion, Vandaele *et al.*, 1996). If, on the other hand, the gully contribution is inferior to 50 per cent, one may expect valley aggradation to take place. If one looks at the figures for central Belgium, one may expect a slight valley aggradation to take place in central Belgium and a deepening of the valley bottoms in south-east Portugal assuming no other erosion process produces sediment on the slopes. Land abandonment induces valley-floor lowering because less sediment is produced by interrill and rill erosion on the slopes, and concentrated overland flow with a low sediment load will tend to incise valley bottoms. The latter situation has been observed on many abandoned lands throughout the Mediterranean where abandoned slopes get covered by matorral and well-developed erosion pavements.

Bank gullies

Few data exist on the relative importance of bank gully erosion. Obviously, the importance of this gully erosion type depends on the density of banks in the landscape, bank height, and the conditions favouring piping and subsequent bank gully initiation among other factors.

In order to assess the contribution of bank gully erosion to annual soil loss, data from two erosion surveys of central Belgium, are used (Table 4). The areas are characterised by a moderate to high bank density (sunken lane banks, lynchets, old quarry banks). The results indicate that in such areas, bank gullies can produce significant amounts of sediment which can be as high as 228 m^3/ha of area draining towards the bank gully. However, since bank gullies are seldom immediately erased after they have formed, they remain visible during subsequent years. Hence, the measured bank gully volumes were assumed to have formed over a period ranging between 10 and 20 years. This results in soil loss rates due to bank gullying of 0.2 to 0.5 m³/ha/year (Table 4), which are one order of magnitude smaller than soil loss rates due to ephemeral gullying (2.3 - 5.1 m³/ha/year).

Table 4. Bank characteristics and sediment production by bank gully erosion in central Belgium

Site	Area (ha)	Bank density (m/ha)	Bank gully volume (m³)	Soil loss (m³/ha)	Soil loss rate * (m³/ha/y)	Source
Ormendaal	126	45.6	641.5	5.1	0.25-0.5	(1)
Kinderveld	100	75.7	404	4.0	0.2-0.4	(2)

* soil loss rates are calculated assuming that the observed bank gullies formed in a period ranging between 20 and 10 years preceding the year of surveying.
(1) Poesen (1989)
(2) Poesen (1993)

Field observations in the Mediterranean reveal that bank gullying can be quite active on terraced land (Poesen, 1995). This is best seen once land abandonment has taken place. However, so far no data are available indicating their contribution to sediment production in such environments.

Gully erosion: modelling needs
In the framework of environmental change studies, there is a need to predict A) where and when (frequency) gullies form in the landscape, and B) the cross-sectional size and shape of the gullies. It is not the purpose of this paper to develop a detailed model strategy which would allow one to answer these questions. However, recent research has provided some insights into gully erosion processes and factors which might help modelling some of these aspects or which, at least, draws the attention to particular points important for modelling gully erosion.

Where do gullies develop in the landscape?
Existing soil erosion models do not predict the location of gullies. Yet this is important for land managers and for predicting the impact of climatic or landuse changes on the spatial

distribution and frequency of gullies. The main question here is where do gullies start and where do they end in the landscape?

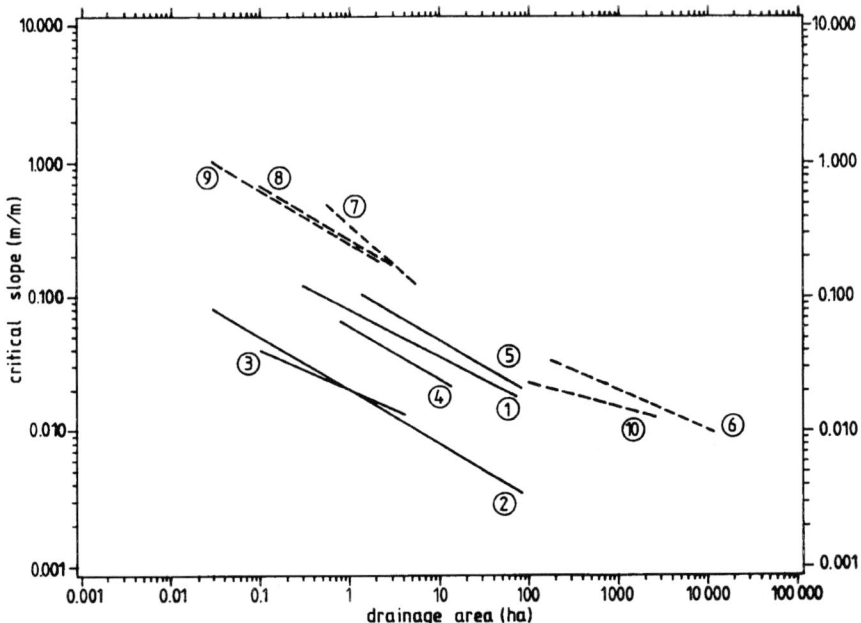

Figure 1. Critical slope-drainage area relationships (threshold lines) for incipient gully (head) development in a variety of environments (partly after Vandaele et al., in press). Solid lines indicate threshold conditions for ephemeral gully development in intensively cultivated lands. Dotted lines indicate threshold conditions for gully head development in uncultivated land (6 = sagebrush and scattered trees; 7 = open oak woodland and grasslands; 8 = coastal prairie; 9 = logged forest and 10 = swampy, reed-covered valley floors).
1. Central Belgium: field survey (Poesen et al., in prep. b);
2. Central Belgium: analysis of aerial photos and topographic maps (Vandaele et al., in press b);
3. Portugal: analysis of aerial photos and topographic maps (Vandaele et al., in press b);
4. France: analysis of aerial photos and topographic maps (Vandaele et al., in press b);
5. UK (South Downs): field survey (Boardman, 1992);
6. USA (Colorado): valley-floor gullies, analysis of aerial photos and topographic maps (Patton and Schumm, 1975);
7. USA (Sierra Nevada): field survey (Montgomery and Dietrich, 1988);
8. USA (California): field survey (Montgomery and Dietrich, 1988);
9. USA (Oregon): field survey (Montgomery and Dietrich, 1988).
10. Australia (New South Wales): valley-bottom gullies, field survey (Nanson and Erskine, 1988)

Where do (ephemeral) gullies start?

A possible approach to predict locations in the landscape where gully heads might develop is represented by the threshold concept, first applied to geomorphic systems by Patton and Schumm (1975). This concept is based on the assumption that in a landscape with a given climate and landuse there exists for a given slope a critical drainage area, necessary to cause gully incision. As slope steepens, this critical drainage area decreases and vice versa. For different environmental conditions and different gully initiating mechanisms, different thresholds apply. This concept permits one to predict where in a landscape and under a given landuse gullying may occur by providing a physical basis for the initiation of gullies. Little research has been conducted on threshold conditions for ephemeral gully development in cultivated lands and the few data available are highly site specific (e.g. Thorne *et al.*, 1986; Moore *et al.*, 1988). Threshold conditions for ephemeral gully initiation in central Belgium and in south-east Portugal have recently been established by Vandaele *et al.* (in press b).

Figure 1 shows that the threshold curves for ephemeral gullies in intensively cultivated areas plot well below the threshold curves for gullies in non-cultivated areas (data from USA and Australia). In other words, for a given drainage area, ephemeral gullies start to form in cultivated areas on much gentler slopes than they do in uncultivated, vegetated areas.

These threshold lines can be represented by a power-type equation (Begin and Schumm, 1979; Vandaele *et al.* in press b):

$$S = a A^b \qquad (1)$$

with: S = critical slope for ephemeral gully head development (m/m);
 A = corresponding drainage area (ha)
 a, b = coefficients

Vandaele *et al.* (in press b) noticed that the b-values corresponding to most threshold curves for ephemeral gullying in cultivated lands range between -0.3 and -0.4.

However, the corresponding a-values vary considerably: i.e. between 0.02 for ephemeral gullies in Portugal (curve 3) and 0.12 for the gullies in the UK (curve 5) (Vandaele *et al.*, in press b). One of the factors controlling the a-value is the methodology used to establish these threshold lines. For instance, the threshold curve deduced from field data (slopes and drainage area) on ephemeral gullying in central Belgium (curve 1) corresponds to an a-value of 0.08.

On the other hand, the a-value corresponding to the threshold curve, deduced from aerial photos and topographic maps, for ephemeral gullying in central Belgium equals only 0.025 (curve 2). Other factors controlling the position of the threshold lines are climate (Montgomery and Dietrich, 1994; Kirkby, 1994), landuse, soils as well as all other factors controlling the mechanisms of incipient gullying (hydraulic erosion by concentrated overland flow, seepage flow, and mass movement processes) (e.g. Montgomery and Dietrich, 1994). For instance, it might well be that the reason why the position of the threshold line corresponding to incipient ephemeral gullying in the South Downs (UK., curve 5) plots above the threshold line for central Belgium (curve 1) is due to the presence of stony soils in the South Downs. More research is needed to indicate to what extent climatic or landuse changes affect the position of the (ephemeral) gully threshold curves.

The threshold curves for ephemeral gullies reflect a critical flow intensity which needs to be exceeded in order to initiate gullying. This critical flow intensity for gully initiation depends on rainfall regime (climate) and landuse type (infiltration characteristics, erodibility (including stoniness and vegetation cover).

Once the (ephemeral) gully threshold lines are established for a variety of climatic, pedologic and landuse conditions, one could predict the effect of landuse or climatic change (rainfall intensity and frequency) on the gully network density and location using models such as those developed recently by, for instance, Kirkby (1994).

Where do (ephemeral) gullies end?

Massive sedimentation counteracts gully development. Some models have been developed to predict the conditions where sedimentation (colluviation) prevails. For instance, De Ploey (1984) proposed a colluviation model:

$$S_{cr} = \frac{A \cdot C^{0.8}}{q^{0.5}} \qquad (2)$$

with: S_{cr} = critical slope angle below which sedimentation (colluviation) occurs (°);

C = sediment concentration in overland flow (g/l);

q = unit flow discharge (cm²/s);

A = sediment grain-size factor; an empirical coefficient depending on the median sediment grain-size.

Field measurements in intensively cultivated areas reveal that important sediment deposition usually occurs at a narrow range of local slopes along catenas with the same landuse. For instance, field observations indicate that this usually occurs on slopes ranging between 2 to 4 per cent for intensively cultivated soils (main crops are maize, sugar beet, potatoes and chicory) in central Belgium and between 4 to 6 per cent for sparsely vegetated rangeland in south-east Spain (Figure 2). Obviously, landuse changes along the hillslope will also affect gully incision and sedimentation. More research is needed to elucidate how landuse type affects gullying/sedimentation along hillslopes.

From the previous analysis it now becomes possible to indicate where in the landscape ephemeral gullies can start developing and where they end, provided that threshold lines for the area and for a variety of landuses are available. Such predictions can be made using routing algorithms describing how flow and water-borne material will be routed over Digital Elevation Models, such as those tested by Desmet and Govers (1995). These algorithms allow one to calculate drainage area and slope for each point in a catchment. By applying threshold conditions for ephemeral gullying, such as those presented above, one can predict total gully length for a given climate, soil and landuse type. The next step is to predict the cross-sectional size and shape of the gully channel as a function of environmental factors.

Bank gullies

Compared to ephemeral gullies, much less research has been conducted on the location of bank gullies in the landscape. Poesen (1989; 1993) concluded from a field study in central Belgium 1) that the eroded volumes by bank gully erosion were not significantly related to a measure of overland flow intensity (product of local slope and drainage area) and 2) that the location of bank gullies was only in a limited number of cases (i.e. 5 out of the 40 bank gullies in the Kinderveld study area, central Belgium, Table 4) related to topographically-controlled concentrated overland flow crossing a bank. In most other cases, the location of the bank gullies along existing banks could not easily be explained and depended often on the presence of macropores (biopores and tension cracks) and the occurrence of mass movement processes (slumping and soil fall).

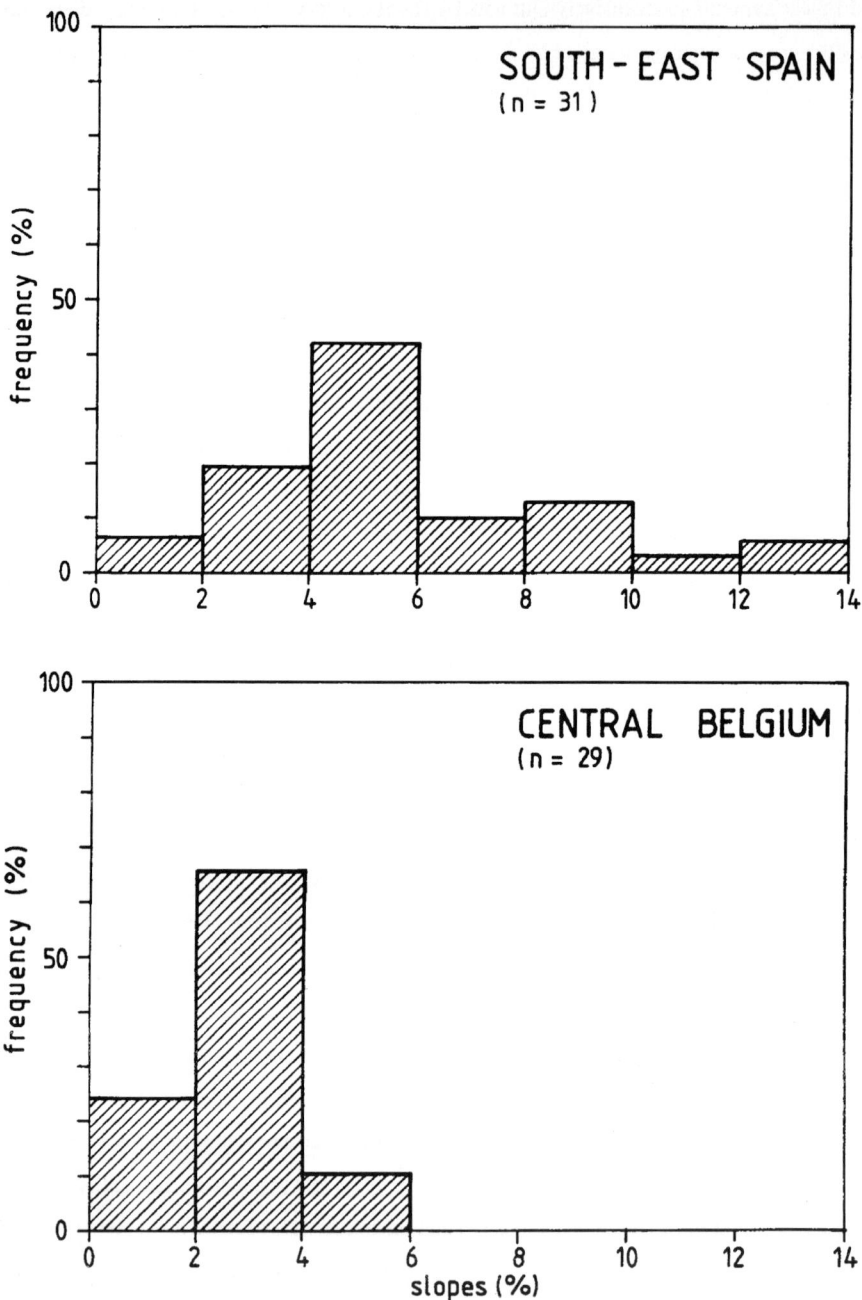

Figure 2. Histograms of critical slopes below which (ephemeral) gullies end and sedimentation in concentrated flow becomes dominant for rangeland on stony soils in south-east Spain and for intensively cultivated silt loam soils in central Belgium (main crops are maize, sugar beet, potatoes and chicory). n = number of observations

Recent findings from central Belgium indicate that the probability of bank gully development in a given spot of a bank increases significantly if dead furrows on parcel borders concentrate overland flow and discharge it on to the bank (Vandaele, 1995). In the Kinderveld study area (Table 4) the location of 16 out of the 35 bank gullies and pipes which are not controlled by topographic-induced concentrated overland flow lines, coincides with the position of old field borders (mapped from aerial photos taken in 1947). Parcel borders are often marked by the presence of a dead furrow which intercepts overland flow and discharges this runoff on to the bank. This runoff will then either flow through pre-existing pipes and cause considerable pipe erosion leading to bank gully formation when the roof of the pipe collapses, or saturate the bank which will lead to mass movement. These results may help to better locate sites along banks where bank gullies might develop.

Obviously, the spatial pattern of bank gully development also depends on the density of the banks in the landscape, their maintenance, the biological activity (e.g. burrowing animals) in the banks, the frequency of runoff events, etc.

Field observations in central Belgium reveal that bank gullies do not always form during high-intensity rainstorms. During low intensity rains of long duration, bank gullies can also develop, particularly where very erodible soil horizons (e.g. unweathered calcareous loess or sandy deposits) are present at shallow depth in the bank.

From these data and observations we learn that the prediction of the location of major pipes and bank gullies is enhanced by looking at the location of actual or old parcel borders coinciding with dead furrows through which overland flow could be discharged on to the bank.

What is the cross-sectional size and shape of the (ephemeral) gullies?
The cross-sectional size of ephemeral gullies determines total soil loss, whereas cross-sectional shape (w/d) controls what proportion of total soil loss consists of fertile topsoil. The understanding of gully cross-section is a first step in evaluating gully processes and factors (Imeson and Kwaad, 1980). Field observations on gully morphology in a variety of agricultural environments in Europe have revealed that beside rainstorm characteristics, a number of site factors play a crucial role in the development of gully cross-section: i.e.

landuse (vegetation type and cover), topography (drainage area and local slope) and soil profile characteristics. The w/d of ephemeral gullies is controlled by a combination of factors affecting flow width, flow intensity as well as resistance properties of soil horizons to concentrated flow erosion. In this section, we briefly discuss the effect of rainfall, topography and soil profile characteristics on the w/d of ephemeral gullies.

High intensity, low-frequency rainstorms precipitating on an initially dry and freshly cultivated loamy top soil are likely to cause wide ephemeral gullies with a w/d > 1 up to a w/d = 20 which, in northern Europe, usually occur in spring and early summer. The large width, compared to their depth, of these gullies is due to the fact that the high overland flow discharges generated during these storms will flow over a considerable width in the flat valley-bottoms. Since the strength of the air-dry loamy soil top layer is drastically reduced upon sudden wetting, the erodibility of the loamy top soil is very high (Govers et al., 1990). Concentrated flow incision is reduced at shallow depth if a less erodible layer is exposed by erosion, e.g. a soil layer with a high moisture content. On the other hand, low intensity rainstorms, which in northern Europe are typical in winter and early spring, cause runoff to flow over a limited width in the talwegs. Since low-intensity rains are more frequent than high intensity storms, cumulative shear stresses produced by these flows on the soil surface and on the channel bed in the concentrated flow zone are expected to be higher than cumulative shear stresses produced during low-frequency, high-intensity rainstorms. Therefore, gullies with a w/d = or < 1 are often formed during winter in Northern Europe. A change in rainfall regime (due to a climatic change) will also affect the frequency of gullies with a given w/d.

The effects of local talweg slope gradient and of a resistant soil horizon (a plough pan) on the width, the depth and the w/d ratio of an ephemeral gully in central Belgium were studied by Poesen (1990). If no resistant soil horizon occurs at shallow depth in the profile, the talweg slope gradient does not seem to affect gully width (w) significantly, but has a significant positive effect on the depth (d) and therefore on the w/d ratio of the gully. The depth of an ephemeral gully increases rapidly beyond a critical slope of 0.03 - 0.04 which is in line with findings of Savat and De Ploey (1982) and Govers (1985). If, however, an erosion-resistant horizon, such as for instance a Bt horizon, a plough pan or a fragipan is present at shallow depth in the profile, gully depth remains small and therefore gully width and w/d are high, even if the talweg gradient is steep. In Mediterranean environments, the presence of a calcrete or of high concentrations of rock fragments in the top soil, often hampers the vertical development of ephemeral gullies (Poesen et al., in prep. b). If, on the other hand, an erodible horizon is present at shallow depth, such as for instance calcareous loess (C horizon), typical deep ephemeral gullies with a w/d <<1 develop.

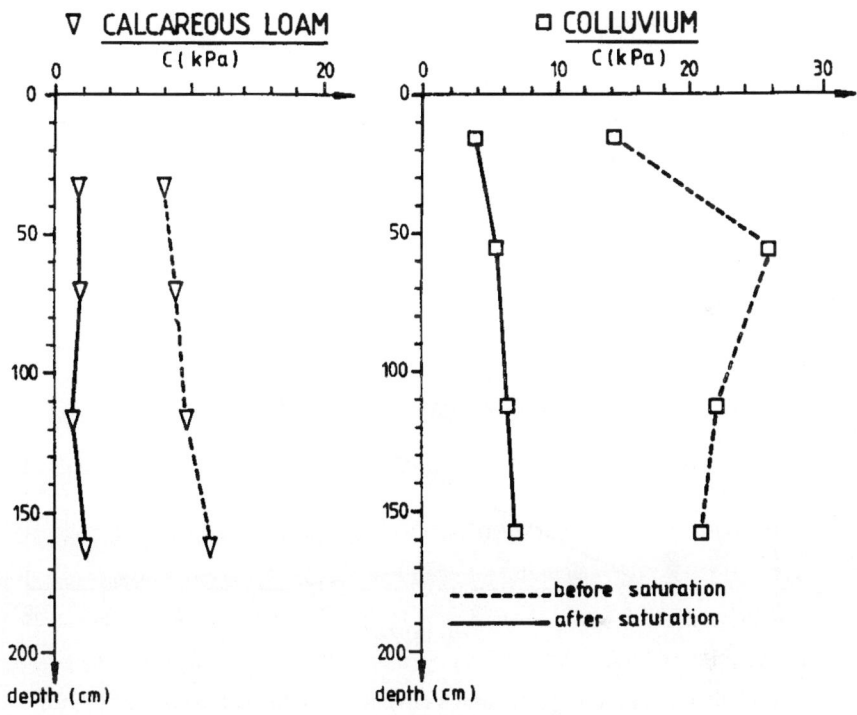

Figure 3. In situ soil shear strength (C) measured with a torvane, before and after artificial saturation by flooding, versus soil depth for unweathered calcareous loam and compact colluvium (consisting essentially of textural pores) in central Belgium. Measurements were made on January 14, 1993, when mean antecedent gravimetric moisture content equalled 9 per cent for calcareous loess and 10 per cent for colluvium (Kinderveld, Korbeek-Dijle, Poesen et al., 1993)

Quantification of the erosive power of overland flow is possible through the use of various flow intensity measures such as shear stress ($\tau = \rho gRS$), shear velocity ($u_* = (gRS)^{0.5}$) or stream power ($= \rho gqs$) (where ρ = density of water, g = acceleration due to gravity, R hydraulic radius, S is sine of slope angle and q = unit flow discharge). Little or no quantitative information is available on the resistance of various soil horizons to concentrated flow erosion. Such information, however, is crucial to model A) the cross-sectional size and shape of ephemeral gullies, as well as the related type of environmental damage they cause, and B) the retreat rate of headcuts (De Ploey, 1989) both in ephemeral gullies and in bank gullies. In the literature on ephemeral gully erosion, soil layers are often classified as erodible or non-erodible (e.g. Foster, 1986). However, in field conditions, most horizons have intermediate values of erodibility. The question then is: how can one characterise these horizons in terms of their erodibility or resistance to concentrated flow erosion?

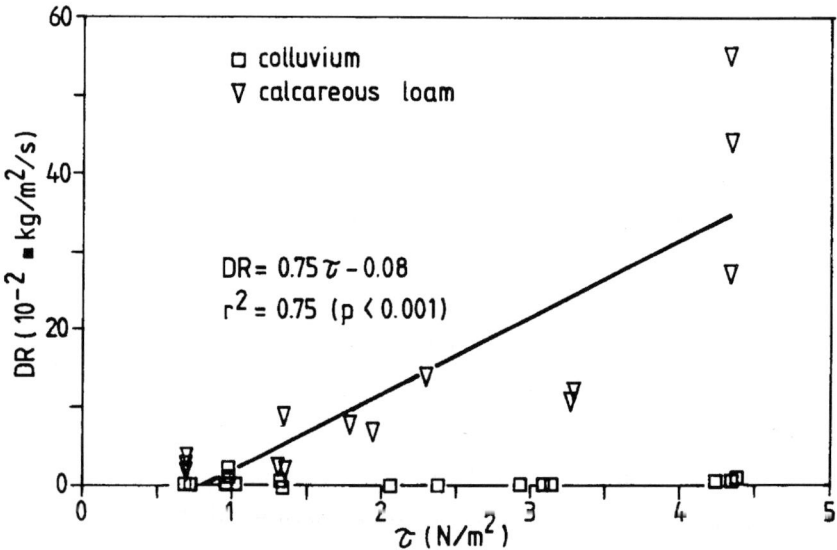

Figure 4. Soil detachment rate (DR) versus concentrated flow shear stress (τ) for calcareous loess (antecedent gravimetric moisture content = 15 per cent) and compact colluvium (antecedent gravimetric moisture content = 23 per cent). Flow detachment measurements were made in the laboratory on undisturbed soil samples from Kinderveld, Korbeek-Dijle, which were installed in a hydraulic flume (Poesen *et al.*, 1993)

Field studies in Central Belgium have been undertaken to detect which soil property best reflects the resistance of loess loam horizons to erosion by concentrated flow. Poesen and Govers (1990) found that out of 14 soil properties investigated, shear strength of the soil horizon, measured under particular conditions, was a reliable and relevant indicator of resistance to concentrated flow erosion. Shear strength of soil horizons, consisting essentially of textural pores (Childs, 1969), was measured with a torvane after artificial saturation by flooding of the undisturbed soil horizon (Figure 3). Shear strength at saturation of the colluvium is significantly higher than that of the calcareous loam (Figure 3). The colluvium is indeed much more resistant to detachment by concentrated flow than is the calcareous loam (Figure 4). Soil shear strength at saturation adequately reflects the condition of the soil horizon when being eroded by concentrated flow. Soil shear strength values obtained in this way are not only a function of intrinsic soil properties but also of the initial soil moisture content. Figure 5 shows that shear strength at saturation is positively and linearly correlated with initial moisture content of a crusted silt loam top soil. Laboratory experiments conducted by Govers *et al.* (1990) indeed show that silt loam soils are most susceptible to concentrated flow erosion when they are dry. Poesen (1993) presented the shear strength profile of an undisturbed soil profile under deciduous forest which developed on calcareous loess. Such information is essential when predicting gully cross-sectional development.

At present, few models are capable of predicting ephemeral gully cross-sectional size and shape: i.e. CREAMS (Chemicals, Runoff, and Erosion from Agricultural Management Systems; Knisel, 1980) - GLEAMS (Groundwater Loading Effects of Agricultural Management Systems; Knisel, 1993), EGEM (Ephemeral Gully Erosion Model; Merkel *et al.*, 1988) and WEPP watershed model (Water Erosion Prediction Project; Flanagan and Nearing, 1995). The channel erosion routines from both EGEM and WEPP watershed model are slightly modified procedures from the CREAMS channel erosion routines (Lane and Foster, 1980). In these models, concentrated flow detachment rate is proportional to the difference between A) flow shear stress exerted on the bed material and the critical shear stress, and B) the transport capacity of the flow and the sediment load. Net detachment occurs when flow shear stress exceeds the critical shear stress of the soil or gully bed material and when sediment load is less than transport capacity. Net deposition occurs when sediment load is greater than transport capacity. For instance, the basic equations used in CREAMS, EGEM and WEPP are (Foster, 1986):

$$D = D_c \cdot (1 - \frac{G}{T_c}) \qquad (3)$$

where: D = detachment rate along the channel boundary (mass/area.time);
D_c = maximum detachment rate or detachment capacity of the flow (mass/area.time);
G = sediment load in the flow (mass/time);
T_c = sediment transport capacity of the flow, calculated using the Yalin equation (mass/time).

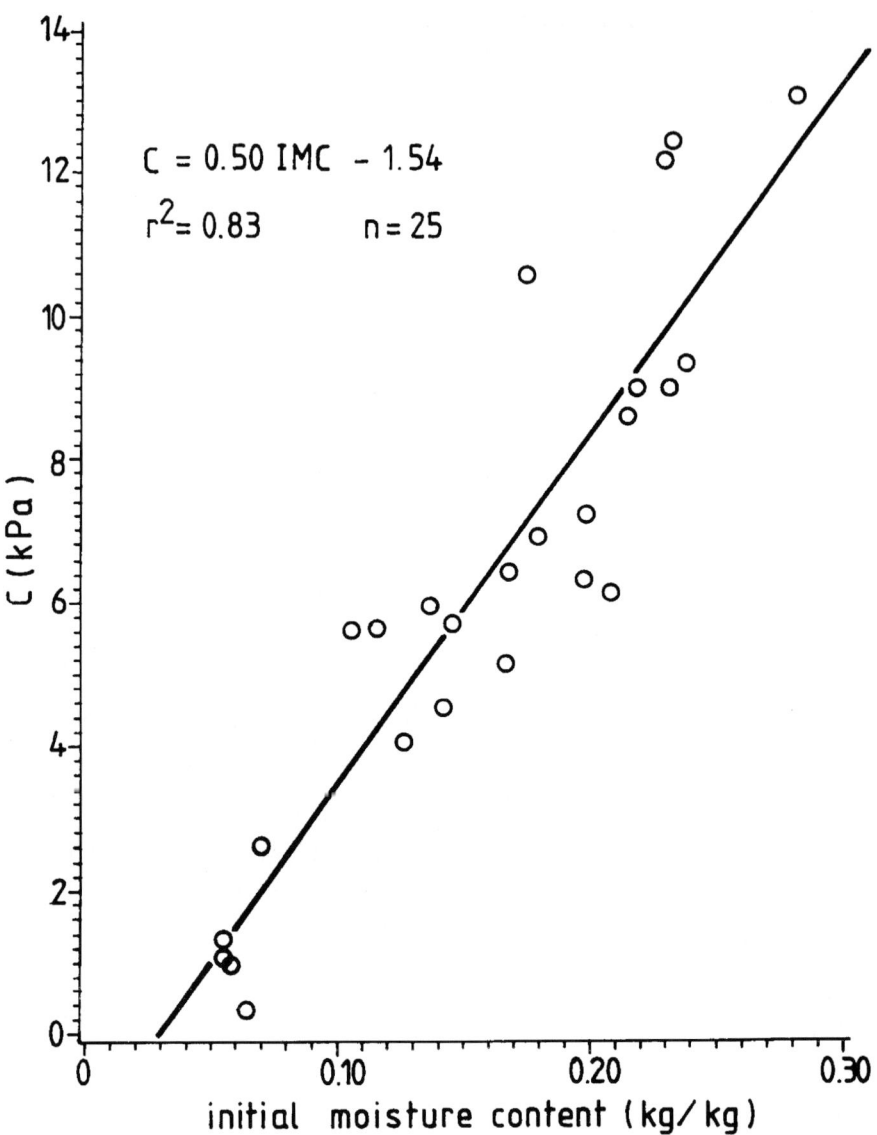

Figure 5. In situ shear strength (C) at saturation (measured with a torvane over a depth of 5 mm after flooding) of structural crusts (consisting essentially of textural pores) which developed on silt loam soils (central Belgium). Shear strength measurements were conducted at different times through the year when the structural crusts had different initial moisture contents (IMC)

$$D_c = K_c \cdot (\tau - \tau_c) \tag{4}$$

where: K_c = channel erodibility factor (mass/time.force);

τ = average shear stress of flowing water (force/area);

τ_c = critical shear stress of soil (force/area).

An alternative procedure, allowing one to predict ephemeral gully width using peak discharge, Manning's roughness, slope and critical tractive stress, was developed by Watson et al. (1986).

These models represent important steps towards modelling ephemeral gully erosion. However, they only simulate entrainment, transport, and deposition of sediment by concentrated flow. Field observations and measurements in a variety of environments (e.g. Bradford and Piest, 1980; Govers and Poesen, 1988; Radoane et al., 1995) reveal that once gullies (either ephemeral gullies or bank gullies) are formed, mass movement processes (i.e. soil sliding and soil fall after undercutting) often supply considerable amounts of sediment into the gully, which is subsequently evacuated during flow events. At present, mass movement processes on the gully walls are not simulated in the existing ephemeral gully models. Also, headcutting extending ephemeral gullies and being quite important in the initiation and further development of bank gullies is not taken into account in existing gully models. De Ploey (1989) produced a simple model predicting the retreat rate for gully headcuts. However, there is still a need for models capable of simulating these processes under a range of environmental conditions.

Conclusions

This study reveals that soil losses due to (ephemeral) gullying in Europe are far from negligible in most agricultural environments (1 - 5 m³/ha/yr) and that in particular environments (e.g. the Mediterranean) (ephemeral) gullying is probably the most important sediment producing erosion process in upland areas. A limited amount of data reveals that although bank gully erosion can produce locally large amounts of sediments, soil loss rates due to this gully type (0.2 - 0.5 m³/ha/yr) measured over areas of about 100 ha are about one order of magnitude smaller than those for ephemeral gullies.

There is an urgent need for more detailed monitoring, experimenting and modelling of the development and infilling of both ephemeral gullies and bank gullies in a variety of (agricultural) environments, in order to better simulate gully erosion subprocesses. More precisely, there is a need to better predict the location, the total length and the cross-section (size and shape) of both ephemeral gullies and bank gullies. The threshold concept could be a useful tool to help locate (ephemeral) gullies in the landscape. However, more research is needed to establish threshold conditions for incipient gullying in a variety of climatological, topographic, pedological and land-use conditions. Existing concentrated flow erosion models need to be refined to incorporate the effects of the resistance of various soil horizons to concentrated flow erosion as well as of other soil detaching mechanisms in gullies, such as soil fall, slumping and headcutting.

Models capable of predicting these aspects of gully erosion are needed in order to predict more accurately the effects of environmental change (e.g. change in rainfall regime, change in landuse such as a shift towards intensive cropping but also towards land abandonment due to Common Agricultural Policy regulations) on soil losses due to gully erosion. Such models are also needed 1) for predicting sediment sources and sediment volumes in upland areas, which often contribute significantly to downstream flooding and sedimentation, and 2) for evaluating the efficiency of gully erosion control measures.

Acknowledgements

Financial support provided by the European Commission (D.G. XII) in the framework of the MEDALUS and the EUROSEM projects is acknowledged. This paper is a contribution to the Soil Erosion Network of the GCTE, which is a Core Research Project of the IGBP.

References

Auzet, A.V., Boiffin, J., Papy, F., Maucorps, J. and Ouvry, J.F. (1990). An approach to the assessment of erosion forms and erosion risk on agricultural land in the Northern Paris Basin, France. In, Boardman, J., Foster, I.D.L. and Dearing, J.A. (eds), *Soil Erosion on Agricultural Land*. Wiley, Chichester. pp. 383-400.

Baade, J., Barsch, D., Mäusbacher, R. and Schukraft, G. (1993). Field experiments on the reduction of sediment yield from arable land to receiving watercourses (N-Kraichgau, SW-Germany). In, Wicherek, S. (ed.), *Farm Land Erosion in Temperate Plains Environment and Hills*. Elsevier, Amsterdam. pp.471-480.

Begin, Z.B. and Schumm, S.A. (1979). Instability of alluvial valley floors: a method for its assessment. *Transactions of the American Society of Agricultural Engineers* **22**, 347-350.

Boardman, J. (1990). Soil erosion on the South Downs: a review. In, Boardman, J., Foster, I.D.L. and Dearing, J.A. (eds), *Soil Erosion on Agricultural Land.* Wiley, Chichester. pp. 87-105.

Boardman, J. (1992). Current erosion on the South Downs: implications for the past. In, Bell, M. and Boardman, J. (eds). *Past and Present Soil Erosion.* Oxbow, Oxford. pp. 9-19.

Boiffin, J., Papy, F. and Eimberck, M. (1988). Influence des systèmes de culture sur les risques d'érosion par ruissellement concentré. I. - Analyse des conditions de déclenchement de l'érosion. *Agronomie* **8**, 663-673.

Bollinne, A. (1982). *Etude et Prévision de l'Érosion des Sols Limoneux Cultivés en Moyenne Belgique.* Ph.D. Thesis, Université de Liège, 356 pp.

Bradford, J. and Piest, R. (1980). Erosional development of valley-bottom gullies in the upper midwestern United States. In, Coates, D.R. and Vitek, J.D. (eds) *Geomorphic Thresholds,* Dowden and Culver, Stroudsburg, Pennsylvania. pp. 75-101.

Brice, J.C., (1966). Erosion and deposition in the loess-mantled Great Plains, Medicine Creek drainage basin, Nebraska. *U.S. Geological Survey Professional Paper* **352-H**:255-339.

Childs, E.C. (1969). *An Introduction to the Physical Basis of Soil Water Phenomena.* John Wiley, London, 475 pp.

De Lima, J.L. (1989). *Overland Flow under Rainfall: Some Aspects related to Modelling and Conditioning Factors.* Ph.D. Thesis, Landbouwuniversiteit Wageningen, 160 p.

De Ploey, J. (1984). Hydraulics of runoff and loess loam deposition. *Earth Surface Processes and Landforms* **9**, 533-539.

De Ploey, J. (1989). A model for headcut retreat in rills and gullies. *Catena Supplement* **14**, 91-86.

Desmet, P. and Govers, G. (1995). Comparison of routing algorithms for digital elevation models and their implications for the prediction of the location of ephemeral gullies. *Proceedings First Joint European Conference and Exhibition on Geographical Information,* The Hague, NL, 27-31 March, pp. 324-329.

Farres, P., Poesen, J. and Wood, S. (1993). Soil erosion landscapes. *Geography Review* **6**, 38-41.

Flanagan, D.C. and Nearing, M. (1995). USDA - *Water Erosion Prediction Project Hillslope Profile and Watershed Model Documentation.* National Soil Erosion Research Laboratory, West Lafayette, Indiana Report No. 10.

Foster, G. (1986). Understanding ephemeral gully erosion. In, *Soil Conservation,* National Academy of Science Press, Washington D.C. 2, pp. 90-125.

Foster, G. and Lane, L. (1983). Erosion by concentrated flow in farm fields. In, Li, R.M., Lagasse, P.F. and Simons, Li and Ass. Inc. (eds), *Proceedings of the D.B. Simons Symposium on Erosion and Sedimentation.* Colorado State University, Fort Collins. pp. 9.65-9.82.

Govers, G. (1985). Selectivity and transport capacity of thin flows in relation to rill erosion. *Catena* **12**, 35-49.

Govers, G. (1991). Rill erosion on arable land in Central Belgium: rates, controls and predictability. *Catena* **18**, 133-155.

Govers, G. and Poesen, J. (1988). Assessment of the interrill and rill contributions to total soil loss from an upland field plot. *Geomorphology* **1**, 343-354.

Govers, G., Everaert, W., Poesen, J., Rauws, G., De Ploey, J. and Lautridou, J.P. (1990). A long-flume study of the dynamic factors affecting the resistance of a loamy soil to concentrated flow erosion. *Earth Surface Processes and Landforms* **15**, 313-328.

23. FIELD DATA AND EROSION MODELS

R. Evans
Division of Geography
Anglia Polytechnic University
East Road
Cambridge CB1 1PT
UK

Abstract

Information collected in farmers' fields provides a sound and realistic foundation upon which to base an understanding of erosion processes and their impacts, and to predict erosion and its impacts in the future. Some results from field-based surveys of erosion in England and Wales are described. Presently, models based on plot experiments do not adequately predict the extent, especially the distribution of eroded fields within the landscape, nor the frequency and rates of water erosion, nor where rills and gullies can occur, and how much soil is transported out of catchments. Compared with field data, models are probably no better at explaining which are the most important factors which govern the occurrence and severity of erosion. Better ways are needed to compare data obtained in the field and from modelling, and to make models more realistic. Thus, models have to generate data which replicate the volumes of soil eroded, and their statistical distributions, which have been measured in farmers' fields. Modellers particularly need to address which are the most important water erosional processes driving their models, as well as the importance of crop cover at the time of erosion.

Introduction

Models used to predict erosion, especially accelerated erosion, are based on rainfall, soil, crop and landform parameters, as well as the impacts of conservation techniques. The data required for models for the most part is obtained from field plots and laboratory experiments. But, it is assumed that rates of erosion predicted by models relate well to actual erosion in farmers' fields, and to the actual extent and frequency of erosion. This may not be so (Evans, 1990a; 1993a; 1995).

In the late 1970s and early 1980s when how to assess the problem of accelerated water erosion in England and Wales was being discussed it was considered, from *ad hoc* evidence collected in the field (Evans, 1980a; 1980b), that plot experiments such as those used in the USA and elsewhere or the Universal Soil Loss Equation (USLE) (Wischmeier and Smith, 1978) type approach were not satisfactory. What needed to be known was what was actually happening in farmers' fields. So erosion was monitored in 17 localities over the five year period 1982-86

(Evans, 1993b). Such field-based techniques give sensible and satisfactorily reliable results (Evans and Boardman, 1994).

The approach described here to assess the incidence and impacts of water erosion comes from a different direction therefore. Whereas most modellers seek to predict what will happen, i.e. how much soil will erode, using empirical data from plot experiments or theoretically-derived formulae, field-workers monitor what has happened, i.e. by locating, mapping and estimating amounts of soil eroded in the field, and then try to deduce the processes or factors contributing to erosion. This field-derived information provides the criteria which should be used to assess the usefulness and validity of models.

Where does accelerated erosion occur?

Predicting where erosion is most likely to occur is not difficult provided there are soil and topographic maps available. For example, field observation in the late 1960s and in the 1970s suggested that certain soil landscapes were more vulnerable to erosion than others. So, by the time the 1:250,000 scale National Soil Map of England and Wales (SSEW, 1983) was published it was already known where water erosion was most likely to occur.

It was predicted for instance that the Newmarket 2 soil association was at slight risk of erosion based on knowledge gained in East Anglia, and so it proved when members of the Soil Association went looking for erosion in Lincolnshire (Evans, 1987). Similarly, the Bridgnorth association, based on knowledge of the West Midlands, was considered at risk, and so it has proved to be in Devon (Evans, unpublished information).

Information collected in an *ad hoc* manner had also indicated which parts of the landscape were vulnerable to rilling — slopes below convexities and valley floors (Evans and Cook, 1986) and so the monitoring scheme showed (Evans, 1993b). In landscapes with clayey or heavier textured soils, or where winter cereals were the dominant crop, rills were often found only in valley floors. Whereas in landscapes with light textured and sandy soils most rills were on slopes only. It is unlikely then that models, particularly sophisticated ones, are needed to predict where erosion is likely to take place.

How extensive is accelerated erosion?

Field observation suggested then (Evans and Cook, 1986; Evans, 1990a), and still does, that rain splash and sheet wash were of little importance in England and Wales in transporting soil downslope, except where soils with more than about 60 per cent silt content occurred. However, although channels may not occur in these silty soils, deposits do, and so erosion can be assessed. As soil is mainly redistributed within a field by rills and gullies, therefore, so those were the processes monitored. If the very small amounts of soil transported in sheet flow are ignored only rilled fields have to be found. This was/is not difficult to do.

Within monitored localities the extent of erosion was related to soil type and land use (Evans, 1993b). On light-textured soils in localities warm enough to grow and ripen crops and not too wet to inhibit their harvesting, fields drilled to different crops can be at risk at different times of the year. Thus, erosion is only widespread and frequent where the land is used in certain ways. Erosion is often wrongly predicted to occur in some localities because these predictions are mostly based on rainfall amount and intensity, not on land use. There are few realistic assessments or predictions of the extent of erosion therefore (Evans, 1993a). For it is assumed that erosion is an ongoing process (see below), and so will be expected to occur throughout the landscape, whereas in fact accelerated erosion is frequently restricted in its occurrence. The distribution of eroded fields within a landscape is rarely considered by modellers.

How frequently does accelerated erosion occur?

In England and Wales erosion occurs in the same field once about every two or three years in many localities (Evans, 1993b). It is more frequent where a wide range of crops is grown, especially if these are irrigated, as in Kent, or where the land is dominantly drilled to one crop (winter cereals) as on the chalk Downs. It is less frequent in localities which are dominantly under grass.

In other countries little good quality information is available on the frequency at which accelerated erosion occurs in the landscape. It seems to be accepted that fields erode frequently, possibly because plot experiments in the USA showed that soil particles were transported even when the vegetation cover was very dense as under sown grassland or natural grassland and woodland (Wischmeier and Smith, 1978). In other words, erosion is a process going on all the time and is predicted by the USLE to occur under any crop, it will be just less serious in some crops than others. However, under natural grassland or woodland or a good sown grass cover, erosion is negligible in amount, and should be ignored. In this way attention is focused on the more easily identified accelerated channel erosion processes. The USLE has a factor in it for cropping (crop rotation) but, in the light of the above comments, how realistic is it?

How severe is accelerated erosion?

Erosion occurred in the 17 localities monitored between 1982-86 in 1705 fields within 83 soil associations. The values of the volumes eroded (m^3/ha) in the fields were very skewed (Figure 1), a similar phenomenon has been noted for amounts eroded from plots (Evans, 1995). In the 17 localities, median values of volumes eroded were generally much less than the means (Evans, 1993b). If such findings from field monitoring apply to plot results, and it is likely they do, they imply that whereas mean values can be used to predict long-term erosion, median values are more relevant for predicting short-term erosion such as when predicting the impacts of different types of conservation schemes, which is one of the aims of the USLE.

In England and Wales, as elsewhere, erosion was generally more severe where soils contained much silt or sand. The relative relationships between mean values of erosion within fields sown to cereals (small grains) and maize, as well as crop rotations, are not unlike those found elsewhere in the world, but the discrepancy between the mean values is large, about an order of magnitude (Evans, 1995). It is likely that using mean rates of erosion from plots as an input to models may well lead to models which seriously over-predict rates of erosion.

The median values of erosion recorded in fields were significantly different in only 28 of the 83 associations; the differences cannot be explained consistently in terms of soil differences. No generalisations can be made, for instance, that soils of a given particle-size class are associated with higher or lower rates of erosion, or with a greater or lesser range of values, or skewness of the data (Figure 1). For these 28 associations the range of median (0.19-2.10 m^3/ha) and mean values (0.92-4.76 m^3/ha) were similar, 11.0 and 9.7 respectively. However, if the monitoring project had continued for a longer period of time, so identifying more eroded fields, although the larger sample numbers could have made more populations statistically significantly different, the range of both median and mean values would likely still be small. Models should produce results which similarly reflect these skewed distributions and the restricted range of median and mean values of erosion.

Field data and erosion models 317

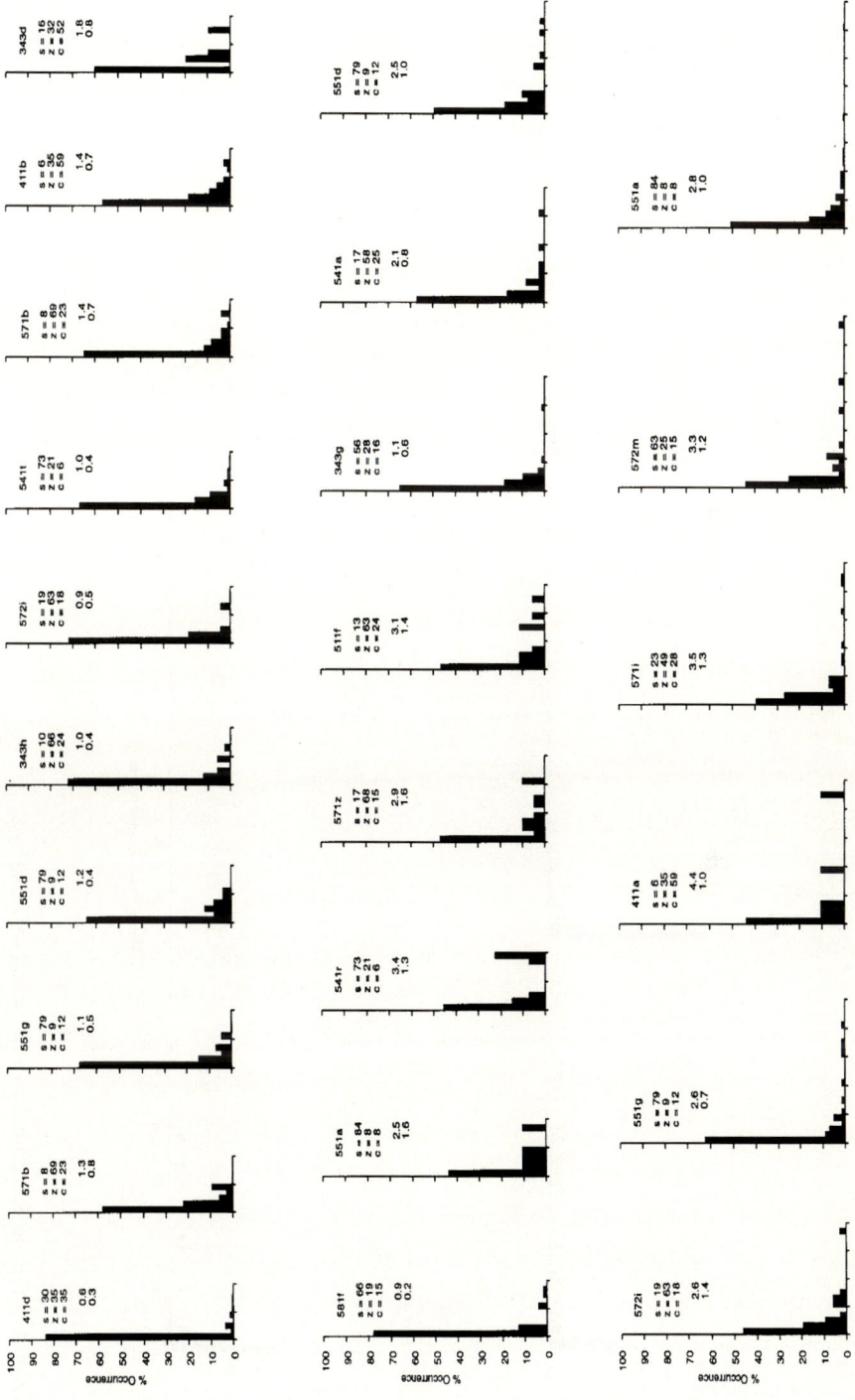

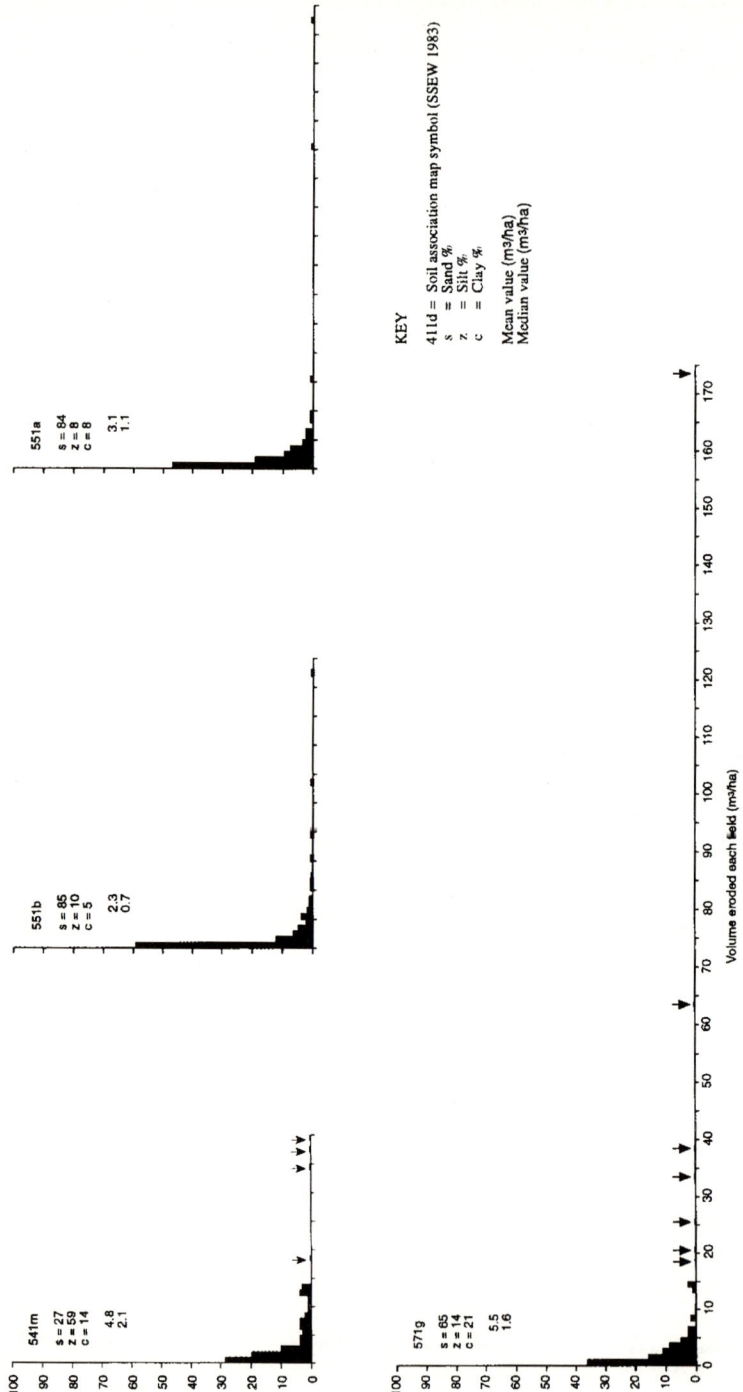

Figure 1. Volumes eroded in fields in 28 soil associations

Volumes eroded and factors contributing to erosion

In most models, erosion rates are a primary output, and are used, for example, to predict the impacts of erosion, how conservation techniques will modify erosion, or to relate to different factors which contribute to erosion. It was envisaged that rates derived from field survey could be used in the same way. However, the small range of values, albeit of about one order of magnitude, and skewness of values (Figure 1) makes difficult any analysis which tries to relate volumes eroded to factors which contribute to erosion.

Table 1. Median volumes (m^3/ha) eroded in fields (and number of fields) sown to different crops

Locality	Winter cereal	Spring cereal	Sugar beet	Potatoes
Somerset 541m* - silty	1.89 (49)	1.21 (13)	4.48 (10)	2.26 (15)
Isle of Wight 571g - coarse loamy	1.70 (61)	-	-	1.80 (10)
Staffordshire 551a - sandy 551g - sandy	1.23 (29) 0.64 (13)	1.02 (25) 0.39 (9)	1.13 (24) 0.92 (20)	0.61 (12) 0.81 (6)
Shropshire 551d - sandy	0.40 (31)	0.94 (13)	1.65 (41)	1.59 (27)
Norfolk West 343g - coarse loamy	0.40 (16)	0.20 (4)	0.37 (27)	-
Norfolk East 551g and 541t - sandy and coarse loamy	0.34 (36)	0.23 (8)	0.59 (37)	0.78 (13)
Nottinghamshire 551b - sandy	0.33 (23)	0.89 (35)	1.07 (70)	0.79 (44)

* Soil map unit, and particle-size subgroup of the most widespread soil in the map unit (SSEW, 1983)

In those soil associations with a sufficient number of eroded fields comparisons can be made of erosion in fields drilled to different crops (Table 1). Generally, fields drilled to sugar beet eroded most severely, those to spring cereals least. The way the land was worked (Evans, 1990a), and the probably greater rainfall intensities in spring, possibly explain the differences in rates between crops. Often winter cereal fields eroded more than spring cereal fields, possibly because within them relief was often greater, and slopes were steeper and longer (see below) and Somerset and the Isle of Wight received more rainfall. However, to explain other differences in values is not easy.

An earlier analysis of the data for the first three years of the monitoring scheme suggested that amounts eroded in a locality were related, once thresholds were exceeded, to the steepness of slope and relief within a field catchment, and soil texture was also a controlling factor (Evans, 1990a). But whilst the median values of erosion in a locality appeared to be related to the median maximum slope steepness and median relief in field catchments in that locality, no statistically valid relationship could be found between these factors in each individual locality.

Table 2. Data related to volumes of soil eroded in fields

Rainfall	- total amount in mm (from 1 to 4 stations per locality) - winter cereal - from October 1st to January 31st - spring cereal - March and April - sugar beet - from April 1st to June 15th - potatoes - from April 1st to May 31st
Particle size	- for sand (60-2000 μm), fine sand (60-200 μm), silt (2-60 μm) and clay (<2 μm); taken from Survey Bulletins describing the soils of England and Wales
Catchment	- area (ha) measured from 1:25,000 scale topographic maps
Slope steepness	- steepest slope in catchment, expressed as per cent, measured from 1:25,000 scale topographic maps
Slope length	- longest contributing slope in catchment, length (m) measured from 1:25,000 or 1:10,000 scale topographic maps, or from 1:10,000 scale aerial photos
Relief	- difference in elevation (m) within the field catchment estimated from 1:25,000 scale topographic maps

Another analysis of the 1982-84 survey data has been made to relate factors which are considered to control water erosion (Table 2) to volumes of soil eroded in field catchments. The catchments were divided into those in which channels had incised only into slopes and those where channels were cut only in valley floors. The information was compiled for the first three years (1982-84) of the monitoring project. Most of the catchments eroded in 1983, whereas there was little erosion in 1984.

The variances explained by the regression analyses are low: 30 per cent for spring cereals; 25 per cent for sugar beet; and 24 per cent for winter cereals and potatoes on eroded slopes; and in eroded valley floors in sugar beet (28 per cent) and winter and spring cereal (6 per cent) fields. For individual localities, more of the variance is explained by the regressions, for example, for eroded slopes in winter cereal fields in Dorset (56 per cent) and Somerset (50 per cent); but again the variances explained are lower in those fields where only valley floors eroded, for example Gwent (42 per cent), Bedfordshire and Staffordshire (36 per cent), and Herefordshire (34 per cent).

Table 3. Median values of erosion and of some factors contributing to erosion on slopes

	Winter cereal	Spring cereal	Sugar beet	Potatoes
No. of catchments	123	60	93	61
Erosion (m³/ha)	3.6	3.3	3.9	3.5
Topsoil texture -				
Sand (60-2000 μm)	56^{abc}	79^a	79^b	79^c
Fine sand (60-200 μm)	24	21^d	24^d	21
Silt (2-60 μm)	28^{abc}	10^a	10^b	10^c
Clay (<2 μm)	15^{abc}	8^a	12^b	8^c
Catchment area (ha)	1.8	1.1^{de}	2.2^d	2.2^e
Steepest slope (per cent)	10.2^b	10.1	7.6^b	8.7
Slope length (m)	185^a	150^{ad}	175^d	170
Relief (m)	15.0^{abc}	10.7^a	9.1^b	10.7^c

Statistically significantly different at 0.05 per cent (Mann-Whitney test):
a Winter cereal to spring cereal
b Winter cereal to sugar beet
c Winter cereal to potatoes
d Spring cereal to sugar beet
e Spring cereal to potatoes

Parameters considered important in explaining the variance of the regressions are those which explain more than 5 per cent of the variance, when regressed against the volume eroded. However, except for one parameter — catchment area — there is no consistency in the

Soils in rilled or gullied valley floors in sugar beet fields generally contained not only more fine sand, but less silt than did soils in eroded valley floors under spring cereals. Catchments under sugar beet with rilled valley floors were larger with longer, but less steep slopes than those under cereals, results similar to those in catchments where erosion had occurred only on slopes. There were few catchments in potato fields with rilled valley floors, which can be partly attributed to the potato ridges crossing valley floors and being resistant to down-valley erosion.

Erosion on slopes in winter cereal fields was more severe than it was in valley floors (Table 5), and where rilling took place only on slopes soil textures were considerably lighter. Although slopes adjacent to rilled valley floors were steeper, the length of slope contributing runoff was shorter. In eroded spring cereal fields, morphological factors did not differ significantly where rills cut into either slopes or valley floors, but where erosion only took place on slopes, soils were lighter textured (Table 6).

Table 6. Comparison of values of erosion, and of some factors contributing to erosion, on slopes and in valley floors in catchments in spring cereal fields

	Slope		Valley floor	
	60 catchments		53 catchments	
	Median	Mean	Median	Mean
Erosion (m^3/ha)	3.3*	7.6	1.5	2.2
Topsoil texture -				
Sand (60-2000 µm)	79	67	65	51
Fine sand (60-200 µm)	21*	22	19	21
Silt (2-60 µm)	10*	21	21	33
Clay (<2 µm)	8*	11	14	16
Catchment area (ha)	1.1	2.0	1.4	1.7
Steepest slope - per cent	10.1	11.7	10.2	12.1
Slope length (m)	150	166	125	143
Relief (m)	11	11	10	12

* Statistically significantly different at 0.05 per cent (Mann-Whitney test)

In rilled sugar beet fields (Table 7), where only slopes were eroded soils were little different in texture to those where only valley floors were eroded. However, rilled slopes were steeper and had smaller catchments.

Erosion is more likely to take place on lighter textured soils, therefore, and on slopes rather than in valley floors. Soils containing more fine sand are more likely to slake and erode. Sugar beet fields are more vulnerable to rilling than fields drilled to other crops, but rates of erosion are similar regardless of crop type. Erosional thresholds are higher in fields drilled to winter cereals.

Table 7. Comparison of values of erosion, and of some factors contributing to erosion, on slopes and in valley floors in catchments in sugar beet fields

	Slope		Valley floor	
	93 catchments		26 catchments	
	Median	Mean	Median	Mean
Erosion (m³/ha)	3.3*	12.0	1.0	2.3
Topsoil texture -				
Sand (60-2000 µm)	79	73	79	72
Fine sand (60-200 µm)	24	25	24	25
Silt (2-60 µm)	10	16	10	16
Clay (<2 µm)	12	11	12	12
Catchment area (ha)	2.2*	3.2	4.0	4.6
Steepest slope - per cent	7.6*	9.3	5.2	7.8
Slope length (m)	175	196	200	225
Relief (m)	9	10	10	11

* Statistically significantly different at 0.05 per cent (Mann-Whitney test)

Information on the extent of crop cover when erosion took place was not gathered in the monitoring scheme and could have aided in explaining the differences in erosion rates from year to year. To illustrate this, there appeared a reasonable exponential relationship between rate of erosion and extent of erosion for four out of the five years of the monitoring scheme. However, in the fifth year (1986) of the scheme erosion was similarly widespread but not as severe as that in the second year (1983). In 1986 erosive storms occurred later in the growing season in fields drilled to both winter cereals and spring-sown crops. In that year then, crop cover at the time of erosion was greater than it was in the other years and so erosion was impeded more.

Conclusions

Monitoring erosion in the field is quick and easy to do and can give useful information to predict and quantify erosion risk (Evans, 1990b). It gives data which models presently cannot satisfactorily mimic, such as the frequency of erosion, and its extent, i.e. the distribution of eroded fields within the landscape.

Also, and importantly, when rates of erosion are compared at different scales, i.e. from the very small areas directly affected by erosion to the area of the monitored transect, erosion is put into perspective within the landscape (Evans, 1992a). This again emphasises the distributional aspect of erosion.

Field observation suggests that splash and sheet erosion are much less important than channel erosion in redistributing soil in farmers' fields. Modellers need to be sure that their models reflect this reality.

A factor of primary importance for predicting and explaining the incidence and severity of erosion is the extent of crop cover at the time of the erosive storm. Amounts eroded will only be accurately predicted when good quality data on crop cover is incorporated into the models.

Erosional thresholds have to be overcome before rilling takes place, but soil and morphological thresholds may vary with crop type (Evans, 1992b). These thresholds need to be built into models.

Field monitoring suggests that a combination of factors such as rainfall amount and intensity, land use, soil type, soil surface characteristics, crop cover and morphology result in erosion rates which may be unique to a particular locality.

Models do what you want them to. Data collected in the field to allow the prediction of erosion risk do that. But analysis of this field data to arrive at useful information with regard to factors contributing to erosion, i.e. which factors are more important than others in contributing to the incidence of rilling, is difficult. As a corollary, it seems that models based on plot data which predict how erosional processes work will do that, but the predicted rates of erosion may have little relevance to the field where rates of erosion seem to fall within a limited range.

Acknowledgements

Jackie Taylor and Melanie Legg drew Figure 1. This paper is a contribution to the Soil Erosion Network of the GCTE, which is a Core Research Project of the IGBP.

References

Evans, R. (1980a). Characteristics of water-eroded fields in lowland England. In, De Boodt, M. and Gabriels, D. (eds), *Assessment of Erosion*, Wiley, Chichester. pp.77-87.

Evans, R. (1980b). Mechanics of water erosion and their spatial and temporal controls: an empirical viewpoint. In, Kirkby, M.J. and Morgan, R.P.C. (eds.). *Soil Erosion*, Wiley, Chichester. pp. 109-128.

Evans, R. (1987). Soilwatch. *Soil Association Membership News*, December issue, 6-7.

Evans, R. (1990a). Water erosion in British farmers' fields - some causes, impacts, predictions. *Progress in Physical Geography* **15**, 199-219.

Evans, R. (1990b). Soils at risk of accelerated erosion in England and Wales. *Soil Use and Management* **6**, 125-131.

Evans, R. (1992a). Assessing erosion in England and Wales. In, Haskins, P.G. and Murphy, B.M. (eds), *People Protecting Their Land. Volume 1*, International Soil Conservation Organisation, Sydney. pp. 82-91.

Evans, R. (1992b). Rill erosion in contrasting landscapes. *Soil Use and Management* **8**, 170-175.

Evans, R. (1993a). On assessing accelerated erosion of arable land by water. *Soils and Fertilizers* **56**, 1285-1293.

Evans, R. (1993b). Extent, frequency and rates of rilling of arable land in localities in England and Wales. In, Wicherek, S. (ed.), *Farm Land erosion: In Temperate Plains Environment and Hills*, Elsevier Science Publishers, Amsterdam. pp. 177-190.

Evans, R. (1995). Some methods of directly assessing water erosion of cultivated land - a comparison of measurements made on plots and in fields. *Progress in Physical Geography* **19**, 115-129.

Evans, R. and Boardman, J. (1994). Assessment of water erosion in farmers' fields in the UK. In, Rickson, J. (ed.), *Conserving Soil Resources: European Perspectives*, CAB International, Wallingford. pp. 13-24.

Evans, R. and Cook, S. (1986). Soil erosion in Britain. *SEESOIL* **3**, 28-58.

SSEW (1983). *Soil Map of England and Wales*. Soil Survey of England and Wales, Harpenden.

Wischmeier, W.H. and Smith, D.D. (1978). *Predicting Rainfall Erosion Losses - A Guide to Conservation Planning*. Handbook No. **537**, U.S. Department of Agriculture, Washington.

24. EFFECTS OF AGRICULTURAL LAND USE ON SPATIAL AND TEMPORAL DISTRIBUTION OF SOIL EROSION IN SMALL CATCHMENTS: IMPLICATIONS FOR MODELLING

Anne-Véronique Auzet[1], Jean Boiffin[2], Bruno Ludwig[2] and Jérôme Guérif[2]

[1] CEREG-URA 95 CNRS
Université Louis Pasteur
3 rue de l'Argonne
F-67083 Strasbourg cédex
France

[2] INRA
Unité d'Agronomie de Laon-Péronne
rue Fernand Christ
F-02007 Laon cédex
France

Abstract

Agricultural land use has important effects on states of the soil surface and on spatial organisation of the catchment, which determines runoff generation and routing. It interacts with geomorphic and climatic conditions to influence the main physical processes involved in soil erosion. Further, it induces great spatial and temporal changes in soil conditions. This paper proposes, in the light of the results of studies that have been carried out in the northern Paris Basin, an approach to take into account these effects and to improve the way to collect input variables for distributed runoff erosion models.

Introduction

Influence of crop management on the dynamics of soil cover has been for long recognised as a major factor controlling erosion. However, agriculture also induces large and fast changes in soil surface structure, which strongly influence runoff generation and determines a spatial organisation of the runoff routing. These last influences were stressed in recent empirical studies that have been carried out in the northern Paris Basin (Boiffin *et al.*, 1988; Papy and Boiffin, 1988; Auzet *et al.*, 1990; Auzet *et al.*, 1993; Auzet *et al.*, 1995; Ludwig *et al.*, 1995; Ludwig *et al.*, in press *a*).

These studies showed that agricultural use of the land interacts with geomorphic and climatic conditions to influence the main physical processes such as infiltration, surface storage, runoff generation, soil particle detachment, runoff concentration and routing. In the present state of the art, however, these dynamic influences and interactions are insufficiently incorporated in erosion models.

The paper proposes an approach to take into account the great spatial and temporal changes which occur and to suggest the collection of input variables for distributed runoff erosion models in this kind of region. After a short presentation on the context of the studies, we will address specific issues for runoff and soil erosion modelling, and discuss the consequences for building and using runoff and soil erosion models.

Context of studies carried out in northern Paris Basin

As in many areas of the loessial belt of north-western Europe occupied by intensive agriculture, rill erosion is widespread even where slope gradients and rainfall intensities are relatively low. Of the twenty catchments studied, areas with slope gradients greater than 5 per cent represented on average 34 per cent of the total area, while 9 per cent had slope gradients over 10 per cent. A large part of the region is covered by loamy soils, with low aggregate stability. Mean annual rainfall ranges from 730 mm to 580 mm from west to east in the studied area, and rainfall intensities >10 mm/h represent less than 20 per cent of the total rainfall amount (Auzet *et al.*, 1995).

Landscapes are characterised by large open fields. Crop rotations are based on annual crops Then, through the year, many farming operations strongly influence the soil surface state (structure, roughness, etc.) and related properties (Boiffin *et al.*, 1988; Papy and Boiffin, 1989; Imeson and Kwaad, 1990) such as hydraulic conductivity and depressional water storage. Each winter, more than half the total area is covered by winter crops. The other fields remain generally untilled after the harvest of spring crops or stubble-ploughed after cereals.

Concentrated flow erosion is the main type of erosion and became a subject of interest because of its frequency and severity of associated off-site damage. In such a context, soil losses are spatially limited to linear features liable to concentrate the surface flow. The runoff contributing areas can differ spatially from the sediment sources.

Specific issues for runoff and soil erosion modelling

Erosion models are usually composed of three main submodels: a hydrological, a detachment and a transport-deposition submodel. The temporal dynamics of input variables and parameters of all these submodels, especially infiltrability, roughness and cohesion, are greatly influenced by the dynamics of vegetation cover and structural states of the surface soil horizons. These dynamics are marked by a succession of progressive phases and discontinuities. Interactions between farming operation sequences, climate and soils induce marked changes in infiltration rate and surface roughness (Boiffin et al., 1988; Papy and Boiffin, 1989; Imeson and Kwaad, 1990). For instance, for a specific field, the runoff risks could be schematically assessed through time, taking into account the temporal distribution of rainfall and farming operations as a function of crop rotation (Figure1: Auzet et al., 1990). The interactions between soil, climate and agricultural land use result in impacts on the structural state of the soil, and surface roughness (Boiffin et al., 1988b) and soil cohesion (Guérif, 1990).

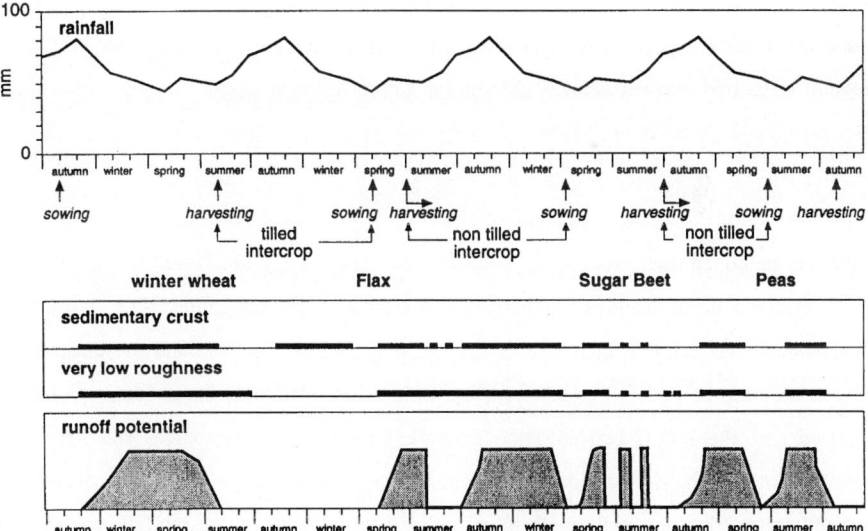

Figure 1. Risk of runoff in relation to crop rotation, farming operations and climatic conditions

Table 1. Runoff potential as a function of the structural state of the soil surface and roughness in the drilling direction

structural state of the soil surface	roughness	runoff potential
sedimentary crust on more than 75 % of the area	very low (0-2 cm)	very high
sedimentary crust on 50% to 75 % of the area	very low (0-2 cm)	high
sedimentary crust on more than 75 % of the area	low (2-5 cm)	high
sedimentary crust on 50% to 75 % of the area	low (2-5 cm)	low
all situations	5 to 20 cm	very low
no sedimentary crust	all situations	very low

The spatial distribution of those variables and parameters is mainly due to the distribution of soil physical properties and the juxtaposition of fields having different histories. A three winter survey of 20 cultivated catchments in different regions in the northern Paris Basin (Ludwig, 1992; Auzet et al., 1995) demonstrates the relationships between the rill erosion rate and the extension of runoff-contributing areas evaluated by a criterion combining the area covered by sedimentary crusts or by wheelings and low roughness in the drilling direction (Table 1). Results obtained for all the three winters showed the same relationships, and could be fitted to a single relation (Figure 2), confirming that the proportion of runoff contributing areas is the primary determinant of differences in rill erosion between-catchments. The residual variability, particularly between catchments located close to each other in the same region, suggested that variables related to the runoff collector length influence rill volumes (Ludwig, 1992; Auzet et al., 1995).

Also the rills do not have a random spatial distribution in a catchment: location of the rill network remained the same from year to year (Figure 3) and used for a large part predetermined linear features (Ludwig, 1992) related to the topography (e.g. talwegs and hollows) or agricultural land use (e.g. wheelings, dead furrows, and headlands). The characterisation of each segment of rill by its location relative to linear features (Table 2) showed that the runoff concentration network formed by topographical and agricultural features controls the rill location and determines the potential length of the rill network. The incised part of this potential length in any given year depends on the rainfall characteristics and on the spatial relationships between collectors and runoff contributing areas.

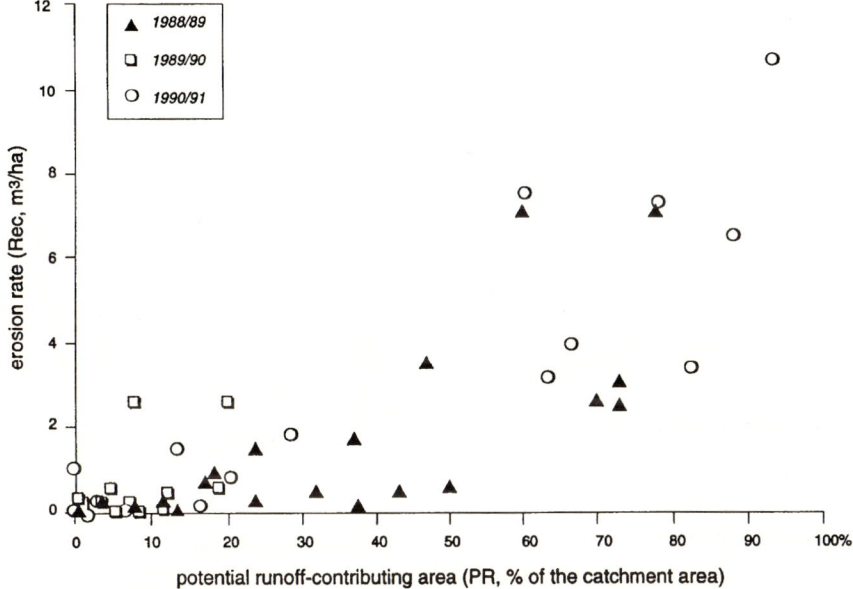

Figure 2. The relationship between erosion rate (Rec, m^3ha^{-1}) and potential runoff contributing area (PR, percentage of total catchment area)

Table 2. Volume of rills and distribution of the part located on predetermined linear features, as a function of feature type

	winter seasons		
	1988/89	1989/90	1990/91
total rill volume (a)	1198 m^3	353 m^3	2185 m^3
volumes of rills located on predetermined linear features (b)	1138 m^3	337 m^3	1746 m^3
percentage of (b) associated with			
topographical features only	61 %	58 %	52 %
both topographical and agricultural features*	12 %	21 %	28 %
agricultural features	27 %	21 %	20 %

*agricultural features could be located also on a topographical feature, e.g. a talweg

Ludwig et al. (1995) found that a better explanation of the variability of rill erosion rate could be obtained when taking into account the internal hydrological structure of the catchment. Their approach consisted of (a) identifying the runoff collectors and splitting them into homogeneous segments; (b) characterising each segment by its length, slope, soil susceptibility to rill incision and the upslope runoff contributing area connected to it, identified by the structural state of their soil surface; (c) studying the relationships between these characteristics and the frequency of rill occurrence and the rill cross-sectional areas.

Such a typology is applicable in time and space: but it implies agronomical monitoring of farming operations and, possibly, soil surface states (Papy and Douyer, 1991; Ludwig and Boiffin, 1994). Then, in a catchment, areas that give rise to runoff could be determined at each visit, by applying the typology to each land unit, corresponding to portions of fields having the same topsoil texture (Auzet *et al.*, 1995; Ludwig *et al.*, 1995).

The catchment spatial structure, which is greatly influenced by geomorphology, can be used to identify the various functional elements, which are, in addition to potential runoff contributing areas, collectors and hydrological connections. The spatial organisation of these elements is largely dependent on the location of field boundaries, headlands, dead furrows, wheel tracks and the spatial distribution of structural states of the soil surface (Figure 5).

Hydrological models need a complementary model of the collector network and the hydrological connections to the runoff contributing areas. This submodel should include collectors which are topographical or agricultural linear features and should take into account the influence of tillage direction on runoff. It could be built using the flow chart (Figure 6) proposed by Ludwig *et al.* (in press *b*).

Conclusions

The use of land for crop production in a cultivated catchment has major effects on the spatial organisation of potential runoff contributing areas, collectors and connections, via the location of field boundaries, headlands, dead furrows, wheel tracks and the spatial distribution of structural states of the soil surface. The temporal dynamics of the conditions of these elements depends greatly on the interaction between farm operations timetable and practices, soil characteristics and climatic conditions. Assessment of the climatic change effects on runoff and erosion in regions of intensive agriculture requires both process models and a system that accurately describes the parameters of cultivated land. The descriptive system should provide information in a form compatible to the model inputs and accurately describe the spatial and temporal dynamics.

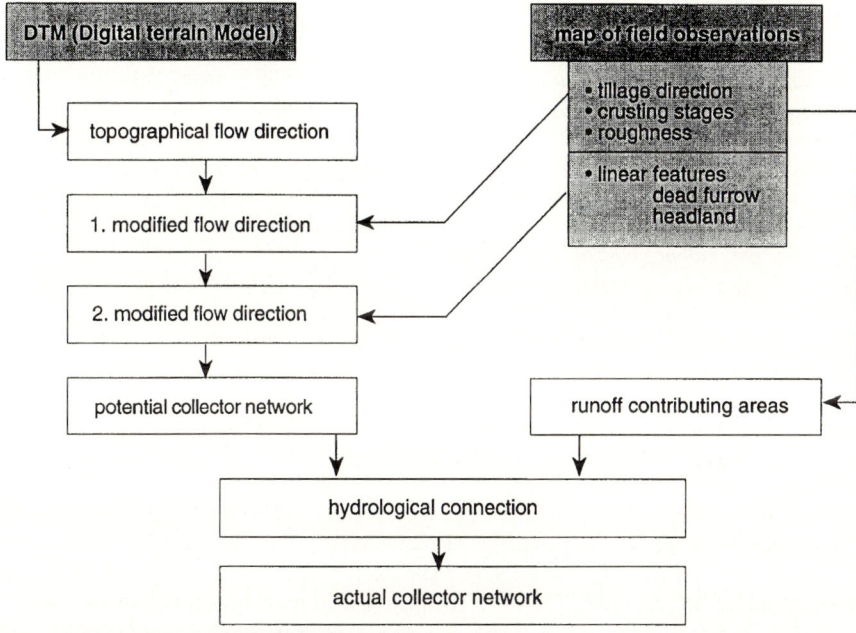

Figure 6. Flow chart proposed to take into account the effects of agricultural land use on runoff contributing areas and runoff routing

References

Auzet, A.V., Boiffin, J., Papy, F., Maucorps, J. and Ouvry, J.F. (1990). An approach to the assessment of erosion forms and erosion risk on agricultural land in the Northern Paris Basin, France. In, Boardman, J., Foster, I. D. L. and Dearing, J. A. (eds), *Soil Erosion on Agricultural Land*, Wiley, Chichester. pp. 383-400.

Auzet, A.V., Boiffin, J. and Ludwig, B. (1995). Concentrated flow erosion in cultivated catchments: influence of soil surface state. *Earth Surface Processes and Landforms* **20**, 759-767.

Auzet, A.V., Boiffin, J., Papy, F., Ludwig, B. and Maucorps, J. (1993). Rill erosion as a function of the characteristics of cultivated catchments in the north of France. *Catena* **20**, 41-62.

Boiffin, J., Papy, F. and Eimberck, M. (1988a). Influence des systèmes de culture sur les risques d'érosion par ruissellement concentré. I - Analyse des conditions de déclenchement de l'érosion. *Agronomie* **8**(8), 663-673.

Boiffin, J., Papy, F. and Monnier, J. (1988b). Some reflexions on the prospect of modelling the influence of cropping systems on soil erosion. In, Morgan, R. P. C. and Rickson, J. (eds), *Agriculture: Erosion Assessment and Modelling*. CEE Report 10860, 215-214.

Guérif, J. (1990). Conséquence de l'état structural sur les propriétés et les comportements physiques et mécaniques. In, Boiffin, J. and Marin-Laflèche, A. (eds), *La Structure du Sol et son Évolution: Conséquences Agronomiques, Maîtrise par l'Agriculteur*. Les Colloques de l'INRA, 53, INRA, Paris. pp 71-89.

Imeson, A.C. and Kwaad, J. P. M. (1990). The response of tilled soils to wetting by rainfall and the dynamic character of soil erodibility. In, Boardman, J., Foster, I. D. L. and Dearing, J. A. (eds), *Soil Erosion on Agricultural Land*, Wiley, Chichester. pp. 3-14.

Ludwig, B. (1992). *L'Érosion par Ruissellement Concentré des Terres Cultivées du Nord du Bassin Parisien: Analyse de la Variabilité des Symptômes d'Érosion à l'Échelle du Bassin Versant Élémentaire*. Thèse doctorat d'université, Université Louis Pasteur Strasbourg I. 201p:

Ludwig, B. and Boiffin, J. (1994). Simulation of the influence of protection measures on the genesis of ephemeral gullies in cultivated catchments. In, Jensen, H.E., Schjønning, P., Mikkelsen, S.A. and Madsen, K.B. (eds), *Proceedings of 13th International ISTRO Conference, Aalborg DK*. pp. 1169-1174

Ludwig, B., Boiffin, J., Chadoeuf, J. and Auzet, A. V. (1995). Hydrological structure and erosion damage caused by concentrated flow in cultivated catchments. *Catena* **25**, 227-252.

Ludwig., B., Auzet, A. V., Boiffin, J., Papy, F., King, D. and Chadoeuf, J. (in press *a*). L'érosion par ruissellement rôle des états de surface et de la structure hydrologique des bassins versants. *Etude et Gestion des Sols*.

Ludwig, B., Daroussin, J., King, D. and Souchère, V. (in press *b*).Using GIS to predict concentrated flow erosion in cultivated catchments. *IAHS Proceedings,* HydroGIS'96, Vienna 1996.

Papy, F. and Boiffin, J. (1988). Influence des systèmes de culture sur les risques d'érosion par ruissellement concentré. II - Evaluation des possibilités de maîtrise du phénomène dans les exploitations agricoles. *Agronomie* **8**(9), 745-756.

Papy, F. and Boiffin, J. (1989). The use of farming systems for the control of runoff and erosion. In, Schwertmann, U., Rickson, R. J. and Auerswald, K. (eds), *Soil Erosion Protection Measures in Europe*, Soil Technology Series 1, 29-38.

Papy, F. and Douyer, C. (1991). Influence des états de surface du territoire agricole sur le déclenchement des inondations catastrophiques. *Agronomie* **11**, 201-215.

25. SENSITIVITY OF THE MODEL LISEM TO VARIABLES RELATED TO AGRICULTURE

Victor Jetten[1], Ad de Roo[2] and Jerome Guérif[1]

[1] INRA - Station d'Agronomie de Laon
rue Fernand Christ
02007 Laon cédex
France

[2] Department of Physical Geography
Utrecht University
PO Box 80.115
3508 TC Utrecht
The Netherlands

Abstract

A sensitivity analysis of the model LISEM was performed for those variables that are directly related to arable farming in Northern France. LISEM is a physically based spatial erosion model integrated into a raster GIS. The variables are: saturated hydraulic conductivity, random roughness, Manning's n, cohesion and aggregate stability. The sensitivity analysis was done on an artificial dataset, first by a direct comparison of input and output, second by calculating a normalised sensitivity factor, and third by coupling the responses of the main algorithms in LISEM to analyse the chain of events. It appeared that changing one variable and comparing it with a baseline value is not sufficient to understand the model response. Combinations of different input values gave the same output. LISEM is sensitive to changes in Ksat in all processes, but the sensitivity varies with different levels of surface roughness. Discharge and net erosion are very sensitive to random roughness, but only when Ksat is relatively high and there is not much water for overland flow. Flow detachment is moderately sensitive to Manning's n (through velocity and transport capacity) and to cohesion. Aggregate stability only played a role when there was not much water at the surface. The response of overland flow and flow detachment to a change in flow depth is highly non-linear with thresholds and compensating effects, which poses constraints to the accuracy of the input variables that has to be obtained.

Introduction

Erosion research in areas of intensive agriculture in northern France, focuses on off-site damages including mud flows and surface water pollution. The water and sediment discharges from these catchments are strongly influenced by agriculture. Agricultural activities, together with soil and climate, induce a large spatial and temporal variability of soil properties and surface drainage direction (e.g. Boiffin and Monnier, 1987; Auzet *et al.*, 1995). Infiltrability of the soil and surface roughness undergo rapid changes caused by tillage operations, followed by a slow decrease of soil structure caused by rainfall. Apart from these, soil cover by plants or residue, aggregate stability and soil cohesion are also determined by the interaction of climate with agricultural activities. The erosion model LISEM (De Roo *et al.*, 1995) is used in a research project of the Institut National de Recherche Agronomique (INRA) in northern France, that investigates the effects of a wide range of agricultural management practices on overland flow and erosion. A sensitivity analysis was done as part of this research, (i) to see how sensitive the model is to the range of values that the input variables take under the conditions described above, and to detect thresholds and non-linear behaviour; and (ii) to give an indication of the accuracy with which the variables should be obtained. A sensitivity analysis can be done in combination with calibration and validation using a test dataset (see e.g. testing of the models SHE by Bathurst, 1986; ANSWERS by De Roo, 1993; EUROSEM by Quinton, 1994), or with the emphasis on the theoretical response of the model (see e.g. EPIC by Sharpley and Williams, 1990; WEPP by Nearing *et al.*, 1989). Since the calibration and validation of LISEM has been done elsewhere (De Roo and Offermans, 1995), the latter approach is used here. The sensitivity analysis is done with a simple small catchment in three parts: first a direct comparison of input with output is done, second a normalised sensitivity factor is calculated, and third the response curves of the most important algorithms in LISEM are analysed.

Methodology

In a complex spatial deterministic model such as LISEM the effect of a change of a input parameter may be different for various combinations of the other input parameters. Therefore it is not sufficient to vary only one parameter at a time and compare the output to a baseline output, one should investigate all possible combinations. This was done in the first part of this sensitivity analysis. However, to avoid an unmanageable dataset, only a limited number of input parameters are considered (Table 1): those that are influenced directly by arable farming. The minimum and maximum values are derived from an erosion research project in Limburg, The Netherlands (De Roo *et al.*, 1994), where soil type (loess) and agriculture are similar to Northern France. The sensitivity of the model is checked against the catchment output: total discharge (Qtot) and total sediment (Etot), as well as the accumulated splash and flow detachment and deposition in the field. A direct comparison of input and output has the disadvantage that the effect of a change in input may be obscured because the input variables have different orders of magnitude. Therefore a relative 'sensitivity parameter' S is calculated in the second part (Nearing *et al.*, 1989):

$$S = \frac{(O_2 - O_1)/O_{12}}{(I_2 - I_1)/I_{12}} \tag{1}$$

where I_1 and I_2 are the least and greatest of the input values used for a parameter, I_{12} is the average of both values, O_1 and O_2 and O_{12} are the associated outputs. S represents a normalised change in output to a normalised change in input. In this paper, only the response to a difference of minimum and maximum values is analysed. Most processes in LISEM however, produce highly non-linear curves or have certain thresholds. Therefore in the third part of this analysis, the main algorithms are placed in a spreadsheet and the response curves of the internal variables are made. In this way the 'one-dimensional' chain of processes that take place within a grid cell is calculated.

Table 1. Minimum and maximum values of variables used in the sensitivity analysis of LISEM

Variable		units	min	max
Sat. hydr. conductivity	Ksat	mm/h	5.0	15.0
Random Roughness	RR	cm	0.5	4.5
Manning's n	n	-	0.07	0.27
Cohesion	COH	kPa	0.7	7.0
Aggregate Stability	AS	-	5	23

A catchment with complex relief and land use will have a large spatial variability of the input and output variables. To avoid averaging out and internal compensating effects, a simple 4.2 ha artificial catchment, with a convex-concave slope and one channel, was used for this sensitivity analysis. The soil is considered homogeneous for all its properties and assumed to be near saturation. To avoid further complexity a single 30 mm/h rainfall 'pulse' of 30 min was used instead of a real rainstorm.

Absolute effects of input variable changes

Figure 1 shows the absolute values of the five output variables for different combinations of input variable values (each bar represents the results of one simulation). The input variables are organised in a hierarchical fashion, as shown in the last graph.

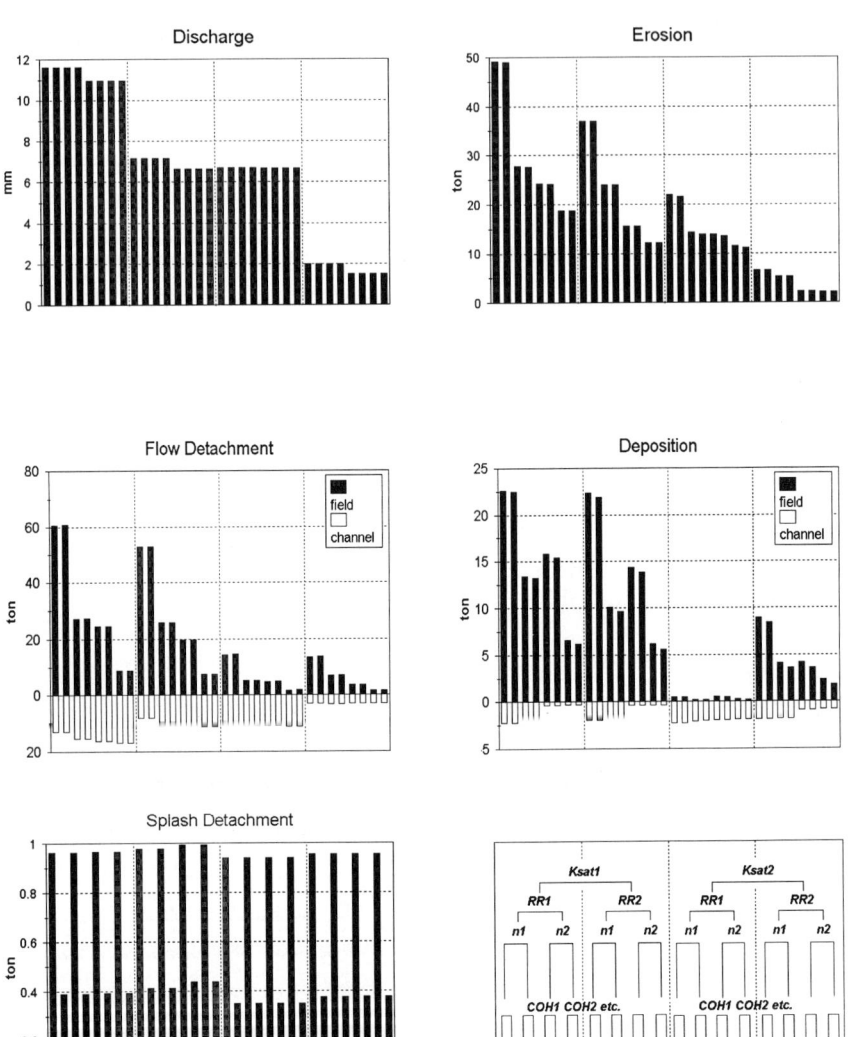

Figure 1. Response of five output variables to all combinations of 5 input variables, arranged according to the configuration diagram. Discharge and net erosion are net amounts leaving the catchment, deposition, flow and splash erosion are spatial and temporal cumulative values

Qtot   With a low conductivity (Ksat1) the difference in RR decreases the discharge by about 40 per cent, while if Ksat is high the effect is even larger. A combination of low conductivity-high roughness (Ksat1/RR2) has the same effect as a high conductivity-low roughness (Ksat2/RR1). There is a small effect of Manning's n.

Etot   Net erosion follows roughly the same pattern as Qtot and the combination Ksat/RR again plays a large role. On a second level, Etot is sensitive to Manning's n and cohesion: there are large differences between pairs of 4-bar groups (n1/n2) and between pairs of 2-bar groups (COH1/COH2).

DETF   The variables influence the output in the order of Ksat-n-COH. RR hardly plays a role and AS is of no influence. DETF seems to be equally sensitive to n and COH. Because the channel characteristics are not altered, channel erosion becomes more important if there is more flow resistance on the field (all combinations with n2), or less water on the field than in the channel (Ksat2/RR1).

DEP   Deposition is determined by Ksat in the first place, while on a second level RR either has no influence (Ksat1) and n and COH determine the differences in output, or RR determines DEP for Ksat2.

DETS   Splash detachment is influenced by AS only, with a small effect of RR.

Relative effects of input variable changes

Figure 2 shows the sensitivity factor S of the 5 output variables to each combination of input variables. Because S is calculated as the difference between the minimum and maximum of the variable in question, the number of points is halved compared to Figure 1. The graphs are more difficult to interpret because a different hierarchical order of combinations is used for each variable: the variable of interest moves to the top of the tree structure in each subsequent graph. For example the points in the third graph all signify a difference in Manning's n, whereby points 1-8 correspond to Ksat1 and points 9-16 to Ksat2, and e.g. point 6 corresponds to the combination Ksat1/RR2/COH1/AS2. This leads to the following behaviour of the variables:

Ksat   Sensitivity of Qtot, Etot to a change in saturated hydraulic conductivity increases when increases from RR1 to RR2 (left and right half of the graph), while sensitivity of DEP decreases. DETF and DETS remain equally sensitive to Ksat for all combinations of other variables.

RR   Sensitivity to random roughness of Qtot, Etot and DEP increases sharply when Ksat changes from low to high, probably because there is less water available. Also a small influence of n is noticeable for these combinations.

n   DETF is most sensitive to n as it is determined directly by the flow velocity. The sensitivity generally increases when there is less overland flow (Ksat2/RR2).

COH   Only DETF and DEP (and to some extent Etot) are sensible to cohesion, without much interference from the other variables. There is a small variation in sensibility according to Manning's n.

AS   Only DETS is determined by AS. Indirectly the influence on DEP is larger when there is less water, because the flow detachment decreases fast with water depth.

The results of both the absolute and the relative sensitivity analysis show that changing one variable while keeping the others at some baseline value is not sufficient to understand the response of LISEM. There are several very different combinations that have the same net effect.

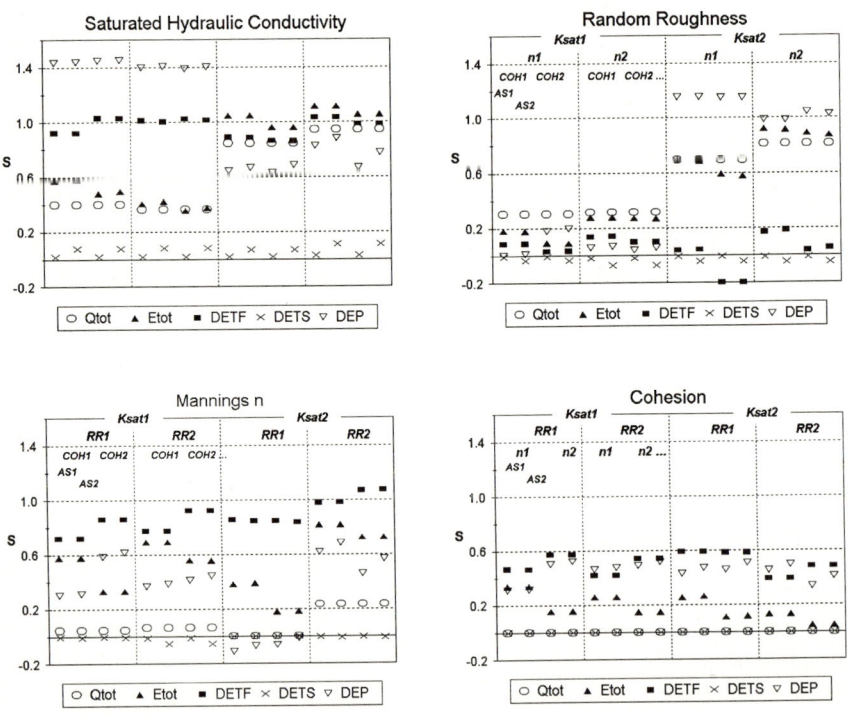

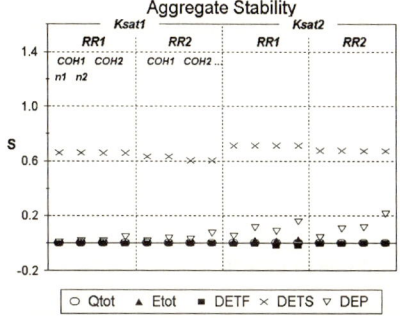
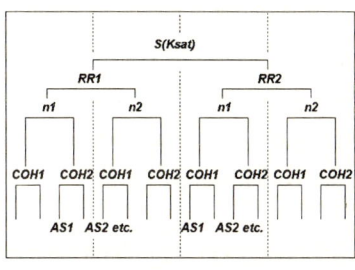

Figure 2. Normalised sensitivity S of the output variables to different combinations of the input variables. The diagram (bottom right) shows the configuration for Ksat. For the other input variables this configuration changes (see text)

One dimensional simulation

To investigate the non-linearity of the main processes in LISEM, as well as the influence of thresholds, response curves for the corresponding algorithms are made. These curves represent more or less the range of values obtained on a grid cell level. Infiltration is not calculated but it is assumed that the water depth at the surface ranges between 0 and 2 cm. To avoid discussing again a series of combinations, three variables are combined in such a way that their influence is maximised: it is assumed that a high slope gradient S corresponds to a low surface roughness. This means that S, RR and n are varied simultaneously with S = 9-1 per cent, from RR = 0.5-4.5 cm and n = 0.07-0.27.

The first process after infiltration calculated for each cell is surface depressional storage (microdepressions only). Two important internal variables are calculated: the flow depth FD (i.e. total water depth minus surface retention), and the fraction of wetted/ponded surface FWA. RR and S are used to derive minimum and maximum retention depths with which FD is calculated (Onstad, 1984). It can be seen from Figure 3a that the FD varies between 100 and 50 per cent of the water depth, while the threshold retention is at most 0.25 cm. Also the maximum fraction of ponded area corresponding to the maximum retention is derived from RR and S. This is used to determine the non-linear relation between FD and fraction of wet area (FWA, Figure 3b). Table 2 shows the sequence of processes calculated in a grid cell. Both surface storage variables influence the overland flow. The flow velocity V, calculated with Manning's equation for turbulent flow, depends on the FD, n and S. Figure 3c shows that for a water depth of 2 cm the velocity varies between 0.03 and 0.33 m.s^{-1}. The discharge Q (Figure 3d) depends both on V and

produced, but the associated FWA is small and little is transported. At a large depth the transport capacity is high but there is not much sediment available.

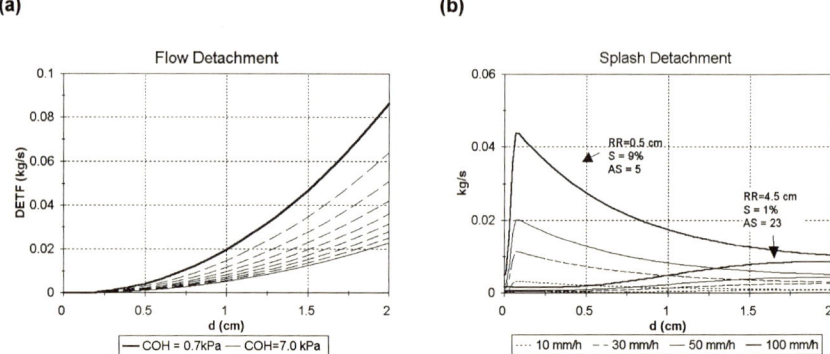

Figure 4. a) Flow detachment (DETF in kg.s^{-1}) for COH = 0.7 to 7 kPa, using the maximum velocity (Figure 3c); b) Splash detachment (DETS in kg.s^{-1}) for 4 rainfall intensities and combinations of RR, S and AS that minimise and maximise their effect

Conclusions

The results show that changing one variable while keeping the others at some baseline value is not sufficient to understand the response of LISEM. Different combinations of input values may have the same net effect. The input variable for which LISEM is most sensitive, is Ksat (or in fact the infiltration it represents) for almost all processes. Nevertheless the sensitivity changes considerably for different levels of surface roughness. RR influences the net output of the basin (Qtot and Etot), especially when Ksat is high and there is not much water for overland flow. Flow detachment is sensitive to Manning's n (through velocity and transport capacity) and of course to cohesion. It should be noted that n and RR are partly correlated (together with a crop if present). Splash detachment is sensitive to aggregate stability and surface roughness but only when there is not much water at the surface. There are only a few real thresholds in the model. Apart from the infiltration which determines whether or not there is water at the surface, the unit stream power threshold in the transport capacity has the largest influence. The minimum retention threshold plays a minor role. More important is a cumulative effect of the non-linear equations in Table 2 which causes a small change in water depth to have a large effect. For instance decrease in water depth of several millimetres (10 to 5 mm) causes an 80 per cent decrease in flow detachment (0.0066 to 0.0014 kg.s^{-1}). This indicates that great accuracy is needed in the determination of the water depth at the surface.

References

Auzet, A.V., Boiffin, J. and Ludwig, B. (1995). Concentrated flow erosion in cultivated catchments: influence of soil surface state. *Earth Surface Processes. and Landforms* **20**, 759-767.

Bathurst, J.C. (1986). Sensitivity analysis of the Systeme Hydrologique Europeen for an upland catchment. *Journal of Hydrology* **87**, 103-123.

De Roo, A.P.J., van Dijk, P.M., Ritsema, C.J., Cremers, N., Stolte, J., Offermans, R., Kwaad, F. and Verzandvoort, M. (1994). *Erosienormeringsonderzoek Zuid-Limburg*. Dept. of Physical Geography, University of Utrecht.

De Roo, A.P.J. and Offermans, R.J.E. (1995). LISEM: a physically based hydrological and soil erosion model for basin-scale water and sediment management. In, *Modelling and Management of Sustainable Basin-scale Water Resource Systems* (Proc. Boulder Symp.). IAHS Publication 231, pp 399-407.

De Roo A.P.J., Wesseling, C.G., Jetten, V.G., Offermans, R.J.E. and Ritsema, C.J. (1995). *LISEM, Limburg Soil Erosion Model, User Manual*, Dept. of Physical Geography, Utrecht University.

De Roo A.P.J. (1993). Modelling surface runoff and soil erosion in catchments using Geographical Information Systems. *Netherlands Geogr. Studies* **157**, 105-122.

Govers G. (1990). Empirical relationships for the transport capacity of overland flow. In, *Erosion, Transport and Deposition Processes*, IAHS Publivation 189, pp 45-63.

Nearing, M.A., Ascough, L.D. and Chaves H.M.L. (1989). WEPP model sensitivity analysis. In, *WEPP-USDA Water Erosion Prediction Project Documentation*, USDA-ARS NSERL Report 2, pp 14.1-14.33.

Morgan, R.P.C., Quinton, J.N. and Rickson, R.J. (1992) *EUROSEM Documentation Manual, Version 1: June 1992*, Silsoe college, Silsoe..

Onstad, C.A. (1984). Depressional storage on tilled soil surfaces. *Transactions of the American Society of Agricultural Engineers* **27**(3), 729-732.

Quinton, J.N. (1994). Validation of physically based erosion models, with particular reference to EUROSEM. In, Rickson, R.J. (ed.), *Conserving Soil Resources*, CAB International, pp 300-313.

Sharpley, A.N. and Williams, J.R. (1990). *EPIC: Erosion Productivity Impact Calculator 1. Model Documentation*. USDA-ARS Technical Bulletin **1768**.

26. APPLYING GIS TO CATCHMENT-SCALE SOIL EROSION MODELLING

Rachael McDonnell

School of Geography
University of Oxford
Mansfield Road
Oxford OX1 3TB
UK

Abstract

Most catchment-scale soil erosion models require an explicit spatial representation of the heterogeneity of inputs such as topography, soil conditions and land cover, to capture the within-area variations in movements as well as outflows. The basic spatial units adopted for this range from fields to grid cells. There is an increasing number of examples of soil erosion models being linked to or embedded within both commercial and in-house written Geographical Information Systems (GIS). The functionality of these systems support the generation and storage of data for the models, as well as allowing the spatial relationships between soil erosion processes and form to be explored. There are a number limitations with using GIS in this work. Many of the problems are associated with the lack of spatial data available to support the modelling.

Introduction

The increasing costs of soil erosion in both environmental and monetary terms have encouraged both data collection and model development efforts, at a range of spatial and temporal scales. Catchment-scale fieldwork has highlighted the spatial complexity of the sediment detachment and transport processes resulting from the variability of soils, hydraulic conditions, slope and to a lesser extent land use and precipitation. Even within sub-catchment areas antecedent hydrological conditions — as well as changes in soil properties during a storm — give rise to dynamic temporal and spatial variabilities in saturated conditions, overland flow, aggregate size and stability, and the resulting erosion. In addition, spatially-limited soil conditions — such as tractor wheelings or surface crusting — bring about locally concentrated overland flow and erosion which contributes significantly to the total movement of sediment in a catchment.

Models have represented soil erosion processes at many different levels of complexity and at varying spatial and temporal scales. They have been used most widely as generators of quantitative predictions of sediment movement for single slopes with the Universal Soil Loss Equation (USLE) (Wischmeier and Smith, 1978), for all its acknowledged limitations, forming the basis of many models. Recently a more spatially distributed approach has been adopted in modelling work (e.g. the AGNPS model of Young et al., 1987), which attempts to embody the spatial variability in the erosion and deposition processes acting within a catchment. With these models an area is divided into a series of basic units and detachment/erosion values are calculated for each. Various routing functions, based on the topological structure defined for the catchment, are then used to determine the amount of soil actually entering and leaving the individual units.

The move to a more distributed approach has been supported by a better understanding of the variability of detachment and erosion processes, as well as by increasing resources of digital map and sensor data. In addition computing technology in the form of GIS has been developed specifically for handling spatial data. This paper details the possibilities and limitations of using current GIS in soil erosion modelling at the catchment-scale. The work is set in context by reviewing briefly the spatial data requirements of two existing soil-erosion models. The functionality and modelling capabilities of GIS are then detailed and their previous application in soil erosion modelling are explored. The benefits highlighted by these examples are then weighed against the limitations of using current GIS. The possibilities and developments in the broader context of spatial data handling and their implications for new modelling are then elucidated.

Modelling soil erosion at the catchment scale

Soil erosion models have been developed to extend the understanding of physical process as well as for generating quantitative predictions of losses often in the management context. The resulting models differ widely in their representation of processes, and temporal and spatial scales. The GAMES model (Dickinson et al., 1985; Snell, 1985; Dickinson et al., 1986; 1990a and 1990b), for example, is designed to estimate seasonal or annual soil loss at the catchment scale using field-sized individual units. The USLE is used to calculate soil loss for each field and then delivery ratio functions are used to determine the amount of sediment delivered downslope and into streams (Dickinson et al., 1990a and b). The model requires values for the land use, slope, soil erodibility and a hydrologic coefficient for each of the field units and the flow paths for the catchment. The results of the calculations may be then used to gain estimates of the potential soil loss for the catchment and importantly to indicate erosional hotspots within (Favis-Mortlock, 1994).

In recent years a number of models developed have attempted to incorporate the physical dynamics of soil erosion processes within the catchment. LISEM (de Roo et al., 1994), for example, represents in physical terms both hydrological and soil erosion processes within an area during a precipitation event. The catchment is divided into a series of cells; routing through the catchment is based on the resolution of a four-point finite-difference solution of the kinematic wave equation. Calculations are made for the processes of interception, infiltration, water storage, overland and channel flow, splash detachment and rill and inter-rill erosion requiring data such as rainfall, leaf area-index, hydraulic conductivity and soil water-retention, slope gradient and roughness. Depending upon the resolution of the cells, local features such as tractor wheelings may be included in the model calculations.

As the two examples of the GAMES and LISEM models illustrate, catchment modelling brings with it demands for scales and types of data not required previously by single field or slope models. Data for much larger areas though at a coarser resolution are usually required (Lammers et al., 1995). Input data sets may also be different with the heterogeneity of variables such as topography, soils and land use tending to be of greater importance in determining the calculated erosion rates. Topological information of the individual units are also required in the modelling for flow routing and for deriving areal values above and below a point. A resulting problem facing catchment modellers is that these new demands for model inputs may not be met from existing data sets.

Concomitant with these new demands is the need to be able to store and manipulate increasing quantities of spatially-referenced data. The models require variable or parameter values for the individual basic units either direct or derived from the various datasets. This brings increasing demands on computer technology to handle, manipulate and integrate large, geographically-referenced data. GIS, specialised computer software designed for inputting, storing, analysing and displaying spatial data, are being used increasingly in soil erosion modelling work to this end. Their functionality has been used in the storage and generation of input data; more recently, it has even been used as the software development base for actual models. This functionality offered by GIS will now be explored.

GIS data handling and functionality

The term GIS has been used to describe a variety of software systems ranging from major commercial packages such as Arc/Info, GRASS, and Intergraph to in-house developed software for specific data handling (e.g. Chakroun *et al.*, 1993). In common is the fact that they are specialised computer software for handling, displaying and analysing spatial data.

These data, from sources such as paper maps, remote sensors, field surveys or existing databases are held in a GIS according to either the raster or vector storage models. These define the way the basic spatial units of the data are defined in terms of either points, lines and polygons for the vector or as contiguous sets of grid squares for the raster data model. This will also determine the possibilities for defining the basic spatial units to be used in any modelling exercise although most GIS today support conversions from one to the other.

Once the data is within the GIS various functions may be used to manipulate and analyse the data. This functionality may be categorised as mapping, querying and modelling (Nyerges, 1991), although the availability of specific operations will vary between the different software systems. Mapping and querying capabilities are present in most systems with specific modelling capabilities being developed for many of the new GIS systems, such as Smallworld and PCRaster (van Deursen and Wesseling, 1992). Most soil erosion model research has used a combination of these, although the first two have been most used to date.

Mapping - this provides referential and browse information, and answers 'what' and 'where' type queries. Answers to queries such as 'On what slopes/aspects/soil associations is most soil erosion found?' may be derived by integrating various different data layers using overlay analysis. Boolean operations are used to find the occurrence of particular combinations of variables. This has been used in some modelling work to divide catchments into a series of sufficiently homogeneous land units, which form the basic units of a model of soil erosion. Other operations may be used to explore the spatial relationships between instances or patterns of variables so as for example to highlight areas of soil erosion risk (e.g. Fernandez *et al.*, 1993; Viet and Phuong, 1993). Many GIS are also capable of generating spatial information using various geostatistical techniques for interpolation or for statistically defining the variance of one variable relative to another (Band and Moore, 1995).

Querying - this addresses specific requests for information such as where all instances of a phenomenon occur, measurements, proximity/neighbourhood questions and network analysis; and it essentially answers the 'how much/where' type queries. This allows questions to be answered such as how much contributing area is there above a particular point or what is the length of the flow path to the channel from that same point. Many GIS also support network analysis so providing the routing capabilities needed in catchment scale modelling.

Modelling - this uses the spatial, temporal and attribute information of phenomena held in the database to investigate and simulate hypotheses of processes and variable interactions. The capabilities available within individual GIS for this type of work varies widely with some allowing mathematical calculations to be made across the different data layers to generate a prediction of soil loss for defined basic units. The modelling languages supported by many GIS, such as Arc/Info or SPANS, are proprietorial and in many cases are limited in terms of mathematical functionality. In other GIS, such as GRASS the actual programming code of the software may be accessed allowing relatively complex models to be built within the system. This has lead to two main approaches being adopted in integrating models and GIS which are known as loosely- and tightly-coupled linkages. These will now be described in more detail in the context of soil erosion modelling.

Catchment soil erosion modelling using GIS

The use of GIS in soil erosion modelling may be classified according to the extent of integration between the temporal and spatial dynamics of the model. The loosely- and tightly-coupled linkages approaches mentioned previously reflect essentially the ability of a particular GIS to support spatio-temporal modelling. The types of integration are illustrated in Figure 1.

a) Loosely-coupled models

Loosely-coupled modelling (also known as low-level integration) is used where a model already exists in computer form, or where the available programming language of the systems is limited in terms of data handling and functionality. The soil erosion model itself is developed outside of the system using some dynamic modelling language but the GIS is used to provide input data to it, for displaying particular time shots of the calculated results and for comparing them with observed data. The data are transferred between the model program and the GIS using some conversion program or batch file transfers.

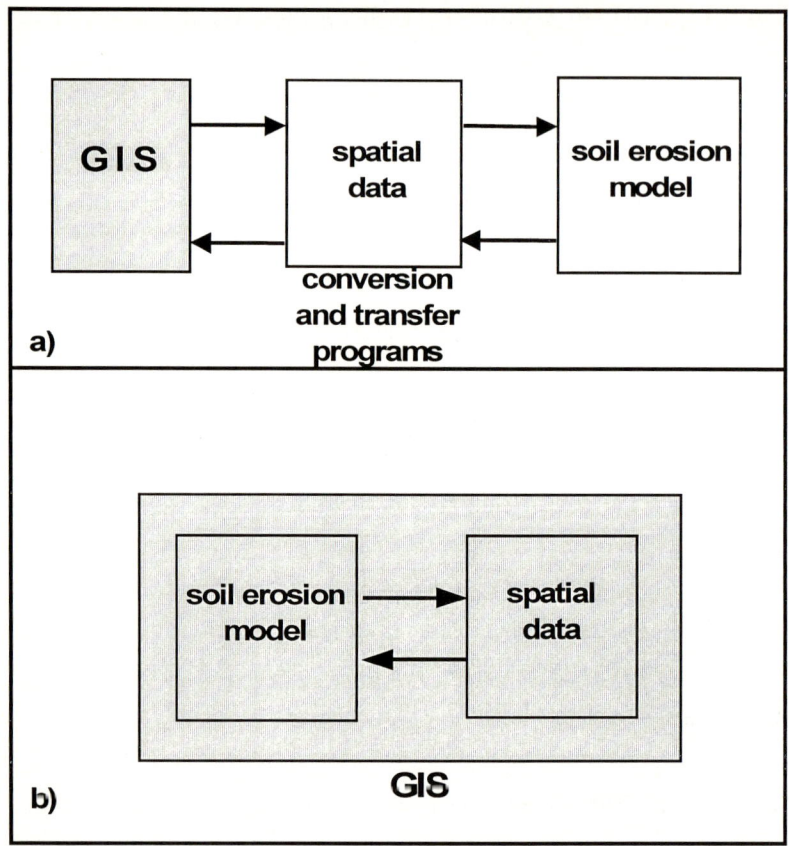

Figure 1. Loosely- a) and tightly- b) coupled GIS and dynamic models

In this type of work the GIS's mapping and querying functionality is used to generate model inputs either directly from the physical data held or indirectly from manipulating or analysing them. For catchment soil erosion modelling the main variables required are derived from topographical, land use, hydrological, soils and meteorological data. One of the important GIS functions used in such work is the ability to generate a 'continuous' surface for a variable from discrete information held at point or line locations within the catchment. Various statistical interpolation techniques such as distance-weighted averages, spline fitting or kriging algorithms are available in many GIS to generate estimates of the variables.

Topographic data, whilst readily obtainable in Digital Elevation Model (DEM) form for a limited number of areas (e.g. United States Geological Survey DEM), is often only available as contour or spot height data for many others. Similarly soils, hydrological and meteorological data are often only held for a series of field points. Using non-linear interpolation algorithms such as Thiessen polygons or distance-weighted averages, values may be generated for the whole of the catchment. Whilst there are obvious limitations to using data derived in this way, particularly for highly variable data sets such as hydraulic conductivity or where few data points exist, as long as these are acknowledged, simulations of soil erosion for the catchment may be based on these statistical surfaces.

Data sets may be further manipulated to generate new information for model inputs. For example, topographic variables such as angle and length of slope, profile curvature, upslope drainage area are required by many models. These may be derived from numerical elevation surfaces through the solution of first and second-order derivative equations by the GIS. More complex parameters such as surface saturation conditions may also be derived through computations integrating elevation and other spatial data (Moore *et al.*, 1991).

Elevation surfaces may also be used to partition catchments and to extract stream networks and drainage areas based on local slope change. Once this is generated stream orders, length and longitudinal slope, and starting and ending co-ordinates for individual channel segments may be derived prior to routing and modelling work. The end result is a capability for flow direction, flow accumulation and flow intensity data to be derived by the GIS for inputting into a soil erosion model.

Many of the examples of GIS-based soil erosion modelling in the literature have adopted a loosely-coupled approach. Prato *et al.* (1989) used a GIS to generate input data for the USLE model to examine the economic efficiency of various soil management systems for a catchment in Idaho. Lo *et al.* (1993) and Lo (1994) used a GIS to combine and integrate various map data, to convert them to a grid-based values and then to generate inputs for the AGNPS model in a soil erosion study in Taiwan. In turn, Brown *et al.* (1993) and de Roo (1989) used GIS to generate variable inputs for the ANSWERS model (Beasley and Huggins, 1982).

process. Probabilistic or spatially-nested hierarchical approaches are needed to provide the modellers with the flexibility in data handling needed for modelling erosional processes.

The existing functionality of the software brings problems with respect to the actual modelling as well as in generating input values. In many of the more physically-based models there is a need to solve complex equations and to adopt an iterative approach in the calculations. This is particularly important in the routing of the water and sediment through the soil profile and the catchment. Most GIS are unable to support such mathematical and data holding demands so requiring the models to simplify the solutions to the various differential and partial differential equations of the processes.

The important role of GIS generating variable values for models has already been highlighted. However, this ability is currently restricted by the small range of mathematical and in particular statistical analytical capabilities available in most GIS. This means that techniques may be used that are not wholly appropriate to the datasets available. Attempts recently to add functionality to the systems through the linking of spatial statistical packages to, or the use of computer programs outside the GIS go some way to address this but are cumbersome in terms of data transfer and use.

The limitations of the software are not the only cause for concern in GIS-based soil erosion modelling. There are also problems associated with the injudicious use of the technology. As highlighted in previous sections, GIS allow data to be integrated and manipulated in many different ways but the user needs to be wary of making simplistic assumptions when using the various available functions (Nemani *et al.*, 1993). For example in some modelling based on grid structures, calculations are made across the various data layers using values held for the same geographically-referenced cell ignoring any spatial interactions with the surrounding area or upslope contributions. Similar problems result when various data layers are combined using overlay analysis prior to model parameterisation. Simple overlay combinations are not able to account for spatial co-variances in the data or multi-variable interaction which are important in modelling or interpreting processes. Nemani *et al.* (1993: 296) warn that:

> "*since process models are inherently complex and non-linear, the way in which continuous geographic information is aggregated has far-reaching effects on model output*".

It is also important for users to be aware of the nature, limitations and error of the data used. For example, when combining map detail of many different scales there are problems associated with varying levels of generalisation within each. Boundaries between the entities within the different data sets will vary giving spurious results when combined in overlay or modelling exercises.

Awareness of scale is also important when defining the basic unit sizes for the model. GIS permits the user to define the map scale of the database but choosing an appropriate one to work at is difficult given the varying sources and availability of information, the nature of erosion processes, the inherent variation in the sub-scale phenomena such as rills, tractor wheelings and surface crusting, as well as computationally-enforced limitations (Grayson *et al.*, 1993). Brown *et al.* (1993) have highlighted the influence of different grid-square sizes on results from using the ANSWERS model. They found that variations in the magnitude of the model's outputs closely matched changes in the spatial dependence of the input variables.

Conclusions

GIS have been shown to offer important benefits for modellers working at the catchment scale. The ability to store, integrate and provide data for soil erosion models is already being exploited in both the research and management context. Advances in GIS programming languages will ensure they will play an increasing role in model development where a more spatially distributed approach is adopted.

The main limitation to their use will result not from the technology itself but from lack of availability of spatial data to support these models. The conversion of existing data into digital form, whilst laborious and costly is beginning to ease this. However, it will only be with the development of new data sources, such as from remote sensing, automatic telemetry, or Global Position Systems that the demand for *spatial* data will begin to ease. That said, information on variables such as soil surface conditions will continue to limit the modelling of soil erosion processes at the catchment-scale.

References

Band, L.E. and Moore, I.D. (1995). Scale: landscape attributes and geographical information systems. In, Kalma, J.D. and Sivapalan, M. (eds), *Scale Issues in Hydrological Modelling*, Wiley, Chichester, pp. 159-180.

Beasley, D.B. and Huggins, L.F. (1982). *ANSWERS - User's Manual*, Dept of Agricultural Engineering, Purdue University, Lafayette.

Brown, D.G., Bian, L. and Walsh, S.J. (1993). Response of a distributed watershed erosion model to variations in input data aggregation levels. *Computers and Geosciences* **19**, 499-509.

Burrough, P.A. (1990). Methods of spatial analysis in GIS. *International Journal of Geographical Information Systems* **4**, 221-223.

Burrough, P.A. (1993). Spatial data quality and error analysis issues: GIS functions and environmental modelling. *Proceedings, Second International Conference/Workshop on Integrating Geographic Information Systems and Environmental Modelling*. Breckenridge, Colorado.

Chakroun, H., Bonn, F, and Fortin, J.P. (1993). Combination of single storm erosion and hydrological models into a geographic information system. In, Wicherek, S. (ed.), *Farm Land Erosion: In Temperate Plains Environment and Hills*, Elsevier, Amsterdam. pp. 261-270.

Dickinson, W.T., Rudra, R.P. and Wall, G.J. (1985). Discrimination of soil erosion and fluvial sediment areas. *Canadian Journal of Earth Sciences* **22**, 1112-1117.

Dickinson, W.T., Rudra, R.P. and Wall, G.J. (1986). Identification of soil erosion and fluvial sediment problems. *Hydrological Processes* **1**, 111-124.

Dickinson, W.T., Rudra, R.P. and Wall, G.J. (1990a). Targetting remedial measures to control nonpoint source pollution. *Water Resources Bulletin* **26**, 499-507.

Dickinson, W.T., Rudra, R.P. and Wall, G.J. (1990b). Model building for predicting and managing soil erosion and transport. In, Boardman, J., Foster, I.D.L. and Dearing, J.A. (eds), *Soil Erosion on Agricultural Land*, Wiley, Chichester, pp. 415-428.

Favis-Mortlock, D.T. (1994). *Use and Abuse of Erosion Models In Southern England*. Unpublished PhD thesis, University of Brighton.

Fernandez, R.N., Ruskinkiewicz, M., da Silva, L.M. and Hohannse, C.J. (1993). Design and implementation of a soil geographic database for rural planning and management. *Journal of Soil and Water Conservation* **48**, 140-144.

Grayson, R.B., Bloschl, G., Barling, R.D. and Moore, I.D. (1993). Process, scale and constraints to hydrological modelling in GIS. In Kovar, K. and Nachtnebel, H.P.(eds), *Application of Geographic Information Systems in Hydrology and Water Resources Management*. IAHS No 211, Wallingford. pp. 83-92.

Lammers, R.B., Band, L.E., Kremer, R.G., and Baron, J.S. (1995). *Scaling behaviour of variables in a hydro-ecological model over heterogeneous topography*. Unpublished paper presented at American Geophysical Union Spring Meeting, Baltimore.

Langran, G. (1992). *Time in Geographic Information Systems*. Taylor and Francis, London.

Liengsakul, M., Mekpaiboonwatana, S., Pramojanee, P., Bronsveld, K. and Huizing, H. (1993). Use of GIS and remote sensing for soil mapping and for locating new sites for permanent cropland - A case study in the 'highlands' of northern Thailand. *Geoderma* **60**, 293-307.

Lo, K.F.A., Chiang, S.H. and Tsai, B.W. (1993). Soil erosion evaluation on hillslopes in Taiwan. In, Wicherek, S. (ed.), *Farm Land Erosion: In Temperate Plains Environment and Hills*, Elsevier, Amsterdam. pp. 451-462.

Lo, K.F.A. (1994). Quantifying soil erosion for the Shihmen Reservoir Watershed, Taiwan. *Agricultural Systems* **45**, 105-116.

Mason, D.C., O'Conaill, M.A. and Bell, S.B.M. (1994). Handling four-dimensional geo-referenced data in environmental GIS. *International Journal of Geographic Information Systems* **8**, 191-215.

Moore, I.D., Grayson, R.B., and Ladson, A.R. (1991). Digital terrain modelling: a review of hydrological, geomorphological and biological applications. *Hydrological Processes* **5**, 3-30.

Nemani, R., Running, S.W., Band, L.E. and Peterson. D.L. (1993). Regional Hydroecological Simulation System: an illustration of the integration of ecosystem models in a GIS. In, Goodchild, M.F., Parks, B.O. and Steyaert, L., (eds), *Environmental Modelling with GIS*. Oxford University Press, New York. pp. 296-304.

Nyerges, T. L. (1991). Analytical map use. *Cartography and Geographic Information Systems* **18**, 11-22.

Peuquet, D., Davis, J.R. and Cuddy, S. (1993) Geographic Information Systems and environmental modelling. In, Jakeman, A.J., Beck, M. B. and McAleer, M.J., (eds), *Modelling Change in Environmental Systems*, Wiley, Chichester. pp. 543-556.

Prato, T., Shi, H-Q, Rhew, R. and Brusven, M. (1989). Soil erosion and nonpoint-source pollution control in an Idaho watershed. *Journal of Soil and Water Conservation* **44**, 323-328.

de Roo, A.P.J., Hazelhoff, L. and Burrough, P.A. (1989). Soil erosion modelling using 'ANSWERS' and Geographical Information Systems. *Earth Surface Processes and Landforms* **14**, 517-532.

de Roo, A.P.J., Wesseling, C.G., Cremers, N.H.D.T., Offermans, R.J.E., Ritsema, C.J. and van Oostindie, K. (1994). LISEM: a new physically-based hydrological and soil erosion model in a GIS-environment, theory and implementation. In, Olive, L.J., Loughran, R.J. and Kesby, J.A. (eds), *Variability in Stream Erosion and Sediment Transport*, Proceedings of Symposium, Canberra. IAHS Publication 224, Wallingford. pp. 439-448.

Snell, E.A. (1985). Regional targeting of potential soil erosion and nonpoint-source sediment loading. *Journal of Soil and Water Conservation* **40**, 520-524.

van Deursen, W.P.A. and Wesseling, C.G. (1992). *The PCRaster Package*. Department of Physical Geography, Utrecht University, Netherlands.

Viet, C.P. and Phuong, M.N. (1993). Natural resources evaluation by the use of remote sensing and GIS technology for agricultural development. *Advances in Space Research* **13**, 117-121.

Wischmeier, W.H. and Smith, D.D. (1978). *Predicting Rainfall Erosion Losses*, US Department of Agriculture, Agricultural Research Service Handbook 537.

Young, R.A., Onstad, C.A., Bosch, D.D. and Anderson, W.P. (1987). *AGNPS, Agricultural Non-Point Source Pollution Model: a watershed analysis tool*. Conservation Research Report 35, US Department of Agriculture, Washington D.C.

27. SNOWMELT AND FROZEN SOILS IN SIMULATION MODELS

Peter Botterweg

Centre for Soil and Environmental Research
N-1432 Ås
Norway

Abstract

Large parts of the world are temporarily or permanently covered with snow. In areas with regular snowmelt in the spring season, meltwater can be the major cause of erosion. Freezing and thawing reduces soil shear strength and infiltration rates can be very low because of ice layers in the soil. Thawing soil over a frozen layer may be oversaturated and susceptible to erosion. Deterministic models dealing with winter hydrology, snowmelt and erosion exist but their data requirements are large, which makes it difficult to apply them to field situations. Modelling the erosion process for cold climate conditions has been tried with EUROSEM with reasonable success. The most difficult task is estimating initial conditions, which often depend on circumstances a long time before the moment of interest. Probably more empirically based models will better succeed in simulating erosion for winter conditions, but the applicability of these models is restricted.

Introduction

About one third of the earth's surface is permanently covered by ice or snow and there is stored about 90 per cent of the available fresh water on earth. Of the land surface about 60 per cent is seasonally covered by snow and frozen soil. Based on regional climate conditions four different situations can be described with decreasing average winter temperatures and increasing latitude:

1. Areas without snow and frozen soil during winter (e.g. the lowlands of Spain and the south of France).

2. Areas with snow and frozen soil but only for a short time (1-10 days) but several periods during winter (e.g. The Netherlands, northern Germany and Denmark).

3. Areas with a nearly permanent snow cover and frozen soil during a long winter season (3-4 months) (e.g. Scandinavian countries).

4. Areas as category 3, but with permafrost (northern Siberia, and northern Canada).

This paper concentrates on the situation in areas of category 2 and 3. Simulation models for the hydrological cycle and nutrient cycles in the soil, for transport processes in the soil or for soil surface processes, applicable for these areas, should include routines for snow and ice formation and for snow and ice melting. Where ice formation takes place in the topsoil, routines should be available to quantify permanent and temporal changes in physical soil characteristics caused by ice formation. The transition of water into or from the solid state is triggered by temperature: therefore modules for calculation of energy fluxes and energy status are needed in the models.

A global climate change with a rise in mean temperature causes the four areas listed to move into a lower category. In addition, a change in total precipitation or in the distribution of precipitation over the seasons may cause changes in the average depth and distribution of the snow cover. Understanding and being able to model satisfactorily the present state is a condition for being able to predict changes caused by climate change as related to processes in and on the soil. This paper summarises the questions that have to be raised when modelling soil processes for regions of category 2 and 3.

The physical basis for modelling snow and ice
At temperatures below $0°C$ water goes from the liquid state into the solid phase, creating solid ice or snow crystals and releasing energy. The temperature of freezing decreases with an increasing concentration of dissolved ions as e.g. found in soil water. A characteristic, unique to water is that it has its maximum density at $3.98°C$. As a consequence ice has a larger volume (about 10 per cent) than the corresponding amount (g) of water resulting in a mechanical force. When snow and ice are supplied with sufficient energy the liquid stage is reached again. The transition of water from one state to the other not only has an effect on the hydrological processes at the soil surface but also affects local climate.

The snow cover
From a hydrological point of view a snow cover on the soil surface can be considered as a temporary storage of precipitation that will be released during snowmelt. Depending on local climate conditions (e.g. areas of category 3 and 4) the precipitation of several months is stored and released in a relatively short period of 2-3 weeks, resulting in high infiltration and/or runoff values. Further the snow pack affects the energy fluxes between the soil surface and the atmosphere. First of all the snow functions as an insulator and in addition snow reflects more incoming radiation than bare soil (93 per cent compared with 25 per cent), while the white snow radiates more heat then a dark soil surface.

A snow cover is a dynamic system. In a snow column density (g/l) increases over time through changes in the snow crystal forms and by wind pressure. Because of the temperature gradient in the snow, melt or rain water from the top freezes again in the snow layer and new snow falls on top. Finally this results in a snow profile where different layers can be recognised that differ in temperature, density, thermodynamic properties and hydrologic conductivity. In the horizontal dimension wind causes an irregular (re)distribution of the snow cover depending on a combination of wind force, wind direction, topography and vegetation elements.

Frozen soil

A soil becomes frozen when the temperature falls so low that soil water changes into the solid state. Because of freezing point depression, soil temperature will be lower then $0^{\circ}C$ when freezing. Besides dissolved ions, freezing point depression also depends on pore sizes. During freezing, capillary transport of water to the frost zone occurs. Freezing affects the hydraulic conductivity, soil cohesion and aggregate size distribution.

To estimate the hydraulic conductivity of a frozen soil is difficult. It depends on a combination of past conditions: 1) soil water content at the time of freezing, 2) periods of thawing in only the top layer but not through the whole frozen layer, 3) infiltration of meltwater or precipitation into the frozen soil. Depending on a combination of these three aspects, the hydraulic conductivity may vary between a value higher than for the unfrozen soil caused by the formation of macro pores and a value close to zero when infiltrating water has reached a frozen layer and blocks the pores. For more details about hydraulic conductivity in frozen soils the reader is referred to Lundin (1990) and Engelemark (1993).

The volume increase at the transition from water to ice affects soil structure. Bindings between soil particles are broken by ice crystals and the mean aggregate size is reduced. The level of disturbance will depend on aggregate type, soil type and the water content at the time of freezing will be important, too. The cohesion of frozen soil may be high when the soil particles are bound strongly together by ice. However, when thawing, soil cohesion decreases to a level lower then existed before freezing. This is caused by the mechanical processes described, but also by the fact that thawing soil can contain more water than the pore size volume; the soil is then over-saturated and forms a slurry. Experimental work has shown that soil cohesion is negatively correlated with the number of freeze-thaw cycles the soil has gone through (Kok, 1989).

Snowmelt models

In disciplines other than soil science the need to predict snowmelt has existed for a long time; e.g. for predicting runoff and floods in river basins. In IAHS Publication No.155 (Morris, 1986) nearly all papers deal with river basins or large scale snowmelt/runoff models and only one paper by Tregubov (1986) quantitatively describes the relation between soil frost, snowmelt and erosion. Another interest is when calculating the potential electricity production for hydro-power plants, which depend on annual snowmelt in high mountain areas. The Norwegian Water Resources and Energy Administration applied the large scale model HVB (Bergström, 1976; 1990) for this purpose. The increasing interest in non-point source pollution from agriculture during the last decades has forced soil scientists to study and describe processes in the soil and on the soil surface also outside the main growing season (Cooley, 1990). Botterweg (1995) has shown that there exists a clear relation between the structure and contents of a simulation model and the type of problems that it aims to address. It can therefore be expected that snowmelt models dealing with areas the size of river basins and a time scale of weeks or more are formulated differently compared to models used for single fields and a time scale of one day or less as e.g. needed for event-based erosion models.

The principles of snowmelt modelling

This section is primarily based on the thesis published by Sand (1990). A complete snowmelt model with short time steps (< one day) should combine three different processes in the snow pack: *A)* melting of snow at the surface, *B)* meltwater refreezing in the snow pack, *C)* the flow of meltwater through the snow pack. For processes *A* and *B* energy fluxes have to be modelled and process *C* is modelling flow through a heterogeneous porous medium, only vertical for a point model and both vertical and horizontal for a field or basin model.

A. Snowmelt at the surface.

Estimation of available energy for snowmelt can be derived from the energy balance equation for the snowcover with the following energy fluxes:

    net short wave radiation flux absorbed (SR)

    net long wave radiation flux at the snow-air interface (LR)

    sensible heat flux from the air at the snow-air interface (SH)

    flux of latent heat at the snow-air interface (LH)

    flux of heat at the snow-ground interface by conduction (CH)

    flux of heat from rain (RH)

    rate of change of internal energy per unit area (IE)

The energy flux available for snowmelt (ES) = SR+LR+SH+LH+CH+RH-IE. The physical mechanisms and quantification of the fluxes are described from laboratory experiments but parameter values are difficult to estimate in the field; e.g. net long wave radiation is affected by clouds, canopy and bare soil.

A simpler type of model to estimate snowmelt at the snow surface is a temperature index model. The basic assumption in these models is that above a given threshold temperature, a linear relation exists between air temperature and snowmelt. Although these type of models do not describe the physics of snowmelt they have been applied successfully. Best results are achieved for large areas under stable weather conditions with minor air movement. As there does not exist a relation between air temperature and distribution of short range radiation during one day, the temperature index models cannot be used for a time resolution less then one day. This lack has been compensated by introducing a factor that accounts for short wave radiation and a time resolution down to one hour gave reasonable results. In applications of a simple model for snowmelt at the surface, the processes *B* and *C* mentioned above are often not modelled separately.

B. Refreezing of meltwater

Melting of a snowpack is a discontinuous process with melting during day time and refreezing at night. An example of this is shown in Figure 2, section 5. Depending on how much energy is lost during the night a freezing front moves downwards into the snow pack. When the energy flux reverses in the morning, the internal cold of the snowpack must be overcome before snowmelt starts. The depth of the freezing front can be calculated with an energy balance equation. Refreezing has been taken into account in some temperature index models by introducing one or more extra coefficients.

C. Water flow through a snow pack

The physics of water flow through snow are not completely understood. Compared to water flow in a porous medium like soil, it is more complex because the matrix of the snow pack is changed by the water flow, which has also an energy dimension. The theoretical basis for flow through a homogeneous snow pack is based on the work by Colbeck (1972) and Colbeck and Davidson (1973) who applied Darcy's flow theory for both unsaturated and saturated flow. Because of a rapid metamorphism of snow crystals in a saturated layer, the intrinsic permeability increases and the permeability increases several orders of magnitude. On slopes, saturated flow near the soil surface has a horizontal dimension too. An application of Colbeck's model to a homogeneous snowpack gave good results, but a natural snowpack is seldom homogeneous. A natural snowpack will show layers of different grain sizes and density, ice layers and crusts. It was found that drainage channels develop in a snowpackage and multipath flow was introduced by Colbeck (1978). However, the model's demand for input variables increases considerable. Several other models have been developed based on the principles given by Colbeck. Bengtsson (1982) included refreezing; Jordan (1983) takes also into account capillary pressure gradients; Marsh and Woo (1985) introduced another solution for distributing flow among the different pathways, and reduced the number of necessary input parameters. Anderson (1976) has developed an empirical model for water flow in a snowpack that needs information about the liquid water content, snow depth and snow density.

In conclusion, it can be stated that the physical basis for the processes related to snow and ice is well understood, but modelling the processes behind ideal situations still is difficult because of the temporal and spatial variation in required parameter values, if they are measurable at all.

Snowmelt and frozen soil in complex models

Modelling processes in the soil ecosystem has to be based on an adequate hydrology model to estimate the magnitude, paths and direction of the water flows functioning as transport medium, and to estimate the hydrological and heat status affecting biochemical and chemical transformation of biotic and abiotic substances.

The main input variable in hydrological models is precipitation or snowmelt water and the most critical parameter is hydraulic conductivity. When the time scale used is less then the average period a snowcover is present in the area simulated, a hydrology model must include modules for snow pack dynamics and frost in the soil. Botterweg (1995) states that in models built up with several modules there often exists a difference in complexity between the modules. Hydrology models developed primarily for «summer» conditions include simple or lack routines for snowmelt and frozen soil. Typical winter hydrology models will include a simple solution for infiltration, because the distribution between infiltration and surface runoff is not important. No general rule can be given about what type of solution in a hydrology model is best. It strongly depends on the type of application.

In transport models a long time scale is used when dealing with pesticides or other 'foreign' chemicals. However a good description of the expected flow paths is needed (Jarvis, 1995) and also these models should take into account the changes occurring in the soil matrix during freezing-thawing.

The plant nutrients nitrogen (N) and phosphorus (P) are in focus when modelling non-point source pollution in agricultural districts. Losses of N during the growing season but also outside have been estimated with simulation models (Groot, 1991). It is accepted that N losses outside the growing season represent an important flux and has to be quantified. The nitrogen transformations in the soil are driven by microbiological activity. So, for plant nutrient models soil temperature as well as water flows and water content are important input variables also for the winter period.

Figure 1 shows a simplified flow chart for an erosion model as it may look when winter conditions are included. An effect of low temperature is found in the driving input (❶), the distribution between surface runoff and infiltration (❷) and in the soil surface conditions(❸). These aspects are incorporated in the deterministic erosion models WEPP (Flanagan and Nearing, 1995) and EUROSEM (Morgan et al., 1997) recently developed in the USA and

Europe respectively. A more detailed discussion of these aspects is presented using EUROSEM as an example.

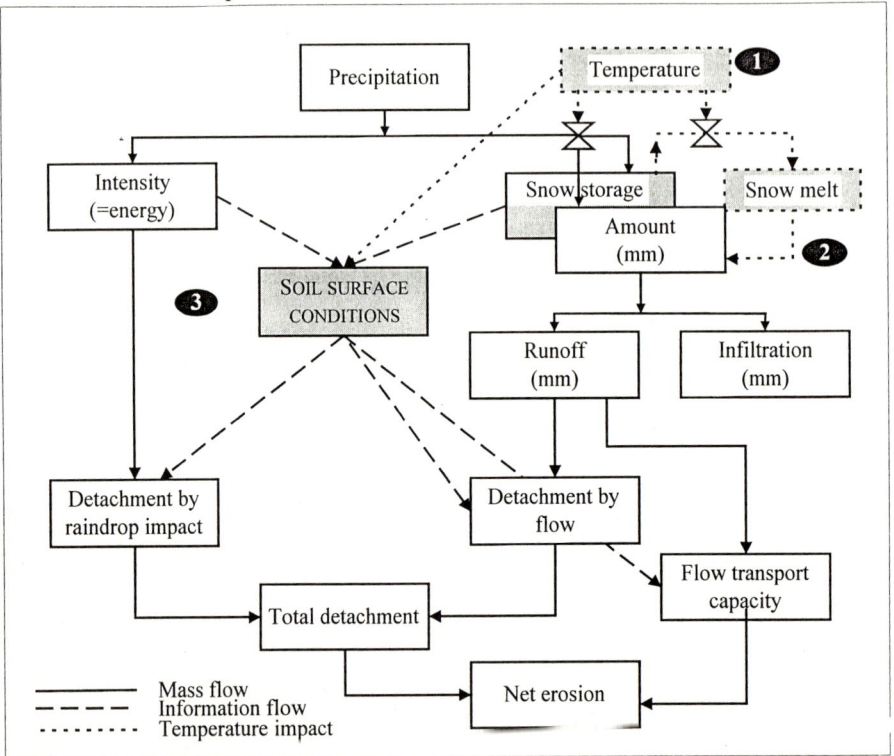

Figure 1. Flow diagram for an erosion model to be used for cold climate conditions. Explanation in main text

Application of EUROSEM under winter conditions

The European soil erosion model (EUROSEM) is a process-based erosion prediction model designed to predict erosion in individual events and to evaluate soil protection measures. A more detailed description of the model is found in Morgan et al. (1998). To be able to use EUROSEM for cold climate conditions it has to be used in connection with a continuous hydrology model as e.g. SOIL (Jansson,1991). SOIL is a continuous, process based one-dimensional hydrology model that simulates water and heat flow through a layered soil profile. Compartments for snow, intercepted water and surface ponding are included to account for processes at the upper soil boundary. Weather input variables for the model are daily values for temperature, precipitation, wind velocity, relative humidity, and cloud cover. SOIL has been shown to satisfactory simulate hydrology for a wide range of soil types and vegetation covers in different climate zones (Jansson,1994). Output variables from SOIL needed for running EUROSEM under snowmelt with or without frozen soil are given in Table 1.

Table 1. SOIL output variables from the hydrology model SOIL needed for running EUROSEM with snowmelt with or without frozen soil

Variable	Dimension	Deduced information	Use in EUROSEM
❶,❷ª⁾ Snowmelt	mm/day or mm/hr		driving input instead of precipitation
❸ª Snow cover	depth in m	initial condition	if snow depth>0 then splash erosion=0
❸ª Soil surface temperature	degrees Celsius per day or per hour	number of frost thaw cycles	estimating parameter value for cohesion and erodibility
		initial condition	depth non-erodible layer (0 cm or >0 cm)

A) Numbers refer to Figure 1

Soil surface, initial conditions (❸)

Initial conditions for erosion simulation for an area partly covered with snow are based on both a development over time and the actual status. There exists a negative relation between the number of frost-thaw cycles and cohesion and erodibility of thawed soil. Further for a frozen soil the depth of a non-erodible layer is assumed to be zero, and for thawed soil a depth of more than 0.5 cm is used. As the thawed layer may be over saturated with water because of no infiltration and continuous thawing, soil cohesion may be very low (< 1.0 kPa).

The points described above open the possibility of using EUROSEM for cold climate conditions. However parameterisation of the model may even be more difficult here than for conditions not affected by frost or snow (Quinton and Morgan, 1998). At the end of a stable winter, snowmelt runoff may follow a regular pattern with long lasting events of 10-12 hours (Figure 2), and because of thawing a dramatic change in cohesion of the surface soil occurs. The parameterisation of EUROSEM to simulate the observed erosion given the measured snowmelt and sediment concentrations showed that somewhere after April 8 erosion changes from being limited by erodibility of the soil into being transport limited. It was not possible to predict the time of this change exactly through modelling with SOIL when run with a time resolution of one day. EUROSEM does not include dynamic updating of parameter values as would be needed in this case.

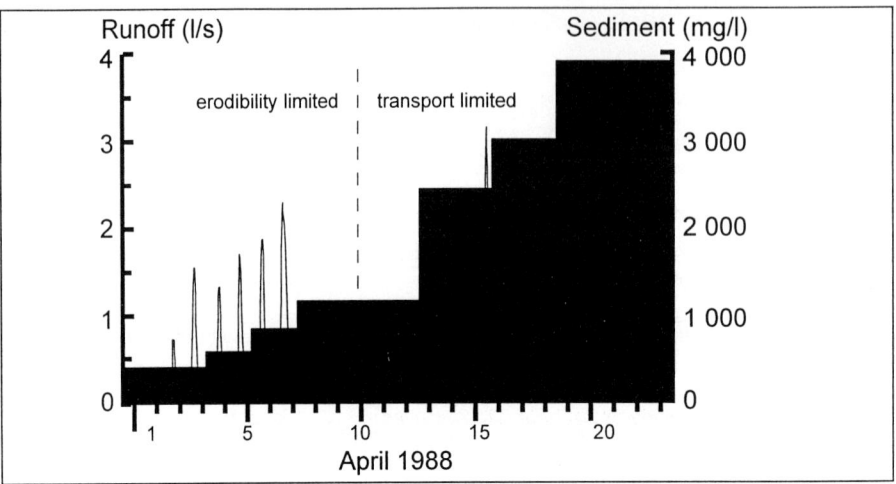

Figure 2. The runoff hydrograph for a 0.82 ha large clay field in southern Norway April 1988 during snowmelt. Sediment concentrations are based on flow proportional combined samples for 7 periods of different length. Data from Øygarden (1989)

It will be clear that in the case of an unstable winter with frequent changes in temperature around zero, estimation of the initial conditions is a difficult task. In addition knowing that event based models are very sensitive to initial conditions makes modelling individual erosion events during such periods nearly impossible.

Conclusions

Separate, deterministic process-based models describing erosion processes, or snowmelt processes or infiltration into frozen soils are available, but one model combining all these processes with all the details is not available. The deterministic models require a large number of parameter values i.e. quantification of initial conditions. What is missing are methods (models or measurement instruments) to estimate these parameters or conditions over a long period of time before the event of interest. So, even if individual processes are well described and modelled, modelling the total system does not seem to be possible yet. However, working with the available models will increase our knowledge about what information is needed and it may show how to simplify the system in a way that still makes sense. In relation to climate change it is concluded that soil processes under cold climate conditions are important to know, because there will be new areas with unstable winters and temperatures fluctuating around zero, while others will no longer have such a situation. To be able to predict the effect of climate change on erosion levels for these areas compared to the present situation, demands an understanding of the processes and a predictive tool e.g. a model to quantify the changes that can be expected. It can be questioned if process-based deterministic models can be developed that will have the potential to give the answers needed. It may be that empirical models will come earlier with a solution for estimating the effect of climate change on erosion with special emphasis on cold climate conditions. These empirical models should however be based on the experience and knowledge achieved by applying deterministic models on individual processes. Today empirical-based erosion models do not yet include cold climate conditions satisfactorily.

Acknowledgements

The modelling work described in this paper was supported by the EC (contract STEP-0053). I thank Lillian Øygarden for help and stimulating discussions about the measured runoff data. The comments made by the referee on a preliminary version of the manuscript are appreciated.

References

Anderson, E.A. (1976). A point energy and mass balance model of a snow cover. *NOAA Technical report NWS* **19**. US Department of Commerce, Washington, 1976.

Bengtsson, L. (1982). Percolation of meltwater through a snowpack. *Cold Regions Science and Technology* **6**, 73-81.

Bergström, S. (1976). Development and application of a conceptual runoff model for Scandinavian catchments. *SMHI RHO* 7, Norrköping.

Bergström, S. (1990). Parametervärden för HBV-modellen in Sverige. *SMHI Hydrologi* **28**, Norrköping.

Botterweg, P. (1995). The user's influence on model calibration results: an example of the model SOIL, independently calibrated by two users. *Ecological Modelling* **81**, 71-81.

Colbeck, S.C. (1972). A theory of water percolation in snow. *Journal of glaciology* **11**, 369-385.

Colbeck, S.C. (1978). The physical aspects of water flow through snow. *Advances in Hydroscience* **11**, 165-206.

Colbeck, S.C. and Davidson, G. (1973). Water percolation through homogeneous snow. *IAHS Publication* **107**, Theme 1, 242-257. UNESCO, Paris.

Cooley, K.R. (ed.) (1990). Frozen soil impacts on agricultural, range, and forest lands. Proceedings of an International Symposium, March 21-22, 1990, Spokane, Washington. *CRREL Special Report* **90-1**.

Engelmark, H. (1993). *Heat and Water Flows in Freezing and Thawing Soils*. Ph.D. thesis, Luleå University of Technology, 1993:128D, pp 102. Luleå, 1993

Flanagan, D.C. and Nearing, M.A. (eds) (1995). *USDA-Water Erosion Prediction Project WEPP; Technical Documentation*. NSERL Report No. **10**, USDA-ARS-MWA, West Lafayette, USA.

Groot, J.J.R., de Willegen, P. and Verberne, E.L.J. (eds) (1991). Nitrogen turnover in the soil-crop system. Modelling of biological transformations, transport of nitrogen and nitrogen use efficiency. *Fertilizer Research* **27**, 141-383.

Jansson, P-E. (1991). *Simulation Model for Soil Water and Heat Conditions. Description of the SOIL Model*. Report 165, Department of Soil Science, Swedish University of Agricultural sciences, Uppsala. 74 pp.

Jansson, P-E. (1994). *SOIL model, Users Manual. Communications 94-3*. Department of Soil Science, Swedish University of Agricultural Sciences, Uppsala. 67 pp.

Jarvis, N. (1995). Simulation of soil water dynamicsand herbicide persistence in a silt loam soil using the MACRO model. *Ecological Modelling* **81**, 97-110.

Jordan, P. (1983). Meltwater movement in a deep snowpack. 2. Simulation model. *Water Resources Research* **19**, 979-985.

Kok, H. (1989). *Freeze-Thaw Induced Variability of Soil Shear Strength*. Ph.D thesis, University of Idaho, USA, 183 pp.

Lundin, L.-C. (1990). Hydraulic properties in an operational model of frozen soil. *Journal of Hydrology* **18**, 289-310.

Marsh, P. and Woo, M-K. (1985). Meltwater movement in natural heterogeneous snow covers. *Water Resources Research* **21**, 1710-1716.

Morgan, R.P.C., Quinton, J.N., Smith, R.E., Govers, G., Poesen, J.W.A., Chisci, G. and Torri, D. (1998). The EUROSEM Model. In, Boardman, J. and Favis-Mortlock, D.T. (eds), *Modelling Soil Erosion by Water*, Springer-Verlag NATO-ASI Global Change Series, Heidelberg.

Morris, E.M. (ed.) (1986). *Modelling Snowmelt-Induced Processes*. IAHS Publication No. 155, IAHS Press, Wallingford, UK, 1996.

Quinton, J.N. and Morgan, R.P.C. (1998), EUROSEM: an evaluation with single event data from the C5 watershed, Oklahoma, USA. In, Boardman, J. and Favis-Mortlock, D.T. (eds), *Modelling Soil Erosion by Water*, Springer-Verlag NATO-ASI Global Change Series, Heidelberg.

Sand, K. (1990). *Modeling Snowmelt Runoff Processes in Temperate and Arctic Environments*. Ph.D. thesis, The University of Trondheim, The Norwegian institute of technology, IVB-rapport B-2-1990-1. 181 pp.

Tregubov, P.S. (1986). Mechanisms and spatial variability of erosion caused by meltwater in the USSR. In, Morris, E.M. (ed.) (1986). *Modelling Snowmelt-Induced Processes*. IAHS Publication No. 155., IAHS Press, Wallingford, UK.

Øygarden L. (1989). *Utprøving av tiltak mot arealavrenning i Akershus. Handlingsplan mot landbruksforurensninger*.(In Norwegian) Rapport nr 6, 118 pp, GEFO, Ås.

28. THE USE OF USLE COMPONENTS IN MODELS

A.D. Nicks[a]

USDA-Agricultural Research Service
National Agricultural Water Quality Laboratory
P.O. Box 1430
Durant, OK 74702
USA

Abstract

The Universal Soil Loss Equation (USLE) has been in use during the past several decades. Developed from small erosion plots, it has become the standard for erosion prediction in the U.S. and other parts of the world. There have been few changes in the model since its conception. Attempts to improve the predictive power of the model have largely been the development of databases for soils, cropping and management practices. Users of the model have mixed reactions to the results obtained. USLE technology has been adapted to continuous simulation such as the CREAMS, EPIC, and SWRRB models. This paper reviews some of the uses of USLE in other models and presents some findings from a recent USLE evaluation with the erosion plot data used in its development.

Introduction

As early as 1915, erosion plot studies were started at the University of Missouri. These erosion research activities described by Woodruff (1987), were crude by today's standards of instrumentation. However, the scientific interest aroused by the results of these and other plot studies led to establishing a network of ten soil erosion experiment stations in 1928 (Bennett, 1928). Located in various geographical areas of the U.S., the research from the erosion plot experiments supplied the data that were used in the development of the USLE (Wischmeier and Smith, 1978). Later, the USLE technology was used in the development of continuous simulation models such as CREAMS (Chemicals, Runoff, Erosion, and Agricultural Management Systems) (Knisel, 1980); EPIC (Erosion-Productivity Impact Calculator) (Williams *et al.*, 1984); SWRRB (Simulator for Water Resources in Rural Basins) (Williams *et al.*, 1985); and WEPP (Water Erosion Prediction Project) (Lane and Nearing, 1989; Flanagan, and Nearing, 1995).

The Red Plains Conservation Experiment Station near Guthrie, Oklahoma, was one of the ten stations established to investigate erosion from agricultural lands in the Red Plains area of

---

[a] Deceased

Kansas, Oklahoma, and Texas (Daniel *et al.*, 1943). After opening for settlement in 1889-1892, pioneers broke the virgin sod and plowed parallel to section lines established for property boundaries. This resulted in cultivation up and down hill with no consideration given to direction of slope (McDonald, 1938). After 35 to 45 years, most of the shallow, more sloping soils were forced out of cultivation by the use of these practices. By 1935, in a survey made of the 14.5 million ha region, more than 0.4 million ha were found to have been completely destroyed by gullies, 5.6 million ha had lost 75 percent of the topsoil, and 4.5 million acres had lost 25 to 75 percent of topsoil. Abandoned farm land was left to revert back to native grasses without reclamation and was utilized mostly for livestock grazing.

Figure 1. Location of the 10 original USDA Erosion Research Stations

Similarly, nine other stations were established on other lands representative of large problem areas of eroding land in other geographic regions of the U.S. (Figure 1). Stations were established near Temple, Texas (prairie soils from chalk and marl); Tyler, Texas (from sedimentary deposits of the coastal plain); Statesville, North Carolina (Piedmont soils from igneous rocks); LaCrosse, Wisconsin (from loessial deposits); Bethany, Missouri (prairie soils from glacial deposits); Hays, Kansas (plains soils from limestone, sandstone and loess); Zanesville, Ohio (timbered soils from sandstone, shale and conglomerate); Pullman, Washington

(Palouse soils from loessial deposits); and Clarinda, Iowa (Missouri Valley loess area soils). As the research programs of USDA expanded from 1930 to 1960, erosion plots and small watershed studies were established at other locations across the U.S. bringing the total number of sites to 46. Erosion plots constructed at these locations have been used to bench-mark soil loss on slopes up to 30 per cent and slope lengths ranging from 11 to 104 m, with various crop rotations ranging from continuous fallow to double crops. Wischmeier and Smith (1965) used these data in the development of the USLE.

The USLE and data

The Universal Soil Loss equation derived by Wischmeier and Smith (1978) is expressed as the product of a series of factors:

$$A = R \times K \times LS \times C \times P \quad (1)$$

where A is the annual soil loss, R is the rainfall and runoff factor, K is the soil erodibility factor, LS is the slope-length-steepness factor, C is the cover and management factor, and P is the practice factor.

The USLE was designed to produce a mean annual estimate of soil loss. It also can be used to estimate soil loss on a storm by storm basis where incremental rainfall is available. For each storm rainfall event the rainfall erosivity index (EI) is calculated by

$$EI = R_5 \sum (210 + 89 \, \text{Log}_{10} I) \quad (2)$$

where I is the incremental rainfall intensity and R_5 the maximum storm 30-minute rainfall. Individual storm erosion amounts were calculated with USLE using this EI estimate to replace the R factor in equation (1), summed to give a yearly soil loss, and then averaged to produce a mean annual erosion estimate.

Although the USLE has been used for a variety of purposes at many locations in the U.S. and around the world because it seems to give acceptable results, Wischmeier (1976) warned users about the use and misuse of the equation. Since its development, much research has been carried out to improve and further develop factors and data used by the equation. Today, the USLE is considered mature technology and limited in further development. However, the plot data used in its development are still the basis for comparisons when using newer models.

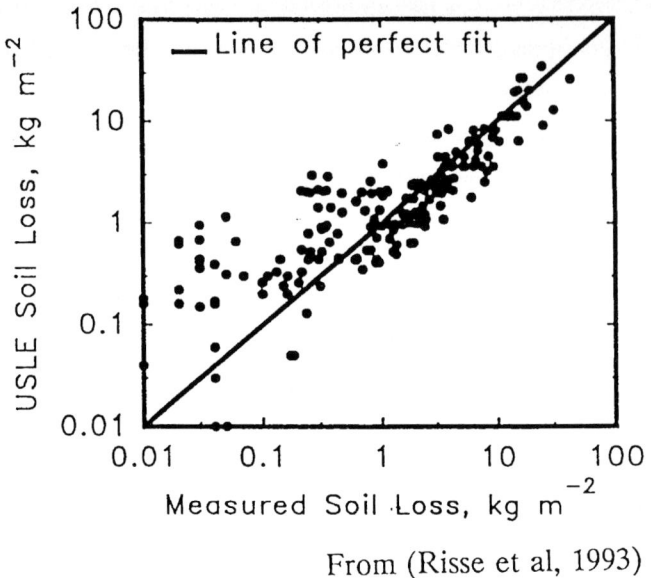

From (Risse et al, 1993)

Figure 2. Average annual soil loss measured and predicted by the Universal Soil Loss Equation

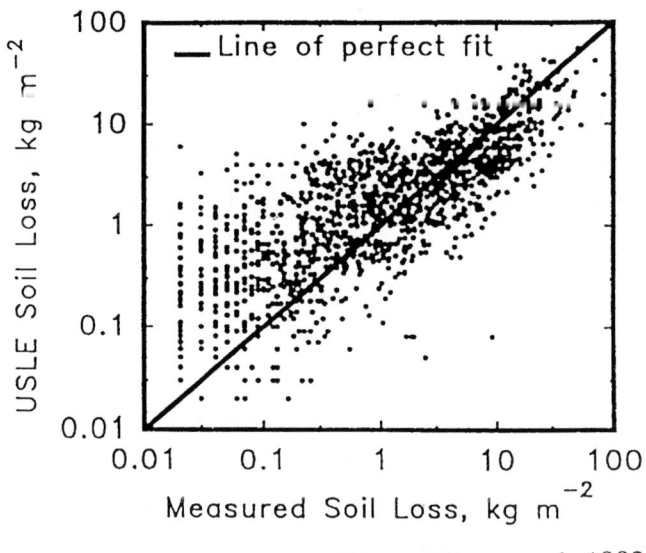

From (Risse et al, 1993)

Figure 3. Yearly soil loss measured and predicted by the Universal Soil Loss Equation

Risse et al. (1993) compiled the USLE erosion database for 1700 plot-years from 220 plots at 22 sites including the plot data from the original erosion research station listed above. Figure 2 is a plot of predicted annual average soil loss using USLE compared with the measured soil loss (r^2=0.75). Figure 3 is a similar plot with soil loss predicted on a yearly basis compared with yearly measured soil loss (r^2=0.58). In both figures USLE tends to overpredict soil loss on plots with lower erosion rates.

USLE technology in other models

The enactment in the 1970s of federal legislation (Federal Water Pollution Control Act of 1972, Public Law No. 92-500 and the Soil and Water Resources Conservation Act) led to the development of continuous simulation models which employ USLE technology. Furthermore, models such as CREAMS, EPIC and SWRRB each use modifications of USLE as intrinsic parts of their erosion component submodels.

Brief descriptions of the principal equations used to estimate interrill and rill erosion by each of the models are given in the following sections. Only the erosion equations are presented so that the technology used in each model can be compared. The reader is directed to the references cited for complete documentation of each model.

The CREAMS model

The CREAMS model is made up of three separate component submodels, for hydrology, erosion, and chemicals. Each model may be run separately with output from the hydrology model being input to the erosion model and then output from the erosion input to the chemical model. Runoff is computed in the hydrology using the NRCS curve number method. The erosion component of CREAMS requires storm runoff volume and peak flow rate passed from the hydrology component. Erosion by detachment is calculated for both interrill and rill areas using two modifications to USLE (Foster et al., 1980). Interrill erosion is calculated using:

$$D_{Li} = 0.210 \text{ EI } (s + 0.014) \text{ KCP } (q_p/V_u) \tag{3}$$

and rill erosion using

$$D_{fr} = 37983 \text{ mV}_u \text{ } q_p^{1/3} \text{ } (x/72.6)^{m-1} \text{ } s^2 \text{ KCP } (q_p/V_u) \tag{4}$$

where D_{Li} is interrill detachment rate, D_{fr} is rill detachment capacity, EI is rainfall erosivity, x is the distance down slope, s is the sine of the slope angle, m is a slope length exponent, K is the USLE erodibility factor, C is the USLE cover and management factor, P is the USLE practice factor, V_u is the runoff volume, and q_p is the peak flow rate.

The storm erosivity, EI, calculated in the hydrology component and passed to the erosion model is estimated from daily rainfall volume, V_R, by

$$EI = 8.0 \, V_R^{1.51} \tag{5}$$

The EPIC model

The EPIC model produces several estimates of erosion using the USLE; Modified Universal Soil Loss Equation (MUSLE), (Williams, 1975); Onstad-Foster model, (Onstad and Foster, 1975); and Theoretical Modified Soil Loss Equation (MUSLT), (Williams et al., 1990). Similarly to CREAMS, EPIC is driven by a daily weather file, a soils characteristics file and a crop management file (Sharpley and Williams, 1990). The hydrology model component also uses the NRCS curve number method to partition daily precipitation into runoff. The model has several options for simulating or reading observed weather data and other options for selecting calculated daily, monthly or annual output variables. The following sections list the erosion equation used in EPIC.

EUSLE model

This model is essentially the USLE model given in equation (1) with the annual R value replaced with an estimated storm EI value derived from

$$EI = R \, [12.1 + 8.9 \, (\log r_p - 0.434)] \, (r_s) / 1000 \tag{6}$$

where EI is the storm erosivity, R is the daily rainfall amount, r_p is peak rainfall rate, and r_s is the maximum 0.5 hour intensity. Then for storm erosion estimates

$$A = EI \times K \times LS \times C \times P \tag{7}$$

where A, K, LS, C, and P are the USLE factors listed previously. Storm erosion estimates are then summed to produce monthly, annual, or mean annual erosion estimates.

AOF Model

The Onstad-Foster model, designated here as the AOF model, estimates soil loss by:

$$A = XI \times K \times LS \times C \times P \tag{8}$$

where XI rainfall-runoff-energy factor is

$$XI = 0.646 \, EI + 0.45 \, (Q \times q_p)^{0.33} \tag{9}$$

and Q is the runoff volume, q_p is the peak flow rate, and A, K, LS, C, and P are as given previously.

MUSLE Model

The MUSLE model uses a runoff-energy factor to replace the EI factor in USLE. The runoff-energy factor is

$$RI = 1.586 (Q * q_p)^{0.56} DA^{0.12} \qquad (10)$$

where Q is the runoff volume, q_p is the peak runoff rate, and DA is the drainage area. Erosion is calculated on a storm by storm basis as

$$A = RI \times K \times LS \times C \times P. \qquad (11)$$

MUSLT Model

The MUSLT model is another form of the MUSLE model with a different coefficient and exponent in the runoff-energy factor

$$RI = 2.5 (Q * q_p)^{0.5} \qquad (12)$$

Erosion estimates are then calculated in the same manner as with MUSLE.

Replacement of USLE

USLE was considered to be a mature technology, because it was a lumped model with empirical origins and structure that limited its improvement for increased accuracy of soil loss prediction. In 1985, a team of U.S. engineers and scientists from three USDA agencies — Agricultural Research Service (ARS), Natural Resource Conservation Service (NRCS), and Forest Service (FS) and the USDI-Bureau of Land Management (BLM) — was formed to develop user specifications and requirements, computer code, and databases for a series of models to replace USLE (Foster, 1987). This effort was named the Water Erosion Prediction Project (WEPP).

The WEPP utilizes new technology for runoff and erosion simulation. Precipitation is partitioned into runoff and infiltration by disaggregating daily precipitation into a storm intensity pattern. A stochastic weather generator, CLIGEN (Nicks and Lane, 1989) is used to generate the required storm precipitation inputs of amount, duration, time to peak intensity and maximum storm intensity. WEPP then disaggregates these variables into a single peak storm intensity pattern. The model is driven by four input files, climate, soil characteristic, slope, and crop management.

WEPP technology differs from the other models listed above by having two separate soil erodibility factors, K_i for interrill, and K_r for rill erosion. Suspended sediment in a rill is calculated by:

$$dG/dx = D_f + D_i \tag{13}$$

where G is the sediment load, x is the distance down slope, D_f is the rill erosion, and D_i interrill erosion. Both D_i and D_f are computed on a per rill area basis. Rill erosion, D_f, is calculated by:

$$D_f = D_c (1 - G/T_c) \tag{14}$$

where D_c is detachment capacity, and T_c is sediment transport capacity and:

$$D_c = K_r(\tau_f - \tau_c) \tag{15}$$

where K_r is a rill erodibility parameter, τ_f is flow shear stress acting on a soil particle, and τ_c is the critical shear stress. Interrill erosion is given as:

$$D_i = K_i I_s^2 C_s G_s (R_s/\omega) \tag{16}$$

where K_i is baseline interrill erodibility, I_s is effective rainfall intensity, C_s is the effect of canopy cover on interrill erosion, G_s is the effect of ground cover on interrill erosion, R_s is the rill spacing, and ω is the width.

Because of the theory developed for the WEPP technology, the USLE erosion database could not be used directly for parameterisation. Therefore, three sets of field experiments, cropland, rangeland and forestland, were initiated to determine parameters for calculating K_r and K_i given in equations (15) and (16), respectively. Starting in 1987, cropland erosion plot data were collected at 36 sites (Elliot *et al.*, 1989); rangeland at 20 sites and forestland at 21 sites (Nearing *et al.*, 1989) across the U.S. A rotating boom rainfall simulator was used to supply fixed rainfall intensity to the plots. Measurements of infiltration, soils properties, roughness, etc. were made before and after experimental runs. Thus, another erosion database was created from these studies.

Conclusions

The need for soil loss prediction became evident to US agriculturalists in the early decades of this century. Their interest led to the establishment of benchmark erosion studies at 10 locations across the country. Data from the early studies and those established later became the basis for development of erosion prediction relationships which are still in use today.

While the USLE is not a perfect predictor of soil loss on plots (r^2=0.75), it does represent the technology prevalent during the time of its development. More recent efforts, such as WEPP, may have benefited from more advanced measurement techniques, computer technology and better analytical tools available today, but the increase in predictive power of this newer technology is yet to be determined. However, it can be concluded that concentrated efforts in erosion research such as the USLE and WEPP experiments perhaps occur only once in every 20 to 30 years. Therefore, the use of the database developed should be maximized.

References

Bennett, H.H. (1928). The geographical relation of soil erosion to land productivity. *Geographical Review* **18**, 579-605.

Daniel, H.A., Elwell, H.M. and Cox, M.B. (1943). *Investigation in Erosion Control and Reclamation of Eroded Land at the Red Plains Conservation Experiment Station Guthrie, Oklahoma 1930-40.* USDA-SCS Technical Bulletin No. 837.

Elliot, W.J., Liebenow, A.M., Laflen, J.M. and Kohl, K.D. (1980). *A Compendium of Soil Erodibility Data From WEPP Cropland Soil Field Erodibility Experiments 1987 and 88.* NSERL Report No. 3. The Ohio State University and USDA - ARS National Soil Erosion Laboratory, West Lafayette, IN 47907.

Flanagan, D.C. and Nearing, M.A. (eds) (1995). *USDA - Water Erosion Prediction Project Hillslope Profile and Watershed Model Documentation.* NSERL Report No. 10. USDA-ARS National Soil Erosion Research Laboratory, West Lafayette, IN 47907.

Foster, G.R., Lane, L.J., Nowlin, J.D., Laflen, J.M. and Young, R.A. (1980). A model to estimate sediment yield from field-sized areas: development of model. In, Knisel, W.G. (ed.), *CREAMS — A Field Scale Model for Chemicals, Runoff, Erosion and Agricultural Management Systems.* U.S. Department of Agriculture, Conservation Research Report No.26. pp. 36-64.

Foster, G.R. (Compiler). (1987). *User Requirements USDA-Water Erosion Prediction Project.* NSERL Report No. 1. USDA - ARS National Soil Erosion Research Laboratory, West Lafayette, IN. 47907.

Knisel, W, G. (ed.) (1980). *CREAMS — A Field Scale Model for Chemicals, Runoff, and Erosion from Agricultural Management Systems.* USDA Conservation Report No. 26.

Lane, L. J. and Nearing, M. A. (eds) (1989). *USDA-Water Erosion Prediction Project: Hillslope Profile Model Documentation.* NSERL Report No. 2. USDA-ARS National Soil Erosion Research Laboratory, West Lafayette, IN 47907.

McDonald, A. (1938). *Erosion and its Control in Oklahoma Territory.* USDA Miscellaneous Publication 301.

Nearing, M.A., Weltz, M.A., Finkner, S.C., Stone, J.J. and West, L.T. (1989). Parameter identification for plot data. In, Lane L.J. and Nearing, M.A. (eds), *USDA Water Erosion Prediction Project: Hillslope Profile Model Documentation.* NSERL Report No. 2, USDA-ARS National Soil Erosion Research Laboratory, West Lafayette, IN. 47907. pp. 11.1-11.15.

Nicks, A.D. and Lane, L.J. (1989). Weather Generator. In, Lane L.J. and Nearing, M.A. (eds), *USDA Water Erosion Prediction Project: Hillslope Profile Model Documentation.* NSERL Report No. 2. USDA-ARS National Soil Erosion Research Laboratory, West Lafayette, IN. 47907. pp. 2.1-2.19.

Onstad, C.A. and Foster, G.R. (1975). Erosion modeling on a watershed. *Transactions of the American Society of Agricultural Engineers* **18**(2), 288-292.

Risse, L.M., Nearing, M.A., Nicks, A.D. and Laflen, J.M. (1993). Error assessment in the Universal Soil Loss Equation. *Soil Science Society of America Journal* **57**(3), 825-823.

Sharpley, A.N. and Williams, J.R. (eds) (1990). *EPIC — Erosion/Productivity Impact Calculator: 1 Model Documentation.* USDA Technical Bulletin No. 1768.

Williams, J.R. (1975). *Sediment Yield Prediction with Universal Equation Using Runoff Energy Factor.* USDA-ARS, ARS-S-40.

Williams, J.R., Jones, C.A. and Dyke, P.T. (1984). A modeling approach to determining the relationship between erosion and soil productivity. *Transactions of the American Society of Agricultural Engineers* **27**(1), 129-144.

Williams, J. R., Nicks, A. D., and Arnold, J. A. (1985). Simulator for water resources in rural basins. *Journal of Hydrologicalc Engineering, American Society of Civil Engineers* **111**(6), 970-976.

Williams, J.R., Dyke, P.T., Fuchs, W.W., Benson, V.W., Rice, O.W. and Taylor, E.D. (1990). *EPIC — Erosion/Productivity Impact Calculator. 2. User Manual.* USDA Technical Bulletin No. 1768.

Wischmeier, W.H. and Smith, D.D. (1965). *Predicting Rainfall Erosion Losses from Farm Land East of the Rocky Mountains.* USDA Agricultural Handbook 282. U.S. Government Printing Office, Washington, D.C.

Wischmeier, W.H. (1976). Use and misuse of the Universal Soil Loss Equation. In, *Soil Erosion: Prediction and Control.* Soil Conservation Society Publication 21, Soil Conservation Society of America, Ankeny, IA. pp. 371-378.

Wischmeier, W. H. and Smith, D. D. (1978). *Predicting Rainfall Erosion Losses — a Guide to Conservation Planning.* USDA-ARS Agricultural Handbook No. 537. U.S. Government Printing Office, Washington, D.C.

Woodruff, C. M. (1987). Pioneering erosion research that paid. *Journal of Soil and Water Conservation,* March-April. pp. 90-91.

SECTION 5. MODEL DESCRIPTIONS

29. THE EUROSEM MODEL

R.P.C. Morgan[1], J.N. Quinton[1], R.E. Smith[2], G. Govers[3], J.W.A. Poesen[3], G. Chisci[4] and D. Torri[5]

[1]*Silsoe College*
Cranfield University
Silsoe
Bedford MK45 4DT
UK

[2]*USDA-ARS, North Plains Area*
AERC, Colorado State University
Fort Collins CO 80523
USA

[3]*National Fund for Scientific Research and Laboratorium voor Experimentele Geomorfologie*
Katholieke Universiteit Leuven
Redingenstraat 16 bis
3000 Leuven
Belgium

[4]*Dipartimento di Agronomia e Produzione Erbacee*
Università degli Studi di Firenze
Piazzale delle Cascine 18
50144 Firenze
Italy

[5]*CNR Centro di Studio per la Genesi Classificazione e Cartografia del Suolo*
Piazzale delle Cascine 15
50144 Firenze
Italy

Abstract

The European Soil Erosion Model (EUROSEM) is a dynamic distributed model for simulating erosion, transport and deposition of sediment over the land surface by interrill and rill processes. It is designed as an event-based model for both individual fields and small catchments. Model outputs include total runoff, total soil loss, the storm hydrograph and the storm sediment graph. EUROSEM provides for explicit simulation of interrill and rill flow; the effects of plant cover on rainfall interception, infiltration, rainfall energy and flow velocity; and the effects of rock fragment cover on infiltration, flow velocity and splash erosion. Catchments are represented as a simplified cascading network of elements, which may be either planes (for hillslope segments) or channels. Each plane is considered uniform in its soil, slope, surface microtopography and land cover.

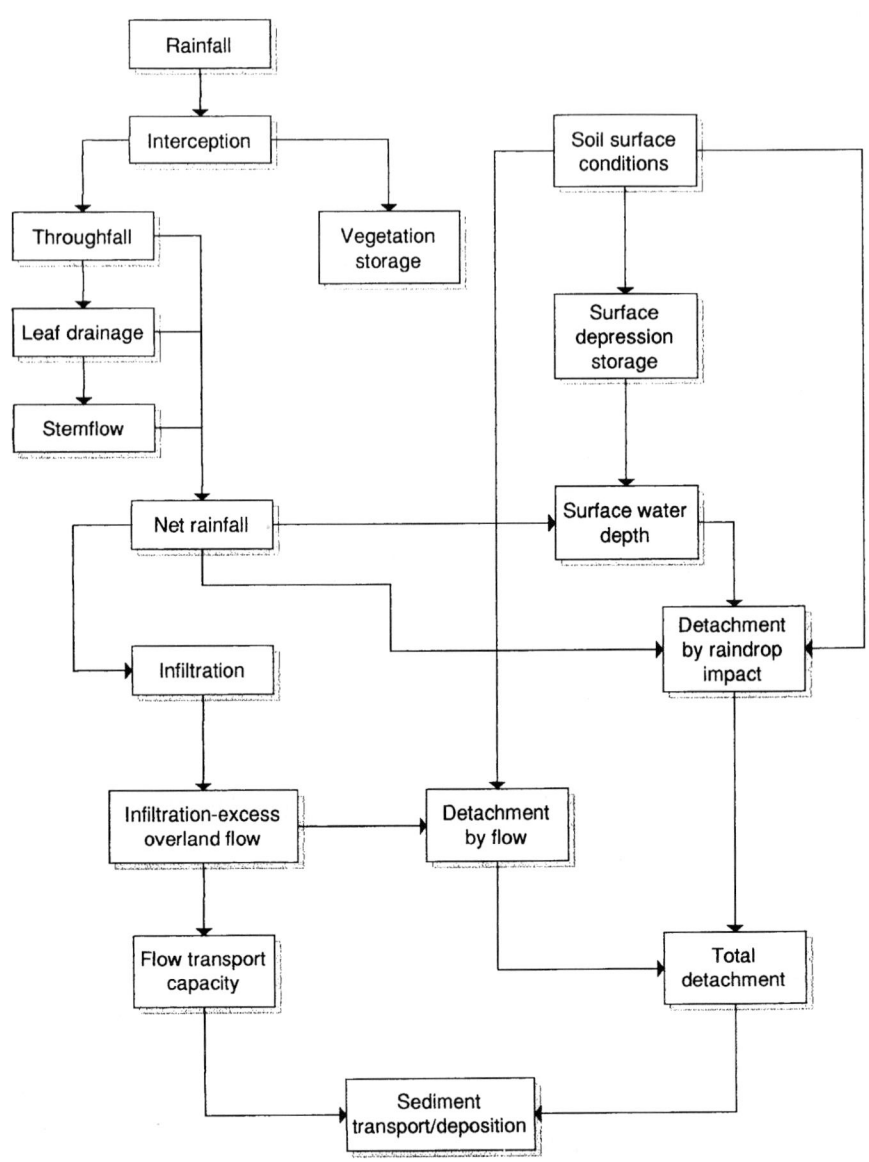

Figure 1. Flow chart of EUROSEM

Introduction

The European Soil Erosion Model (EUROSEM) is a dynamic distributed model designed to simulate the erosion, transport and deposition of sediment over the land surface by interrill and rill processes. It can be applied to individual storm events and to spatial scales ranging from a small field to a small catchment. It has been developed over the last six years by more than twenty scientists from twelve countries as a state-of-art erosion model for the assessment of soil erosion risk and the evaluation of soil protection measures. Existing models were considered inadequate for these purposes. Empirical models like the Universal Soil Loss Equation (Wischmeier and Smith, 1978) require much research to parameterise them for Europe, while physically-based models such as CREAMS (Foster et al., 1981) and WEPP (Nearing et al., 1989) adopt a steady-state approach which falls short of what is required for understanding and predicting the timing and magnitude of peak transfers of sediment to water courses. In order to obtain a good approximation of what is happening at different points over the land surface at different times during a storm, a fully dynamic approach is required.

Model structure

EUROSEM requires break-point rainfall data, ideally on a one-minute resolution, for the storm and computes, in turn, the interception of the rain by the plant cover, the generation of runoff as infiltration excess, soil detachment by raindrop impact, soil detachment by runoff, transport capacity of the runoff and deposition of sediment (Figure 1). EUROSEM uses the runoff generator and the water and sediment routing routines from KINEROS (Woolhiser et al., 1990). A mass balance equation (equation 1) is used to compute the volume (or mass) of sediment passing a given point on the land surface at a given time:

$$\frac{\partial (AC)}{\partial t} + \frac{\partial (QC)}{\partial x} - e(x,t) = q_s(x,t) \tag{1}$$

in which C = sediment concentration (m³/m³), A = cross sectional area of the flow (m²), Q = discharge (m³/s), q_S = external input or extraction of sediment per unit length of flow (m³ s⁻¹ m⁻¹), e = net detachment rate or rate of erosion of the bed per unit length of flow (m³ s⁻¹ m⁻¹), x = horizontal distance (m), and t = time (s).

The term, e, in equation (1) is defined by two major components:

$$e = DR + DF \qquad (2)$$

where DR = the rate of soil particle detachment by raindrop impact ($m^3\ s^{-1}\ m^{-1}$), and DF = the balance between the rate of soil particle detachment by the flow and the particle deposition rate ($m^3\ s^{-1}\ m^{-1}$).

Simulation of erosion and deposition

Soil particle detachment by raindrop impact

Soil particle detachment by raindrop impact for each time step (t) is modelled as a function of the kinetic energy of the rainfall at the ground surface, the detachability of the soil and the surface water depth. This is done in several stages within the model but the equations used can be simplified into the relationship:

$$DR = \{ (KE)\ e^{-bh} \} / \rho_s \qquad (3)$$

where k = an index of the detachability of the soil (m^3/J), KE is the kinetic energy of the rainfall at the ground surface (J/m^2), b is an exponent taken as equal to 2.0 (Torri et al., 1987), h is the depth of the water surface layer (m) and ρ_s is the sediment particle density (= 2.65 Mg/m^3).

The kinetic energy of the rainfall represents the combined energy of the rain reaching the ground surface as direct throughfall and that of leaf drainage. The energy of the direct throughfall is modelled assuming that its drop size distribution follows that described by Marshall and Palmer (1948) and that the raindrops are at terminal velocity. The energy of the leaf drainage is computed from the equation developed by Brandt (1990) which considers the height of fall of the drops and assumes that the drop size distribution follows that described by Brandt (1989).

Soil particle detachment by runoff

Soil particle detachment by runoff is modelled in terms of a generalised erosion-deposition theory proposed by Smith et al. (1995). This assumes that the concentration of sediment in the flow at transport capacity reflects a balance between the counteracting processes of erosion and deposition. It implies that the ability of running water to erode its bed is a function of the energy expended by the flow, particularly in shear between the water and the bed and in turbulence, and is independent of the amount of material it is carrying. The general equation for soil particle detachment and/or deposition by flow is:

$$DF = \beta w v_s (TC - C) \tag{4}$$

where DF is the net rate of detachment of soil particles by the flow (negative values represent deposition), β is a flow detachment efficiency coefficient, w is the width of the flow (m), v_s is the settling velocity of the particles in the flow (m/s), TC is the sediment concentration in the flow at transport capacity and C is the actual sediment concentration in the flow. By definition, $\beta = 1$ when deposition is taking place and $\beta < 1$ for cohesive soils when DF is positive. EUROSEM defines β as a function of the cohesion of the soil as measured by a torvane under saturated conditions. Such a measure has been shown to relate to the detachment of soil particles by concentrated overland flow (Poesen and Govers, 1990) and to the likelihood of rill initiation (Rauws and Govers, 1988). An approach based on soil cohesion would appear to be the most appropriate one at present because it enables detachability to be expressed by a parameter which can be easily measured in the field. In reality, there will be variability around the detachability value depending upon the initial moisture content (Govers and Loch, 1993) and initial structural conditions (Govers et al., 1990).

Transport capacity of the flow
EUROSEM uses two somewhat different transport capacity relationships for rill and interrill flows. Together they summarise data from over 500 experimental observations of sediment transport by shallow surface flows carried out on materials ranging in median particle size from 50 to 250 μm, slopes from 1 to 15 per cent and discharges from 2 to 100 cm^3 cm^{-1} s^{-1}. For flow in rills, transport capacity (TC) is modelled as a function of unit stream power using the equation (Govers, 1990):

$$TC = c (\omega - \omega_c)^\eta \tag{5}$$

where ω is unit stream power (the product of slope and flow velocity), ω_c is a critical value of unit stream power = 0.4 cm/s, and c and η are coefficients related to the median particle size of the soil and obtained from laboratory experiments.

For interrill flow, EUROSEM uses an equation based on the work of Everaert (1991), which included transport experiments with loose sediments with median particle sizes ranging from 33 to 390 μm:

$$TC = \frac{b}{\rho_s q}\left[(\Omega - \Omega_c)^{0.7/n} - 1\right]^n \tag{6}$$

where b is a function of particle size, ρ_s is the sediment density, Ω is Bagnold's modified stream power, Ω_c is a critical value of Bagnold's modified stream power, and $n = 5$. Compared with other erosion models which generally assume that all the sediment detached by raindrop impact and flow on interrill areas is delivered to the rills, EUROSEM computes delivery based on the transport capacity of the interrill flow. For interrill flow paths longer than 1 m, EUROSEM explicitly routes the interrill flow, allowing deposition of sediment to occur within the interrill areas.

Factors affecting erosion

Soil erodibility

Soil erodibility is expressed by two parameters, the detachability of the soil by raindrop impact and soil cohesion, used to describe the detachability of the soil by flow. The parameters were chosen from extensive studies of splash erosion by Poesen (1985) and Govers (1991) and research on initiation of soil particle movement by overland flow (Govers, 1987; Rauws, 1988) and generation of rills (Rauws and Govers, 1988).

Soil surface characteristics

Four main characteristics of the soil surface can be simulated by EUROSEM: the roughness of the surface, resulting either naturally or from a range of tillage practices; the presence of impermeable or paved surfaces; the effect of rock fragments; and the presence of rills. For rilled surfaces, the user needs to define the number, width, depth and side slope of the rills or microchannels.

Compared with other erosion models, EUROSEM provides for explicit simulation of rock fragment effects on the relative volume of the soil not acting as a porous medium, the reduction in the area of fine earth exposed to raindrop impact, and a change in the effective saturated hydraulic conductivity of the soil. The user is able to determine whether the effect on saturated hydraulic conductivity is positive or negative, based on whether the rock fragments are embedded within the soil or sit on the surface and on whether the soil matrix is characterised by structural or textural porosity (Poesen et al., 1994).

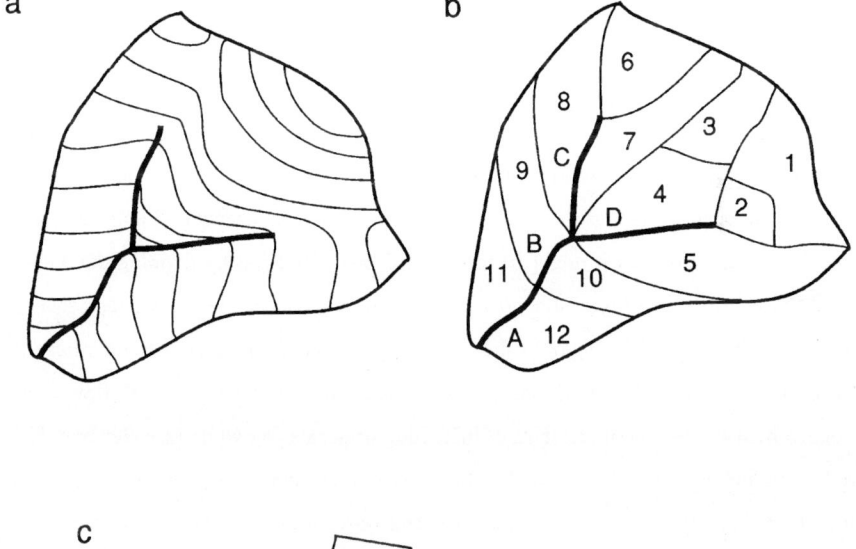

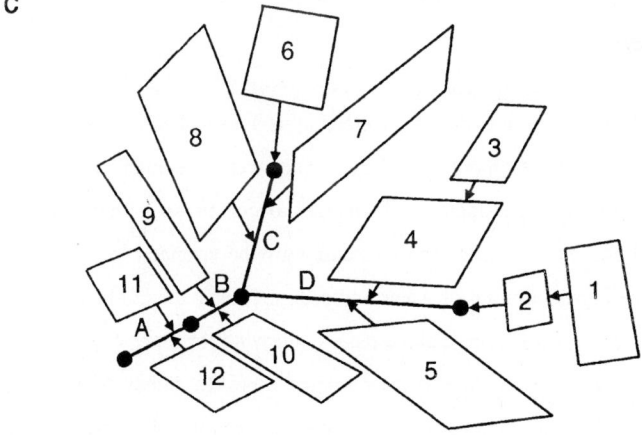

Figure 2. Schematic representation of a catchment by elements

Plant cover

EUROSEM uses direct measurements of plant properties to model explicitly the influence of the plant cover on rainfall amount and energy at the ground surface, saturated hydraulic conductivity and cohesion of the soil, and roughness imparted to flow. The plant properties are percentage canopy cover, effective canopy height, maximum interception storage, average acute angle between the ground surface and the stems or shoots, an index of leaf shape and percentage basal area. To allow for the influence of the plant cover, the user can adjust the values of soil cohesion and Manning's *n*.

Applications

EUROSEM can be applied at a range of scales from a small catchment to a hillslope and an individual field. The land surface is represented, following KINEROS, as a series of interlinked elements which are either uniform slope planes or channels. Figure 2 shows a simple catchment and how it might be represented in EUROSEM by dividing the area into elements along flow lines taken orthogonal to the contours. Each slope plane is represented by a rectangle whose length is equal to the average flow path of the element and whose area matches that of the element. Surfaces with convergence or divergence of flow can be simulated by a succession of elements of increasing or decreasing width (see elements 1, 2, 3 and 4). Channel elements should be identified based on changes in the length of contributing slope planes (see elements 9, 10, 11 and 12) and on changes in channel characteristics of width, depth and side slope.

Care should be taken to ensure that each element is reasonably homogeneous in all its characteristics. New elements should be chosen whenever an element cannot be properly represented by a mean slope angle, and changes occur in soil type, land cover, soil management or the number of rills. The elements must be arranged in a natural cascading sequence to enable correct routing of water and sediment over the land surface. A hillslope can be represented by a succession of slope planes, allowing for representation of changes in slope steepness along, for example, a convexo-concave profile. Subdivisions into additional planes should be made where changes occur in soils, land cover, soil management or number of rills. The effects of live contour barrier strips, riparian barriers and bench terraces can be simulated with this approach (Rickson, 1994).

Three types of surface topography can be simulated on the individual slope planes: unrilled surfaces but characterised by some surface irregularities or roughness; rilled surfaces with interrill flow routed towards the rills; and furrowed surfaces, ranging from plough furrows to ridges, where interrill routing is not used because of the short distance traversed by interrill flows. Changes in the dimensions (widths and depths) of the flow paths (rills and furrows) as a result of erosion and deposition are modelled throughout the storm.

Model outputs

Model outputs from EUROSEM include total runoff, total soil loss, storm hydrograph and storm sediment graph for each element of the land surface. By interpreting the output, information is obtained on the timing of peak runoff and sediment delivery to the water courses and the location of major sources and sinks of sediment.

Conclusions

EUROSEM is a dynamic distributed state-of-art erosion model which can be used to simulate the erosion, transport and deposition of sediment in storm events over areas ranging in size from fields to small catchments. EUROSEM provides for explicit simulation of interrill and rill flow and the effects of plant cover and rock fragments. By using a dynamic rather than steady-state approach, EUROSEM gives a better understanding of the spatial and temporal distribution of runoff and erosion within a catchment during an individual storm than can be achieved by most other models.

Acknowledgements

The work on EUROSEM was funded by Directorate General XII of the Commission of European Communities under their Research and Development Programme in the Field of the Environment (1986-1990) and the Science and Technology for Environmental Protection (STEP) Programme (1989-1992). This paper is a contribution to the Soil Erosion Network of the GCTE, which is a Core Research Project of the IBGP.

References

Brandt, C.J. (1989). The size distribution of throughfall drops under vegetation canopies. *Catena* **16**, 507-524.

Brandt, C.J. (1990). Simulation of the size distribution and erosivity of raindrops and throughfall drops. *Earth Surface Processes and Landforms* **15**, 687-698.

Everaert, W. (1991). Empirical relations for the sediment transport capacity of interrill flows. *Earth Surface Processes and Landforms* **16**, 513-532.

Foster, G.R., Lane, L.J., Nowlin, J.D., Laflen, J.M. and Young, R.A. (1981). Estimating erosion and sediment yield on field-sized areas. *Transactions of the American Society of Agricultural Engineers* **24**, 1253-1263.

Govers, G. (1987). Initiation of motion in overland flow. *Sedimentology* **34**, 1157-1164.

Govers, G. (1990). Empirical relationships on the transporting capacity of overland flow. *International Association of Hydrological Sciences Publication* **189**, 45-63.

Govers, G. (1991). Spatial and temporal variations in splash detachment: a field study. *Catena Supplement* **20**, 15-24.

Govers, G. and Loch, R.J. (1993). Effects of initial water content and soil mechanical strength on the runoff erosion resistance of clay soils. *Australian Journal of Soil Research* **31**, 549-587.

Govers, G., Everaert, W., Poesen, J., Rauws, G., De Ploey, J. and Lautridou, J.P. (1990). A long-flume study of the dynamic factors affecting the resistance of a loamy soil to concentrated flow erosion. *Earth Surface Processes and Landforms* **15**, 313-328.

Marshall, I.S. and Palmer, W.M. (1948). The distribution of raindrops with size. *Journal of Meteorology* **5**, 165-166.

Nearing, M.A., Foster, G.R., Lane, L.J. and Finckner, S.C. (1989). A process-based soil erosion model for USDA-Water Erosion Prediction Project technology. *Transactions of the American Society of Agricultural Engineers* **32**, 1587-1593.

Poesen, J. (1985). An improved splash transport model. *Zeitschrift für Geomorphologie* **29**, 193-211.

Poesen, J. and Govers, G. (1990). Gully erosion in the loam belt of Belgium: typology and control measures. In, Boardman, J., Foster, I.D.L. and Dearing, J.A. (eds), *Soil erosion on agricultural land*. Wiley, Chichester. pp. 513-530.

Poesen, J.W., Torri, D. and Bunte, K. (1994). Effects of rock fragments on soil erosion by water at different spatial scales: a review. *Catena* **23**, 141-166.

Rauws, G. (1988). Laboratory experiments on the resistance of overland flow due to composite roughness. *Journal of Hydrology* **103**, 37-52.

Rauws, G. and Govers, G. (1988). Hydraulic and soil mechanical aspects of rill generation on agricultural soils. *Journal of Soil Science* **39**, 111-124.

Rickson, R.J. (1994). Potential applications of the European Soil Erosion Model (EUROSEM) for evaluating soil conservation measures. In, Rickson, R.J. (ed.), *Conserving soil resources: European perspectives*. CAB International, Wallingford. pp. 326-335.

Smith, R.E., Goodrich, D. and Quinton, J.N. (1995). Dynamic distributed simulation of watershed erosion: the KINEROS2 and EUROSEM models. *Journal of Soil and Water Conservation* **50**, 517-520.

Torri, D., Sfalanga, M. and Del Sette, M. (1987). Splash detachment: runoff depth and soil cohesion. *Catena* **14**, 149-155.

Wischmeier, W.H. and Smith, D.D. (1978). *Predicting rainfall erosion losses*. USDA Agricultural Handbook No. 537.

Woolhiser, D.A., Smith, R.E. and Goodrich, D.C. (1990). *KINEROS: A kinematic runoff and erosion model: documentation and user manual*. USDA Agricultural Research Service Publication No. ARS-77.

30. GRIFFITH UNIVERSITY EROSION SYSTEM TEMPLATE (GUEST)

Calvin W. Rose, Keppel J. Coughlan and Banti Fentie

Faculty of Environmental Sciences
Griffith University
Nathan Campus
Kessels Road
Queensland
Australia 4111

Abstract
Excluding gully processes and mass movement, the rate of erosion of bare soil depends on the rate of overland flow and rainfall, on the erodibility and depositability of surface soil, and on the features of rilling if this occurs. The program GUEST is designed to analyse data collected from runoff plots of simple form, and to yield an approximate non-dimensional erodibility parameter denoted by β. The parameter β has a theoretical basis, and is more physically meaningful if flow-driven erosion processes dominate those due to rainfall impact.

Evaluation of β requires a value of the depositability of the soil, which can be readily determined by a variety of laboratory methods. Conversion of such laboratory data into depositability, which is the mean settling velocity of sediment, is assisted by the Griffith University Depositability Program (or GUDPRO).

Evaluation of erodibility β using GUEST also requires the following information:
- geometry and slope of the runoff plot;
- runoff rate measured as a function of time during the erosion event;
- total net loss of sediment from the runoff plot during the event;
- rill frequency and geometry (if rills occur).

The utility of GUEST is illustrated by its application to data collected in a multi-country study of soil erosion and conservation methodologies involving Australia, the Philippines, Malaysia and Thailand. The erodibility of a given soil type is found to vary significantly with soil management, and with the extent or duration of erosion.

Introduction

Before it is feasible to adequately predict soil erosion under a variety of conditions, a measure of the soil's resistance to erosion processes is required. Whilst there is a great variety of management methods which can be utilised to reduce soil loss, a benchmark is the erosion which will occur on bare soil, though even here soil loss can be significantly affected by management, particularly tillage.

Program GUEST (Griffith University Erosion System Template) (Misra and Rose, 1990) was developed in order to derive an erodibility parameter for bare soil, based on measurements of soil loss and runoff rate. Major applications of this programme have been in tropical and semi-tropical countries where the scale of agricultural operations is small, so that the runoff unit can be approximated by a plane land surface, which may, however, be rilled. GUEST assumes that runoff rate is a measured quantity. With erodibility determined using GUEST, erosion prediction is a separate and subsequent step, though utilising the same type of erosion theory.

In this paper, application of program GUEST is described for a soil conservation project involving four countries: Malaysia, Thailand, the Philippines, and Australia. This project was supported by agencies and Universities in these countries, but funds for co-ordination came from the Australian Centre for International Agricultural Research (ACIAR). Aims of this project included the demonstration of acceptable cropping systems which also substantially reduce soil loss, and the development of methods which assist in the extrapolation of results obtained. A description of the common methodology employed, and examples of outcomes from an earlier stage in the project are in press in a special issue of *Soil Technology* edited by Rose (1995).

Methodology
A common experimental methodology in the project is illustrated in Figure 1 for crops other than tree crops, where larger plot areas and a weir for flow measurement were used.

Figure 1 illustrates the use of tipping-bucket technology described by Bonell and Williams (1987) in measuring rate of runoff. Water and sediment from the runoff plot is first collected on a settling basin of low slope, where coarser sediment (referred to as 'bedload') deposits. Clearer water with the remaining sediment (referred to as 'suspended load') then drops vertically through a slot in the extended collection system to fall into one side or the other of a split-container bucket, which tips when adequately full. The tipping motion induces an electric pulse when time of occurrence is recorded by an electronic recorder. This data logger also records tips from a tipping-bucket raingauge adjacent to the plot.

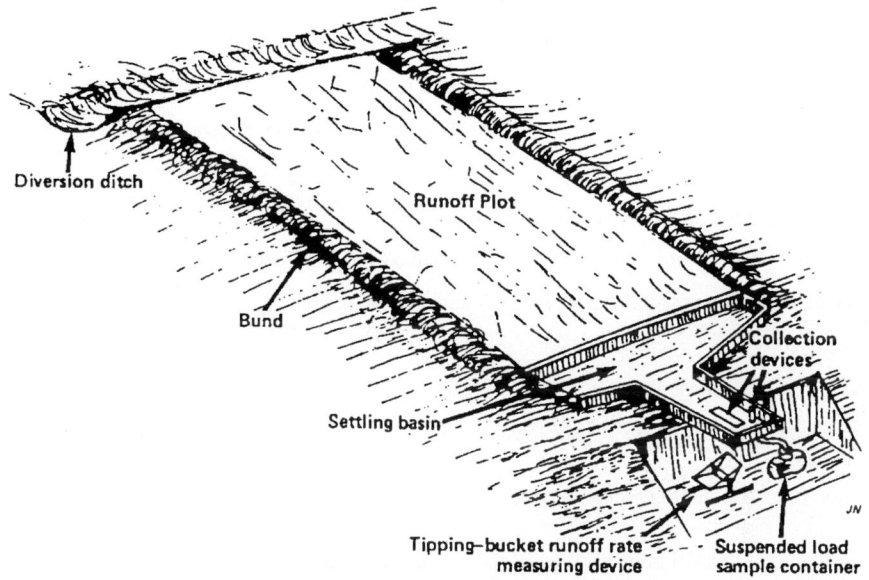

Figure 1. Illustrating a hydrologically defined runoff plot. Water entry at its upper end is prevented by the diversion ditch. Runon or runoff can be prevented from occurring laterally to or from the plot by a low bund (illustrated) or galvanised iron sheeting, for example. In the ACIAR experiments referred to in the text, the settling basin was concrete and of lower slope than the plot, allowing coarser sediment (or 'bed-load') to settle out on it for later collection. A (flow-weighted) sample of 'suspended sediment', defined as that which is lost from the settling basin, can be collected from a split pipe shown located at the end of the settling basin, and stored in a container shown located beneath the settling basin. The other collection device shown is a slot in the end of the settling basin through which runoff falls into a tipping-bucket flow rate measuring device. The time of the bucket tipping is recorded on an electronic datalogger. (The end of the settling basin need not be cantilevered as shown; beneath it is shown cut away for clarity). (From Rose, 1993)

A sub-sample of water with its suspended load is taken as it leaves the tipping bucket. Since this sub-sample is approximately flow weighted, the loss of sediment as suspended load can be calculated as the product of the total runoff volume and the concentration of suspended load in the sub-sample collector. Addition to this mass of soil of that which deposited in the Gerlach-trough settling basin yields total soil loss for the event. Division of this total mass of eroded soil by the total runoff for the event (corrected for rainfall falling on the settling basin) gives the average sediment concentration for the erosion event, denoted $\bar{c}$. This procedure is the same whatever management and cropping system is used in the plot.

Program GUEST is like a template which analyses data on both total soil loss per event, and runoff rate during the event, to yield an empirical parameter β which is a measure of the erodibility of the soil. With experience built up in this way on how β varies for any particular soil type and management system, the same theory is used in longer term predictions of soil erosion.

The theoretical bases on which GUEST depends makes a fundamental distinction between the two types of sediment characteristics which control soil erosion in a bare soil situation. These are:

- Soil erodibility, described by β, and
- Soil depositability (Rose et al., 1990) denoted here by ϕ, where ϕ is the mean settling velocity of sediment given by $\sum_{i=1}^{I} v_i / I$, where v_i is the settling velocity of any arbitrary size class i, with the total number of equal mass size classes being denoted by I.

The magnitude of the depositability depends on the settling velocity characteristic of the sediment. However, with the shallow flows quite common in the context of soil erosion from plots of modest size, soil aggregates of size up to some 4 mm which contribute to ϕ may not be fully immersed, and so cannot be considered as contributing to deposition. Hence there is a reduction in value of depositability, ϕ, to the 'effective depositability', denoted ϕ_e, calculated using the Griffith University Depositability Program called GUDPRO (Lisle et al., 1995).

Another consequence of non-immersion of larger aggregates is that flow occurs only around such aggregates (assuming they remain in place and do not shift by rolling). Thus some of the shear stress exerted by flowing water is dissipated on these larger stationary sedimentary units, and only the remainder of shear stress is available to erode other immersed sediment. This reduction in effective shear stress has consequences for soil erosion theory (Rose et al., 1996). Program GUDPRO also provides an estimate, for any depth of water flow, of the fraction of the soil surface which is occupied by these larger aggregates.

Data on sediment depositability can be obtained using a variety of experimental methods. These methods include wet sieving, a modified bottom withdrawal tube (Lovell and Rose, 1988) and a top entry tube (Hairsine and McTainsh, 1986). Each of these methods has its advantages and disadvantages. For example, whilst being cheap to construct, the bottom withdrawal tube appears to require more training or experience than the other method in order to obtain reliable results.

The program GUDPRO accepts data from any of the three experimental techniques listed above and provides, as output to program GUEST, the effective depositability, ϕ_e, and the fraction (1-C) of the surface exposed to erosion, where C is the fraction covered by larger unimmersed aggregates. Both ϕ_e and (1-C) are provided as functions of the depth of water, D. An example of this functional dependence is given in Figure 2.

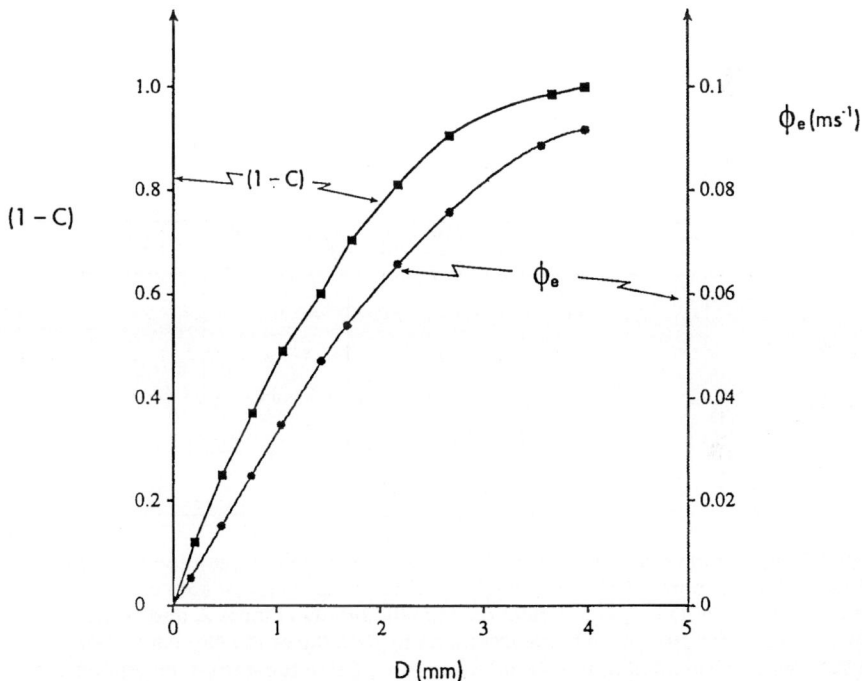

Figure 2. Illustration of the dependence of the effective depositability, ϕ_e and the exposure factor, (1-C) on water depth D for a particular soil. (From Rose et al., 1995)

A lap-top computer is used to extract data from loggers, and to provide on-site a graphical display of the recorded data in cumulative form. This allows some check on equipment operation. Total number of bucket tips per event, measured mechanically, is also compared with the electronic record as a further check.

Program DATALOG converts runoff measurements into average flow rate per minute, Q and it does the same for rainfall rate. GUEST accepts the input from DATALOG, together with information from GUDPRO, and information on soil loss, plot geometry, instrument and rill configuration, (from what is called a configuration file). GUEST then computes the erodibility parameter β for the recorded erosion event. The interactions or relationships between the various programs is illustrated in Figure 3.

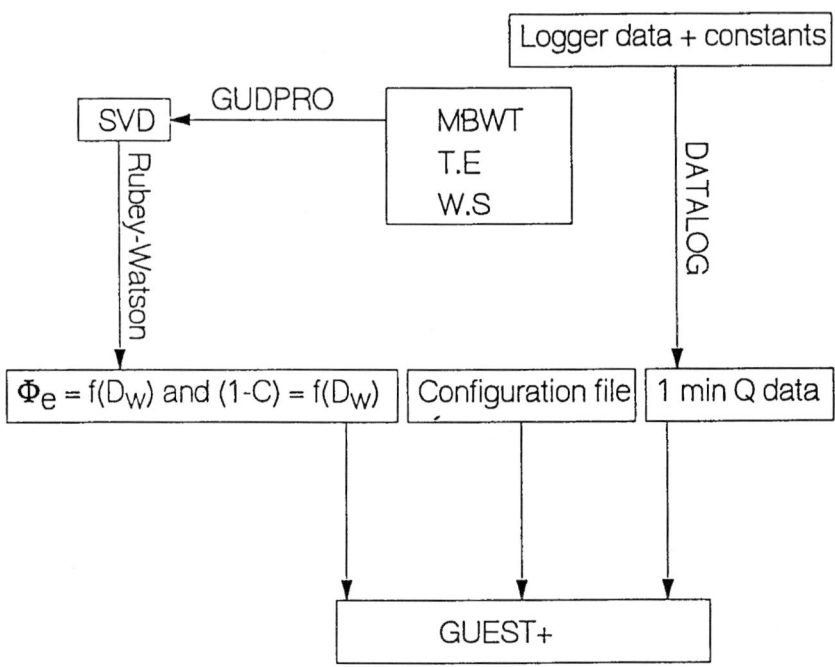

Figure 3. Illustrating the flow relationship between the data preparation program, DATALOG, the program GUDPRO which processes experimental data of soil from the experimental site to give the relationship illustrated in Figure 2, and GUEST+, which uses output from both these programs to yield the erodibility parameter β. MBWT refers to modified bottom-withdrawal tube, T.E to top-entry-type equipment, and W.S to wet sieving, which are alternative experimental techniques from which the settling velocity distribution (SVD) can be derived. D_w is water depth

In Figure 3 the program GUEST is shown as GUEST+ to indicate recent additions to the original program, partly due to the current common availability of greater computing power. The output of GUEST+ is the erodibility parameter β. The theoretical basis on which the derivation of β depends will now be outlined, a more complete description being given by Rose (1993).

Indication of theoretical basis for β

Foster (1982) amongst others, recognised experimentally that in the absence of mass movement, there is an upper limit to the observed concentration of sediment in any given soil and flow context. This has been called the 'transport limit', or the 'transport capacity', and is denoted by c_t. Use is made of this limit in many soil erosion models, for example WEPP (Nearing et al., 1989), and EUROSEM (Morgan et al., 1992). These erosion models use experimentally-based relationships to estimate c_t in any particular flow circumstance.

Program GUEST, however, uses a theoretically-derived expression to estimate c_t (Rose and Hairsine, 1988, and Rose, 1993). The theoretically-derived equation for c_t is in good agreement with the experimental database used for c_t where available data makes it possible to make that comparison in quantitative terms (e.g. the data of Yang, 1972).

This theoretically-derived expression for c_t uses the concept of stream power, Ω, introduced by Bagnold (1977), and defined as the rate of working of the mutual shear stresses which exist between water and the soil surface over which it is flowing. Thus:

$$\Omega = \tau V \ (Wm^{-2}) \tag{1}$$

where τ is the mutual shear stress between flowing water and the soil surface, and V is the bulk velocity of the flow.

The theoretical expression used in GUEST to estimate c_t is based on the following set of assumptions:

- At the transport limit the eroding surface is completely covered by sediment previously eroded in the same erosion event. Such sediment is termed a 'deposited layer'.
- The mechanical strength of sediment in this deposited layer is negligible. (This assumption is plausible, since the typically short dwell time for eroded particles is unlikely to allow bonds of significant strength to develop between neighbouring sedimentary units in the deposited layer).
- A fraction, F, of stream power, Ω, is used in the process of erosion by flowing water. The magnitude of the fraction F has been found to be commonly in the range 0.1-0.2 (Proffitt et

al., 1993), though in some circumstances F may be a little higher (Misra and Rose, 1993). (The fraction (1-F) Ω is dissipated as heat and noise).

- No sediment is assumed transported if $\Omega < \Omega o$, where Ωo is the threshold streampower, and the term $F(\Omega - \Omega o)$ can be called the 'effective excess streampower'. This component of streampower is assumed to be consumed in lifting sediment from the deposited layer into the flow against its (downward) immersed weight in water.
- In a steady state (or steady rate) situation, the rate of re-entrainment of sediment from the deposited layer is equal to the rate of deposition. This leads to an equation for c_t which is assumed to hold generally for the sediment concentration at the transport limit, even if conditions are not steady.

It follows from these physical assumptions (and for simplicity neglecting Ωo relative to Ω, and the correction for unimmersed large units) that the maximum sediment concentration due to flow (the concentration of the transport limit, c_t) is given for sheet flow by:

$$c_t = \frac{F}{\Sigma v_i / I} \left[\frac{\sigma}{\sigma - \rho} \right] \rho S V \qquad (2)$$

where: F   = fraction of streampower used in erosion of sediment,
        Ev_i/I = mean settling velocity or depositability of the sediment,
        σ = wet density of sediment,
        ρ = fluid density,
        S   = sine of the angle of land inclination (the land slope),
and: V   = flow velocity (related to volumetric flux using Manning's equation with measured roughness coefficient)

Should the flow occur in rills, as is quite common, equation (2) is modified to recognise this altered geometry, but the physical assumptions behind the equation for c_t in rill flow are basically the same as those outlined earlier.

The implication of equation (2) for c_t in a plane or sheet flow situation is illustrated in Figure 4. This figure shows c_t increases with Ω, but the rate of increase with Ω declines as Ω itself becomes larger.

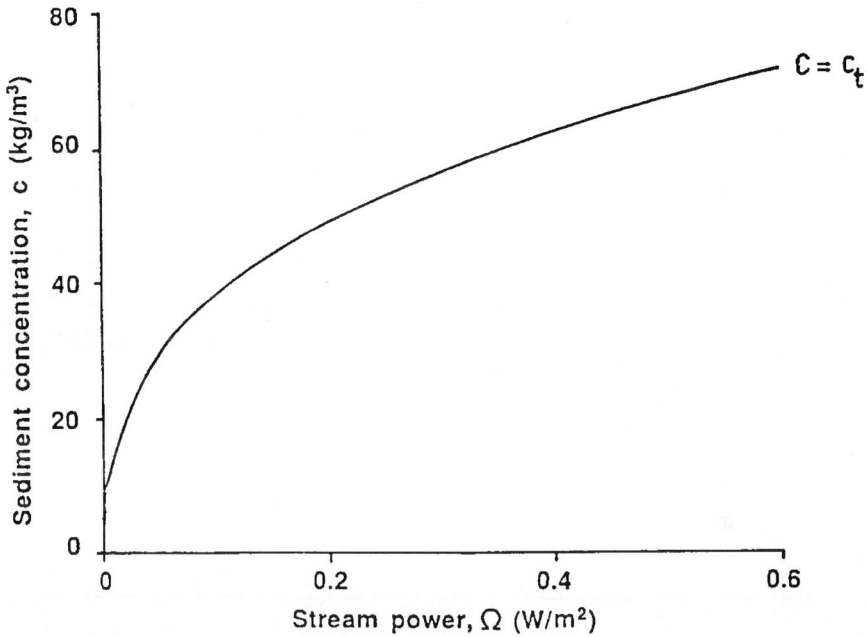

Figure 4. Illustrating the form of variation with streampower in sediment concentration at the transport limit, c_t for sheet flow. (For simplicity, threshold streampower Ωo has been taken to be zero)

Whilst it is possible for the sediment concentration to equal c_t, commonly the actual observed sediment concentration, c, is less than c_t. Theory of how c is expected to vary with Ω when $c<c_t$ is given by Hairsine and Rose (1992a,b) and by Rose (1993). This theory is based on the assumption that for flow to entrain unit mass of a cohesive soil, a certain energy per unit mass, J (Joules/kg) must be expended. This energy per unit mass must be related to the strength of the soil matrix. The theory referred to predicts that c will vary with Ω in the form of a unique curve for any given value of J (shown as solid lines in Figure 5). As is discussed more fully in Rose (1993), these curved relationships for different values of J are very well fitted by the other geometrical forms shown as dashed curves in Figure 5. These dashed curves are for the erodibility parameter β, where β is defined by the simple equation

$$c = c_t^\beta \tag{3}$$

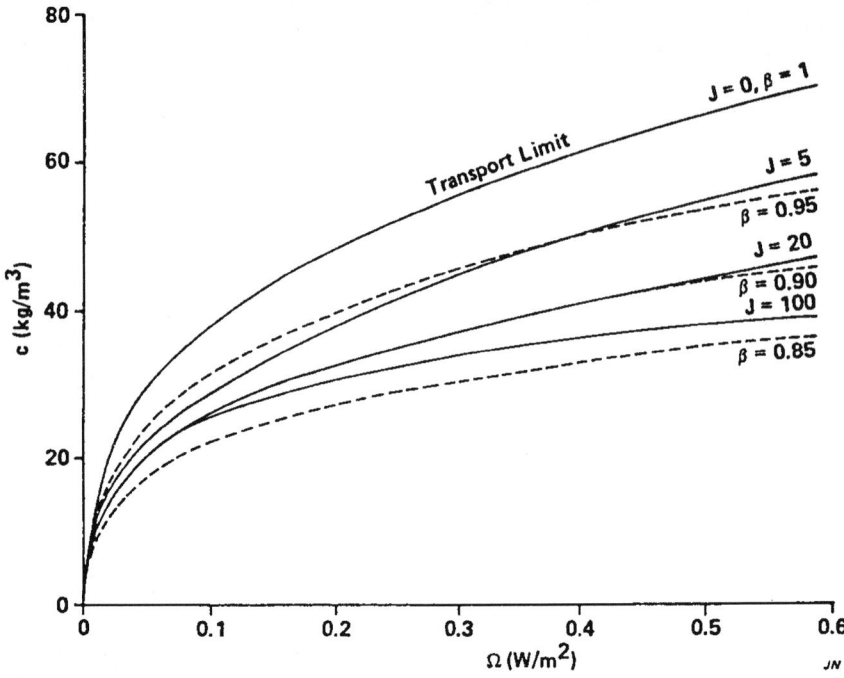

Figure 5. Sediment concentration (c) calculated as a function of streampower (Ω) for various values of a parameter J, related to soil strength, and to the soil erodibility parameter β defined in the text. The uppermost curve describes the transport limit situation for which $c=c_t$, $J=0$ and $\beta=1$. Other solid curves apply to increasing values of J (indicating stronger soils), and dashed curves to values of β indicated. Comparison of solid and dashed curves show that the form of variation of c with Ω calculated using the simple Equation (3) using β are similar in form to the results of a more complex equation involving J. (From Rose, 1993)

Thus β is the power to which the calculated value of c_t must be raised to equal the measured value of c.

Program GUEST+ calculates an average value of c_t for the entire duration of the event, denoted by $\overline{c_t}$. This is then related to the average sediment concentration, $\underline{c}$, measured for the entire event, and then β is evaluated from

$$\overline{c} = \overline{c_t}^{\beta} \quad \text{or} \quad \beta = \frac{\ln \overline{c}}{\ln \overline{c_t}} \tag{4}$$

It follows from its definition that the lower the value of β, the lower the erodibility of the soil. There is a dependence of β on soil strength, a soil characteristic which can be measured in the field, but theory ensures β is not dependent on sediment concentration (Rose, 1993).

With low streampowers (e.g. low slopes) and for compacted soils, erosion rate commonly depends more on the detachability of soil to rainfall, which is also affected by soil strength, and by rainfall characteristics. With low slopes and streampowers and tilled soils, β is commonly close to 1. Whatever the relative contribution of processes to erosion may be, they are effectively incorporated into the value of β determined as described earlier.

Some examples of the values of β and their variation through time for bare soil plots will now be given for the ACIAR project referred to in the Introduction.

Results for β

The time-trend in values of β depend very much on soil management practices. In the context of fairly frequent cultivation associated with keeping weeds in check, β tends to vary about some mean value dependent on soil type and cultivation method. This is illustrated by data collected in the island of Leyte, the Philippines (latitude 10°44' north, longitude 124°48' east, 30 m above sea level). Average annual rainfall is 2238 mm at the site. Experimental plots were constructed to have a uniform slope of 10 per cent, with some soil taken from the surrounding area, soil type being Oxic Dystropepts. Measurement methods were as described earlier in this paper.

Data is from a plot kept bare by hand-hoe weeding (Presbitero *et al.*, 1995). Following each erosion event the plot was observed for the presence or absence of rills. The event number was preceded by the letter R if rills were observed; no evidence of rills, with the surface appearing close to planar, led to P being placed before the event number.

Figure 6 shows the value of β calculated using Equation 4 for a series of erosion events labelled either R or P. The value of β was found to increase following cultivation, with a decline during subsequent erosion events. This illustrates the common finding that soil condition, and in particular soil strength as affected by tillage in particular, can affect soil erodibility as indicated by the value of β. Thus there is a possibility that β might be able to be approximately estimated using surrogate measurements, perhaps to an accuracy of ±0.1, reducing or avoiding the necessity of establishing and monitoring runoff plots, which is somewhat resource intensive.

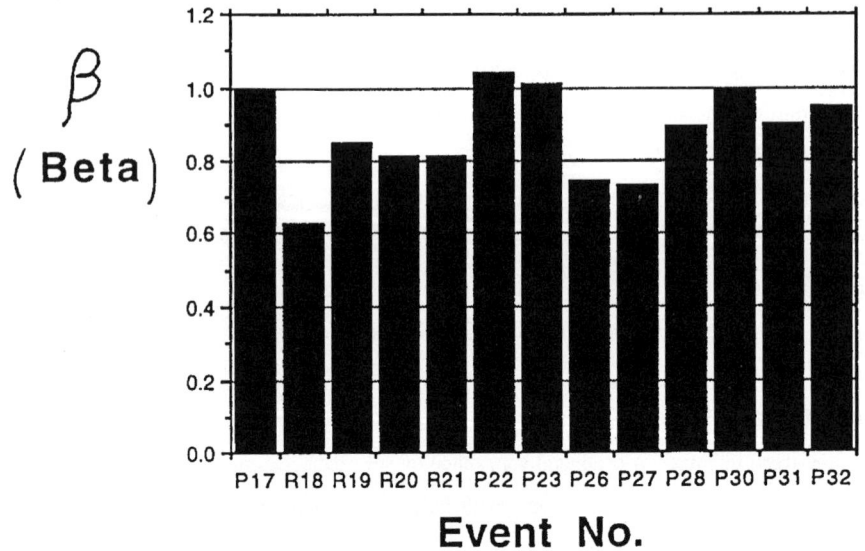

Figure 6. Variation in erodibility parameter β with successive significant erosion events on bare soil plots of 10% slope in the island of Leyte, the Philippines (From Presbitero *et al.*, 1995)

If cropping practices provide a significant amount of cover in close contact with the soil surface, the resultant reduction in sediment concentration, and thus in soil loss, decreases dramatically. For example, Paningbatan *et al.* (1995) found the ratio sediment concentration with fractional surface contact cover to that without decreased approximately exponentially with contact cover fraction, with the exponential rate coefficient being about ten. Thus dramatic improvement in soil conservation can be achieved if a management system is developed which maintains good levels of cover in contact with the soil surface, especially during periods of high erosion potential.

Acknowledgements

Research described in this paper was supported by the Australian Centre for International Agricultural Research, ACIAR (Project 8551), and by institutions in collaborating countries.

References

Bagnold, R.A. (1977). Bedload transport by natural rivers. *Water Resources Research* **13**, 303-311.

Bonell, M., and Williams, J. (1987). Infiltration and redistribution of overland flow and sediment on a low relief landscape in semi-arid tropical Queensland. In, Swanson, R.H., Bernier, P.Y. and Woodward, P.D. (eds), *Forest Hydrology and Watershed Management*. JAHS Publication No. 167; 199-211.

Foster, G.R. (1982). Modelling the erosion process. In, Hann, C.T. (ed.), *Hydrologic Modelling of Small Watersheds*. American Society of Agricultural Engineers Monograph No. 5, St. Joseph, Michigan, USA. pp. 297-399.

Hairsine, P.B. and McTainsh, G. (1986). *The Griffith Tube: a Simple Settling Tube for the Measurement of Settling Velocity of Soil Aggregates*. AES Working Paper 3/86 (Griffith University, Nathan, Queensland, Australia 4111).

Hairsine, P.B. and Rose, C.W. (1992a). Modelling water erosion due to overland flow using physical principles. I. Sheet flow. *Water Resources Research* **28**, 237-243.

Hairsine, P.B. and Rose, C.W. (1992b). Modelling water erosion due to overland flow using physical principles. II. Rill flow. *Water Resources Research* **28**, 245-250.

Lisle, I., Coughlan, K. and Rose, C.W. (1995). *Gudpro 3.1 User Guide and Reference Manual*. Faculty of Environmental Sciences, Griffith University, Queensland, Australia 4111.

Lovell, C.J. and Rose, C.W. (1988). Measurement of soil aggregate settling velocities I. A modified bottom withdrawal tube method. *Australian Journal of Soil Research* **26**, 55-71.

Misra, R.K. and Rose, C.W. (1990). *Manual for use of Program GUEST*. Division of Australian Environmental Studies Report, Griffith University, Brisbane, Australia, 4111.

Misra, R.K. and Rose, C.W. (1995). An examination of the relationship between erodibility parameters and soil strength. *Australian Journal of Soil Research* **33**, 715-732.

Morgan, R.P.C., Quinton, J.N., and Rickson, R.J. (1992). *EUROSEM: Documentation Manual*. Silsoe College, Silsoe, UK.

Nearing, M.A., Foster, G.R., Lane, L.J., and Finkner, S.C. (1989). A process based erosion model for USDA water erosion prediction project technology. *Transactions of the American Society of Agricultural Engineers* **32**, 1587-1593.

Paningbatan, E.P., Ciesiolka, C.A., Coughlan, K.J., and Rose, C.W. (1995). Alley cropping for managing soil erosion of hilly lands in the Philippines. *Soil Technology* (in press).

Presbitero, A.L., Escalante, M.C., Rose, C.W., Coughlan, K.J., and Ciesiolka, C.A. (1995). Erodibility evaluation and the effect of land management practices on soil erosion from steep slopes in Leyte, the Philippines. *Soil Technology*. (in press).

Proffitt, A.P.B., Hairsine, P.B., and Rose, C.W. (1993). Modelling soil erosion by overland flow: application over a range of hydraulic conditions. *Transactions of the American Society of Agricultural Engineers* **36**, 1743-1753.

Rose, C.W. (1993). Erosion and sedimentation. In, Bonnell, M., Hufschmidt, M.M. and Gladwell, J.S. (eds), *Hydrology and Water Management in the Humid Tropics — Hydrological Research Issues and Strategies for Water Management*. Cambridge University Press, Cambridge, UK. pp. 301-343.

Rose, C.W. (1995). Introduction by the guest editor. *Soil Technology* **8**, 177-178.

Rose, C.W. and Hairsine, P.B. (1988). Process of water erosion. In, Steffen, W.L. and Denmead, O.T. (eds), *Flow and Transport in the Natural Environment: Advances and Applications*, Springer Verlag, Berlin. pp. 312-326.

Rose, C.W. and Hairsine, P.B., Proffitt, A.P.B. and Misra, R.K. (1990). Interpreting the role of soil strength in erosion processes. *Catena Supplement* **17**, 153-165.

Rose, C.W., Presbitero, A.L., Coughlan, K.J., Lisle, I., Fentie, B. and Cresiolka, C.A. (1996). Saltation and effective shear stresses in soil erosion due to overland water flow. (in preparation).

Yang, C.T. (1972) Unit stream power and sediment transport. *Journal of the Hydraulics Division, American Society of Civil Engineers* **78**, (HY10), 1805-1826.

31. A CONTINUOUS CATCHMENT-SCALE EROSION MODEL

J.G. Arnold[1] and R. Srinivasan[2]

[1] US Department of Agriculture — Agricultural Research Service
Grassland, Soil and Water Research Laboratory
808 East Blackland Road
Temple
Texas 76502
USA

[2] Texas Agricultural Extension Service
808 East Blackland Road
Temple
Texas 76502
USA

Abstract

A spatially distributed catchment model called SWAT (Soil and Water Assessment Tool) was developed that operates on a daily time step. The model allows considerable flexibility in watershed configuration and discretisation. Watersheds can be subdivided into cells and/or subwatersheds. A command structure is used for routing runoff, sediment, nutrients, and pesticides through the watershed. Commands are included for simulating outputs from cells/subwatersheds, routing through channel reaches, routing through reservoirs, and adding outputs. The command structure file can be generated automatically from the drainage patterns. The model is being applied on a small watershed within the Indian Pines Experimental Watershed near West Lafayette, Indiana. The 329 hectare watershed is subdivided into 100 x 100 m cells for model simulation and validation. The model is also applied to the Lower Colorado River Basin (8,927 km^2) in Texas.

Introduction

Water quality is becoming an increasing national concern in the United States. With past emphasis focusing on point sources, non-point source pollution is attaining a higher relative importance. Damage from soil erosion alone is estimated at tens of billions of dollars per year (Committee on Conservation Needs and Opportunities, 1986), which does not include nutrient and pesticide contamination of ground and surface water supplies.

There is an increasing concern for the offsite impacts of nonpoint source pollution including lake water quality and instream nutrient and toxic concentrations. The development of integrated systems consisting of comprehensive models is crucial in solving complex catchment management problems. To estimate offsite loadings, the ability to simulate catchments with heterogeneous soils, land use, and topography is required. Several basin scale water quality models have recently been developed that simulate spatial variability within a watershed. However, these models have been limited by several factors including: 1) computer speed; 2) computer memory; and 3) availability of inputs. These limitations have produced models falling into one of the following three categories:

- Continuous time models with natural subwatershed boundaries that require considerable lumping of subwatershed inputs
- Single event models subdivided into grid cells that allow more spatial detail
- Continuous time, spatial models that are so complex that obtaining required model inputs inhibits their general use.

The challenge is to develop a catchment scale model that: 1) is computationally efficient; 2) allows considerable spatial detail; 3) requires readily available inputs; 4) is continuous time, capable of simulating land management scenarios; and 5) gives reasonable results. The objective of this paper is to describe the SWAT model, which is a continuous time routing model for water, sediment, and agricultural chemicals. The model will be demonstrated on a small watershed within the Indian Pines Experimental Watershed at West Lafayette, Indiana, and on a large river basin in Texas.

Model operation

SWAT is a continuous time model that operates on a daily time step (Figure 1). The objective in model development was to predict the impact of management on water, sediment and agricultural chemical yields in large ungauged basins. To satisfy the objective, the model (a) is physically based (calibration is not possible on ungauged basins); (b) uses readily available inputs; (c) is computationally efficient to operate on large basins in a reasonable time, and (d) is continuous time and capable of simulating long periods for computing the effects of management changes. SWAT uses a command structure for routing runoff and chemicals through a catchment. Commands are included for routing flows through streams and reservoirs, adding flows, and inputting measured data or point sources. Using a routing command language, the model can simulate a catchment subdivided into grid cells or subwatersheds. Additional commands have been developed to allow measured and point source data to be input to the model and routed with simulated flows. Also, output data from other simulation models can be input to SWAT. Using the transfer command, water can be transferred from any reach or reservoir to any other reach or reservoir within the basin. The user can specify the fraction of flow to divert, the minimum flow remaining in the channel or reservoir, or a daily amount to divert. The user can also apply water directly to a subbasin for irrigation. Although the model operates on a daily time step and is efficient enough to run for many years, it is intended as a long term yield model and is not capable of detailed, single event, flood routing.

Model components

The sub-catchment components of SWAT can be placed into eight major divisions — hydrology, weather, sedimentation, soil temperature, crop growth, nutrients, pesticides, and agricultural management. A complete description of all components can be found in Arnold *et al.* (1995). The components directly impacting erosion/sedimentation are briefly described here.

Hydrology

Surface runoff. Surface runoff from daily rainfall is predicted using a procedure similar to the CREAMS runoff model, option one (Knisel, 1980; Williams and Nicks, 1982). Like the CREAMS model, runoff volume is estimated with a modification of the SCS curve number method (USDA Soil Conservation Service, 1972). The curve number varies non-linearly from the 1 (dry) condition at wilting point to the 3 (wet) condition at field capacity and approaches 100 at saturation. The SWAT model also includes a provision for estimating runoff from frozen soil. Peak runoff rate predictions are based on modification of the Rational Formula. The runoff coefficient is calculated as the ratio of runoff volume to rainfall. The rainfall intensity during the watershed time of concentration is estimated for each storm as a function of total rainfall using a stochastic technique. The watershed time of concentration is estimated using Manning's Formula, considering both overland and channel flow.

Percolation. The percolation component of SWAT uses a storage routing technique to predict flow through each soil layer in the root zone. Downward flow occurs when field capacity of a soil layer is exceeded if the layer below is not saturated. The downward flow rate is governed by the saturated conductivity of the soil layer. Upward flow may occur when a lower layer exceeds field capacity. Movement from a lower layer to an adjoining upper layer is regulated by the soil water to field capacity ratios of the two layers. Percolation is also affected by soil temperature. If the temperature in a particular layer is 0°C or below, no percolation is allowed from that layer.

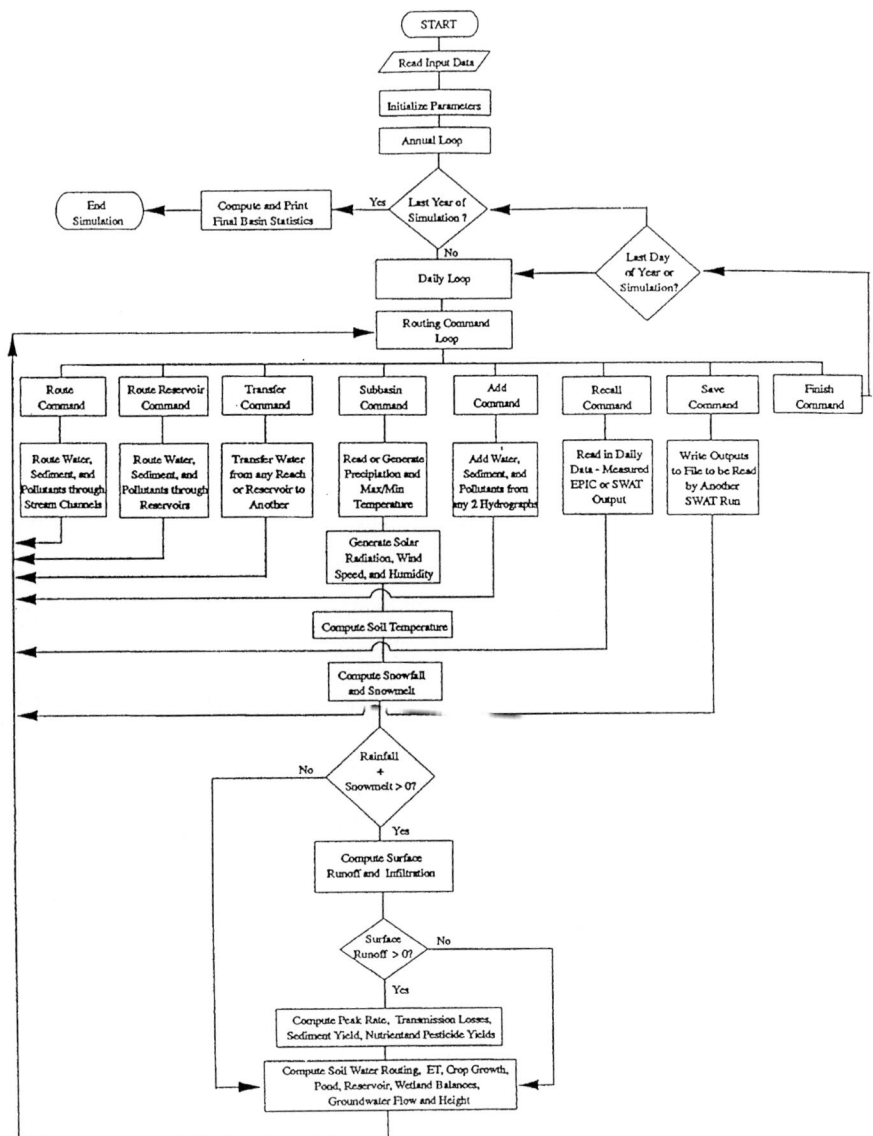

Figure 1. A flowchart of SWAT

Groundwater flow. Groundwater flow contribution to total streamflow is simulated by creating a shallow aquifer storage (Arnold *et al.*, 1993). Percolate from the bottom of the root zone is recharge to the shallow aquifer. A recession constant, derived from daily streamflow records, is used to lag flow from the aquifer to the stream. Other components include evaporation, pumping withdrawals, and seepage to the deep aquifer.

Evapotranspiration. The model offers three options for estimating potential ET — Hargreaves and Samani, 1985), Priestley-Taylor (Priestley and Taylor, 1972), and Penman-Monteith (Monteith, 1965). The Penman-Monteith method required solar radiation, air temperature, wind speed, and relative humidity as input. If wind speed, relative humidity, and solar radiation data are not available (daily values can be generated from average monthly values), the Hargreaves or Priestley-Taylor methods provide options that give realistic results in most cases. The model computes evaporation from soils and plants separately as described by Ritchie (1972). Potential soil water evaporation is estimated as a function of potential ET and leaf area index (area of plant leaves relative to the soil surface area). Actual soil water evaporation is estimated by using exponential functions of soil depth and water content. Plant water evaporation is simulated as a linear function of potential ET and leaf area index.

Erosion

Sediment yield. Sediment yield is estimated for each subcatchment with the Modified Universal Soil Loss Equation (MUSLE) (Williams, 1975.). The hydrology model supplies estimates of runoff volume and peak runoff rate. The crop management factor is evaluated as a function of above-ground biomass, crop residue on the surface, and the minimum C factor for the crop. Other factors of the erosion equation are evaluated as described by Wischmeier and Smith (1978).

Crop growth model

A single model is used in SWAT for simulating all crops. Energy interception is estimated as a function of solar radiation and the crop's leaf area index. The potential increase in biomass for a day is estimated as the product of intercepted energy and a crop parameter for converting energy to biomass. The leaf area index is simulated with equations dependent upon heat units. Crop yield is estimated using the harvest index concept. Harvest index increases as a non-linear function of heat units, from zero at planting to the optimal value at maturity. The harvest index may be reduced by water stress during critical crop stages (usually between 30 and 90 per cent of maturity).

Nutrients and pesticides

Nitrogen. Amounts of NO_3-N contained in runoff, lateral flow, and percolation are estimated as the products of the volume of water and the average concentration. Leaching and lateral subsurface flow in lower layers are treated with the same approach used in the upper layer, except that surface runoff is not considered. A loading function developed by McElroy *et al.* (1976) and modified by Williams and Hann (1978) for application to individual runoff events is used to estimate organic N loss. The loading function estimates the daily organic N runoff loss based on the concentration of organic N in the top soil layer, the sediment yield, and the enrichment ratio. Also, crop use of N is estimated using a supply and demand approach.

Phosphorus. The SWAT approach to estimating soluble P loss in surface runoff is based on the concept of partitioning pesticides into the solution and sediment phases, as described by Leonard and Wauchope (Knisel, 1980). Because P is mostly associated with the sediment phase, the soluble P runoff is predicted using labile P concentration in the top soil layer, runoff volume, and a partitioning factor. Sediment transport of P is simulated with a loading function as described in organic N transport. Crop use of P is also estimated with the supply and demand approach.

Pesticides. GLEAMS (Knisel, 1980) technology for simulating pesticide transport by runoff, percolate, soil evaporation, and sediment was added to SWAT. Pesticides may be applied at any time and rate to plant foliage or below the soil surface at any depth. The plant leaf-area-index determines what fraction of foliar applied pesticide reaches the soil surface. Also, a fraction of the application rate (called application efficiency) is lost to the atmosphere. Each pesticide has a unique set of parameters including solubility, half absorption coefficient, and cost. Pesticide on plant foliage and in the soil degrade exponentially according to the appropriate half lives. Pesticide transported by water and sediment is calculated for each runoff event and pesticide leaching is estimated for each soil layer when percolation occurs.

Agricultural management

SWAT allows for unlimited years of crop rotations and up to three crops per year. The user can also input irrigation, nutrient, and pesticide application dates and amounts.

Tillage and residue. The SWAT tillage component was designed to partition the above-ground biomass at harvest. Part of the biomass is removed as yield, part is incorporated into the soil, and the remainder is left on the soil surface as residue. The model has no process interactions with incorporated residue. Also, tillage does not affect soil properties.

Irrigation. The user has the option to simulate dryland or irrigated agriculture. If irrigation is selected, the runoff ratio (volume of water leaving the field/volume applied) and a plant water stress level to trigger irrigation must be specified. The plant water stress factor ranges from 0 to 1.0 (1 means no stress and 0 means no growth). When the user-specified stress level is reached, enough water is applied to fill the root zone to field capacity.

Routing components

Channel flood routing. Channel routing uses a variable storage coefficient method developed by Williams (1975). Channel inputs include the reach length, channel slope, bankfull width and depth, channel side slope, flood plain slope, and Manning's n for channel and floodplain. Flow rate and average velocity are calculated using Manning's equation and travel time is computed by dividing channel length by velocity. Outflow from a channel is also adjusted for transmission losses, evaporation, diversions, and return flow.

Channel sediment routing

The sediment routing model consists of two components operating simultaneously (deposition and degradation). The deposition component is based on fall velocity and the degradation component is based on Bagnold's stream power concept (Williams, 1980). Deposition in the channel and floodplain from the subbasin to the basin outlet is based on sediment particle fall velocity. Fall velocity is calculated as a function of particle diameter squared using Stokes Law. The depth of fall through a routing reach is the product of fall velocity and reach travel time. The delivery ratio is estimated for each particle size as a linear function of fall velocity, travel time, and flow depth. Stream power is used to predict degradation in the routing reaches. Bagnold (1977) defined stream power as the product of water density, flow rate, and water surface slope. Williams (1980) modified Bagnold's equation to place more weight on high values of stream power — stream power raised to 1.5. Available stream power causes bed degradation: this is adjusted by the USLE soil erodibility and cover factors of the channel and floodplain.

Reservoir routing

The water balance for reservoirs includes inflow, outflow, rainfall on the surface, evaporation, seepage from the reservoir bottom, and diversions and return flow. There are currently three methods to estimate outflow. The first method simply reads in measured outflow and allows the model to simulate the other components of the water balance. The second method is for small uncontrolled reservoirs, and outflow occurs at a specified release rate when volume exceeds the principle storage. The volume exceeding the emergency spillway is released within one day. For larger managed reservoirs, a monthly target volume approach is used.

Reservoir sediment routing

Inflow sediment yield to ponds and reservoirs (P/R) is computed with MUSLE. The outflow from P/R is calculated as the product of outflow volume and sediment concentration. Outflow P/R concentration is estimated using a simple continuity equation based on volumes and concentrations of inflow, outflow, and pond storage. Initial pond concentration is input and between storm concentration decreases as a function of time and median particle size of inflow sediment.

GIS interface

In recent years, there has been considerable effort devoted to utilising GIS to extract inputs (soils, land use, and topography) for comprehensive simulation models and spatially display model outputs. Much of the initial research was devoted to linking single-event, grid models with raster-based GIS (Srinivasan and Engel, 1991; Rewerts and Engel, 1991). An interface was developed for SWAT (Srinivasan and Arnold, 1994) using the Graphical Resources Analysis Support System (GRASS) (U.S. Army, 1988). The input interface will automatically subdivide a basin (grids or subwatersheds) and then extract model input data from map layers and the associated relational data bases for each subbasin. Soils, land use, weather, management, and topographic data are collected and written to appropriate model input files.

The output interface allows the user to display output maps and graph output data by selecting a subbasin from GIS map.

Model application

Small watershed

The Animal Sciences watershed is located at Purdue University's Indian Pine Natural Resources Field Station. A gauge is located at the outlet of the watershed where runoff, sediment, and nutrients are sampled. Land use on the 329 ha watershed is predominantly corn. Soils were digitised from the SCS county soil survey (1:24,000) and elevation was digitised from the USGS quadrangle maps (1:24,000). Soils data for the corresponding soil series were obtained from the Soils-5 database, while missing values and local interpretations were made by county SCS soil scientists. Since the watershed is relatively small, and the data had been digitised and was already in GIS format, the watershed was subdivided into one hectare grid cells. This allows for significant spatial detail. Model input data including soils, topography, land use, and the routing structure were developed automatically using the GRASS GIS (Srinivasan and Arnold, 1994).

Simulation results

Measured water, sediment and nutrient data has been sampled at the watershed outlet for nearly two years (Engel *et al.*, 1993). However, the data for only a few events have been processed to evaluate the model. Model output includes water, sediment, nutrient, pesticide, and crop yields for each cell. Also output by the model are the water sediment, chemical balances within the channel of each cell. Measured and SWAT predicted values for four storm events are shown in Table 1. Although the rainfall and runoff are extremely low, the model was able to reproduce the values reasonable well, considering calibration was not performed. To demonstrate long term capabilities, the model was run for ten years with a generated average annual rainfall of 960 mm. Only one weather station was simulated and each cell used the identical simulated weather. Simulated annual runoff at the watershed outlet (surface runoff plus lateral flow in the soil profile) was 104 mm and annual sediment yield at the outlet was 0.5 t/ha.

Table 1. Simulated (SWAT) and observed runoff data at the Animal Sciences Catchment

Event	Rainfall (mm)	Runoff (mm)		Sediment Yield (t/ha)		Nitrate (kg/ha)		Phosphate (kg/ha)	
		Meas	Pred	Meas	Pred	Meas	Pred	Meas	Pred
May 11, 1991	10.9	0.45	0.69						
May 18, 1991	18.5	2.3	2.7						
May 23, 1991	9.4	1.4	0.5	0.0291	0.0031	0.15	0.20	0.009	0.016
Oct. 28, 1991	35.6	5.6	4.8	0.0143	0.0216	0.13	0.19	0.020	0.028

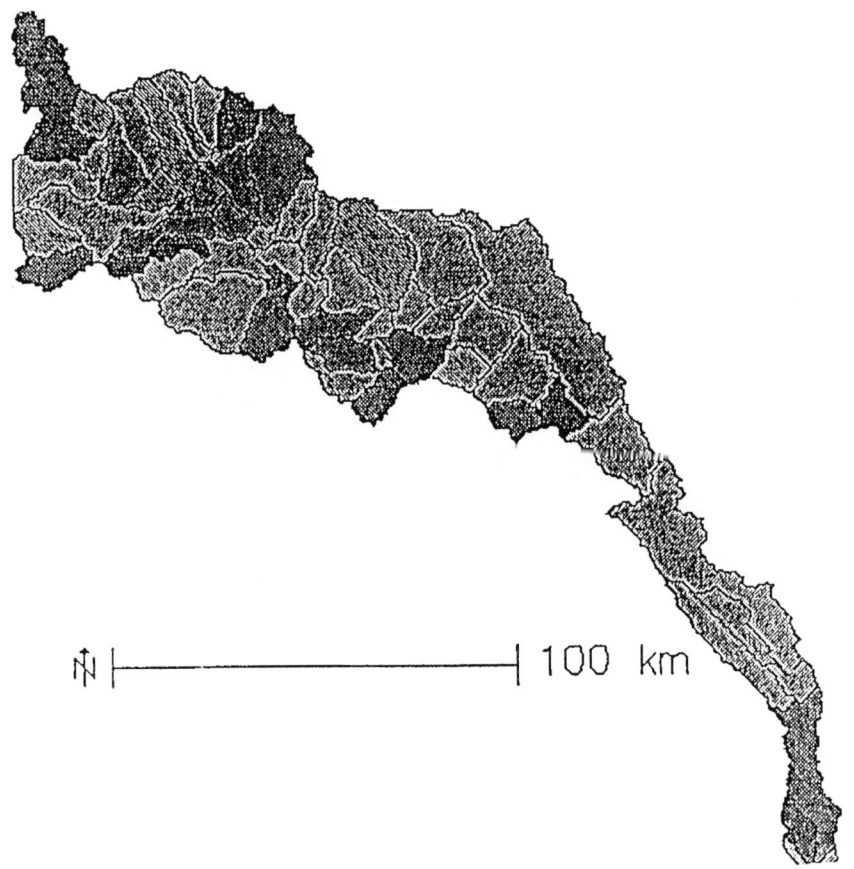

Figure 2. Subwatershed map for the Lower Colorado River Basin simulation

River Basin Scale

The river basin selected for model validation is the Lower Colorado River basin in Texas. Only the downstream portion of the basin was simulated. The entire basin area is 109,401 km^2 while the simulated area was 8,927 km^2. At the upstream end of the simulated area, measured outflows from Lake Travis (west of Austin) were input to the model and flow was routed through the basin until it reached the Gulf of Mexico at Matagora Bay. Figure 2 shows the basin and subwatersheds used in modelling the basin. Four basic data sets are required to develop model input files. They are topography, soils, land use, and measured weather and streamflow. Elevation and slope files were extracted from the USGS Digital Elevation Models (DEM). The data is in digital form and the vertical resolution on the 1:250,000 maps is one metre. Sixty subwatersheds were simulated with an average size of 150 km^2 or approximately 50 mi^2. Soils data for the Lower Colorado were derived from the STATSGO data base (USDA, 1991). STATSGO is spatially divided into state soil identifiers, which include percentages of several soil series. The GIS link used to develop the soils inputs (Srinivasan and Arnold, 1994) currently selects the dominant soil series within a subwatershed. It then searches the Soils-5 data base for this series and places the required model input data for that series into the proper model input file and format. Land use data was obtained from the USGS Land Use/Land Cover (LULC) data base. The spatial resolution was comparable to the topographic maps (1:250,000) with a cell size on the digitised maps of 10 ha. Measured daily precipitation and maximum and minimum temperatures were obtained from the National Weather Service. Six gauging stations were located within the basin that had significant lengths of record. The weather stations are located at Austin, Matagora, Bay City, Columbus, Lagrange, and Redrock. The weather station for each subwatershed was determined simply by selecting the one nearest to the subwatershed. Ten years of data (1980-1989) were collected for each of the six stations. Measured streamflow data were obtained from USGS records that are readily available on CD. The gauging station nearest the simulated outlet is the Colorado River near Bay City (USGS record number 08162500). Ten years of daily flow data were obtained directly from the CD for the identical period of record (1980-1989) for use in model validation.

Validation results

The only measured data available for validation is streamflow data. Sediment and nutrient loadings are not available for the basin. A comparison of monthly and annual and measured and predicted streamflows at Bay City are shown in Table 2. The predicted results shown in Table 2 are uncalibrated. The input data were obtained directly from the GIS and were not modified. Given the fact that no calibration was performed, model results appear adequate, with measured and predicted means within 5 percent and the Nash-Sutcliffe coefficient of determination above 0.60. Statistics are valuable criterion but often a graph sheds considerable insight to the goodness-of-fit. Measured and predicted monthly water yields are plotted in Figures 3 and 4 to show model performance.

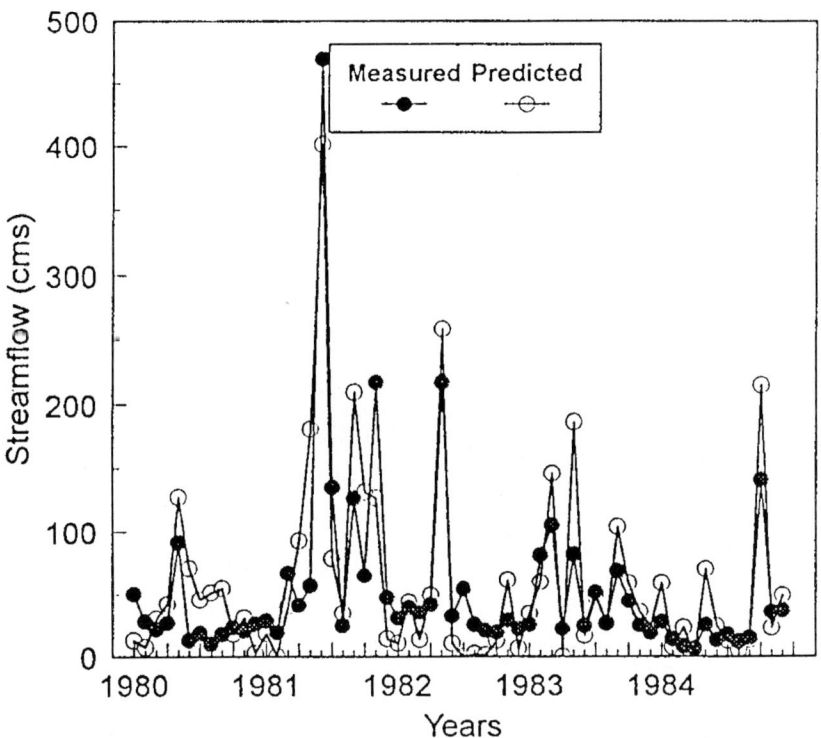

Figure 3. Measured and predicted monthly streamflow for the Lower Colorado River Basin from 1980 to 1984

Table 2. Comparison of measured and predicted streamflow at Bay City in the lower Colorado River basin

	Annual		Monthly	
	Measured	Predicted	Measured	Predicted
Mean Flow ($m^3 s^{-1} d$)	726.4	693.8	60.5	57.8
Standard Dev. ($m^3 s^{-1} d$)	524	317.9	97.6	74.3
R^2	0.72			0.6
Slope	1.44			1.01
Nash-Sutcliffe	0.65			0.6

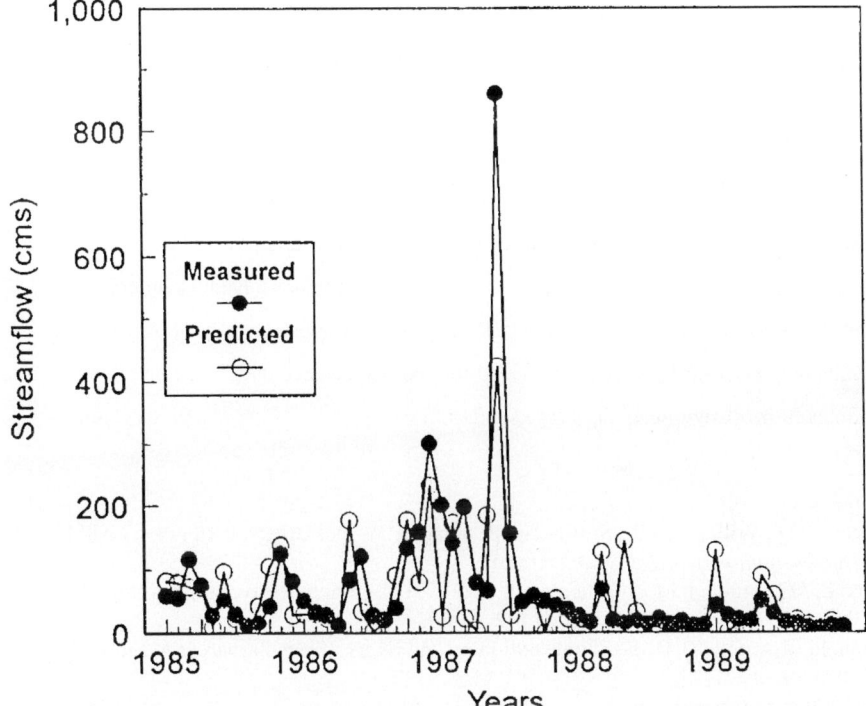

Figure 4. Measured and predicted monthly streamflow for the Lower Colorado River Basin from 1985 to 1989

Selected applications

The SWAT model is currently being used throughout the world by consulting engineers, government agencies, universities, and chemical companies. Several applications follow to demonstrate the model's potential capabilities.

Hydrology Unit Model of the United States (HUMUS)

The SCS is using the SWAT model as part of the 1997 Resource Conservation Assessment. The models will be linked to national economic models and used for national planning of water supply and quality on the 18 major river basins in the U.S. This system of models allows water, sediment, and attached pollutants to be tracked from their point or origin to major rivers, reservoirs, and coastal zones. Geographic information systems are utilised to integrate the models with national soils, land use, and digital elevation databases. The GIS automatically extracts model input from the map layers and displays model output.

Coastal Pollutant Discharge Inventory

As a part of the National Coastal Pollutant Discharge Inventory, the National Oceanic and Atmospheric Administration (NOAA) is using SWAT to estimate non-point source loadings from non-urban lands in all coastal counties of the U.S. (Singer *et al.*, 1988). Site-specific data are obtained from the SCS's National Resources Inventory and Soils-5 data bases, NOAA weather stations, U.S. Geological Survey digital land use data tapes, and other local sources. Simulations have been run for cropland, rangeland, and forest land in approximately 770 subwatersheds comprising the Gulf Coast, Eastern, and Western coastal zones of the U.S. Results are compiled by season and added to a comprehensive data base containing pollutant loadings from all significant discharge sources.

References

Arnold, J.G., Williams, J.R., Srinivasan, R., King, K.W. and Griggs, R.H. (1995). *SWAT Documentation and User Manual* (draft).

Arnold, J.G., Allen, P.M. and Bernhardt, G. (1993). A comprehensive surface-groundwater flow model. *Journal of Hydrology* **142**, 47-69.

Bagnold, R.A. (1977). Bedload transport in natural rivers. *Water Resources Research* **13**(2), 303-312.

Committee on Conservation Needs and Opportunities. (1986). *Soil Conservation: Assessing the National Resource Inventory*. National Academy Press 1, Washington D.C., 114 pp.

Engel, B.A., Srinivasan, R., Arnold, J.G., Rewerts, C. and Brown, S.J. (1993). Nonpoint Source (NPS) pollution modeling using models integrated with Geographic Information Systems (GIS). *Water Science Tech*nology **28**(3-5), 685-690.

Hargreaves, G.H. and Samani, Z.A. (1985). Reference crop evapotranspiration from temperature. *Applied Engineering in Agriculture* **1**, 96-99.

Knisel, W.G. (ed.) (1980). *CREAMS: A Field Scale Model for Chemicals, Runoff and Erosion from Agricultural Management Systems*. USDA Conservation Research Report 26. 643 pp.

McElroy, A.D., Chiu, S.Y., Nebgen, J.W., Aleti, A. and Bennet, R.W. (1976). *Loading Functions for Assessment of Water Pollution from Nonpoint Sources*. EPA-600/2-76-151 (NTIS PB-253325), U.S. Environmental Protection Agency, Washington, D.C.

Monteith, J.L. (1965). Evaporation and environment. Symposia of the Society for Experimental Biology **19**, 205-234.

Priestley, C.H.B. and Taylor, R.J. (1972). On the assessment of surface heat flux and evaporation using large-scale parameters. *Monthly Weather Review* **100**, 81-92.

Ritchie, J.T. (1972). A model for predicting evaporation from a row crop with incomplete cover. *Water Resources Research* **16**(5), 1204-1213.

Srinivasan, R. and Engel, B.A. (1991). *A Knowledge Based Approach to Extract Input Data from GIS*. American Society of Agricultural Engineers Paper No. 91-7045. ASAE Summer Meeting, Albuquerque, New Mexico.

Srinivasan, R. and Arnold, J.G. (1994). Integration of a basin-scale water quality model with GIS. *Water Resources Bulletin* **30**(3), 453-462

USDA (1991). *STATSGO - State Soils Geographic Data Base*. Soil Conservation Service, Pub. No. 1492, Washington, D.C.

USDA Soil Conservation Service. (1983). *National Engineering Handbook.* Hydrology Section 4, Chapter 19.

U.S. Army (1978). *GRASS Reference Manual*. USA CERL, Champaign, IL.

Williams, J.R. (1975). Sediment routing for agricultural watersheds. *Water Resources Bulletin,* AWRA **11**(5), 965-974.

Williams, J.R. (1980). SPNM, a model for predicting sediment, phosphorus, and nitrogends from agricultural basins. *Water Resources Bulletin* **16**(5), 843-848.

Williams, J. R. and Hann, R.W. (1978). *Optimal Operation of Large Agricultural Watersheds with Water Quality Constraints.* Texas Water Resources Institute, Texas A&M University, TR-96. 152 pp.

Williams, J.R. and Nicks, A.D. (1982). CREAMS hydrology model — option one. In, Singh, V.P. (ed.), *Applied Modeling Catchment Hydrology.* Proceediongs of the International Symposium on Rainfall-Runoff Modeling, May 18-21, 1981, Mississippi State MS. pp. 69-86

Wischmeier, W.H. and Smith, D.D. (1978). *Predicting Rainfall Erosion Losses — A Guide to Conservation Planning*. USDA Agricultural Handbook No. 537. 58 pp.

32. LISEM: A PHYSICALLY-BASED HYDROLOGIC AND SOIL EROSION CATCHMENT MODEL

Ad de Roo[1], Victor Jetten[2], Cees Wesseling[1] and Coen Ritsema[3]

[1] Department of Physical Geography
Utrecht University
P.O.Box 80.115
3508 TC Utrecht
The Netherlands

[2] Station d'Agronomie, INRA
Rue Fernand Christ
02007 Laon
France

[3] The Winand Staring Centre
P.O.Box 125
6700 AC Wageningen
The Netherlands

Abstract

The LImburg Soil Erosion Model (LISEM) is a physically-based hydrological and soil erosion model, which can be used for research, planning and conservation purposes in drainage basins. Processes incorporated in the grid-based LISEM model are rainfall, interception, surface storage in micro-depressions, infiltration, vertical movement of water in the soil, overland flow, channel flow, detachment by rainfall and throughfall, detachment by overland flow, and transport capacity of the flow. Also, the influence of tractor wheelings, small paved roads (smaller than the pixel size), stones and surface sealing on the hydrological and soil erosion processes is taken into account. Vertical movement of water in the soil is simulated using the Richards equation. Optionally, the user can choose a one or two layer Green and Ampt or the Holtan infiltration model as an alternative to Richards. LISEM is completely incorporated in a raster Geographical Information System. From the first validation results it is clear that, although the model has several advantages over other models and the qualitative erosion and runoff patterns appear plausible, the quantitative prediction of erosion and discharge is still far from perfect.

Soil Erosion Models

Quantitative simulation models of surface runoff and soil erosion are useful in areas with soil erosion problems to provide spatially distributed estimates. The model results can be used to evaluate alternative strategies for improved land management, not only in the monitored areas, but also in ungauged catchments. Until recently, the Universal Soil Loss Equation (Wischmeier and Smith, 1978) has been widely used as a soil erosion model on the field scale. Since then, more process based soil erosion models have been developed such as CREAMS (Knisel, 1980), ANSWERS (Beasley *et al.*, 1980), Morgan/ Morgan/Finney (Morgan *et al.*, 1984), WEPP (Lane *et al.*, 1992) and EUROSEM (Morgan, 1994). Although some of these models use spatial input variables that are made with a GIS, only a few have been really integrated in a raster GIS. Examples are the ANSWERS model (De Roo *et al.*, 1989) and the Morgan model that has been modified and integrated in a GIS: the SEMMED model (De Jong, 1994). However, both in ANSWERS and SEMMED hydrological and erosion processes are grossly simplified. The LISEM model discussed below has been developed as a next step, i.e. a process based soil erosion model at the catchment scale, fully integrated in a GIS. A number of reasons existed to develop a new model:

- improve process descriptions for e.g. infiltration and detachment;
- integrate the model with GIS to prevent unnecessary lumping — simplification — of e.g. topography;
- build a model that allows input from remotely sensed data, because data availability is a major problem in physically-based modelling and remote sensing might be a cost-effective alternative. The use of remote sensing data in soil erosion has advantages but has difficulties at the same time (De Jong, 1994). However, one might expect useful future improvements such as imaging spectroscopy.

The LISEM model

Recently, a physically-based hydrological and soil erosion model has been developed by the Department of Physical Geography of the Utrecht University, the Soil Physics Division of the Winand Staring Centre in Wageningen, the Netherlands, and the Agronomy unit of INRA, Laon, France: the LImburg Soil Erosion Model (LISEM). LISEM is event based and can be used for planning and conservation purposes. The model is named after the region for which it has been developed originally (Limburg, The Netherlands), but its physical basis allows application in other areas. The LISEM model is one of the first examples of a physically based model that is completely incorporated in a raster Geographical Information System (PCRaster: Van Deursen and Wesseling, 1992). The main reason for using a GIS is that runoff and soil erosion processes vary spatially, so that cell sizes should be used that allow spatial variation to be taken into account.

Description of the processes simulated by LISEM

Processes incorporated in the model, which are described below, are rainfall, interception, surface storage in micro-depressions, infiltration, vertical movement of water in the soil, overland flow, channel flow, detachment by rainfall and throughfall, detachment by overland flow, and transport capacity of the flow. Also, the influence of tractor wheelings and small paved roads (smaller than the pixel size) on the hydrological and soil erosion processes is taken into account. A flowchart of LISEM is presented in Figure 1.

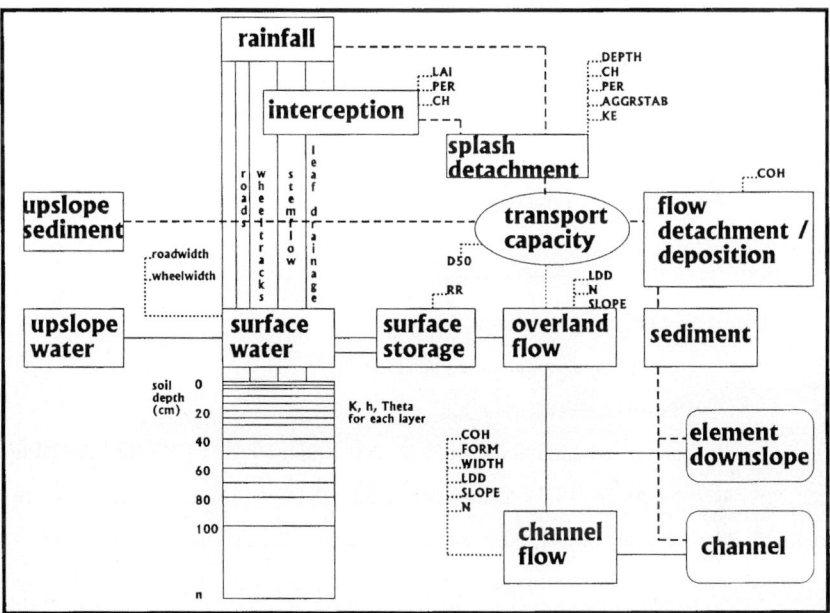

Figure 1. Flowchart of the LISEM model

Rainfall

Data from multiple rain gauges can be entered in an input data file of the type time series (Van Deursen, 1995). Breakpoint rainfall data may be used down to a resolution of one second. A map is used as input to define for each pixel which raingauge must be used. Thus, the model allows for spatial and temporal variability of rainfall.

Interception

Interception by crops and/or natural vegetation is simulated by calculating a maximum storage capacity, which is filled during rainfall. The Maximum Interception Storage Capacity is

$$SMAX = 0.935 + 0.498 * LAI - 0.00575 * LAI^2 \qquad (1)$$

estimated using an equation developed by Von Hoyningen-Huene (1981):

with: SMAX = Maximum Storage Capacity (mm)
LAI = Leaf Area Index (-)

Cumulative interception during rainfall is simulated using an equation developed by Aston (1979), which is modified from Merriam (1960):

$$CINT = SMAX * \left[1 - e^{-(1-p) * \frac{PCUM}{SMAX}}\right] \qquad (2)$$

with: CINT = Cumulative Interception (mm)
PCUM = Cumulative Rainfall (mm)
p = Correction factor, equals (1-0.046*LAI)

This equation simulates throughfall already before SMAX is reached. The factor k, equal to (1-p), was introduced by Aston (1979) to incorporate the effect of slower interception when the vegetation is dense. Simultaneously, this factor incorporates the fact that only that part of the cumulative rainfall which falls on the vegetation (PCUM*PER, where PER is the soil coverage by vegetation) can contribute to the interception storage. From the cumulative interception equation, interception rate is calculated by subtracting the CINT at time = (t-1) from CINT at time = t.

Infiltration and soil water transport

Infiltration and soil water transport in soils are simulated by a solution of the well known Richards equation, which combines the Darcy equation and the continuity equation. The Mualem/Van Genuchten equations (Mualem, 1976; Van Genuchten, 1980) are used to predict the soil-water retention curves and the unsaturated hydraulic conductivity, which are needed to solve the equation above. Soil profile types are defined, and for each characteristic soil horizon tables with the measured K-θ-h relations are required. The submodel operates with a variable time increment depending on pressure head changes. For areas without detailed knowledge of the soil physics the user can choose also a one or two layer Green-Ampt infiltration model, or the Holtan model.

Storage in micro depressions

Storage in micro-depressions and related variables are simulated by a set of equations developed by Onstad (1984) and Linden *et al.* (1988). The variable Random Roughness is used as a measure of microrelief. Moore and Larson (1979) identified three possible stages during a rainfall event:

a) Micro-relief storage building up, no surface runoff;
b) Additional micro-relief storage, accompanied by runoff;
c) Runoff only with the micro-relief storage at maximum.

Using the equations, the starting point of runoff, the maximum fraction of the surface covered with water, the actual fraction surface covered with water, and the fraction of the isolated depressions, not contributing to overland flow, are calculated (De Roo *et al.*, 1996a).

Overland flow and channel flow

For the distributed overland and channel flow routing, a four-point finite-difference solution of the kinematic wave is used together with Manning's equation. Procedures of the numerical solution can be found in Chow *et al.* (1988) and Moore and Foster (1990).

Splash detachment

Splash detachment is simulated as a function of soil aggregate stability, rainfall kinetic energy and the depth of the surface water layer. The kinetic energy can arise from both direct throughfall and drainage from leaves. The equation used here is adapted from the EUROSEM model (Morgan *et al.*, 1992). However, aggregate stability is used instead of the soil detachability index, and the equation is calibrated for loess conditions using splash field data:

$$DETR = \left[\frac{2.82}{AGGRSTAB} * KE * \exp^{-1.48*DEPTH} + 2.96 \right] * (P-I) * \frac{(dx)^2}{dt} \quad (3)$$

with: DETR = Splash detachment ($g.s^{-1}$);
AGGRSTAB = Soil aggregate stability (median number of drops);
KE = Rainfall kinetic energy ($J.m^{-2}$);

DEPTH	= Depth of the surface water layer (mm);
P	= Rainfall (mm);
I	= Interception (mm);
dx	= Size of an element (m);
dt	= Time increment (s).

Transport capacity

The transport capacity of overland flow is modelled as a function of unit stream power (Govers, 1990):

$$TC = C1 * [S*V - 0.4]^{D1} \quad (4)$$

with:	TC	= Volumetric transport capacity (cm^3.cm^{-3});
	S	= Slope gradient (m.m^{-1});
	V	= Mean flow velocity (cm.s^{-1});
	C1,D1	= Empirically derived coefficients.

The values of C1 and D1 depend on the D50 of the upper soil layer (Govers, 1990), which is an input map of the model.

Rill and interrill erosion

Flow detachment and deposition are simulated using equations from the EUROSEM model (Morgan *et al.*, 1992; Morgan, 1994). Whenever the transporting capacity, calculated using equation 4, is less than the available sediment from splash, from upslope areas and from previous time steps, deposition occurs at the following rate:

$$DEP = w * v_s * [TC - C] \quad (5)$$

with:	DEP	= Deposition rate (kg.m^{-3});
	w	= Width of the flow (m);
	v_s	= Settling velocity of the particles (m.s^{-1});
	TC	= Transport capacity (kg.m^{-3});
	C	= Sediment concentration in the flow (kg.m^{-3}).

If the transporting capacity of the flow exceeds the sediment concentration in the flow, detachment by the flow takes place and is calculated using the following equation (Morgan et al., 1992; Morgan, 1994):

$$DF = y * w * v_s * [TC - C] \qquad (6)$$

with: DF = Flow detachment rate (kg.m^{-3});
       y = Efficiency coefficient (-).

The efficiency coefficient in equation 6 is determined by (Morgan et al., 1992; Rauws and Govers, 1988):

$$y = \frac{u_{gmin}}{u_{gcrit}} = \frac{1}{0.89 + 0.56 * COH} \qquad (7)$$

with: u_{gmin} = Minimum value required for critical grain shear velocity (cm.s^{-1});
      u_{gcrit} = Critical grain shear velocity for rill initiation (cm.s^{-1});
      COH = Cohesion of the soil at saturation (kPa).

In the LISEM model, the user can enter both the cohesion of the bare soil (COH.MAP) as well as the additional cohesion caused by vegetation or crops (COHADD.MAP). These two cohesion values are added and then used in equation 7.

Roads, wheel tracks and channels
Roads and tractor wheel tracks smaller than the pixel size are simulated using information on their width. According to their width, interception on a percentage of the pixel is zero for both roads and wheel tracks. Infiltration and splash detachment are also zero on roads. Soil water movement under wheel tracks is simulated separately by using a special soil physics table for wheel tracks, which is different because of the soil compaction. Water and sediment flow in channels is simulated separately, using Manning's n for the channel bed, the width, the channel gradient, channel form and width, and the channel bed cohesion.

Crusts and surface stones

Infiltration through crusted soils can be simulated using separate conductivity tables for crusts. In a map, the percentage of crusted soil within a pixel is entered. The Richards equation is solved for both 'normal' soils and crusted soils. After the infiltration equations, the water is summed and the other processes are simulated. The fraction of stone cover within a pixel can also be given in a separate map. The current version of the model simulates no splash or flow detachment on the stone covered part. Infiltration effects are not (yet) taken into account.

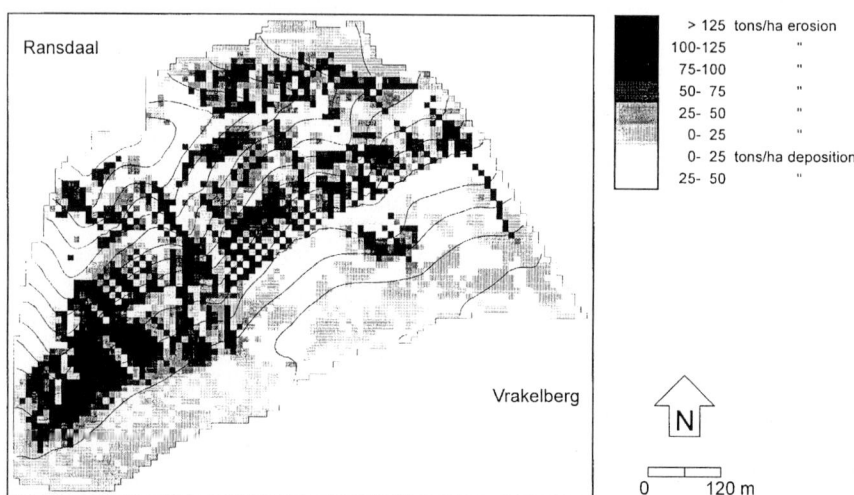

Figure 2. Example of soil erosion and deposition simulated with the LISEM model in the Ransdaal catchment for a summer rainstorm with a 25 year return period and current land use

LISEM input and output

For a detailed description of the input and output of the LISEM model, the reader is referred to De Roo *et al.* (1996a). Basically, much of the input and output are GIS maps. The results of the LISEM model consist first of all of a file with totals (total rainfall, total discharge, peak discharge, time of peak discharge, total soil loss, total infiltration, storage in micro depressions etc.). Furthermore, an ASCII data file which can be used to plot hydrographs and sedigraphs is provided. Also, PCRaster maps of soil erosion and deposition, as caused by the event (Figure 2), are produced. Finally, PCRaster maps of overland flow at desired time intervals during the event are calculated (Figure 3).

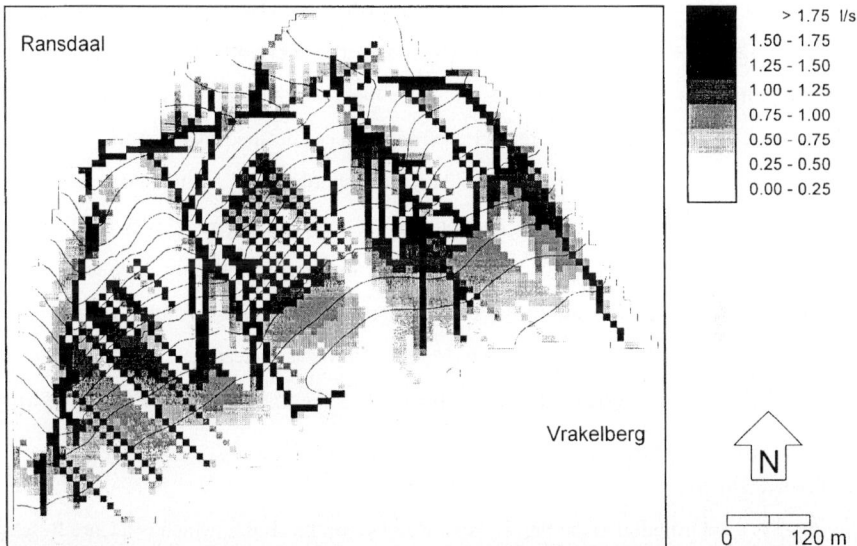

Figure 3. Example of surface and channel runoff simulated with the LISEM model in the Ransdaal catchment for a summer rainstorm with a 25 year return period and current land use

Sensitivity analysis, calibration, validation and applications of LISEM

For a detailed sensitivity analysis of LISEM, the reader is referred to Jetten *et al.* (1998). Calibration, validation and the applications of LISEM in three catchments in South-Limburg (The Netherlands) with a total of about 35 events are discussed in De Roo *et al.* (1996b). The preliminary results of early versions of LISEM are far from perfect. Based on the sensitivity analysis and field observations, the main reasons for these differences seems to be the spatial and temporal variability of the soil hydraulic conductivity and the initial pressure head at the basin scale. Another reason for the differences between measured and simulated results is the lack in our theoretical understanding of the hydrological and soil erosion processes. It is clear from our data that summer and winter response to rainfall are quite different. But even within the main seasons there are different responses to rainfall due to tillage operations and biological activity.

Conclusions

LISEM is a powerful model which simulates hydrological and soil erosion processes during single rainfall events on a catchment scale. Using LISEM it is possible to calculate the effects of land use changes and to explore soil conservation scenarios. Driven by hypothetical rainstorms of known probability of return, LISEM is a valuable tool for planning cost-effective measures to mitigate the effects of runoff and erosion. LISEM produces detailed maps of soil erosion and overland flow that are useful for planners. The integration of LISEM in a raster-based GIS, which holds the many data on the distributions of land attributes, is very useful. Other advantages of LISEM are the use of physically-based mathematical relationships, the ease with which newly developed relationships can be incorporated and the incorporation of information about the spatial variability of land characteristics. However it is clear that, although the model has several advantages over other models, the preliminary results of LISEM are far from perfect.

Acknowledgements

This paper is a contribution to the Soil Erosion Network of the GCTE, which is a Core Research Project of the IGBP. The Province of Limburg, the Waterboard 'Roer en Overmaas', the Ministry of Agriculture and 14 municipalities of South-Limburg are greatly acknowledged for funding a large part of the research which lead to the development of the LISEM model. In addition to the authors, several researchers have contributed to the model development. Mr. N.H.D.T. Cremers, Mr. M.A. Verzandvoort, Mr. H. Kolenbrander and Mr. R.J.E. Offermans of the Utrecht University and Mr. J. Stolte and Mr. K. Oostindie of the Winand Staring Centre Wageningen are all thanked for their contributions.

References

Aston, A.R. (1979). Rainfall Interception by eight small trees. *Journal of Hydrology* **42**, 383-396.

Beasley, D.B., Huggins, L.F. and Monke, E.J. (1980). ANSWERS: a model for watershed planning. *Transactions of the American Society of Agricultural Engineers* **23**(4), 938-944.

Chow, V.T., Maidment, D.R. and Mays, L.W. (1988). *Applied Hydrology*. McGraw-Hill, 572 pp.

De Jong, S.M. (1994). Applications of reflective remote sensing for land degradation studies in a Mediterranean environment. *Netherlands Geographical Studies* **177**.

De Roo, A.P.J., Hazelhoff, L. and Burrough, P.A. (1989). Soil erosion modelling using 'ANS-WERS' and Geographical Information Systems. *Earth Surface Processes and Landforms* **14**, 517-532.

De Roo, A.P.J., Wesseling, C.G. and Ritsema, C.J. (1996a). LISEM: a single event physically-based hydrologic and soil erosion model for drainage basins. I: Theory, input and output. *Hydrological Processes*, in press.

De Roo, A.P.J., Offermans, R.J.E. and Cremers, N.H.D.T. (1996b). LISEM: a single event physically-based hydrologic and soil erosion model for drainage basins. II: Sensitivity analysis, validation and application. *Hydrological Processes*, in press.

Govers, G. (1990). *Empirical Relationships on the Transporting Capacity of Overland Flow*. IAHS Publication 189, 45-63.

Howes, S. and Anderson M.G. (1988). Computer simulation in geomorphology. In, Anderson, M.G. (ed.), *Modelling Geomorphological Systems*. Wiley, Chichester. pp. 421-440.

Jetten, V., de Roo, A.P.J. and Guérif, J. (1998). Sensitivity of the model LISEM to variables related to agriculture. In, Boardman, J. and Favis-Mortlock, D.T. (eds), *Modelling Soil Erosion by Water*, Springer-Verlag NATO-ASI Global Change Series, Heidelberg.

Knisel, W.G. (ed.) (1980). *CREAMS: A Field Scale Model for Chemicals, Runoff and Erosion from Agricultural Management Systems*. USDA Conservation Research Report 26. 643 pp.

Lane, L.J., Nearing, M.A., Laflen, J.M., Foster, G.R. and Nichols, M.H. (1992). Description of the US Department of Agriculture Water Erosion Prediction Project (WEPP) Model. In, Parsons, A.J. and Abrahams, A.D. (eds), *Overland Flow: Hydraulics and Erosion Mechanics*. UCL Press, London, 377-391.

Merriam, R.A. (1960). A note on the interception loss equation. *Journal of Geophysical Research* **65**, 3850-3851.

Moore, I.D. and Larson, C.L. (1979). Estimating micro-relief surface storage from point data. *Transactions of the American Society of Agricultural Engineers* **20**(5), 1073-1077.

Moore, I.D. and Foster, G.R. (1990). Hydraulics and overland flow. In, Anderson, M.G. and Burt, T.P. (eds), *Process Studies in Hillslope Hydrology*. Wiley, Chichester. pp. 215-254.

Morgan, R.P.C., Morgan, D.D.V. and Finney, H.J. (1984). A predictive model for the assessment of soil erosion risk. *Journal of Agricultural Engineering Research* **30**, 245-253.

Morgan, R.P.C., Quinton, J.N. and Rickson, R.J. (1992). *EUROSEM Documentation manual. Version 1*: June 1992. Silsoe College, Silsoe.

Morgan, R.P.C. (1994). The European Soil Erosion Model: an update on its structure and research base. In, Rickson, R.J. (ed.), *Conserving Soil Resources: European Perspectives*. CAB International, Cambridge, pp. 286-299.

Mualem, Y. (1976). A new model for predicting the hydrologic conductivity of unsaturated porous media. *Water Resources Research* **12**, 513-522.

Onstad, C.A. (1984). Depressional storage on tilled soil surfaces. *Transactions of the American Society of Agricultural Engineers* **27**, 729-732.

Rauws, G. and Govers, G. (1988). Hydraulic and soil mechanical aspects of rill generation on agricultural soils. *Journal of Soil Science* **39**, 111-124.

Van Deursen, W.P.A. (1995). Geographical Information Systems and dynamic models. *Netherlands Geographical Studies* **190**, 198 p.

Van Deursen, W.P.A. and Wesseling, C.G. (1992). *The PCRaster Package*. Department of Physical Geography, Utrecht University.

Van Genuchten, M.Th. (1980). A closed-form equation for predicting the hydraulic conductivity of unsaturated soils. *Soil Science Society of America Journal* **44**, 892-898.

Von Hoyningen-Huene, J. (1981). *Die Interzeption des Niederschlags in landwirtschaftlichen Pflanzenbeständen*. Arbeitsbericht Deutscher Verband für Wasserwirtschaft und Kulturbau, DVWK, Braunschweig, 63 pp.

Wischmeier, W.H. and Smith, D.D. (1978). *Predicting Rainfall Erosion Losses — a Guide to Conservation Planning*. U.S. Department of Agriculture, Agricultural Handbook No. 537, Science and Education Administration USDA, Washington D.C. 58 pp.

33. APEX: A NEW TOOL FOR PREDICTING THE EFFECTS OF CLIMATE AND CO_2 CHANGES ON EROSION AND WATER QUALITY

Jimmy R. Williams[1], Jeffrey G. Arnold[1], Raghavan Srinivasan[2] and Tharacad S. Ramanarayanan[2]

[1] US Department of Agriculture — Agricultural Research Service
Grassland, Soil and Water Research Laboratory
808 East Blackland Road
Temple, TX 76502
USA

[2] Texas Agricultural Experiment Station
808 East Blackland Road
Temple, TX 76502
USA

Abstract

Several field scale hydrologic/water quality models have been developed to study the impacts of agricultural management practices. The EPIC (Environment Policy Integrated Climate — previously the Erosion Productivity Impact Calculator) model is one of the more popular models, which has been widely applied in the United States and around the world. Such models are limited to small field size areas, where the soil, management, crop, and topography are assumed to be homogeneous. To extend the capabilities of EPIC to simulate large complex farming systems (multiple fields, soils, rotations, management, etc.), a model called APEX (Agricultural Policy/Environmental eXtender) was developed. In addition to the capabilities of EPIC, APEX has components for routing water, sediment, and chemicals (nutrients and pesticides) across complex landscapes and channel systems to the watershed outlet. The subsurface routing routine to APEX is enhanced from that in EPIC and can be used to simulate subsurface processes to a depth of 30 m. In this paper we present an overview of EPIC and APEX, and describe in detail the recently added CO_2 component of the model.

Introduction

Most recent development of the EPIC model has focused on problems involving water quality and global climate/CO_2 change. Example additions include the GLEAMS (Leonard et al., 1987) pesticide fate component, nitrification and volatilisation submodels; a new more physically based wind erosion component, optional SCS technology for estimating peak runoff rates; and newly developed sediment yield equations. Stockle et al. (1992) modified EPIC to simulate the effects of CO_2 on plant growth and water use efficiency. In this paper we present an overview of the EPIC (Environmental Policy Integrated Climate) model and describe the newly added CO_2 component.

Williams *et al.* (1984) developed EPIC to assess the effect of soil erosion on soil productivity. EPIC was used for that purpose as part of the 1985 RCA (1977 Soil and Water Resources Conservation Act) analysis. Subsequently, the model has been expanded and refined to allow simulation of many processes important in agricultural management (Sharpley and Williams, 1990; Williams, 1995). To better address these changes the model name was changed from Erosion Productivity Impact Calculator to Environmental Policy Integrated Climate, but the acronym EPIC remains the same.

Water balance
The runoff model simulates surface runoff volumes and peak runoff rates, given daily rainfall amounts. Runoff volume is estimated by using a modification of the Soil Conservation Service (SCS) curve number technique (USDA-SCS, 1972).

There are two options for estimating the peak runoff rate — the modified Rational formula and the SCS TR-55 method (USDA-SCS, 1986). A stochastic element is included in the Rational equation to allow realistic simulation of peak runoff rates, given only daily rainfall and monthly rainfall intensity.

The model offers four options for estimating potential evaporation — Hargreaves and Samani (1985), Penman (1948), Priestley-Taylor (1972), and Penman-Monteith (Monteith, 1977). The Penman and Penman-Monteith methods require solar radiation, air temperature, wind speed, and relative humidity as input. If wind speed, relative humidity, and solar radiation data are not available, the Hargreaves or Priestley-Taylor methods provide options that give realistic results in most cases.

Weather
The weather variables necessary for driving the EPIC model are precipitation, air temperature, and solar radiation. If the Penman methods are used to estimate potential evaporation, wind speed and relative humidity are also required. Wind speed is also needed when wind-induced erosion is simulated. If daily precipitation, air temperature, and solar radiation data are available,

they can be input directly into EPIC. Rainfall and temperature data are available for many areas of the United States, but solar radiation, relative humidity, and wind data are scarce. Even rainfall and temperature data are generally not adequate for long-term (more than 100 years) EPIC simulations. Thus, EPIC provides options for simulating various combinations of the five weather variables.

Crop growth and tillage

A single model is used in EPIC for simulating all the crops considered (corn, grain sorghum, wheat, barley, oats, sunflower, soybean, alfalfa, cotton, peanuts, potatoes, Durham wheat, winter peas, faba beans, rapeseed, sugarcane, sorghum hay, range grass, rice, cassava, lentils, and pine trees). Each crop has unique values for the model parameters. EPIC is capable of simulating growth for both annual and perennial crops. Annual crops grow from planting date to harvest date or until the accumulated heat units equal the potential heat units for the crop.

Perennial crops maintain their root systems throughout the year, although they may become dormant after frost. They start growing when the average daily air temperature exceeds their base temperature. Potential crop growth and yield are usually not achieved because of constraints imposed by the plant environment. The model estimates stresses caused by water, nutrients, temperature, aeration, and radiation.

The EPIC tillage component was designed to mix nutrients and crop residues within the plough depth, simulate the change in bulk density, and convert standing residue to flat residue. Other functions of the tillage component include simulating ridge height and surface roughness.

Water erosion

The EPIC component for water-induced erosion simulates erosion caused by rainfall and runoff and by irrigation (sprinkler and furrow). To simulate rainfall/runoff erosion, EPIC contains six equations — USLE (Wischmeier and Smith, 1978), the Onstad-Foster modification of the USLE (Onstad and Foster, 1975), MUSLE (Williams, 1975), two recently developed variations of MUSLE, and a MUSLE structure that accepts input coefficients. Only one of the equations (user specified) interacts with other EPIC components. The six equations are identical except for their energy components. The USLE depends strictly upon rainfall as an indicator of erosive energy (EI). The MUSLE and its variations use only runoff variables to simulate erosion and sediment yield. The Onstad-Foster equation contains a combination of the USLE and MUSLE energy factors.

Thus, the water erosion model uses an equation of the form:

$$Y = \chi (K)(CE)(PE)(LS)(ROKF) \tag{1}$$

and:

$$\begin{aligned}
\chi &= EI \quad \text{for USLE} \\
\chi &= 0.646\ EI + 0.45\ (Q \cdot q_p)^{0.33} \quad \text{for Onstad - Foster} \\
\chi &= 1.586(Q\ q_p)^{0.56}\ A^{0.12} \quad \text{for MUSLE} \\
\chi &= 2.5\ (Q\ q_p)^{0.5} \quad \text{for MUST} \\
\chi &= 0.79\ (Q\ q_p)^{0.65}\ A^{0.009} \quad \text{for MUSS} \\
\chi &= by_1\ Q^{by_2}\ q_p^{by_3}\ A^{by_4} \quad \text{for MUSI}
\end{aligned} \tag{2}$$

where Y is the sediment yield in t ha^{-1}, K is the soil erodibility factor, CE is the crop management factor, PE is the erosion control practice factor, LS is the slope length and steepness factor, ROKF is the coarse fragment factor, Q is the runoff volume in mm, q_p is the peak runoff rate in mm h^{-1}, and A is the watershed area in ha. MUST is a new equation theoretically developed from sediment concentration bases, MUSS is a new equation developed by fitting small watershed data (no channel erosion), and MUSI allows user input of four coefficients (by$_i$).

Wind erosion

The original EPIC wind erosion model (WEQ) required daily mean wind speed as a driving variable. The new EPIC wind erosion model, WECS (Wind Erosion Continuous Simulation) requires the daily distribution of wind speed to take advantage of the more mechanistic erosion equation. The new approach estimates potential wind erosion for a smooth bare soil by integrating the erosion equation through a day using the wind speed distribution. Potential erosion is adjusted using four factors based on soil properties, surface roughness, cover, and distance across the field in the wind direction.

The basic WECS wind erosion equation is

$$YW = (FI)(FR)(FV)(FD) \int_0^{DW} YWR \, dt \tag{3}$$

where YW is the wind erosion in kg m^{-1}, FI is the soil erodibility factor, FR is the surface roughness factor, FV is the vegetative cover factor, FD is the mean unsheltered travel distance of wind across the field, YWR is the wind erosion rate in kg m^{-1} s^{-1} at time t, and DW is the duration of wind greater than threshold velocity in s.

CO_2 relationships

The CO_2 developments of Stockle et al. (1992) were added to the EPIC potential evaporation and crop growth components. The Penman-Monteith method (Monteith, 1965) is one of four potential ET equations available for use in EPIC. The Penman-Monteith equation is expressed as

$$E_p = \frac{\delta(h_o - G) + 86.7 \, AD(e_a - e_d) / AR}{HV(\delta + \gamma(1. + CR / AR))} \tag{4}$$

where E_p is the potential evaporation in mm, δ is the slope of the saturation vapour pressure curve in kPa °C^{-1}, h_o is the net radiation in MJ m^{-2}, G is the soil heat flux in MJ m^{-2}, HV is the latent heat of vaporisation in MJ kg^{-1}, e_a is the saturation vapour pressure at mean air temperature in kPa, e_d is the vapour pressure at mean air temperature in kPa, AD is the air density in g m^{-3}, AR is the aerodynamic resistance for heat and vapour transfer in s m^{-1}, and CR is the canopy resistance for vapour transfer in s m^{-1}.

The CO_2 effect on canopy resistance is computed with the equation

$$CR = \frac{p_1}{(LAI)(g_o^*)(1.4 - 0.00121 CO_2)} \tag{5}$$

where p_1 is a parameter ranging from 1.0 to 2.0, LAI is the leaf-area-index of the crop, g_o^* is the leaf conductance in m s^{-1}, and CO_2 is the carbon dioxide level in the atmosphere in ppm. Leaf conductance is estimated from the crop input rate adjusted for vapour pressure deficit (VPD).

$$g_o^* = (g_o)(FV) \tag{6}$$

where g_o is the crops leaf resistance when VPD is less than the crop's threshold VPD and FV is the VPD correction factor given by

$$FV = 1. - b_v(VPD - VPD_t) \geq 0.1 \tag{7}$$

where b_v is a crop coefficient and VPD_t is threshold VPD for the crop.

valuable tool for long-term fertility management. For example, though manure may be rich in phosphorus, inorganic sources of nitrogen may be needed to provide the nutrients needed for maximum crop production. APEX can be used to also evaluate their effects on nutrient movement to surface and ground waters.

APEX is currently operational and is being used to evaluate alternative animal waste management strategies in the Upper North Bosque watershed of Texas. It is also undergoing rapid development, testing, and improvement. One of the more useful improvements expected in the next year is integration of the model with a user interface utilising a Geographic Information System (GIS) to facilitate input and visualise model output. This interface will be similar to that currently used by the SWAT model.

Conclusion
APEX is an operational model and decision tool designed to simulate the behaviour of complex farming systems at the whole-farm or small watershed scale. Its treatment of individual fields or areas of homogeneous soils, topography, management, etc. is taken from the EPIC model. Like the SWAT model, movements of water, sediment, nutrients, and pesticides are used to link fields within the farm or watershed. Future improvements are expected to include integration of APEX with a GIS and user interface like that used by the SWAT watershed hydrology model.

References
Arnold, J.G., Williams, J. R., Nicks, A.D., and Sammons, N.B. (1990). *SWRRB - A Basin Scale Simulation Model for Soil and Water Resource Management.* Texas A&M Press, College Station, Texas.
Arnold, J.G., Allen, P.M. and Bernhardt, G. (1993). A comprehensive surface-groundwater flow model. *Journal of Hydrol*ogy **142**, 47-69.
Easterling, W.E., Rosenberg, N.J., McKenney, M.S., Jones, C.A., Dyke, P.T., and Williams, J.R. (1992). Preparing the Erosion Productivity Impact Calculator (EPIC) model to simulate crop response to climate change and the direct effects of CO_2. *Agriculture and Forest Meteorology* **59**, 17-34.
Hargreaves, G.H. and Samani, Z.A. (1985). Reference crop evapotranspiration from temperature. *Applied Engineering in Agriculture* **1**, 96-99.
Leonard, R.A., Knisel, W.G. and Still, D.A. (1987). GLEAMS: Groundwater loading effects on agricultural management systems. *Transactions of the American Society of Agricultural Engineers* **30**, 1403-1428.
Monsi, M. and Saeki, T. (1953). Uber den Lictfaktor in den Pflanzengesellschaften und sein Bedeutung für die Stoffproduktion. *Japanese Journal of Botany* **14**, 22-52.
Monteith, J.L. (1977). Climate and the efficiency of crop production in Britain. *Philosophical Transactions of the Royal Society of London Series B.* **281**, 277-329

Onstad, C.A. and Foster, G.R. (1975). Erosion modeling on a watershed. *Transactions of the American Society of Agricultural Engineers* **18**, 288-292.

Penman, H.L. (1948). Natural evaporation from open, bare soil and grass. *Proceedings of the Royal Society of London Series* A**193**, 120-145

Priestley, C.H.B. and Taylor, R.J. (1972). On the assessment of surface heat flux and evaporation using large-scale parameters. *Monthly Weather Rev*iew **100**, 81-92.

Robertson, T., Benson, V.W., Williams, J.R., Jones, C.A. and Kiniry, J.R. (1987). The impact of climate changes on yields and erosion for selected crops in the southern United State. In, *Proceedings of the Symposium on Climate Change in the United States: Future Impacts and Present Policy Issues*, New Orleans, Louisiana. pp 89-134.

Robertson, T., Rosenzweig, C., Benson, V. and Williams, J.R. (1990). Projected impacts of carbon dioxide and climate change on agriculture in the Great Plains. In, Unger, P.W., Sneed, T.V., Jordan, W.R. and Jensen, R. (eds), *Proceedings of the International Conference on Dryland Farming: Challenges in Dryland Agriculture — A Global Perspective*, Amarillo/Bushland, Texas. pp. 675-677.

Shapiro, M., Westervelt, J., Gerdes, D., Larson, M. and Brownfield. (1992). *GRASS 4.0 Reference Manual.* USA CERL, Champaign, IL.

Sharpley, A.N. and Williams, J.R. (eds) (1990). *EPIC-Erosion/Productivity Impact Calculator: 1. Model Documentation.* U.S. Department of Agriculture Technical Bulletin No. 1768.

Stockle, C.O., Williams, J.R., Rosenberg, N.J. and Jones, C.A. (1992). A method for estimating the direct and climatic effects of rising atmospheric carbon dioxide on growth and yield of crops: Part I — Modification of the EPIC model for climate change analysis. *Agricultural Systems* **38**, 225-238.

USDA-SCS (1972). *National Engineering Handbook, Hydrology Section 4, Chapters 4-10.*

USDA-SCS (1986). *Urban Hydrology for Small Watersheds.* U.S. Department of Agriculture Technical Release No. 55.

Williams, J.R. (1975). Sediment routing for agricultural watersheds. *Water Resources Bulletin* **11**, 965-974.

Williams, J.R., Jones, C.A. and Dyke, P.T. (1984). A modeling approach to determining the relationship between erosion and productivity. *Transactions of the American Society of Agricultural Engineers* **27**, 129-144.

Williams, J.R. (1995). The EPIC model. In, Singh, V.P. (ed.), *Computer Models of Watershed Hydrology*, Water Resources Publication, Highlands Ranch, Colorado. pp. 909-1000.

Wischmeier, W.H. and Smith, D.D. (1978). *Predicting Rainfall Erosion Losses — a Guide to Conservation Planning.* U.S. Department of Agriculture Handbook No. 537.

34. A DYNAMIC MODEL OF GULLY EROSION

A. Sidorchuk
Laboratory of Soil Erosion and Fluvial Processes
Geographical Faculty
Moscow State University
Moscow, 119899
Russia.

Abstract
The main causes of gully formation are anthropogenic: the clearing of native forests, tilling of fallow lands and associated change of the hydrological conditions in the rainfall-runoff system. Gully formation is very intensive during the period of initiation, when morphological characteristics (length, depth, width, area, and volume) are far from stable. About 80 per cent of a gully's length, 60 per cent of its area and 35 per cent of its volume are formed in only 5 per cent of its lifetime. This stage of development can be described by a dynamic model to predict rapid changes of gully morphology.

The dynamic gully model is based on the solution of the equations of mass conservation and gully bed and wall deformation. An analysis of experimental results shows that the rate of soil particle detachment is linearly correlated with the product of bed shear stress and mean flow velocity. Basic equations were written in the form of a transport equation and solved with the use of an explicit predictor-corrector scheme of the Lax-Wendroff type. The side walls of gullies become practically straight after rapid sliding following its incision. A straight stable slope model was used for prediction of gully-side wall inclination.

This dynamic gully erosion model has been verified with data on gully morphology and dynamics from the Yamal peninsula (Russia) and New South Wales (Australia).

Introduction

The significance of gully erosion has been well documented. The volume of gullies on the Russian Plain is about 4×10^9 m^3, i.e. about 4 per cent of the whole volume of erosion since 1700 AD (Sidorchuk, 1995). In south-east Australia with mainly pasture lands, the volume of gully erosion amounts to 37 per cent of the whole erosion volume (Graham, 1987). Gullies completely destroy the fertile topsoil layer, and surrounding lands are damaged by severe sheet and rill erosion.

There are two main stages of gully development which are controlled by different sets of geomorphic processes. During the first stage — initiation — hydraulic erosion predominates at the gully bottom; rapid mass movement occurs on the gully sides. During the second stage — gully stabilised — sediment transport and sedimentation are the main processes in the gully's bottom, its width increases due to lateral erosion, and slow mass movement transforms the gully sides. Gully channel formation is very intense during the period of gully initiation when the morphological characteristics of the gully (length, depth, width, area, and volume) are far from stable. This stage is relatively short and occupies about 5 per cent of the gully's lifetime, but 80 per cent of its length, 60 per cent of its area and 35 per cent of its volume are formed in this period (Kosov et al., 1978).

The dynamic gully erosion model
The model describes the first, quick stage of gully development. At this stage the following main processes occur:

a) During the snowmelt or rainstorm event flowing water erodes a rectangular channel in the topsoil or at the gully bottom if the flow velocity exceeds some critical value for erosion initiation. Sediment concentration in the flow is controlled by lateral inflow from the gully's catchment, detachment of the particles from the bottom and banks, and by sedimentation on the gully bottom.

b) The vertical walls of this trench are unstable. Shallow landslides quickly transform the gully's cross-sectional shape to trapezoidal in the period between water flow events.

Processes of gully incision
Theoretical framework
The rate of gully erosion is controlled by water flow velocity, depth and turbulence, as well as soil texture, mechanical pattern and protection by vegetation. These characteristics are combined in equations of mass conservation and deformation, which can be written in the form:

$$\frac{\partial Q_s}{\partial X} + \frac{\partial AC}{\partial t} = C_w q_w + M_0 W + M_b D - CV_f W \qquad (1)$$

$$-W\frac{\partial Z}{\partial t} = \frac{\partial Q_s}{\partial X} - M_b D - C_w q_w \qquad (2)$$

Here $Q_s = Q\,C$ is sediment discharge (m³/s), Q = water discharge (m³/s); X = longitudinal co-ordinate (m); t = time (s); C = mean volumetric sediment concentration; A = flow cross-sectional area (m²); C_w = sediment concentration of the lateral input; q_w = specific lateral discharge; M_0 = upward sediment flux (m/s); M_b = sediment flux from the channel banks (m/s); Z = gully bottom elevation (m); W = flow width (m); D = flow depth (m); V_f = sediment particle fall velocity in the turbulent flow (m/s).

The first term in the left part of the equation of mass conservation (1) defines the sediment budget in the channel reach; the second term is sediment storage in the flow. The right part of (1) defines the sediment flux: the first term is lateral flux, the second is upward flux, the third is sediment flux from the banks, and the fourth is the downward flux. The equation of deformation (2) defines the change of gully bottom elevation and bank co-ordinates according to the sediment budget.

The solution of the equations with the main assumptions and simplifications

The sediment storage in the flow is usually very small and can be neglected. In this case the equation (1) is a first order ordinary differential equation, and equation (2) is a first order partial differential equation with variable coefficients. The solution of these equations depends on the form of the terms, which describe sediment fluxes.

For a given section the lateral specific discharge q_w may be assumed to be constant for the length L and water discharge in the flow may be assumed to increase linearly with the distance X from the initial value Q_o: $Q = Q_o + q_w X$. The sediment concentration in the lateral flow C_w is controlled by the conditions within the basin and also assumed to be constant for the section L.

The upward sediment flux is the product of volumetric bottom sediment concentration C_0 and vertical bottom velocity of the sediment particles $U_\uparrow$: $M_0 = \cdot U_\uparrow C_0$. Vertical bottom velocity of sediment particles is about 0.3 U (Rossinskiy and Debolskiy, 1980), where U (m/s) is the mean flow velocity. The near bed sediment concentration (or probability of particle detachment) after Einstein (1942) is a function of the measure of the transport rate $C_0 = f_1(\tau/\tau_{cr})$. Here $\tau = g\rho DS$ is the bed shear stress, τ_{cr} is its value for sediment detachment initiation, g is the acceleration due to gravity (m/s²), ρ is water density (kg/m³), and S is flow surface slope (for gullies this is equal to bed slope).

Mirtskhulava (1988) showed that critical shear stress is mainly controlled by the forces of friction and cohesion: $\tau_{cr} = 1.2\lambda \left(m/n\right)\left[(\rho_s - \rho)gd + 1.25 C_f^n K\right]$. Here λ is a coefficient of flow resistance: $\lambda = 0.18*(d/D)^{1/3}$; m is equal to 1 for clean water flows, and is equal to 1.4 for flows with colloidal particle content more than 0.1 kg/m³; the parameter of turbulence n is usually about 4; ρ_s is sediment density (kg/m³); d is the mean diameter of soil aggregates (m); K is a coefficient of variability of soil mechanical pattern, usually 0.5; $C_f^n = f_2(C_h)$ is the soil fatigue strength to rupture, which is a function of soil cohesion C_h (Pa). The first term in square brackets represents the influence of friction on particle stability, and is of most importance for non-cohesive soils, the second term represents the influence of cohesion on particle stability, and is of most importance for cohesive soils.

Field experiments were run to determine the functions $C_0 = f_1(\tau/\tau_{cr})$ and $C_f^n = f_2(C_h)$ for gully erosion conditions. Flumes 1, 2 and 3 of length 9.7, 3.5 and 6.0 m respectively were prepared in natural soils on the sides of Brook gully (Yass River basin, NSW, Australia) with different inclinations. The water entered the head of the flume from a reservoir (tank) with a volume of 15 m³ through a transportable weir that permitted constant discharge up to 12 l/s for not less than 15 minutes. The water samples for sediment concentration determination were taken several times during each run at the head and at the end of the flume in order to estimate the sediment budget $\Delta Q_s/\Delta X$ along the flume, and particle detachment rate $M_0 = (1/W) \cdot \Delta Q_s/\Delta X$ (sedimentation and bank erosion was negligible). The main hydraulic parameters of the flow were measured during the run, and soil cohesion was measured with a tore vane after each run (Table 1). The soil was composed mainly of silt particles and had cohesion 3.0-7.0 10^4 Pa at saturation.

Analysis of experimental results shows that for the steep slopes typical of gullies, the rate of soil particle detachment is linearly correlated with the product of bed shear stress and mean flow velocity:

$$M_0 = k_e UDS \qquad (3)$$

This formula was also validated on the basis of field experiments in the gullies of the Yamal peninsula (north-western Siberia, Russia) which showed satisfactory conformity to these data

(Sidorchuk, in press). The erodibility coefficient k_e is equal to 6.46 $10^{-2}/\tau_{cr}$. To calculate the critical bed shear stress the formula of Mirtskhulava can be used with $C_f'' = 6.7 \cdot 10^{-7} C_h^2$. The latter expression is based on a limited set of data and additional experiments must be carried out for its verification.

Table 1. The parameters of the flow, sediment budget and soils in experimental flumes in the Brook gully (Yass River basin, NSW, Australia)

flume	run	discharge Q m³/s	velocity U m/s	width W m	depth D m	slope S	C kg/m³	C_h	τ_{cr} Pa	(1/W)* dQs/dX Pa 10^5 m/s 10^{-5}	$\rho g U D S$ m/s 10^2
1	1	0.0014	0.420	0.180	0.019	0.063	2.7	0.29	10.0	0.081	0.049
1	2	0.0022	0.370	0.170	0.034	0.063	3.0	0.29	8.3	0.145	0.078
1	3	0.0042	0.580	0.177	0.041	0.063	3.5	0.29	7.8	0.316	0.015
1	5	0.0086	0.640	0.208	0.064	0.068	2.5	0.29	6.7	0.368	0.273
1	6	0.0078	0.600	0.220	0.060	0.064	3.3	0.29	6.9	0.433	0.224
2	3	0.0015	0.534	0.120	0.023	0.285	4.5	0.31	11.0	0.576	0.349
2	4	0.0023	0.644	0.130	0.028	0.285	3.7	0.31	10.0	0.680	0.499
2	5	0.0041	0.587	0.165	0.042	0.240	2.4	0.31	8.7	0.566	0.578
2	6	0.0057	0.641	0.190	0.047	0.240	5.6	0.31	8.4	1.770	0.709
2	7	0.0086	0.730	0.233	0.034	0.248	3.6	0.31	9.3	1.370	0.893
2	9	0.0110	1.127	0.277	0.035	0.252	6.0	0.31	9.2	2.490	0.981
3	3	0.0011	0.830	0.120	0.008	0.577	6.4	0.50	36.0	0.351	0.509
3	4	0.0015	1.726	0.132	0.007	0.578	4.9	0.52	42.0	0.341	0.646
3	5	0.0019	1.709	0.142	0.008	0.579	5.7	0.52	40.0	0.460	0.747
3	6	0.0023	1.817	0.151	0.009	0.580	5.8	0.53	40.0	0.542	0.879
3	7	0.0032	1.700	0.195	0.009	0.582	6.5	0.54	40.0	0.650	0.922
3	8	0.0041	1.900	0.177	0.012	0.584	3.2	0.59	44.0	0.431	1.330
3	9	0.0049	1.900	0.195	0.013	0.588	4.5	0.61	45.0	0.697	1.460
3	10	0.0063	0.720	0.202	0.015	0.590	4.3	0.64	48.0	0.839	1.810
3	11	0.0083	0.800	0.226	0.018	0.594	4.6	0.74	60.0	1.030	2.150

The width of the flow in gullies can be calculated with the empirical formula: $W = 0.3 * Q^{0.4}$ (based on data from Yamal peninsula), and depth and velocity with the Chezy formula.

The process of flow bank erosion in the gullies has not been satisfactorily investigated. It is assumed that the rate of bank erosion dW_b/dt is equal to sediment flux from the banks M_b. Using an analogy with estimations of the river bank erosion the expression $M_b = M_0 \cdot V / U$ can be suggested as a first approximation. Here V is lateral velocity, W_b is gully bottom width. In large channels M_b is usually about 5 per cent of M_0 (Vikulova, 1972). For a curved channel Rozovzkiy (1957) obtained a simple formula: $V = 11.0 \cdot U \cdot D / R$. At the narrow incised gully bottom with $W_b < 10.0W$ the radius R of confined meanders will decrease when W_b increases due to bank erosion: $R = 50.0 \ W(W/W_b)$. When W_b becomes $> 10.0W$ the flow forms free meanders with R = 5.0W. At the same time curved flow can wash only part of the side walls and this part P_e decreases when the relative bottom width increases. The investigations in the gullies of Yamal peninsula show that $P_e = W/W_b$. After combining all the above formulae, the expression for calculation of gully bank erosion rate takes the form: $\frac{dW_b}{dt} = k_b \cdot M_0$. Here $k_b = 0.22 \cdot D/W$ when $W_b < 10.0 \cdot W$ and $k_b = 2.2 \cdot D/W_b$ when $W_b > 10.0 \cdot W$.

The expression for downward flux is rather simple and includes the product of fall velocity in turbulent flow and depth-averaged sediment concentration in the flow. The fall velocity in the turbulent flow is lower than Stocks fall velocity in laminar flow or in steady water V_{st}, and in the case of thin particles and high turbulence can be 0.

After substitution of equations (1) and (3) into (2) the expression takes the form of a transport equation in terms of bottom elevation Z:

$$\frac{\partial Z}{\partial t} - a \frac{\partial Z}{\partial x} - V_f C = 0 \tag{4}$$

Here $a = k_e UD$. The equation (4) can be numerically solved with the use of an explicit predictor-corrector scheme of the Lax-Wendroff type.

$$Z_i^{j+1/2} = (1-\beta)Z_i^j + \beta Z_{i+1}^j - \alpha \frac{\Delta t}{\Delta x}\left[\frac{(aq)_{i+2}^j + (aq)_{i+1}^j}{2} Z_{i+1}^j - \frac{(aq)_{i+1}^j + (aq)_i^j}{2} Z_i^j\right]$$

$$Z_i^{j+1} = Z_i^j - \frac{\Delta t}{2\alpha \Delta x}\{(\alpha-\beta)\frac{(aq)_{i+2}^j + (aq)_{i+1}^j}{2} Z_{i+1}^j - (2\beta-1)\frac{(aq)_{i+1}^j + (aq)_i^j}{2} Z_i^j +$$

$$(1-\alpha-\beta)\frac{(aq)_i^j + (aq)_{i-1}^j}{2} Z_{i-1}^j + \frac{(aq)_{i+1}^j + (aq)_i^j}{2} Z_i^{j+1/2} - \frac{(aq)_i^j + (aq)_{i-1}^j}{2} Z_{i-1}^{j+1/2}\} + V_f C_i \Delta t$$

The symbol i represents change in length, and the symbol j change in time. For a sediment concentration C_i the solution of (1) on the flow reach with length Δx may be used:

$$C_i = \left(C_{i-1} - \frac{(k_e + k_b) * Q_{i-1} S}{q_w(Y+1)} - \frac{C_w}{Y}\right) * \left(\frac{Q_{i-1}}{Q_i}\right)^Y + \frac{(k_e + k_b) * Q_i S}{q_w(Y+1)} + \frac{C_w}{Y}$$

Here C_o is the sediment concentration in channel flow at the beginning of the reach, $Y = (q_w + V_f * W)/q_w$. Best-fit values of net numbers α and β are: β = 0.75 — 1.0; α = 0.25

— 0.5. For the explicit scheme stability the Courant number must be less than 1.0: $aq\,\Delta t/\Delta x \leq 1$. The same approach was used for numerical solution of the equations of bank erosion rate.

The process of the side wall transformation

The side walls of the gully becomes practically straight after rapid sliding following incision. In this case a straight stable slope model can be used for prediction of gully side wall inclination. If the depth of incision D_v is more than $D_v = \dfrac{2.0 * C_h}{g * \rho_s} \cos(\varphi) \Big/ \sin^2 \dfrac{1}{2}\left(\varphi + \dfrac{\pi}{2}\right)$, then gully wall inclination ϕ can be calculated with the help of the formula

$$\dfrac{C_h}{g * \rho_s * D_v} = \dfrac{\rho_s - w * \rho}{\rho}\tan(\varphi) * \cos^2(\phi) - \dfrac{\sin(2\phi)}{2}.$$ Here w is volumetric water content in the soil and φ is the angle of internal friction.

When the bottom width, wall inclination and volume of incision V_0 are known, the shape of the gully cross-section can be transformed into a trapezium with bottom width W_b, depth

$$D_t = \left(\sqrt{W_b^2 + \dfrac{4V_0}{\tan(\phi)}} - W_b\right)\dfrac{\tan(\phi)}{2}$$ and top width $W_t = W_b + 2.0 D_t / [\tan(\phi)]$.

Algorithm of the dynamic model of gully erosion

Input to the model includes topographical, hydrological and lithological data. Topography is described by elevations and distances from the gully mouth at N points along the longitudinal profile of each flowline on the initial slope (including existing gullies). Water discharge change in time (hydrograph) has to be calculated for all these points with a hydrological model (which must be linked with the gully erosion model). Multi-layered soil properties are used in the model: for each layer an input is needed for the elevation of the base of the layer at the same N points, as well as soil density, cohesion, the angle of internal friction, the diameter of stable aggregates, water content, and the root content of vegetation.

Longitudinal profile transformation in space and time and gully bottom widening are calculated with the Lax-Wendroff predictor-corrector scheme. A stability criterion is determined for each calculation step. Stability of the numerical scheme stability is attained by changing the duration of the model's time step. At each timestep, the length of the flow width, depth, velocity and critical shear stress for the soils in the gully bottom and banks are calculated. After each flood event the rectangular bottom trench is transformed using the straight slope model to a trapezoidal shape, and longitudinal distributions of gully width (top and bottom), depth and bottom elevations are estimated.

Dynamic model verification

The model was used for prediction of gully erosion on the Yamal peninsula (Sidorchuk, in press), and of the Brook gully in the Yass River basin, NSW, Australia. Table 2 shows calculated and observed elevations of longitudinal profile of Brook gully. The erodibility coefficient k_e was determined from the data in Table 1, and equals $2.0 \cdot 10^{-3}$ for the loam soils in the upper section of the gully. The gully profile in 1932 was reconstructed from the plan of lot 64 in Parish Purrorumba, County of Murray and air photographs from 1941; elevations in 1992 were measured during field work. Runoff for the period 1932-92 was calculated from River Yass discharge data. Comparison of calculated and measured longitudinal profiles is satisfactory for the lower part of the incision into pebbly loams. At the upper part of this section the observed incision was larger than that predicted, due to seepage from the pond at the gully head. Numerical experiments show that the model is sensitive to change of the erodibility coefficient value. Field investigations and careful calibration of the model are necessary for accurate prediction of gully erosion

Table 2. Calculated and observed deformation of the longitudinal profile of Brook gully (River Yass basin, NSW, Australia)

Distance from the gully mouth along the gully bottom (m)	Observed elevations of longitudinal profile in 1932 (m from MSL)	Calculated elevations of longitudinal profile in 1988 (m from MSL)	Observed elevations of longitudinal profile in 1988 (m from MSL)	Soil texture
900.	700.6	700.6	700.6	schist
950.	701.4	700.9	701.2	pebbly loam
1000.	703.4	701.8	701.9	pebbly loam
1050.	705.5	702.4	702.5	pebbly loam
1100.	706.1	702.3	703.0	pebbly loam
1150.	706.4	702.3	703.3	pebbly loam
1200.	706.8	703.0	703.5	pebbly loam
1250.	707.3	704.2	704.2	pebbly loam
1300.	707.8	705.3	704.9	pebbly loam
1350.	708.2	705.5	705.1	pebbly loam
1400.	708.8	706.6	705.9	pebbly loam
1450.	709.4	707.9	706.9	pebbly loam
1500.	709.80	709.20	709.20	pebbly loam

Conclusion

The dynamic gully erosion model describes the first, rapid stage of gully development. During the snowmelt or rainstorm event the flowing water erodes a rectangular channel in the topsoil or at the gully bottom. Change of gully bottom elevations is controlled mainly by upward detachment of the particles from the bed and by sedimentation on the gully bottom. This process is described by the transport equation $\frac{\partial Z}{\partial t} - k_e UD \frac{\partial Z}{\partial x} - V_f C = 0$, which is numerically solved with the an explicit predictor-corrector scheme of Lax-Wendroff type. The vertical walls of this channel are unstable. In the period between adjacent water flow events, shallow landslides quickly transform gully cross-sectional shape to trapezoidal, with bottom width W_b, depth $D_t = \left(\sqrt{W_b^2 + \frac{4V_0}{\tan(\phi)}} - W_b \right) \frac{\tan(\phi)}{2}$, and top width $W_t = W_b + 2.0 D_t / [\tan(\phi)]$. Numerical experiments show that the model is sensitive to change of the coefficient k_e value. Field investigations and that careful calibration of the model are necessary for accurate prediction of gully erosion.

Acknowledgements

This paper was prepared for the NATO Advanced Research Workshop 'Global Change: Modelling Soil Erosion by Water' and is a contribution to the Soil Erosion Network of the GCTE, which is a Core Research Project of the IGBP. I am very grateful to John Boardman and David Favis-Mortlock for the invitation to this meeting and for financial support. This model was developed on the basis of field work in Australia and in the Yamal Peninsula (Russia). I am very much obliged to Bob Wasson and Peter Fogarty, who showed me the gullies of south-eastern Australia, and to Cathy Wilson, who organised the field experiments in Brook gully. The work on the Yamal gullies could not have been done without the helpful assistance of Andrey Alabyan, Boris Vlasov, Sergey Golovenko, Konstantin Voskresenskiy, and Viktor Grigor'yev.

References

Einstein, H.A. (1942). Formulas for the transportation of bed load. *Transactions of the American Society of Civil Engineers* **107**, 561-577.

Graham, O.P. (1988). *Land Degradation Survey of N.S.W*, Soil Conservation Service of N.S.W. Technical Report 7. 47 pp.

Kosov, B.F., Nikolskaya, I.I. and Zorina.Ye.F. (1978). Experimental research of gullies formation. In, Makkaveev, N.I. (ed.), *Experimental Geomorphology*, volume **3**, Moscow University Press, pp 113-140. (in Russian)

Mirtskhulava, Ts.Ye. (1988). *Principles of Physics and Mechanics of Channel Erosion.* Leningrad, Gidrometeoizdat, 303 pp. (in Russian)

Rossinskiy, K.I. and Debol'skiy, V.K. (1980). *River Sediments*. Moscow, Nauka, 216 pp. (in Russian)

Rozovskiy, I.L. (1957). *The Water Motion on the Alluvial Channel Bend*. Kiev, AN UkrSSR, 188 pp.(in Russian)

Sidorchuk, A.Yu. (1995). Erosion-sedimentation processes on the Russian Plain and the problem of aggradation in the small rivers. In, Chalov, R.S., (ed.), *Water Resources Management and Problems of Fluvial Science.* Moscow, Izd AVN, pp. 74-83 (in Russian).

Sidorchuk, A.Yu. (in press). Gully erosion and thermoerosion on the Yamal Peninsula. In, Slaymaker, O. (ed). *Geomorphic Hazards*, Wiley.

Vikulova L.I. (1972). The initial stage of artificial channel formation in sandy soil. In, Rossinskiy, K.I. (ed), *The Dynamic and Thermics of the River Flow*, Moscow, Nauka, pp 63-80 (in Russian).

35. ALTERNATIVE APPROACHES TO SOIL EROSION PREDICTION AND CONSERVATION USING EXPERT SYSTEMS AND NEURAL NETWORKS

Trevor M. Harris[1] and John Boardman[2]

[1]Department of Geology and Geography
West Virginia University
425 White Hall
Morgantown
West Virginia 26506-6300
USA

[2]School of Geography and Environmental Change Unit
Mansfield Road
University of Oxford
Oxford OX1 3TB
UK

Abstract

In response to on-site and off-site water-borne soil erosion impacts, considerable effort has been expended in recent decades in conceptualising and developing quantitative models for soil erosion prediction. Such models have focused on deductive process-based modelling, emphasising replication of the physical laws operational in soil erosion and seeking to encapsulate the conditions, processes, and practices deemed responsible for erosion events. This paper proposes an alternative, inductive, use of two variants of empirical models: those of expert systems and artificial neural networks. Though empirically based, these models offer significant advantages and potential for both soil erosion prediction and conservation. In these models the theoretical understanding of the soil erosion process is expressed in the sample erosion sites selected for rule generation and in the related site attributes. Importantly, the approach is empirical not one of empiricism.

The paper discusses the strengths and weaknesses associated with expert system and neural network models as an alternative paradigm to mathematical process-based erosion modelling. The paper details the results of an inductive rule-based soil erosion expert system which utilises a ten year database of erosion events for the South Downs in Sussex, England. The potential for 'chaining' or linking expert systems beyond a predictive mode toward a conservation component is demonstrated.

Introduction

Considerable effort has been expended in recent decades in conceptualising and developing quantitative models for use in soil erosion estimation and prediction (Foster, 1991). These models are in response to acknowledged on-site and off-site impacts arising from short-term, highly publicised, single event erosion, as well as from longer-term continuous erosion of arable land (Boardman, 1988, 1990, 1992; Evans, 1990a, 1990b; Morgan, 1986a; Skinner, 1986). Such models have sought to encapsulate the conditions, processes, and practices deemed responsible for erosion events. These factors comprise a complex blend of land use and crop cultivation; farm management practices; rainfall magnitude, frequency, and intensity; topographical influences; and soil characteristics. The predictive models vary in terms of their designed application, spatial scale, lumped or distributed form, temporal range, and mathematical construction. The Universal Soil Loss Equation (USLE) continues to be incorporated, either in whole or in part, within many models largely because of its low data requirements and its ease of application. However, even on experimental plots, predictions using USLE are of limited accuracy (Burwell and Kramer, 1983) and the few results from field-scale testing are poor (Govers et al., 1990). Much recent research and development has focused on deductive process-based modelling emphasising replication of the physical laws operational in soil erosion. The chief disadvantage of these models is the quantity of data required to run them and their extreme sensitivity to some input values (Favis-Mortlock, 1994). The use of CREAMS, EPIC and GLEAMS on test sites in southern England have been reported (Morgan 1986b; Favis-Mortlock, 1998).

Somewhat in contrast to the current emphasis on process-based erosion models, this paper focuses on an alternative, inductive, use of two variants of empirical models: those of expert systems and artificial neural networks. Though empirically-based, these models offer significant advantages and potential for both erosion prediction and conservation. In their construction, and especially in their dependence on detailed field-survey generated data, they represent substantive understanding and representation of the processes which influence soil erosion. It is in the selection of the sample erosion sites and related data attributes that the theoretical understanding of the soil erosion process is expressed. In this respect the approach is empirical but not one of empiricism.

This paper extends the results and ideas discussed in a previous paper linking expert systems to soil erosion prediction (Harris and Boardman, 1990). The paper discusses the strengths, weaknesses, and potential of both expert systems and neural networks as an alternative paradigm to mathematical process-modelling erosion prediction. The paper emphasises the results from an inductive rule-based expert system encompassing a ten year database of erosion events. The discussion is extended to include the potential role of neural networks. The paper also considers the application of expert systems beyond a predictive mode and toward a chaining or linking with a powerful conservation component.

Expert systems and neural networks

Inductive rule-based expert systems and neural networks, as used in this research, produce predictive outcomes from a series of rules generated from a set of field-captured or 'trained' soil erosion examples. The models are empirical and deterministic in the sense that the rules and the generated outcomes are dependent upon the specific examples input to the system. Strictly speaking, stochasticity is restricted to interpolation between a limited range of end values, though in essence considerable variability is embedded in the model in terms of how well the examples in the database capture the full range of soil erosion events and the underlying physical and socio-economic processes. The examples form the core of a knowledge-base about erosion events in a particular area or domain. Production rules are generated from this knowledge-base and it is these which provide the system's predictive capability. The examples were taken from a database of several hundred soil erosion events which had been field-surveyed and recorded over a ten year period for an area around Brighton in East Sussex, United Kingdom (Boardman, 1988; 1990). It is worth noting at this point, however, that the examples could have been generated 'hypothetically', or the rules created from the theoretical and experiential knowledge of a soil erosion expert codified into a series of 'if-then' rules (Clancey, 1983). Alternatively, the system rule-base could be a combination of all the above. In such instances these empirical models possess capacity for extension into process-based modelling via the specification, representation, and incorporation of physical laws governing soil erosion processes and events. This latter approach has been tentatively explored by the authors and the initial results suggest the approach is certainly worthy of further attention. Calibrating the data inputs and tracking the sensitivity of the system toward the predicted outcomes is not, however, an easy task.

Empirical rule-based systems such as expert systems and neural networks possess very powerful predictive capabilities, particularly when applied in domain-specific applications (Ignizio, 1991). By domain-specific is meant that the example database from which the rules are generated is limited in extent. This domain limitation may apply to a specific geographical region or to locations which possess underlying commonalties, similar characteristics, or a degree of homogeneity in the erosion experienced. The data used in this research was collected from the South Downs in East Sussex and reflect conditions and processes present in that region. Given the importance of soil type and geomorphologic characteristics, this system could be extended to encompass other chalk dominant sites. This assumes, however, that other characteristics, such as rainfall characteristics or crop types, are somewhat similar between the locations or that the variability has been captured by the examples in the knowledge-base. Attempts to utilise the predictive capability of this system elsewhere, for example on loamy or sandy soils or for areas engaged in very different farm management practices, could result in considerable error generation because the examples and resultant rules poorly capture the differing conditions at the 'new' location. The extent to which the system developed here is valid only for chalk soil types or is 'portable' and extendible to incorporate examples from other locations has yet to be determined.

Importantly, it is assumed that there exists some underlying pattern or structure within the data and that the data is non-random. If the knowledge-base exhibits substantial randomness then the system, as with any other model, has little opportunity to extract a meaningful rule-base. One very significant advantage of these systems is that they are not solely dependent upon numerical notation or form, but can include symbolic representations as well. The use of symbolism permits considerable latitude when representing otherwise very complex phenomena or for dealing with imprecision in the data through the use of additional 'dummy' variables. Thus categorical morphological characteristics of erosion types such as valley bottom or valley side, or the convexity of a site, or drill direction, or crop-type can be incorporated in the original symbolic form.

In essence, these systems seek to uncover logical relationships between the examples contained in the soil erosion database which are then expressed as a series of rules rather than as a series of mathematical equations. The systems are forward chaining, or data-driven, in that they progress from conditions which are known to be true, to conclusions which the examples allow us to establish (Jackson, 1986). At the heart of the expert system is an inference engine which drives it and acts as the interface between user and rule tree. The inference engine induces or constructs a series of rules from the examples. The induction is based on Quinlan's ID3 or Interactive Dichotomizer 3, which sorts and divides the examples based on the attributes which best discriminate between the examples (Quinlan, 1979). These rules comprise a premise, consisting of decision points in the rule tree based upon example attributes, and a conclusion which represents the recorded soil loss. The rules are traditionally displayed in the form of a rule or decision tree, and formalise the empirical associations or patterns which exist within the example knowledge-base.

Neural networks, as their name suggests, draw their structure from biological models of the brain (Dayhoff, 1990). Artificial Neural Networks represent layers of nodes (neurons) in which the varied interconnections between input layer, output layer and intermediate layers, is controlled by a weighted value (Lippmann, 1987). Each node has a threshold value which establishes the 'stimulation' level of that node. A feed-forward neural network commonly has nodes organised into several layers within which every node in one layer is connected to every node in the preceding and succeeding layer. Importantly, interconnections between nodes of the same layer differ from interconnections between nodes of differing layers. In the former, the interconnections have the effect of suppressing the stimulation of nodes to which they are attached to by emitting negative or 'inhibitory' values. In contrast, connections between layers are positive values, or 'excitatory'. Lateral inhibition occurs between nodes in the same layer as they compete with each other to stimulate nodes in the next layer. Only nodes with the strongest stimuli from the layer below will overcome the sum of suppression signals received and be able to issue excitatory signals to the next layer. Multilayered networks were less favoured because of the difficulty of training, estimating the error, and adjusting the node interconnections which exist between the hidden layers deep in the network. A back propagation algorithm (Rumelhart et al., 1986) is most commonly used in feed-forward neural networks to incrementally reduce the error between actual and desired outputs during iterative presentations of a training set.

Neural networks, like expert systems, rely upon domain specific knowledge contained within the input layer. Production rules are propagated through the network layers before outcomes are produced in the final output layer. During propagation each node in the hidden layer(s) calculates a weighted sum of all its inputs and applies the result of that sum to an activation function which produces the node's output. The neural network uses a learning algorithm to change the values of the weighted node connections in order to produce an approximation to the correct output for each example that is input. These weighted adjustments, created during the preliminary learning or training phase, affect the sum calculated for each node and ultimately the node's output or conclusion. Thus, the relationship between input node and output node is specified without ever actually knowing the explicit nature of the relationship itself.

The training set is very important in producing values which correspond to a desired output. Training occurs until a pre-determined error threshold is achieved. The network is trained by applying a series of input patterns and comparing the output pattern to a desired training target pattern. The threshold value is determined by means of a transform function which intercedes between the input and output stimuli. The weights and threshold patterns are adjusted until the network gives the exact response required by the training output pattern. Depending on the level of stimulation from its neighbours, a neural net will either begin contributing output stimuli of its own or will remain inactive. At this point the network is ready to process the real data. Relationships developed within the neural network could in turn be used to construct an inductive expert system as outlined above (Ignizio, 1991).

The examples database

The examples used in this study represent a continuous ten year monitoring study of erosion events in the South Downs of southern England undertaken by John Boardman between 1982 and 1991 (Boardman, 1990). Prior to the Second World War the South Downs was predominantly under permanent grass. During and after the war much of the Downs was ploughed up and large areas brought under cereal cultivation. In the late 1970s there was a movement from the growing of spring to autumn-planted wheat and barley. 'Winter cereals' quickly came to dominate the landscape with cultivation on slopes up to 23 degrees and 60 per cent of some catchments brought under these crops. In the 1970s there were few cases of erosion, even in the wettest month on record, October 1976. When flooding occurred it was generally clean water from grassed surfaces. In the 1980s the South Downs experienced a significant increase in erosion as a result of the change to winter cereals and continued intensification of farming methods (Boardman and Robinson 1985; Boardman 1990).

Between 1982 and 1991 an area of about 36 km^2 of agricultural land in the eastern South Downs was monitored for erosion events. Attribute information collected for each of these events was extensive. For each erosion event a variety of climatic, morphometric, land, soil type, and crop type information was recorded from field survey or published topographic maps along with measured soil loss (Table 1) (Boardman, 1990, 1992). Soil loss was estimated from the volumetric measurement of rills, gullies, and areas of deposition. This collection of 450 recorded soil erosion events comprises a unique database encapsulating specific erosion event information in the South Downs.

Table 1. Attributes recorded for erosion events

a) **Rainfall index**: characterises the rainfall responsible for the measured soil loss and is obtained from the rainfall records of nearby meteorological stations (Boardman and Favis-Mortlock, 1993). Rainfall events of 30mm in two days are allotted an index of one; 30mm in one day a value of two; 60mm in two days a value of three; 60mm in one day a value of four.
b) **Catchment area**: the area contributing runoff to the eroded site, in hectares. Partitioned into 18 classes.
c) **Length of slope**: contributing runoff, in metres. Classified into 15 classes.
d) **Type of erosion**: valley bottom, valley side, or a combination of both (see Evans and Cook, 1986).
e) **Maximum angle**: in the catchment, in degrees. Classified into four classes
f) **Relief of catchment**: the difference in height between the highest and lowest points, in metres. Nine classes.
g) **Morphology**: the presence or absence of convexity or crestal area: none, convex, both (see Evans, 1980).
h) **Direction of drilling**: down or across the direction of maximum slope.
i) **Crop type:** winter cereal, spring cereal, under plough, rape, ley, linseed.
j) **Soil type**: chalky, sandy, or stony.
k) **Whether rolled after drilling**: yes, or no.
l) **Measured soil loss**: based on a seven-fold classification, in m^3ha^{-1} (see Table 2)

Soil losses recorded for each erosion site were placed into one of seven classes representing a continuum from low to high erosion rates (Table 2). These classes were derived from experience obtained from field work over many years of the monitoring study. Aggregations into coarser, but still very valuable categories in the British context, of 'low' (0.0-1.99 m^3/ha^{-1}), 'moderate' (2.0-9.99 m^3/ha^{-1}) and 'high' (>10.00 m^3/ha^{-1}) soil loss easily produced an improvement in the results. 'Low' rates of erosion as defined above would be of little concern either in an on-farm or off-farm context, whereas 'high' rates would certainly be a cause for concern and would require ameliorative action. However, the predictive accuracy of the system was evaluated based on the sevenfold classification. Accuracy to within one or two classes in this fine classification was deemed to be satisfactory and certainly comparable to the coarser categorisation of soil loss magnitude. Field and experimental plot evidence suggest that the rainfall threshold at which rilling is initiated is around 30 mm in 2 days: in the period 1 September to 30 November of the years 1982-91 there were 22 such events (Boardman and Favis-Mortlock, 1993). This observation led to the definition of a 'Rainfall Index' (RI) to characterise the erosion-producing rainfall of a particular growing season (Table 1). The RI for each year shows a close relationship with total and median soil loss in the monitored area (Boardman and Favis-Mortlock, 1993).

Table 2. Soil loss classification

Erosion rate (m^3ha^{-1})	N occurrences in database	%
0.00 - 0.99	166	36.9
1.00 - 1.99	81	18.0
2.00 - 4.99	88	19.5
5.00 - 9.99	48	10.7
10.00 - 19.99	32	7.1
20.00 - 49.99	21	4.7
50.00 +	14	3.1

Generation of the rule-tree was undertaken within the induction section of the Crystal expert system shell, and the rules subsequently imported into the main body of Crystal (Drenth and Morris, 1992; Wallsgrove, 1988). The Crystal expert system shell provides an induction shell with significant advantages in terms of import-export facilities and considerably greater speed and efficiency over previously used induction shells such as Expert Ease.

Predicting soil erosion from examples

Previous published results (Harris and Boardman, 1990) were based upon a six year monitoring of erosion sites from the South Downs and bear comparison with the results obtained for the ten year span. The former comprised 334 erosion events recorded between 1982-87. Using a cross-validation approach 1.5 per cent and 5 per cent samples (5 and 17 examples respectively) were iteratively extracted from the knowledge-base and the rule-tree recompiled each time from the remaining examples. In this way it was possible to query the resultant rules using the specific details of the extracted erosion cases and compare the known soil loss of the extracted cases against the predicted results of the system as a means of estimating system performance and accuracy. Based on a seven-fold soil loss classification and 15 runs of the 1.5 per cent resample, the system predicted soil loss to the exact category in 43 per cent of all instances. Prediction to within ± one class of actual soil loss achieved 76 per cent accuracy, and to within ± two classes 91 per cent accuracy was attained. Based on a threefold classification of low, medium, and high erosion, with divisions at 2 m³ha⁻¹ and 10 m³ha⁻¹, almost two-thirds of all predictions fell within the correct soil-loss category. The 'optimum' size of the knowledge-base at this time was not known, though the effect of reducing the number of example cases, for example from five extracted sample cases to 17 cases, did lessen the predictive accuracy of the system. In these cases it was noted that a relatively small change in the size of the knowledge-base had a significant effect on the levels of accuracy obtained (prediction of 'exact' soil loss accuracy fell from 43 per cent to 32 per cent). Based on this finding it was optimistically expected that enlarging the knowledge-base still further to a ten year study and 450 sites would improve system accuracy markedly.

A number of changes were made to the data by the authors in the time between the 1990 study and the current one. In the former the attributes for each example were left unclassified. A number of data transformations and classifications were subsequently undertaken on the original 334 examples to see if the 'noise' which arose from the incorporation of specific attribute values could be lessened and the prediction rates improved. Based on a revised attribute classification as detailed in Table 1, the correction of transcription errors, and the refinement of the Rainfall Index for individual site records, a small overall improvement was achieved using the original 334 examples. However, the addition of another 116 samples for the years 1988-1991 produced somewhat disappointing results in that they only maintained the level of accuracy already achieved.

The examples collected between 1988 and 1991 completed a ten year span of erosion monitoring. The quality of data collected was undertaken to the same standards as for the earlier years though soil type information was considered to be somewhat crude. These latter examples tended to reflect relatively low rainfall index values and attention became focused on aspects of rainfall intensity and the impact of cumulative rainfall on erosion events. In this respect these factors indicated a weakness of the RI; a failing that was not apparent in the earlier data analysis. Furthermore, between 1989 and 1990 the trend of putting land under plough in this region became more pronounced. Thus, experience with the data and the erosion events from the latter years suggested a possible earlier oversimplification of the natural system and the climatic regime. The selection and collection of data is clearly not a static affair, but is dynamic and priorities change through time. Nonetheless, despite these perceived problems, the database represents an excellent resource for soil erosion prediction and conservation.

By increasing the number of erosion examples in the knowledge-base it was fully expected that an improvement in the overall accuracy of the predicted results would occur. In reality the predictive accuracy remained comparable, if not slightly less accurate, than the results produced in the more limited seven year study. This was disappointing and somewhat confusing given earlier experiences. Using a similar testing approach to that employed in the 1990 study, the ten year database achieved 'exact' predictions (using the detailed seven-fold soil loss classification) in 40 per cent of cases; attained 67 per cent accuracy to within ± one class; 90 per cent accuracy to within ± two classes; and 98 per cent accuracy to within ± 3 classes. Using a three-fold soil loss aggregation of low, medium and high, then a 65 per cent predictive accuracy was achieved. Even though the expected improvement in accuracy was not forthcoming, these figures are still very encouraging from a soil erosion estimation perspective.

Subsequent examination of the data and results suggests two possibilities: that the addition of the extra examples introduced greater noise or variability into the system or that the system had already achieved maximum predictive capability. In the latter case the system could be improved through the addition of more attribute data, or a refinement of the attribute metric to reflect subtleties in the physical or cultural environment. Importantly, and contrary to previous expectations, the addition of more examples did not achieve an improvement in system performance. In an attempt to reduce the amount of noise in the system, several variables were subsequently reclassified. The absolute values for catchment area, slope length, maximum angle, and relief, were classified into a limited series of categories (Table 1). Rather than reduce noise in the system, as theory would suggest, reclassification in this instance produced an overall decrease of a few percentage points in the system's predictive capacity. A second approach, to reclassify the attributes as above and to categorise soil loss outcome to a three-fold classification, produced only marginally better results and achieved an 'exact' prediction in 51 per cent of instances — still a real decrease in performance from the earlier use of absolute figures aggregated into a three-fold soil loss classification. The end result of these experiments was that aggregating example attribute values to dampen noise in the system was not successful and that the best results were achieved with absolute values, albeit these were slightly less accurate for the ten year study than for the seven-year study.

Although the expected improvement in system performance did not arise from the enlarged example knowledge-base, the results are still sufficiently impressive to suggest the approach continues to have merit. A number of refinements are proposed to improve system capacity. As indicated earlier, the impact of rainfall intensity, rainfall accumulation, and rainfall timing are considered important and yet have not been codified within the example database. The use of binary variables to reflect these aspects of the climatic regime could be expected to improve the system's ability to discern erosion outcomes. Similarly, account needs to be taken of the sensitivity of the system to the examples input. A focus on examples that were poorly predicted by the expert system will enable consideration and refinement as to why the system response lacked precision. The difficulty in undertaking sensitivity analysis within the expert system, and in calibrating the input (predicate) data with the conclusion outcome layer, suggests that the neural network approach may provide a more favourable approach because of the ability to train the layers to achieve a desired conclusion or output.

Chaining expert systems: moving from prediction to conservation

One important capability provided by this expert system approach is the ability to link or chain the predictive system to a conservation-oriented component. The iterative nature of the system permits a number of simulations to be run from which high-risk erosion sites can be identified. These are subsequently tracked through a conservation component which prioritises factors for change in order to achieve a maximum reduction in risk status. Two primary parts of the soil loss problem were identified as being crucial in driving the conservation process. One part was that of economics, and the other was that of the topographic and land management component of the erosion process itself.

In the former, economics clearly drives land use in this region for if market conditions and financial returns were not important then much of this area would have remained under permanent grass and erosion would be minimal. Conservation proposals would likely prioritise changing from winter cereal to spring cereal production, for example; in order to decrease soil loss. However, yields would also be reduced and profit margins impacted. Thus, in terms of soil loss mitigation, conservation measures involving land use or crop type changes have obvious pecuniary implications.

An alternative is to focus on the physical properties of site or farm management practices which contribute to the magnitude of soil loss. Thus sites with slope lengths greater than 200m, for example, may be reduced into smaller sections through the construction, or reintroduction, of hedge boundaries between crest and slope to provide break of slope and reduce site catchment area. The optimal placement of these break of slope features would also take into account slope morphology and maximum angle of slope. Alternatively, or in addition, farm management practices such as drill direction or use of rolling would also be important factors to consider. What arises are a series of mitigation measures which can be implemented as conservation rules dependent on a prioritisation of the most important contributors to site erosion. The construction and implementation of a rule base is ideally suited to an expert system. Simulations of site erosion responses can be run in order to identify those conservation measures which best reduce soil loss.

In this study a number of possible responses, or rules, were identified and tested on a series of erosion examples. For one site (site 8617) which recorded a soil loss of between 10 and 20 m^3ha^{-1} arising from a relatively low rainfall index of five, was changed from 'rolled' to 'not rolled' and soil loss prediction fell to below one m^3/ha. However, if the same site were also changed from winter cereal to plough then the high soil loss rate remained unchanged. In another example (site 8727), with a high soil loss of between 20 - 40 m^3ha^{-1} and a rainfall index of 14, changing to non-rolled produced minimal improvements. However, reducing slope length to half reduced soil loss to less than 10 m^3ha^{-1}. In yet another example (site 8787), changing drill direction to across slope and not rolling the field reduced soil loss from 20 - 40 m^3ha^{-1} to less than one m^3ha^{-1}. Furthermore, the ability to divide a slope at different locations in order to minimise soil loss proved particularly useful. Thus the slope for site 8941 was initially divided at mid-slope and subsequent simulations indicated a reduction in soil loss for the two sections from 10 and 20 m^3ha^{-1} to less than 1 m^3ha^{-1} and 2 - 5 m^3ha^{-1} respectively. A division based upon slope morphology at a point one-third down the slope reduced erosion from both sections to less than one m^3ha^{-1}. Similar tests were undertaken on other case records and a series of conservation rules were developed. The resultant conservation rule-tree was sequenced in the following priority order of: crop type, drill direction, whether rolled, slope length, maximum angle of slope, and morphology. Dependent on each case response to these attributes the rule-tree proposed a suitable conservation response which was then simulated through the predictive expert system in order to assess the relative success of the proposed measure.

Despite success in these simulations, the chaining of prediction and conservation modules remains a paper exercise which has yet to be tested rigorously in the field. Furthermore, the simulated predictions are still susceptible to the accuracy ratings discussed earlier in this paper. However, a comparison of the system against responses from an expert long accustomed to the area and its erosion characteristics, does provide considerable qualitative credence and support as to the suitability of the approach and to the likely benefits of the proposed mitigation measures.

Conclusion

Concern for the depletion of the nation's soil resource base and the long-term implications for sustainable arable farming has generated a need for a reliable predictive capacity to identify erosion rates. There is a need to address the off-farm costs of erosion by targeting responses based on the identification of specific at-risk sites. As climatic and land use changes occur and bring the potential of new rainfall patterns, irrigation, and new crops to southern Britain, then a system that can assess the impact of these changes on erosion rates is clearly desirable (Boardman *et al.*, 1990). Expert systems offer one valuable approach which provide a relatively accurate predictive capability though the quality of system extrapolations clearly depends upon how well existing examples capture possible future climatic and land use changes. Current results from this study indicate an accuracy rate that is comparable to any other currently available model for predicting erosion on the South Downs. The potential role of neural networks has also been identified. This study has indicated that the mere addition of more examples to the database does not imply an increase in accuracy. The focus must now be to 'fine-tune' the knowledge-base and calibrate the prediction rules. The ability of neural networks to train the data and system to achieve more accurate outcomes may provide a particularly powerful capability not provided by expert systems. In addition to identifying alternative empirical approaches to soil erosion prediction, the paper has also sought to indicate where these systems may be extended to a consideration of process-based modelling involving the specification and representation of physical laws governing soil erosion events, as well as to their links to conservation applications.

Specifically these approaches have indicated a number positive outcomes. The techniques permit the unknown to be predicted from the known. Importantly, they do so by maintaining the integrity of the actual field survey data. The knowledge-base used here is built on real world data and has demonstrated capabilities for use as a scientific tool, as well as for management purposes. Increasingly, as with the Exploratory Data Analysis techniques of Tukey (1976), there is a need to retain the integrity of the hard-won data during an analysis and not to subject it to a series of data reduction and state altering transformations. Within the expert system the functional relationships between the site examples are developed automatically and the soil loss outcomes are identified without the need for constructing extensive and complex mathematical relationships. This is a powerful capability particularly in situations when it is difficult to explicitly express the relationships between input and output vectors. The systems are also able to integrate a variety of disparate and ancillary data into the soil loss estimation process. The use of symbolic representation, and the lessened reliance on quantification, is one such advantage enabling the incorporation of a variety of site data and 'expert' perspective to be incorporated within the model. Oftentimes, the 'gut' response of an expert, based on assimilated knowledge and experience, is as valuable as any model. The systems are robust and nonparametric and make no assumptions regarding the statistical distribution of data. In the neural network in particular, the systems exhibit 'graceful degradation' in that inaccuracies in one part of the system arising from missing or 'noisy' data do not invalidate the remainder of the system because of its parallel nature. As with many models the development of models to generate complex sets of relationships are susceptible to claims of being a 'black-box' and of not providing feedback as to the nature of the relationships that exist between input data and output categories. This would appear applicable in the case of neural networks, where the functional relationships within the hidden layers are difficult to identify. However, within the expert system these relationships are made explicit in the form of the series of rules which are generated. The expert system approach, of all approaches to soil erosion modelling, probably the least 'opaque' or 'black-box' of all of them in identifying characteristics which contributed to the erosion rate. It does not, of course, provide explanation as to why a certain combination of attributes should result in a certain erosion rate.

The extension of the study to include the data results of a longer ten year study has indicated an inability of the expert system to improve on the predictive capability achieved with a smaller knowledge-base. In the earlier study it was proposed that extending the database by including more examples should continually improve system performance. This study has indicated that subsequent development must focus on reducing data noise in the example database and further refine the data attributes. A focus on outliers or clashes in the example database could also improve system capability. Further development of an expert system approach will need to focus on sensitivity analysis, system calibration, and data attribute refinement. In this respect artificial neural networks may provide a powerful complementary approach to that offered by an expert system.

One important aspect of the expert system or neural net approach is that the models are essentially scale independent. In most process-based distributed models the physical laws, processes, and process interactions that underpin the model differ in their importance and dominance at differing spatial scales (Dickinson *et al.*, 1990). The scaling up, or scaling down, of models for use at plot, field, catchment, or landscape scale thus represents significant problems, not just in the data requirements but in the modelling processes themselves. In the case of these empirical models the structure of the rule induction model remains intact while it is the data examples which are changed to represent the differing spatial foci.

Finally, the ability to link risk assessment with conservation and mitigation measures by chaining systems is a significant advance and contribution from this research. The expert system approach is well suited for development of this capability. The simulation capabilities of expert systems linked to an iterative conservation module could prove extremely valuable not only as a predictive technique but as a powerful management tool. With the need to redress the detrimental impacts of soil erosion the development of a soil erosion prediction and conservation expert system for use by the non-expert becomes an appealing possibility.

References

Boardman, J. (1988). Severe erosion on agricultural land in East Sussex, UK, October 1987. *Soil Technology* **1**, 333-348.

Boardman, J. (1990). Soil erosion on the South Downs: a review. In, Boardman, J., Foster, I. D. L. and Dearing, J. A. (eds), *Soil Erosion on Agricultural Land*, Wiley, Chichester. pp. 87-105.

Boardman, J. (1992). Agriculture and erosion in Britain. *Geography Review* **6**(1), 15-19.

Boardman, J., Evans, R., Favis-Mortlock, D. T. and Harris, T. M. (1990). Climate change and soil erosion on agricultural land in England and Wales. *Land Degradation and Rehabilitation* **2**(2), 95-106

Boardman, J. and Favis-Mortlock, D. T. (1993). Simple methods of characterizing erosive rainfall with reference to the South Downs, southern England. In, Wicherek, S. (ed.), *Farm Land Erosion: in Temperate Environment and Hills.* pp. 17-29.

Boardman J. and Robinson, D. A. (1985). Soil erosion, climatic vagary and agricultural change on the Downs around Lewes and Brighton, Autumn 1982. *Applied Geography* **5**, 243-58.

Burwell, R. E. and Kramer, L. A. (1983). Long term runoff and soil loss from conventional and conservation tillage of corn. *Journal of Soil and Water Conservation* **38**(3), 315-319.

Clancey, W. J. (1983). The epistomology of a rule-based expert system: a framework for explanation, *Artificial Intelligence* **20**, 215-251.

Dayhoff, J. E. (1990). *Neural Network Architectures: An Introduction.* Van Nostrand, New York.

Dickinson, W. T., Wall, G. J., Rudra, R. P. (1990). Model building for predicting and managing soil erosion and transport. In, Boardman, J., Foster, I. D. L. and Dearing, J. A. (eds), *Soil Erosion on Agricultural Land*, Wiley, Chichester. pp. 415-428.

Drenth, H. and Morris, A. (1992). Prototyping expert solutions: an evaluation of Crystal, Leonardo, GURU, and ART-IM. *Expert Systems* **9**(1), 35-45.

Evans, R. (1980). Characteristics of water-eroded fields in lowland England. In, De Boodt, M. and Gabriels, D. (eds), *Assessment of Erosion.* Wiley, Chichester. pp.77-87.

Evans, R. (1990a). Soil erosion: its impact on the English and Welsh landscape since woodland clearance. In, Boardman, J., Foster, I. D. L. and Dearing, J. A. (eds), *Soil Erosion on Agricultural Land,* Wiley, Chichester. pp. 232-254.

Evans, R. (1990b). Water erosion in British farmers' fields — some causes, impacts, predictions, *Progress in Physical Geography* **14**(2), 199-219.

Evans, R. and Cook, S. (1986). Soil erosion in Britain. *SEESOIL* **3**, 28-58.

Favis-Mortlock, D.T. (1994). *Use and Misuse of Soil Erosion Models in Southern England,* unpublished Ph.D. thesis, University of Brighton, England.

Favis-Mortlock, D.T. (1998). Evaluation of field-scale erosion models on the UK South Downs. In, Boardman, J. and Favis-Mortlock, D.T. (eds), *Modelling Soil Erosion by Water*, Springer-Verlag NATO-ASI Global Change Series, Heidelberg.

Favis-Mortlock, D.T. and Smith, R. F. (1990). A sensitivity analysis of EPIC. In, Sharpley, A. N. and Williams, J. R. (eds), *EPIC (Erosion/Productivity Impact Calculator). 1. Model Documentation*, USDA-ARS Technical Bulletin No. 1768, pp. 178-190.

Foster, G. R. (1991). Advances in wind and water erosion prediction. *Journal of Soil and Water Conservation* **46**(1), 27-29.

Govers, G., Everart, W., Poesen, J., Rauws, G., De Ploey, J. and Lautridou, J. P. (1990). A long flume study of the dynamic factors affecting the resistance of a loamy soil to concentrated flow erosion. *Earth Surface Processes and Landforms* **15**, 313-328.

Harris, T.M. and Boardman, J. (1990). A rule-based expert system approach to predicting waterborne soil erosion. In, Boardman, J., Foster, I.D.L., and Dearing, J.A. (eds), *Soil Erosion on Agricultural Land*, Wiley, Chichester. pp.402-412.

Ignizio, J. P. (1991). *Introduction to Expert Systems: the development and implementation of rule-based expert systems.* McGraw-Hill, New York.

Jackson, P. (1986). *Introduction to Expert Systems*. Addison-Wesley, Wokingham.

Lippmann, R. P. (1987). An introduction to computing with neural nets. *IEEE ASSP Magazine*, April.

McClelland, J. L. and Rumelhart, D. E. (1988). *Parallel Distributed Processing: Explorations in the Microstructure of Cognition*. MIT Press. Cambridge.

Morgan, R. P. C. (1986a) Soil erosion in Britain: the loss of a resource. *The Ecologist* **16**(1).

Morgan, R. P. C. (1986b). *Soil Erosion and Conservation*. Longman, London.

Quinlan, J. R. (1979). Discovering rules by induction from large collections of examples. In, Michie, D. (ed.), *Expert Systems in the Micro-Electronic Age*, Edinburgh University Press, pp. 168-201.

Rumelhart, D. E., Hinton, G. E. and Williams, R. J. (1986). Learning internal representation by error propagation. In, Rumelhart, D. E. and McClellend, J. L. (eds), *Parallel Distributed Processing: Explorations in the Microstructure of Cognition*. MIT Press.

Skinner, R. J. (1986). A survey of water erosion problems in England and Wales. *SEESOIL* **3**, 60-61.

Tukey, J. W. (1977). *Exploratory Data Analysis*. Addison-Wesley, Reading, Massachusetts.

Wallsgrove, R. (1988) Crystal. *Personal Computer World*, November, 172-175.

SECTION 6. MODEL APPLICATIONS: ACTUAL AND POTENTIAL

36. SOIL EROSION MODELLING IN HUNGARY

Dénes Lóczy, Ádám Kertész and Tamás Huszár
Geographical Research Institute
Hungarian Academy of Sciences
Budapest
P.O. Box 64
H-1388 Hungary

Abstract
This paper summarises the efforts in Hungary to apply two soil erosion models (USLE and EPIC). The models were applied for various purposes: evaluating erosion and deposition hazards, land capability prediction and landscape sensitivity. The results of modelling and field measurements and perspectives for future applications are also treated.

Introduction

In Hungary the concern for soil erosion grew after World War II and mapping started as early as the 1950's. A forerunner of this activity was Mattyasovszky (1953), who surveyed the western part of the country, Transdanubia, at 1:75,000 scale. Duck and Stefanovits constructed a 1:200,000 scale map of all the mountain and hill regions of Hungary (Duck, 1960). More recent soil erosion surveys in Hungary are often still based on a conventional estimation of soil profile truncation: how deep a layer of soil is missing compared to an intact profile of the same type of soil in the region (Stefanovits, 1964).

First applications of USLE for planning purposes in Hungary

The Universal Soil Loss Equation (USLE) was first used for soil conservation planning in the 1960s and 70s. In the Planning Office for Water Management (VIZITERV), Kamarás elaborated a system based on USLE to promote regional planning for planners (Erõdi *et al.*, 1965, 1974) and this was accepted as the standard approach to soil conservation. Soil erosion estimations from the equation gradually became part of erosion mapping (Máthé, 1974) and the prediction of erosion hazard was founded on meteorological and soil parameters. Recently, process monitoring on a catchment scale (Pinczés *et al.*, 1978; Góczán and Kertész, 1988), testing of soil conservation technologies in agriculture (Birkás and Szabó, 1992) and laboratory experiments with raindrop impact (Kerényi, 1991) have been undertaken.

More recent applications of USLE

1. Erosion hazard mapping in the Lake Balaton catchment

After the publication of Wischmeier and Smith's benchmark paper (1978) further attempts were made to apply the Universal Soil Loss Equation to the conditions of Hungary.

In the early 1980s, Dezsény (1981, 1982) mapped soil erosion hazard over the catchment of the largest lake in Central Europe, Balaton. The lake is largely surrounded by rolling hill landscapes used primarily as arable and pasture and secondarily as orchards and vineyards. There is a high pressure from tourism on the landscape. The major data sources were existing soil charts and land use maps (updated by field checking) and the missing information was generated by the USLE. The various parameters of the USLE were calculated as follows:

R: calculated after Wischmeier and Smith;

K: estimates based on the field survey of humus layer depth and $CaCO_3$ content;

LS: determined from 1 to 10,000 scale topographic maps using Wischmeier's nomograms;

C: average crop rotation for three years from the records of agricultural authorities at local councils;

P: during field surveys 3 categories identified: downslope cultivation; contour cultivation and terraced cultivation after drainage.

Two watersheds in the northern catchment of the lake with variable surface geology were selected for the study. One of them was subdivided into an upper section of steeper slopes and rendzina soils and a lower one partly with alluvial and loess fill and the other into three sections. Rates of soil loss were estimated separately for slope inclination and land use classes.

The USLE predicted soil losses of 3 to 10 tonnes per hectare for flat and 30-40 tonnes per hectare for steep (12-17 per cent slope) arable land and above 100 tonnes per hectare for orchards and vineyards of 12-17 per cent slopes, while in the case of above 25 per cent slope inclination values as high as 400 tonnes per hectare (Table 1).

Table 1. Soil loss on the Örvényes-Séd watershed by land use classes and slope categories based on estimations through USLE, tonnes per hectare (data from VIZITERV, processed by Dezsény, 1982)

Land use classes	Slope categories				
	0-5%	5-12%	12-17%	17-25%	>25%
Arable					
upper section	3.3	12.6	27.3	50.4	-
lower section	4.0	15.4	33.4	61.6	-
Pasture and meadow					
upper section	1.6				
lower section	2.5				
Orchard and vineyard					
upper section	17.8	68.3	148.2	273.2	409.8
lower section	15.0	57.5	124.8	230.0	345.0

The calculated rates of soil erosion served as a basis for identifying categories of soil erosion hazard, which were represented on a 1:100,000 scale map.

Since a sediment delivery ratio from the subcatchment has not been calculated, it is difficult to compare computed and observed data. Dezsény's work was based on mapped and file data and no analysis of daily data on runoff-generating events had been performed. This may explain the significant (*c*. tenfold) overestimation of soil losses.

2. A subsequent survey of the same subcatchment of Lake Balaton

A team of Hungarian and German researchers repeated Dezsény's computations, but with amended input parameters (Kertész, 1993; Kertész *et al.*,1995). They identified the following problems in defining the factors of the USLE:

a) Factor R can only be calculated from a long series of data. In the case of the Örvényes-Séd catchment, for instance, the next meteorological station lies 20-30 km away from the test area. The R factor value was first estimated from these data (R = 55). When other short-interval (here 2-year) rain-gauge records in the immediate neighbourhood were also considered, an R value of 34 was calculated (Table 2). Rainfall kinetic energy was calculated by the Wischmeier-Smith equation.

Table 2. Mean values of the R factor for various stations (after Henzler, unpublished)

Rainfall classes	≤5 mm	5-12.5 mm	>12.5 mm	total
Balatonakali (9 years of measurement)				
average number of events per year	73.7	12.4	5.1	91.2
precipitation (mm)	71.2	97.4	98.2	286.8
R factor	4.54	14.53	30.50	49.57
Balatonszemes (9 years of measurement)				
average number of events	81.8	10.9	4.7	97.3
precipitation (mm)	91.7	84.9	88.5	264.5
R factor	5.48	16.15	37.59	59.21
Visz (2 years of measurement)				
average number of events	60.6	11.0	9.0	80.5
precipitation (mm)	77.6	92.4	180.6	350.6
R factor	2.38	7.14	24.38	33.89

b) The soil erodibility factor (K) should be determined through rainfall simulation and crusting has to be allowed for. Nomograms are unsuitable to give values for all cases. Even the values deriving from experiments differ greatly. Table 3 shows K values for plots numbers 1 to 5 at the Csákvár Research Station. The difference between the original values (column 2 in Table 3) and those corrected for stoniness (percentage of stone cover on the surface) lies between 3 and 23 per cent. A further correction for humus content is also possible.

c) As far as the morphometric factors (L and S) are concerned, the application of a digital elevation model (DEM) enables the user to achieve very high precision. A comparative analysis was carried out by Richter and Kertész (unpublished), which proved the higher precision of DEMs. GIS application seems to be the optimal way for the prediction of soil erosion rates.

Table 3. Computation of K factor values (after Hassel)

plot no	A. modified according to stoniness			B. from rainfall simulation				
	K factor from soil texture	stone cover (%)	corrected K factor	average	K_{min}	K_{max}	range	95% confidence interval
1	0.35	10.8	0.27	0.15	0.07	0.26	0.19	0.05-0.24
2	0.19	7.1	0.17	0.46	0.24	0.57	0.33	0.37-0.55
3	0.24	7.3	0.21	0.23	0.12	0.41	0.29	0.14-0.33
4	0.14	2.7	0.13	0.19	0.04	0.33	0.29	0.10-0.29
5	0.27	2.7	0.26	0.36	0.21	0.54	0.33	0.26-0.45

The results of this repeated survey show considerable variation when compared to the first one (Table 4 — cf. Table 1).

Table 4. Soil losses on the Örvényes-Séd watershed as estimated by a repeated use of USLE (R factor = 36, Richter et al., unpublished)

land use	C value	area (ha)	average soil loss (t/ha)	total annual soil removal (t)
uncultivated (grass and shrubs)	0.03	180.65	1.78	322.20
meadows	0.05	201.05	1.11	222.73
gardens	0.20	8.89	5.11	45.46
arable	0.26	449.73	3.86	1734.58
vineyards	0.40	387.55	10.63	4119.76
total watershed		227.87	5.25	6444.73

3. A demonstration of USLE for urban planning and university education

Recently, increased attention has been paid to the off-site effects of soil erosion. A manuscript paper (Balogh, 1994) summarises the applicability of the USLE for soil erosion estimation and presents an example from which town planners can deduce information for their everyday work.

Some Hungarian towns (e.g. Pécs and Szekszárd) are severely affected by sedimentation. Agricultural areas (primarily vineyards) lie at higher elevations than built-up areas and if land amelioration was not properly implemented, particularly intensive rainfalls cause rapid soil removal and deposition on the streets, blocking sewage. To predict this hazard, USLE is applied and the resulting data are represented on maps.

Maps and nomograms developed by Hungarian soil scientists are used to apply the USLE to the conditions of Hungary (precipitation zoning, erosivity distribution curves, and slope length reduction curve).

In the studied area (the Mecsek Mountains footslopes, north of Pécs, southern Hungary) rendzina soils, equally liable to wetting and drying out, predominate. Their K factor is around 0.33. A detailed slope inclination map was constructed to extend the computed values over the entire area investigated. On gently sloping surfaces soil loss remains below 15 tonnes per hectare in a year, but over large areas in the immediate neighbourhood of settlements the calculated rate of soil erosion exceeds 150 tonnes per hectare annually. Land use in the model is represented by three crop rotation patterns and maize monoculture. The soil conservation applied does not seem to be satisfactory to prevent soil accumulation damage.

Balogh (1994) also raises the need of soil erosion simulation and monitoring for spoil heaps in this region of black coal mining.

This USLE application is also used in university education. With its help undergraduate students of physical geography can get a first impression of the means for the quantitative estimation of soil erosion.

The first attempts to apply EPIC in Hungary

1. A comparative land capability prediction

After the EPIC conferences held in Temple and Colombus (1988), physical geographers in Hungary also became interested in testing the validity of the EPIC model. In international co-operation a hill region in Germany (vineyards in the Mosel valley) and a lowland in southern Hungary (arable fields near Szeged) were selected to test how reliably EPIC (version 3657, dated 1987) is able to predict climatic parameters, water budget processes (infiltration, runoff and evapotranspiration) and soil loss (Richter and Mezősi, 1990; Mezősi and Richter, 1991).

For the German test area the simulation covered the intervals of eight and 30 years. The selected eight years (a relatively dry period) provided the average values for the 30-year simulation. In the case of Szeged averages of climatic parameters from a 6- and a 20-year measurement series were used. The building of the data base involved several problems. Computed wind erosion rates could not be checked properly as there had been no regular wind measurements in the area. For this reason, the authors had to refer to observations during dust storms (with highly variable values of dust deposition). Results of the simulation (2 tonnes per hectare annual soil loss) are realistic.

Crop pattern and agrotechnology were regarded as unchanged over the period of simulation. Maize monoculture is not uncommon in the Hungarian test area. However, in another investigation, several crop rotations were included in the simulation.

Several remarks refer to the modelling of climatic and hydrologic parameters (Table 5). Solar radiation proved to be overestimated by the model. Actual evapotranspiration (on maize fields) was also lower than that estimated either by Penman's or by Priestley and Taylor's formulas, the latter however giving a closer estimate. The model fails to predict potential evapotranspiration properly. The error of monthly mean temperatures is within 10 per cent. The same applies to precipitation with the remark that rainfall during the summer is slightly overestimated, while winter precipitation is underestimated.

Table 5. Comparison of measured (m) and computed (c) climatic parameters for the Szeged test area (after Richter and Mezősi, 1990)

	J	F	M	A	M	J	J	A	S	O	N	D	Year
Global radiation (Ly)													
m	121	183	332	459	608	623	632	503	431	285	114	100	4391
c	150	180	420	600	780	780	780	600	540	360	150	120	5610
Monthly mean temperature (°C)													
m (1970-85)	-2.2	-0.4	4.8	11.2	16.4	19.8	21.8	21.1	17.2	11.1	5.5	0.9	10.6
c	-2.3	-0.1	7.2	9.6	15.8	18.4	22.0	20.1	15.8	12.8	5.0	0.4	10.4
Monthly average precipitation (mm)													
m (1970-85)	22.6	22.2	31.1	43.2	58.3	79.4	48.5	69.3	28.2	37.7	33.5	32.5	491.5
c (ann.)	19.9	18.6	32.7	35.9	44.8	85.3	42.6	88.6	26.8	25.4	39.1	33.4	492.6
c (daily)	21.0	15.2	33.2	30.7	51.1	84.5	46.7	80.2	28.0	28.1	34.8	34.6	488
Actual evapotranspiration (mm)													
m (1951-70)	0	1	22	52	90	106	100	79	59	46	16	2	573
c (after Penman)	5	10	25	34	48	91	123	79	32	17	6	4	474
c (Priestley-Taylor)	13	16	29	36	59	136	137	80	29	15	11	11	574
Potential evapotranspiration (mm)*													
estimate 1	0	1	22	53	98	124	145	127	83	46	16	2	717
estimate 2	6	15	52	95	152	158	169	133	125	89	10	5	1009

* Estimate1 is by means of Varga-Haszonits' formula (Varga-Haszonits, 1977) which is used widely in Hungary; estimate 2 is from Priestley and Taylor's formula

Mezősi and Richter identify a series of parameters to which the model is most sensitive. For runoff they are temperature (the standard deviation values), slope angle and grain size composition of soils. The crucial parameters in the case of soil removal are the cover and management factor (C) in the USLE, the standard deviation of temperature and precipitation values and the SCS curve value. These observations underline the decisive role of climatic (and relief) components as opposed to lithological and soil parameters.

Richter and Mezősi (1990) describe the findings of a cost-benefit analysis taking into account cultivation, seed, pesticide and irrigation costs. The yield and profit analyses affected the two major crops of the area, maize and wheat. In the 6-year simulation for wheat error stays under 5 per cent. The results of the 20-year simulation strongly depend on the application of mineral fertilisers.

The highest profitability can be expected — as suggested by the modelling — from a maize-wheat rotation into which more labour-intensive crops (paprika and beans) are also included.

The paper also compares the outcomes of EPIC with those of CREAMS (Chemicals, Runoff and Erosion from Agricultural Management Systems), but restricted to the German test area. The authors find the EPIC model — even under extreme relief conditions — more reliable in the estimation of soil erosion.

2. EPIC modelling for mapping environmental sensitivity

Very recently, another application of the EPIC model (version 4160, dated 1995) began in Hungary within the framework of the Ph.D. research of Huszár (no publication prepared yet). It is aimed at the mapping of variation in environmental response to intensive land use and a presumed climatic change. The survey covers some subcatchments of Lake Balaton (allowing comparisons with the findings of previous surveys).

At the initial phase of work several problems of data acquisition occurred. Land privatisation led to land allotments, multiplying the number of owners in tracts formerly owned jointly by members of collective farms. Records on cultivation procedures are not available. The extremely varied physical conditions of the area particularly emphasise the difficulties of importing USDA terminology and classifications concerning soil types, pesticides, agricultural crops and the cultivating devices employed. Some climatological data (like wind velocity) are not readily available for the area.

Various climatic scenarios are planned to be employed, using the last 10 years of drought as data basis.

Conclusions

In spite of early attempts to apply the USLE system, soil erosion modelling in Hungary has not yet taken advantage of the opportunities offered by new computer models. Repeated attempts, however, to improve the input parameters of models and to compare the findings, produced data closer to observed values. This is of particular importance in areas where soil erosion endangers the quality of the environment.

At present, the uncertainty in agriculture does not favour progress in this direction, but in the longer term various methods should be applied in parallel in order to decide which can be best adopted to Hungarian conditions.

Acknowledgements

Soil erosion modelling in Hungary is being further developed with international co-operation. The authors express their gratitude to the German Science Foundation (DFG) and the Hungarian Science Foundation (OTKA, project no 7674) for their assistance. The association of a team of Hungarian researchers with the European Union Project Mediterranean Desertification and Land Use (MEDALUS) promises a new opportunity. A catchment on the Danube-Tisza Interfluve has been selected as a test area for the running of the MEDRUSH model. MEDALUS II was funded by the EC under its Environment Programme, contract no. EV5V 0128/0166. The support is gratefully acknowledged.

References

Balogh, J. (1994). *A talajpusztulás és számítása Pécs északi térségében* (Computing Soil Erosion North of Pécs). Manuscript report for contract work, Department of Physical Geography, Janus Pannonius University, Pécs. 35 pp.

Birkás, M. and Dezsény, Z. (1992). Stubble cover - moisture conservation - soil-protecting tillage. In, *Proceedings of the International Symposium INTERPRAEVENT 1992*, Bern. Tagungspublikation, Band 4. pp. 303-312.

Dezsény, Z. (1982). A Balaton részvízgyûjtõinek összahasonlító vizsgálata az erózió-veszélyeztetettség alapján (A comparative study of the subcatchments of Lake Balaton based on erosion hazard survey). *Agrokémia és Talajtan,* Budapest **31**(1-4), 405-425.

Duck, T. (1960). Eróziós területek térképezése és értékelése (Mapping and evaluating areas affected by erosion). *MTA X. Osztály Közleményei,* Budapest **18**. 431-442.

Erõdi, B., Horváth, V. and Kamarás, M. (1965). *Talajvédõ gazdálkodás hegy-és dombvidéken* (Soil Conservation Farming in Mountains and Hills). Mezõgazdasági Kiadó, Budapest.

Erõdi, B. *et al.* (1974). *Irányelvek a hegy-és dombvidéki területek üzemi meliorációs tervezéséhez* (Guidelines for the Planning of Amelioration in the Farms of Mountain and Hill Regions). Ministry for Agriculture and Nutrition (MÉM), Budapest.

Góczán, L. and Kertész, Á. (1988). Some results of soil erosion monitoring at a large-scale farming experimental station in Hungary. In, Imeson, A. and Sala, M. (eds), Geomorphic processes, environments with strong seasonal contrasts. Volume 1. Hillslope processes, *Catena Supplement* **12**, 175-184.

Kerényi, A. (1991). *Talajerózió. Térképezés, laboratóriumi és szabadföldi kísérletek* (Soil Erosion: Mapping, Laboratory and Field Experiments). Akadémiai Kiadó, Budapest 219 pp.

Kertész, Á. (1993). Application of GIS methods in soil erosion modelling. *Computing, Environment and Urban Systems* **17.** 233-238.

Kertész, Á., Márkus, B. and Richter, G. (1995). Assessment of soil erosion in a small watershed covered by loess. *GeoJournal* **36**(2/3), 285-288.

Máthé, F. (1974). Eróziós veszélyeztetettségi térképezés (Mapping erosion hazard). In, *Az MTA TAKI 25 éve* (25 years of MTA TAKI). Research Institute for Soil Science and Agrochemistry, Hungarian Academy of Sciences (MTA TAKI), Budapest. pp. 29-32.

Mattyasovszky, J. (1953). Észak-dunántúli talajok eróziós viszonyai (Erosion conditions on soils northern Transdanubia). *Agrokémia és Talajtan,* Budapest **3,** 333-340.

Mezõsi, G. and Richter, G. (1991). Az EPIC (Erosion-Productivity Impact Calculator) modell tesztelése (Testing the EPIC model). *Agrokémia és Talajtan,* Budapest. **40**(3-4), 461- 468.

Pinczés, Z., Kerényi, A. and Marton-Erdõs, K. (1978). *A talajtakaró pusztulása a bodrogkeresztúri félmedencében* (Soil Erosion in the Bodrogkeresztúr Half-Basin). Publications of Institute of Geography, Kossuth Lajos University, Debrecen. No 129. 26 pp. + 2 tables.

Richter, G. and Mezõsi, G. (1990). Bodenerosion, Winderosion und Bodenfruchtbarkeit. Eine quantitative Näherung mit EPIC Model. *Acta Geographica*, Szeged **28-30.** 67-81.

Stefanovits, P. (1964). *Talajpusztulás Magyarországon. Magyarázatok Magyarország eróziós térképéhez* (Soil loss in Hungary. Memoirs to the map of soil erosion in Hungary). OMMI, Budapest. 58 pp. + map.

Varga-Haszonits, Z. (1977). *Agrometeorológia* (Agrometeorology). Mezõgazdasági Kiadó, Budapest. 224 pp.

Wischmeier, W.H. and Smith, D.D. (1978). *Predicting Rainfall Erosion Losses — a Guide to Conservation Planning.* USDA Agricultural Handbook 537, US Government Printing Office, Washington, D.C. 58 pp.

37. DEFINITION AND MAPPING OF DESERTIFICATION UNITS IN MEDITERRANEAN AREAS UNDER RAINFED CEREALS

C. S. Kosmas and N. G. Danalatos

Agricultural University of Athens
Laboratory of Soils and Agricultural Chemistry
Iera Odos 75
Athens 11855
Greece

Abstract

A methodology is proposed for classifying and mapping desertification units of hilly soils developed in different parent materials, under Mediterranean conditions. The proposed methodology is based first on the actual erosion risk of the soils, and secondly on their potential for biomass production assuming cultivation with rainfed wheat, taking additionally into account the effect of rock fragments on soil water conservation and biomass production. Threshold values of the time to runoff initiation were defined under certain soil surface conditions and used for the establishment of criteria to classify the soils in particular erodibility classes. For the assessment of the biomass production of rainfed wheat, simple models and empirical relationships have been employed. The developed methodology was applied for the compilation of maps for proper land management and protection against desertification, using the available regional Geographic Information Systems (GIS). As a case study, a hilly area of central Greece vulnerable to desertification was selected and studied in detail.

Introduction

Land degradation and desertification through soil erosion comprises the major environmental hazard to the semi-arid areas of the Mediterranean and particularly to the marginal hilly lands of the region. Besides adverse climatic conditions and human action, geology and irregular terrain with steep slopes are the main factors responsible for erosion and desertification. Among the main geological formations occupying the largest part of the substratum of the cultivated hilly Mediterranean landscapes are shale-sandstones, conglomerates and marl.

Large areas of hilly lands in the Mediterranean region are typically cultivated with rainfed cereals. In a number of years, the prevailing weather conditions during the growing period of such crops may be so adverse that the soils remain bare, creating favourable conditions for overland flow and erosion. Any loss of volume from these marginal lands greatly reduces the potential for biomass production, ultimately leading to desertification. Desertification at present threatens only the shallow and severely eroded soils. Global change, however, may threaten the majority of them. The vulnerability of the main units of these soils need to be quantitatively evaluated and the respective soils should be classified into desertification response units. The objective of this work is to develop a methodology for classifying and mapping desertification response units in cultivated Mediterranean soils, such as those formed on shale-sandstones, conglomerate and marl deposits based on the actual erosion risk and the potential for biomass production of rainfed wheat.

Methodology for mapping desertification units

The present methodology is based on the assessment of the actual erosion risk of the soils and their potential for biomass production assuming cultivation with rainfed wheat (Kosmas *et al.*, 1995a).

Actual erosion risk

The assessment of soil erosion risk is based on the general principles and parameters defined by the Universal Soil Loss Equation (Wischmeier and Smith, 1965) such as soil erodibility, topography, management practices and vegetation cover, but in a modified form adapted to the conditions under study.

In order to establish criteria for classifying the soils of different mapping units in particular erodibility classes, threshold values in which runoff is initiated under certain soil surface conditions are defined. A modified Philip equation is used for the assessment of the infiltration parameters, i.e. conductivity and sorptivity, applying the following multiple regression analysis on experimental infiltration data:

$$I = d + S\, t^{-0.5} + K_{tr}\, t \qquad (1)$$

where I is the cumulative intake rate (cm min^{-1}), S is the sorptivity of the soil (cm min$^{-0.5}$) estimated as a function of the soil moisture content, K_{tr} is the transmission zone conductivity (cm min^{-1}) estimated as a function depending on the soil texture, t is the time (min), and d is added to the original Philip equation to assess the effect of the initial disturbances (surface fissures, crotovinas, etc.) (Danalatos, 1993).

Based on these parameters (S and K_{tr}) which are measured with ponded infiltration experiments under field conditions (ring infiltrometers), or they are available from soil survey reports, the time to incipient ponding and runoff initiation for different rainfall applications and initial moisture conditions is estimated from the measured conductivity and sorptivity-soil moisture content relations, as well as the maximum surface storage capacity of the particular piece of land, according to the formulas (Danalatos, 1993):

$$Tp = S^2 / [2 \, Vo \, (Vo - Ktr)]$$
$$Runoff = (Vo - Ksat)(T - Tp) - SSmax \tag{3}$$

where Tp is the time to incipient ponding (min), Ksat is the saturation conductivity (cm min^{-1}), Vo is the rainfall intensity (cm min^{-1}), T is the total duration of the rain (min), and SSmax is the maximum surface storage capacity of the land (cm). The latter is calculated from the slope angle of the land s (degrees), the management variables clod/furrow angle a (°) and the surface roughness RG of the land (cm) using the following formula (Driessen, 1986):

$$SSmax = 0.5 \cdot RG \cdot \frac{\sin 2(a-s)}{\sin(a)} \cdot \frac{\cotan(a+s) + \cotan(a-s)}{2\cos(a) \cdot \cos(s)} \tag{4}$$

Topography is defined in terms of slope grade, and slope length which is defined in terms of landscape position such as summit, shoulder, backslope, footslope, and plain.

Vegetation cover can be readily altered depending on the climatic conditions and the period of the year. Hilly areas under rainfed cereals, like the study areas, remain almost bare after sowing in November and probably during most of December (limited plant cover). During that time, the erosion risk is very high due to the high intensities and long duration of rainfalls occurring under Mediterranean conditions. In the rest of the year, plant cover depends on the amount of soil water available for plant growth and therefore on the prevailing weather conditions. The plant cover is then determined on the basis of biomass production as discussed below.

The actual erosion risk for each mapping unit is then calculated by aggregating the separately defined indices of soil erodibility, rain erosivity, topography and vegetation cover. Classes are defined for each erosion factor with different indices (Table 1). Rainfall erosivity can be considered the same for all mapping units in a certain area existing under the same climatic conditions. Finally, the actual erosion risk is defined as: AER = R S L V. Five classes of actual erosion risk were defined as presented in Table 2.

Table 1. Classes and rating of erosion factors

Factor	Value	Rating	Description
runoff threshold value (R, min)	0-2	4	very high
	2-4	3	high
	4-6	2	moderate
	6-8	1	low
	>8	0	none
slope grade (SG, %)	0-2	0	none
	2-6	1	slight
	6-12	2	moderate
	12-18	3	high
	>18	4	very high
landscape position (LP)	plain	0	none
	footslope	1	slight
	summit	2	moderate
	shoulder	3	high
	backslope	4	very high
vegetation cover (VC, %)	0-5	4	none
	6-30	3	slight
	31-60	2	moderate
	61-85	1	high
	>85	0	very high

Table 2. Classes of actual erosion risk

class	description	abbreviation
0	No erosion risk	NER
1	Low erosion risk	LER
2	Moderate erosion risk	MER
3	High erosion risk	HER
4	Very high erosion risk	VHER

Potential for biomass production

The assessment of reduction in biomass production is based on data obtained from studies on biomass production: a) in rainfall exclusion experiments (Kosmas *et al.*, 1996), b) along catenas on marls, shale-sandstones and conglomerates in successive years under different weather conditions (Kosmas *et al.*, 1993), and c) in experimental plots in which rock fragments were removed from the soil surface and compared with those in which rock fragments remained on the soil (Kosmas, unpublished data) using the empirical relation:

$$RBP = 0.97 + 0.54 \ln(ET_a/ET_m) + 0.035 \ln(RFC) \tag{5}$$

where RBP is the relative biomass production calculated from the actual biomass production measured, divided by its maximum value estimated for each landscape position and parent material under conditions of no water deficit, and RFC is the percentage of rock fragment cover on the soil surface. Actually, despite increasing runoff and erosion, cobbles exert a beneficial effect on soil moisture conservation under conditions of moderate water stress such as those prevailing in spring and early summer (Danalatos et al., 1995).

A simple water balance model is used to calculate the actual evapotranspiration rate (ETa) from the momentary soil moisture data, the prevailing rainfall and potential evapotranspiration rates (Penman), measured soil parameters such as the field capacity and permanent wilting point, and crop parameters (coefficient of crop cover, effective rooting depth, depletion fraction of the total soil water storage, and the actual crop calendar). The model is largely based on the broadly accepted FAO (1977) methodology. The potential evapotranspiration rate (ET_a) is calculated from daily values of maximum and minimum air temperature, sunshine duration, air humidity and wind speed, according to Penman (1948), modified by Frere (1979). The maximum evapotranspiration rate (ET_m) was determined by the potential evapotranspiration rate and the crop (leaf area) coefficient of wheat according to FAO (1977).

Three classes are defined for the assessment of biomass production as follows: **high productivity** (HBP) with total above ground biomass production ranging from 90 per cent to 100 per cent of its maximum value; **moderate productivity** (MBP) with biomass production ranging from 70 per cent to 89 per cent of the maximum value; **low productivity** (LBP) with biomass production ranging from 40 per cent to 69 per cent of the maximum (or lower).

Results and discussion

The study soils

The proposed methodology was tested in a hilly area about 100 km north of Athens. The area has a rolling topography with soils formed on Tertiary and Quaternary deposits of marl, conglomerates, and shale-sandstones. On the shale-sandstone hillsides, despite the existing high gradients, prolonged stability and the lack of calcium carbonate advanced the development of soils with an argillic horizon characterised by strong blocky structure and red colour. Accelerated erosion following the intensive cultivation in the last century resulted in the removal of the great part or all of the original soil and the formation of the present very shallow, very gravelly and stony soils on the upper and middle slopes to moderately deep, slightly gravelly to gravelly soils on the footslopes.

The soils formed on conglomerates are very stony, moderately fine-textured, calcareous and normally contain a red cambic horizon. They are more permeable and resistant to erosion despite the high gradients and therefore they are shallow to moderately deep in the upper slopes and shoulders, and deep in the footslopes. The soils on marl deposits are finer-textured, very calcareous, more compacted and less permeable. They lack gravel and stones and are normally more susceptible to erosion.

Rainfall simulation experiment

In order to calibrate the proposed methodology for estimating of the time to runoff initiation, a series of runoff experiments were conducted using a rainfall simulator. The study was conducted during the dry period (September) when the soils were unploughed and partially covered (0-85 per cent) with the remaining plant residue, and during the wet period (December) when the soils were bare and loose due to ploughing. The experimental plots were situated on catenas, along specific hillslope components (backslope) with certain slope grade (13.5 per cent). The measured runoff rates varied considerably among bare soils with different parent materials and confined to 42.5 per cent, 25.8 per cent and 32.5 per cent of the applied rain (intensity 175 mm h^{-1}, duration 11 min) for soils on marl, shale-sandstones and conglomerates, respectively, measured in September. The highest rate of sediment loss was measured on soils on marl (491.2 g m^{-2}) as compared with soils on shale-sandstone (21.1 g m^{-3}) and conglomerates (175.9 g m^{-2}).

The data obtained during the wet period demonstrated that the soils are the most vulnerable to erosion during that period. The time to runoff initiation was similar to the time measured at the end of the dry period for soils on marl and conglomerates but the runoff rates were higher even though the rainfall intensity was reduced by fifty per cent (Table 3). The soil erodibility during this period was similar due to ploughing (Table 3). However, the greatest sediment loss was measured on the soil on marl due to very short time to runoff initiation, pointing to the great importance of the time to runoff initiation for assessing the erodibility of hilly soils under rainfed cereals. The data obtained with the rainfall simulator showed a good agreement between measured and estimated time to runoff initiation using the above methodology (Figure 1).

Table 3. Selected erosion data measured in the wet period. Rainfall intensity = 87 mm h^{-1}

Site	Slope %	RF cover %	Ksat initiation (cm h^{-1})	Runoff (min)	Runoff rate (mm min^{-1})	Sediment rate (g m^{-2} mm^{-1})
marl	13.2	0	1.1	3.3	1.20	40.2
shale	13.5	10.8	3.5	10.3	0.63	40.3
conglomerate	13.8	9.5	9.4	4.3	0.76	33.8

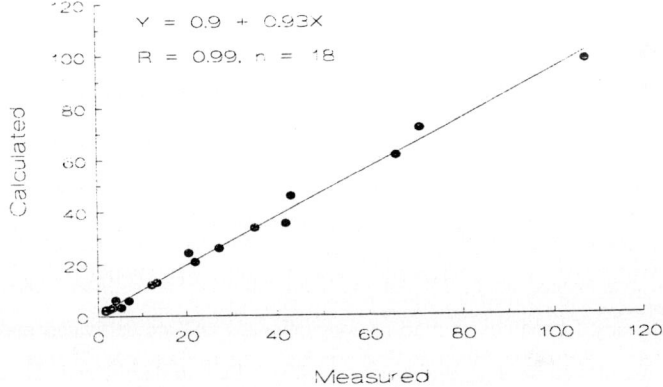

Figure 1. Measured and predicted values of runoff initiation

Evaluation of biomass production

Soil physical properties affecting water availability to the growing plants and total above ground biomass production of wheat were measured along sloping catenas on distinct landscape positions. A fair agreement between measured values of relative biomass production and those calculated with equation (5) can be seen in Figure 2, indicating that the above empirical equation satisfactorily describes (R=0.90) the relative biomass production of rainfed wheat under Mediterranean conditions. The above equation is valid only when soil water deficit occurs ($ET_a/ET_m<1$). In the case where there is no water deficit, rock fragments negatively affect biomass production due to the reduction in the rooting depth, less soil volume is available for adsorption of nutrients.

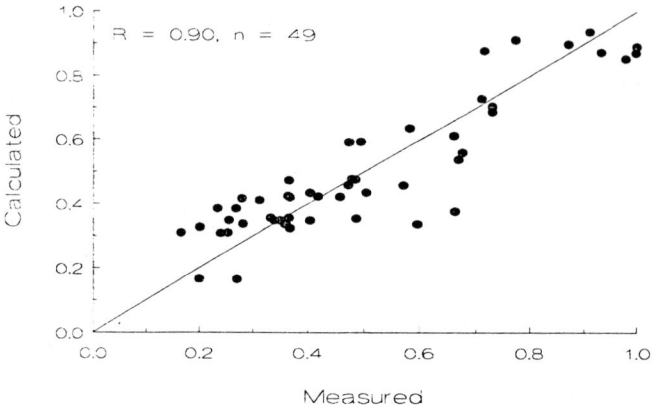

Figure 2. Measured and calculated values of relative biomass production of rainfed wheat

Application

The above methodology was applied as an example of mapping desertification units in the Nikea (Larissa) area located about 350 km north of Athens. The soils are derived from Tertiary to Quaternary formations of marl or conglomerates and are cultivated mainly with rainfed cereals. A soil map (scale 1:10,000) was used as a base to define land characteristics affecting erosion and biomass production of rainfed wheat. Figure 3 shows the defined desertification units for rainfed wheat. The following eight desertification units were defined in terms of actual erosion risk and biomass productivity for rainfed wheat:

Very high erosion risk/low biomass productivity (VHER-LBP). This desertification unit includes soils located mainly on backslopes and shoulders, moderately steep to steep. The actual biomass productivity for rainfed wheat is low due to the very shallow soils (effective rooting depth less than 28 cm). This unit is the most vulnerable to desertification.

Very high erosion risk/moderate biomass productivity (VHER-MBP). This unit has similar characteristics concerning the erosion risk as the previous one. The higher productivity is attributed to the greater effective rooting depth (30-45 cm), giving a higher production potential of rainfed wheat. This land is under very high erosion risk, and the rate of land degradation will be high under restricted plant cover.

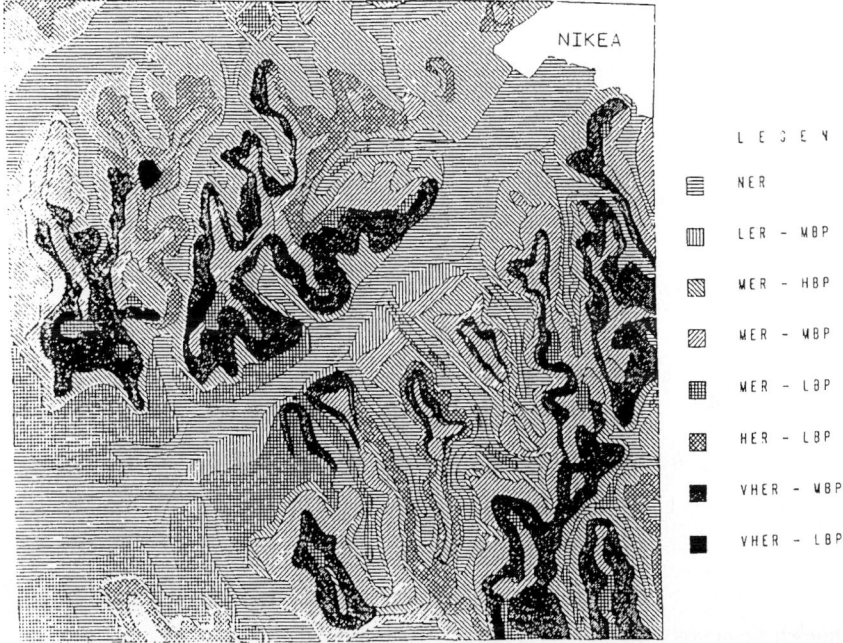

Figure 3. Map of desertification units of the Nikea area cultivated with rainfed wheat

High erosion risk/low biomass productivity (HER-LBP). This unit includes shallow soils (depth less than 38 cm), strongly sloping, located on shoulder or backslope positions. The potential for biomass production is low because of the presence of shallow soils and textures of relatively low available water holding capacity. This unit is very sensitive to erosion and the potential for biomass production is low under dry and hot climatic conditions.

Moderate erosion risk/low biomass productivity (MER-LBP). This unit includes moderately sloping soils located mainly at backslopes and footslopes. The soils are shallow to moderately deep (depth 40-75 cm) formed on marl deposits, free of rock fragments. This unit is very vulnerable to desertification, being unable to support any vegetation under hot and dry climatic conditions.

Moderate erosion risk/moderate biomass productivity (MER-MBP). This unit includes soils with the same characteristics concerning erosion as the previous unit, but the soils are deeper and contain rock fragments positively affecting the potential for biomass production under hot and dry climatic conditions.

Moderate erosion risk/high biomass productivity (MER-HBP). This unit includes soils located mainly on footslopes or on broad summits, gently to moderately sloping. The actual biomass productivity is high due to the presence of deep soils (depth greater than 60 cm). The desertification risk is low.

Low erosion risk/moderate biomass productivity (LER-MBP). This unit occurs on broad summits, footslopes or toeslopes, on gently sloping surfaces. The actual biomass productivity is moderate due to the presence of limiting subsurface layers.

No erosion risk/high biomass productivity (NER-HBP). This unit is confined to plains having no erosion risk and high potential for biomass productivity due to very deep soils, receiving additional runoff water from the surrounding sloping land.

Acknowledgements
This work was partially financed by the EU Research Project MEDALUS II (Mediterranean Desertification and Land Use), contract no EV5V-CT92-012B).

References
Danalatos, N.G. (1993). *Quantified Analysis of Selected Land Use Systems in the Larissa Region, Greece.* Unpublished Ph.D. Thesis, Agricultural University of Wageningen. 370 pp.
Danalatos, N.G., Kosmas, C.S., Moustakas, N.S. and Yassoglou, N. (1995). Rock fragments. II. Their impact on soil physical properties and biomass production under Mediterranean conditions. *Soil Use and Management* **11**, 121-126.
Driessen, P.M., (1986). The water balance of the soil. In, van Keulen, H. and Wolf, J. (eds), *Modelling of Agricultural Production: Weather, Soils and Crops.* Pudoc, Wageneingen, pp. 76-116.
FAO (1977). *Guidelines for Predicting Crop Water Requirements.* Irrigation and Drainage Paper 24, by Doorenbos, J. and Pruitt, W.O. Rome. 144 pp.
Frere, M. (1979). *A Method for the Practical Application of the Penman Formula for the Estimation of Potential Evapotranspiration and Evaporation from a Free Water Surface.* FAO, AGP: Ecol. /1979/1. Rome. 26 pp.

Kosmas, C., Danalatos, N., Moustakas, N., Tsatiris, B., Kallianou, Ch. and Yassoglou, N. (1993). The impacts of parent material and landscape position on drought and biomass production of wheat under semi-arid conditions. *Soil Technology* **6**, 337- 349.

Kosmas C., Danalatos, N., Moustakas, N., Mizara, A., and Yassoglou, N., (1995a). A methodology for mapping desertification units in areas cultivated with rainfed cereals. *Proceedings of Third International meeting on 'Red Mediterranean Soils'*, Chalkidiki, May 21-26, pp. 120-124.

Kosmas, C., Moustakas, N., Danalatos, N. and Yassoglou, N. (1996). The effect of diminishing soil moisture on soil properties and biomass production of wheat under semi-arid conditions. In, Thornes, J. and Brandt, J. (eds), *Mediterranean Desertification and Land Use*. Wiley, Chichester (in press).

Penman, H.L. (1948). Natural evaporation from open water, bare soils and grass. *Proceedings of the Royal Society Series A* **193**, 120-145.

Wischmeier, W.H. and Smith, D.D. (1965). *Predicting Rainfall Erosion Losses from Cropland East of the Rocky Mountains*. USDA Agricultural Handbook 282.

This area was selected for geomorphological and hydrological studies as representative of wide areas in the European Mediterranean middle mountains that were intensely farmed in the recent past and suffer now an extensive and progressive land abandonment. The instrumentation of the research basins started in 1989, in order to analyse the hydrological and sediment yield consequences of these land use changes as well as the dynamics of badland areas (Figure 1).

The aim of this paper is to provide an updated review of the present state of knowledge of the and the kind of data that could be used to test or validate erosion models. hydrological and erosional processes active in this area, as well as the main problems addressed, and the kind of data that could be used to test or validate erosion models.

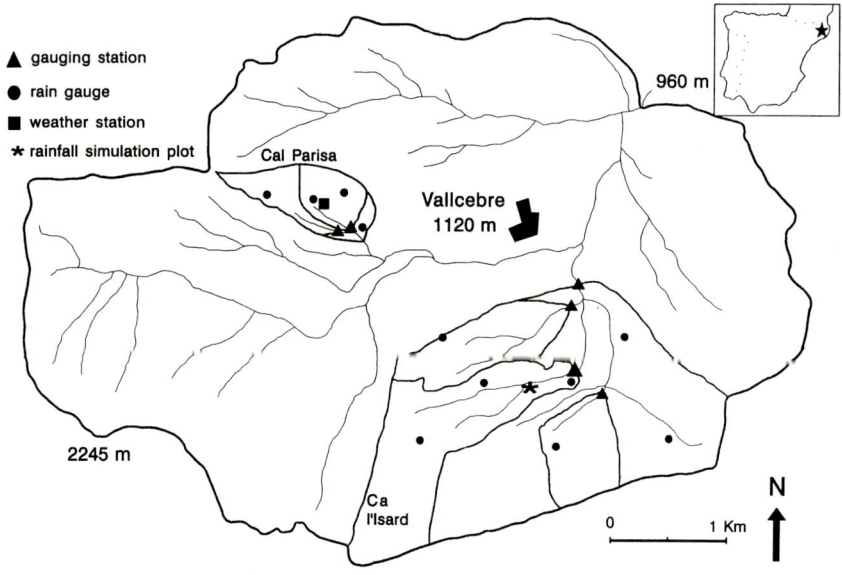

Figure 1. General design of the Vallcebre experimental catchments in the south-eastern Pyrenees (Catalonia, Spain)

Hydrological processes in the terraced areas (Cal Parisa basin)

The terraced old agricultural areas have silty-clayey soils with good structure and aggregate stability because of the high content of Ca^{+2} as well as organic matter. Terraces are typically 3-6 m wide and 4-30 m long flat or gently dipping areas, separated by steep banks 1-2 m high, usually sheltered by man-made stone walls. Infiltration rates are very high in the external part of the terraces, but decrease by one or two orders of magnitude to the internal parts, where deeper soil horizons or even the clayey bedrock outcrop.

The hydrological response of these areas, studied in the Cal Parisa catchment (Llorens and Gallart, 1992) is strongly dominated by antecedent conditions and saturation mechanisms. During the summer these terraces are able to absorb rainstorms of about 54 mm in 4 hours without producing any runoff at the outlet; whereas during the wet season, runoff is generated by the semi-permanent saturation of gentle hollows and the inner parts of some terraces. The terraced microtopography induces premature formation of saturated areas. This, as well as the network of ditches, is inferred to increase the peakedness of the hydrological response (Gallart *et al.*, 1994). Erosion rates in the terraced area are extremely low because the good grass cover and the good condition of the old soil conservation structures.

Hortonian overland flow and significant erosion occurs only in some patches devoid of vegetation in the steeper upslope or the internal part of some terraces. The more significant of these active areas are scars of debris flows that occurred during an extreme event in November 1982 (Gallart and Clotet, 1988). Nevertheless, these areas are disconnected from the drainage network, and runoff and sediment they produce do not reach the outlet of the small basin.

Table 1. Runoff and sediment response of badland regolith to rainfall simulation experiments performed at 50 mm h^{-1}

	runoff delay (sec)	mean runoff coef.	max runoff coeff.	mean sed. conc. (g/l)	max sed. conc. (g/l)	sed. disch. (kg/m² h)
Jul. 92	178	0.36	0.48	6.80	10.02	120.01
Oct. 92	30	0.88	0.98	15.04	20.73	653.19
Nov. 92	120	0.54	0.73	11.67	16.58	287.29
Dec. 92	788	0.19	0.26	17.45	23.35	156.67
Feb. 93	240	0.75	0.98	10.98	22.47	402.33
May. 93	519	0.24	0.32	23.29	31.00	270.52
Aug. 93	1470	0.24	0.42	9.49	14.13	66.93
Sep. 93	738	0.34	0.53	13.72	23.14	243.61
Oct. 93	203	0.67	0.96	16.29	25.13	564.39
Mar. 94	2408	0.06	0.07	14.59	15.88	37.77
Apr. 94	463	0.31	0.48	20.28	34.84	297.71
Jun. 94	658	0.29	0.37	24.47	31.60	332.00
Oct. 94	31	0.77	0.93	17.57	22.66	788.85

Typical suspended sediment concentrations in the runoff from these old agricultural areas is only of some tenths of mg l^{-1}. The sediment yield is therefore very low and the main source for it is the drainage network itself, because the low erosion rates on the terraces and the discontinuities of conveyance of sediment coming from small scattered sediment source patches (Llorens *et al.*, 1990).

Present research activity is focused on the study of the changes in the water balance induced by the spontaneous afforestation of these areas. Experimental data demonstrate the significant increase of water losses due to land cover change (Llorens *et al.*, 1995; Llorens *et al.*, in press), as well as the difficulties for modelling the observed high interception rates (Llorens, in press).

Weathering and erosional processes in the badland areas

Badland surfaces on smectite-rich mudrock are known to suffer erosion rates of about 9 mm per year, as a result of the combination of high physical weathering rates during winter and intense rainstorms during the late summer. It is suggested that erosion rates were limited by weathering rates (Clotet *et al.*, 1988).

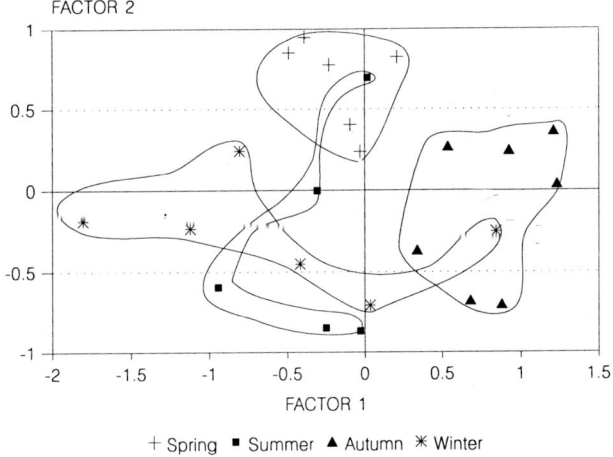

Figure 2. Behaviour of the badland regolith under rainfall simulation experiments

Subsequent experimental research confirmed the important role of winter weathering due to the disruptive role of freezing, and enhanced the seasonal evolution of hydrological and erosional behaviour of the regolith (Regüés *et al.*, 1995). Nevertheless, not all the erodible regolith is removed every year, but there is weathering potential in excess and some interaction occurs between weathering and erosion rates. Erosion seems, therefore, to be mainly limited by the available rainfall energy, the removal rate of the weathered regolith being a control on the weathering rate (Regüés, 1995).

Rainfall simulation experiments performed over three years showed that the infiltration behaviour of the regolith changes with the seasons more than its susceptibility to erosion (Table 1). During spring and autumn the behaviour of the regolith is relatively uniform whereas it shows wide changes during summer and winter, because of respectively high erosion and weathering activity. In Figure 2, the first factor summarises the response to runoff (higher runoff coefficient to the right); the second factor summarises the erosive response (higher sediment concentrations to the top).

Hydrological and sediment yield response of basins with significant badland surfaces (Ca l'Isard basin).

Catchments with significant badland area are known to show a shift in suspended sediment concentrations during late summer and autumn, the more hydrologically active season. Indeed, intense events taking place in summer show typical sediment concentrations higher than 10 g l^{-1}, that decreased more than one order of magnitude in autumn (Balasch *et al.*, 1992). This behaviour was assumed to be the result of sediment depletion and routing.

In October 1994, an ultrasonic beam attenuation sensor able to continuously measure suspended sediment concentrations between 5 and 220 g l^{-1}, coupled to an infrared backscattering sensor that measures sediment concentrations up to 6 g l^{-1} was installed at the gauging station of Ca l'Isard. This basin is 1 Km^2 in area, and the three main landscape units are 12 ha of badland surfaces, 29 ha of old agricultural terraced fields, patchily forested, and 59 ha of forested hard bedrock outcrops with shallow soils.

The analysis of detailed (two-minute step) discharge and suspended sediment concentrations records obtained at this station suggests the important role played by a shift of runoff generation mechanisms during the end of the dry season and the beginning of the wetter one (summer-autumn). During the dry period, Hortonian overland flow occurs only on bare badland surfaces, producing flash floods with high sediment concentrations, whereas the vegetated areas do not contribute to runoff. Later on, subsequent rainfall events share the production of Hortonian overland flow on the badland surfaces with the contribution of saturated areas whose size increases with increasing water reserve in the basin (Latron and Gallart, in press).

Table 2. Main characteristics of three events that occurred in late Summer 1995, measured at the Ca l'Isard station

	29 July	12 September (first peak)	18 September
Total rainfall (mm)	29.6	33.0	22.2
Intensity max. 6' (mm/h)	58.3	57.4	11.9
Kinetic energy (J/m^2)	721.0	797.0	393.0
Peak discharge (l/s)	288.2	680.2	256.9
Runoff volume (m3)	1220.0	4950.0	6220.0
Runoff coefficient	0.04	0.15	0.28
Sediment transport (Tm)	86.0	78.0	23.7
Max. sed. conc. (g/l)	201.0	80.0	28.0
Mean sed. conc. (g/l)	70.6	28.5	3.8

Figure 3 shows the hyetographs, hydrographs and sediment concentration graphs for three characteristic events in summer 1995: the main characteristics are summarised in Table 2. The rainfall event on 29 July occurred over a dry catchment; the sharpness of the hydrograph, the low runoff coefficient and the high sediment concentrations demonstrate that only the degraded areas contributed to runoff. On 12 September, a similar rainfall event (first peak) on a wetter catchment produced a higher peak and lower sediment concentration, but with similar total sediment transport: some clear water from saturated areas was added to the muddy runoff coming from degraded surfaces without any apparent sediment exhaustion. Some sediment depletion effect from badland surfaces could nevertheless be compensated by the removal of the excess sediment deposited on the drainage net during the former events. Finally, on 18 September, a much gentler rainfall event falling on a rather wet catchment produced a wider hydrograph with a higher runoff coefficient, but much lower sediment concentration and transport.

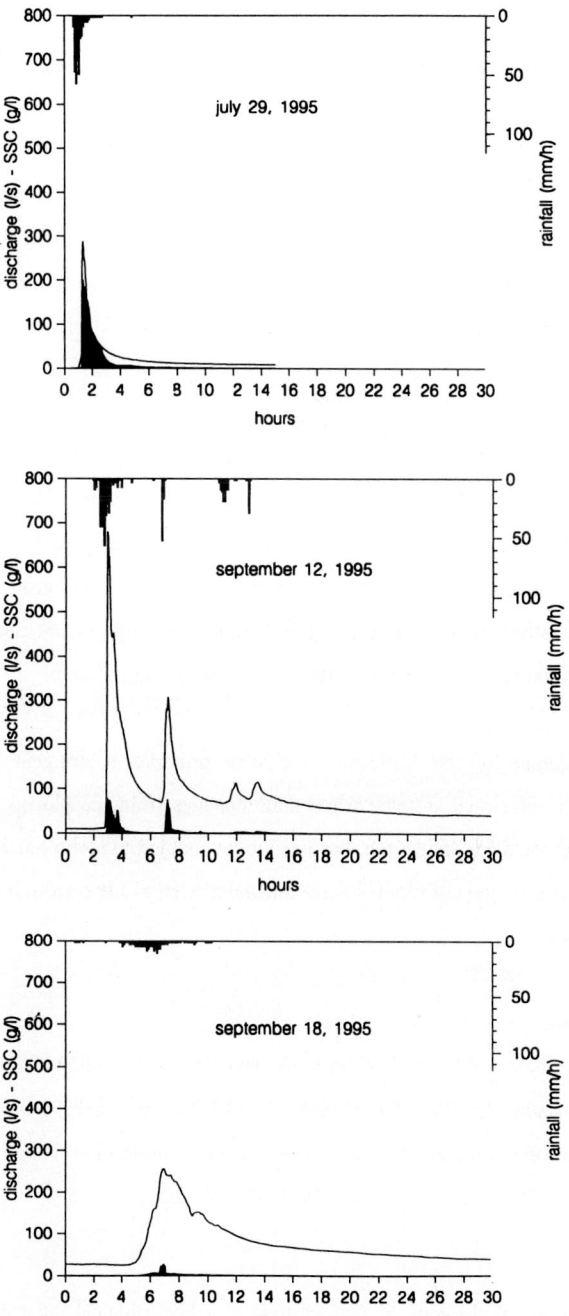

Figure 3. Hyetographs, hydrographs, and sediment concentrations (with black fill) of three events in late Summer 1995

Data available, discussion and conclusions

The use of this ultrasonic device provides us with data of unusual high sediment concentrations at a two-minute time step that can be used to test erosion and sediment transport models at high temporal resolutions. This data set started in October 1994 and contains (on 20 September 1995) more than 10 significant events that represent about 1100 Tm of suspended sediment transport with an average concentration of 80 g l^{-1}. Other data available consist of weather (5-min step) and rainfall (3 rain recorders within the catchment at variable-time step down to 1s and 0.2 mm of resolution). Infiltration characteristics on badland surfaces and soils on terraces are also available.

The results obtained in the Vallcebre catchments emphasise the high differences in erosion rates for different areas with the same climatological and lithological conditions. The abandonment of agricultural activity resulted here in low erosion rates because of the good condition of terraces and ditches and the favourable climatological conditions for vegetation growth. Nevertheless, badland areas — either natural or induced by human activity — are subject to intense degradation and provide major sources of sediment for the drainage system.

Finally, the functioning of the Vallcebre catchment provides a stringent example of the differences found in erosional systems when scale changes from the plot to the small basin. Models of sediment yield for drainage basins need to be coupled to hydrological models able to describe different mechanisms of runoff generation and as well as the corresponding substitution of contributing areas.

Acknowledgements

These research activities have been funded by DGCONA (LUCDEME Project), CYCYT (Projects AMB93-0806, AMB93-0844-C06-05, IN93-0081 and AMB95-0986-C02) and the European Commission (Contract EV5V-CT91-0039). The contribution from the second author was funded by an EC Research Fellowship (EV5V-CT-94-5222).

References
Balasch, C., Castelltort, X., Llorens, P., Gallart, F. (1992). Hydrological and sediment dynamics network design in a Mediterranean mountainous area subject to gully erosion. In, Bogen, J., Walling, D.E. and Day, T. (eds), *Erosion and Sediment Transport Monitoring Programmes in River Basins,* IAHS Publication 210, pp. 433-442.

Clotet, N. Gallart, F. and Balasch, J. (1988). Medium term erosion rates in a small scarcely vegetated catchment in the Pyrenees. *Catena Supplement* **13**, 37-47.

Gallart, F. and Clotet, N. (1988). Some aspects of the geomorphic processes triggered by an extreme rainfall event: the November 1982 flood in the Eastern Pyrenees. *Catena* Supplement **13**, 79-85.

Gallart, F., Llorens, P. and Latron, J. (1994). Studying the role of old agricultural terraces on runoff generation in a Mediterranean small mountainous basin. *Journal of Hydrology* **159**, 291-303.

Latron, J. and Gallart, F. (1995). Hydrological response of two nested small Mediterranean basins presenting various degradation states. *Physics and Chemistry of the Earth* **20**(3-4), 369-374.

Llorens, P. (in press). Rainfall interception by a Pinus sylvestris forest patch overgrown in a Mediterranean mountainous abandoned area. II. Assessment of the applicability of the Gash analytical model. *Journal of Hydrology.*

Llorens, P. and Gallart, F. (1992). Small basin response in a Mediterranean mountainous abandoned farming area: research design and preliminary results. *Catena* **19**, 309-320.

Llorens, P., Latron J. and Gallart, F. (1992). Analysis of the role of agricultural abandoned terraces on the hydrology and sediment dynamics in a small mountainous basin. *Pirineos* **139**, 27-46.

Llorens, P. Poch, R., Rabadà, D. and Gallart, F. (1995). Study of the changes of hydrological processes induced by afforestation in Mediterranean mountainous abandoned fields. *Physics and Chemistry of the Earth* **20** (3-4): 375-383.

Llorens, P., Poch, R. and Gallart, F. (in press). Rainfall interception by a Pinus sylvestris forest patch overgrown in a Mediterranean mountainous abandoned area. I. Monitoring design and results down to the event scale. *Journal of Hydrology.*

Regüés, D. (1995). *Meteorización física en relación con los procesos de producción y transporte de sedimentos en un àrea acarcavada.* Unpublished Ph.D. Dissertation, Facultat de Geologia, Universitat de Barcelona.

Regüés D., Pardini G. and Gallart F. (1995). Regolith behaviour and physical weathering of clayey mudrock in a gullied area, as dependent on seasonal weather conditions. *Catena* **25**, 199-212.

SECTION 7. CONCLUSIONS

39. MODELLING SOIL EROSION BY WATER: SOME CONCLUSIONS

John Boardman[1] and David Favis-Mortlock[2]

[1] School of Geography and Environmental Change Unit
Mansfield Road, University of Oxford
Oxford OX1 3TB
UK

[2] Environmental Change Unit
University of Oxford
5 South Parks Road
Oxford OX1 3UB
UK

The following main conclusions may be drawn.

- It appears that calibration is still essential (or at least desirable) for all current models. In situations close to optimum (for example, when a model is run with data which is very similar to that used to develop it) calibration may be dispensed with. In the 'uncharted territory' of future global change, this represents a problem. There is a need to reduce the extent to which models require calibration.
- Runoff is always better simulated than soil loss. In some ways this is logically comforting: better prediction of erosion would imply that a model functions as a mere 'black box'. This finding also implies, though, that we need to improve the conceptual linkages of runoff to soil loss within current models.
- Long-term average results are generally best simulated. This creates difficulties for model validation, since most measured data is relatively short-term. It is also a problem if any of the impacts of future global change should operate more strongly in the short term than the long term (cf. Favis-Mortlock and Boardman, 1995).
- The model evaluation has focused most strongly on continuous simulation models rather than event-based types. There are two reasons: the more stringent data requirements of event-based models (particularly with regard to initial conditions) render them less able to compete with continuous models when datasets have missing or dubious values. In addition, the requirement for high-quality data on initial conditions disadvantages event

models for global change studies, when knowledge of initial conditions may be little better than speculation.
- The familiarity of the model user with the model and the data appears to be an important factor in some cases. It is however difficult to quantify (but see Botterweg, 1995).
- Many factors or processes affecting erosion are inadequately described by existing models. e.g. crusting, or the influence of soil stoniness. While it is possible to circumvent these limitations to some extent, for example by appropriate choice of input parameters or by calibration, this is likely to be dubious or infeasible under future global change.
- This volume has concentrated on field-scale modelling. However, it is clear that the spatial variability of erosion or its contributing factors is important even at this scale. For estimates of global change impacts at larger scales, modelling approaches which explicitly deal with the spatial aspects of erosion must be developed. For policy purposes, such larger scale forecasts may be essential: a possibly extreme view is that "a field-scale model is only useful as a component of a catchment-scale model" (David Garen, personal communication 1996).
- Model comparison of the kind described here is demanding of resources. There were — inevitably — practical problems in carrying out the model evaluation exercise, both with respect to timetabling and data quality: e.g. the Portuguese soil loss data were originally specified with incorrect units, and were therefore 1000 times too high! Analysis of the common-dataset component of the model evaluation in particular produced a large quantity of data, some of which could not be presented here.

These conclusions suggest many avenues for enhancement of current models, and development of new ones. In addition, further validation of existing models appears desirable in order to reliably evaluate their suitability for global change impact studies (Favis-Mortlock *et al.*, 1996).
- Models should be validated with data from outside the area — and hence the range of conditions — in which they were developed; for example with data from semi-arid regions, or the humid tropics.
- Estimates should be made of each model's sensitivity to change, in particular to the values of those climate and land use parameters which will be affected by global change.

Future global change may be unknown; but with well-tested and reliable predictive tools, we will be in the best possible position to meet its challenges.

References

Botterweg, P. (1995). The user's influence on model calibration results: an example of the model SOIL, independently calibrated by two users. *Ecological Modelling* **81**, 71-81.

Favis-Mortlock, D.T. and Boardman, J. (1995). Nonlinear responses of soil erosion to climate change: a modelling study on the UK South Downs. *Catena* **25**(1-4), 365-387.

Favis-Mortlock, D.T., Quinton, J.N. and Dickinson, W.T. (1996). The GCTE validation of soil erosion models for global change studies. *Journal of Soil and Water Conservation* **51**(5), 397-403.

APPENDICES

APPENDIX A. ACRONYMS AND ABBREVIATIONS

CSEP	Climatic Seasonal Erosion Potential
EPIC	Erosion-Productivity Impact Calculator
EUROSEM	EURopean Soil Erosion Model
GCTE	Global Change and Terrestrial Ecosystems
GLEAMS	Groundwater Loading Effects of Agricultural Management Systems
GUEST	Griffith University Erosion System Template
IBSNAT	International Benchmark Sites Network for Agrotechnology Transfer
IGBP	International Geosphere-Biosphere Programme
MEDALUS	Mediterranean Desertification and Land Use
MEDRUSH	Mediterranean Environmental Degradation and Response Units for Sediment yield and Hydrology
MUSLE	Modified Universal Soil Loss Equation
NRCS	Natural Resources Conservation Service
NSERL	National Soil Erosion Research Laboratory
SCS	Soil Conservation Service
USDA-ARS	United States Department of Agriculture — Agricultural Research Service
USLE	Universal Soil Loss Equation
WEPP	Water Erosion Prediction Project

APPENDIX B. ATTENDEES AT NATO-ARW 'GLOBAL CHANGE: MODELLING SOIL EROSION BY WATER'

Dr Jeff Arnold
USDA-ARS Grassland, Soil and Water
 Research Laboratory
808 East Blackland Road
Temple, Texas 76502
USA
arnold@brcsun0.tamu.edu

Dr Veronique Auzet
CEREG-URA 95 CNRS
Université Louis Pasteur
3 rue de l'Argonne
F-67083 Strasbourg cédex
France
auzet@geographie.u-strasbg.fr

Dr Jussi Baade
Institut für Geographie
Friedrich-Schiller-Universität
Loebdergraben 32
D-07740 Jena
Germany
cub@rz.uni-jena.de

Dr Belgin Bilge
Marmara Arastirma Merkezi Research Centre
P.K. 21
41470 Gebze-Kocaeli
Turkey
belgin@yunus.mam.tubitak.gov.tr

Dr John Boardman
School of Geography/Environmental Change Unit
University of Oxford
Mansfield Road
Oxford OXI 3TB
UK
john.boardman@ecu.ox.ac.uk

Dr Peter Botterweg
Centre for Soil and Environmental Research
Jordforsk
N 1432 Ås
Norway
peter.botterweg@jordforsk.nlh.no

Professor Tim Burt*
Department of Geography
University of Durham
South Road
Durham DH1 3LE
UK
t.p.burt@durham.ac.uk

Professor Miguel Azevedo Coutinho
Instituto Superior Técnico
D.E. Civil-Secçao de Hidràulica
Av. Rovisco Pais 1
P-1096 Lisboa codex
Portugal

Dr Bob Evans
Division of Geography
Anglia Polytechnic University
East Road
Cambridge CB1 1TT
UK
bevans@bridge.anglia.ac.uk

Dr David Favis-Mortlock
Environmental Change Unit
University of Oxford
5 South Parks Road
Oxford OXI 3UB
UK
david.favismortlock@ecu.ox.ac.uk

Ronald Fix
Keble College
University of Oxford
Oxford OXI 3PG
UK
ronald.fix@geog.ox.ac.uk

Dr Francesc Gallart
Institute of Earth Sciences 'Jaume Almera'
PO Box 30102
08080 Barcelona
Spain
fgallart@ija.csic.es

Dr Gerard Govers
Laboratory for Experimental Gemorphology
Katholieke Universiteit Leuven
Redingenstraat 16
B-3000 Leuven
Belgium
gerard.govers@geo.kuleuven.ac.be

Professor Peter Gregory
Department of Soil Science
University of Reading
Whiteknights
Reading RG6 6DW
UK
p.j.gregory@reading.ac.uk

Jerôme Guérif
INRA Unité d'Agronomie de Laon
rue Fernand-Christ
F-02007 Laon cédex
France
jguerif@laon.inra.fr

Dr Trevor Harris
Department of Geology and Geography
University of West Virginia
425 Whitehall
Morgantown, WV 26506
USA
tmh@wvugeo.wvnet.edu

Professor Anton Imeson
Universiteit van Amsterdam
Fysisch Geografisch en Bodemkundig Lab.
Nieuwe Prinzengracht 130
NL-1018 VZ Amsterdam
The Netherlands
ai@fsb.frw.uva.nl

John Ingram
GCTE Focus 3 Offices
NERC Centre for Ecology and Hydrology
Crowmarsh Gifford
Wallingford OX10 8BB
UK
j.ingram@unixa.nerc-wallingford.ac.uk

Dr Victor Jetten
INRA Unité d'Agronomie de Laon
Rue Fernand-Christ
F-02007 Laon cédex
France
jeten@laon.inra.fr

Dr Dominique King
INRA Centre de Recherche d'Orléans
45160 Olivet
France
king@orleans.inra.fr

Professor Mike Kirkby
School of Geography
University of Leeds
Leeds LS2 9TJ
UK
mike@geog.leeds.ac.uk

Dr Contantinos Kosmas
Agricultural University of Athens
Laboratory of Soils and Agricultural Chemistry
Iera Odos 75, Botanikos 11855
Athens
Greece
lsos2kok@auadec.aua.ariadne-t

Dr Frans Kwaad
Universiteit van Amsterdam
Fysisch Geografisch en Bodemkundig Laboratorium
Nieuwe Prinzengracht 130
NL-1018 VZ Amsterdam
The Netherlands
fk@fgb.frw.uva.nl

Dr Yves Le Bissonnais
INRA Centre de Recherche d'Orléans
45160 Olivet
France
lebisson@orleans.inra.fr

Dr Jeff Lee†
US Environmental Protection Agency
Environmental Research Laboratory
200 SW 35 Street
Corvallis, OR 97333
USA

Dr Denés Lóczy
Geographical Research Institute
Hungarian Academy of Sciences
PO Box 64
H-1388 Budapest
Hungary

Dr Simon A. Lorentz
Department of Agricultural Engineering
University of Natal
Box X01
Pietermaritzburg 3209
South Africa
lorentz@aqua.ccwr.ac.za

Dr Rachael McDonnell*
Visiting Fellow, Department of Geography
University of Bristol
University Road
Bristol BS8 1SS
UK
rachael.mcdonnell@compuserve.com

Professor Ts.E. Mirtskhulava
Georgian Research Institute of Water Management
    and Engineering Ecology
60 Avenue I Chavchavadze
3880062 Tbilisi
Georgia

Professor Roy Morgan
Silsoe College
Cranfield University
Silsoe
Bedford MK45 4DT
UK

Dr Mark Nearing
USDA-ARS National Soil Erosion
    Research Laboratory
1196 Building SOIL
West Lafayette
IN 47907-1196
USA
nearing@ecn.purdue.edu

Dr Arlin Nicks†
USDA-ARS National Agricultural
    Water Quality Laboratory
PO Box 1430
Durant, OK 74702
USA

Professor Tony Parsons
Department of Geography
University of Leicester
University Road
Leicester LE1 7RH
UK
ajp16@leicester.ac.uk

Professor Jean Poesen
Laboratory for Experimental Gemorphology
Katholieke Universiteit Leuven
Redingenstraat 16
B-3000 Leuven
Belgium
jean.poesen@geo.kuleuven.ac.be

Dr John Quinton
Silsoe College
Cranfield University
Silsoe
Bedford MK45 4DT
UK
j.quinton@cranfield.ac.uk

Dr Ad de Roo*
AIS Unit Environment and Natural Hazards
Space Applications Institute
Joint Research Centre
I-21020 Ispra (Va)
Italy
ad.de-roo@jrc.it

Professor Calvin Rose
Faculty of Environment sciences
Griffith University
Brisbane
Queensland 4111
Australia.
C.Rose@plato.ens.gu.edu.au

Dr Ramesh Rudra
School of Engineering
University of Guelph
Guelph
Ontario N1G 2W1
Canada
rudra@net2.eus.uoguelph.ca.

Professor Aleksey Sidorchuk
Geography Department
Moscow State University
Lenin Hills GSP-3
Moscow 119899
Russia
sidor@yas.geogr.msu.su

Dr Ed Skidmore
USDA-ARS Wind Erosion Research
Throckmorton Hall
Kansas State University
Manhattan, KS 66506
USA
skidmore@weru.ksu.edu

Dr Raghavan Srinivasan
Texas Agricultural Experiment Station
Texas A&M University
808 East Blackland Road
Temple, TX 76502
USA
srin@brcsun2.tamu.edu

Professor Bernard Tinker
GCTE Focus 3 Offices
NERC Centre for Ecology and Hydrology
Crowmarsh Gifford
Wallingford OX10 8BB
UK

Pedro Tomás
Instituto Superior Técnico
D.E. Civil-Secçao de Hidràulica
Av. Rovisco Pais 1
P-1096 Lisboa codex
Portugal
PT@civil10.ist.utl.pt

Dr Dino Torri
CNR - Soil Genesis and Ecology Institute
Piazzale le Cascine 15
50129 Firenze
Italy
dbtorri@csgccs.fi.cnr.it

Dr Christian Valentin
ORSTOM
rue La Fayette
75450 Paris, cédex 10
France
valentin@orstom.rio.net

Dr John Wainwright
Department of Geography
King's College London
Strand
London WC2R 2LS
UK
j.wainwright@kcl.ac.uk

Professor Des Walling
Department of Geography
University of Exeter
Amory Building
Rennes Drive
Exeter EX4 4RJ
UK

Dr Brad Wilcox*
IAI
Av. dos Astronautas, 1758
12227-010 Sao Jose dos Campos SP
Brazil
bwilcox@dir.iai.int

---

* Address changed since NATO-ARW
† Deceased

INDEX

A

aerial photography 285, 289, 290, 291, 293, 294, 297, 305, 321
aggregate 7, 186, 247, 248, 252, 254, 329, 339, 346, 351, 366, 408, 432, 433, 503
   stability 7, 186, 247, 254, 329, 339, 346, 432, 433, 503
agriculture 14, 329, 336, 339, 340, 366, 417, 437, 446, 481, 488
   barley 287, 442, 464
   crop cover 313, 325, 326, 493
   crop rotation 57, 58, 63, 287, 315, 331, 378, 417, 481, 485
   crop yield 418
   fertiliser 89, 286
   herbicide 372
   maize (corn) 11, 19, 57, 90, 93, 228, 229, 234, 235, 287, 295, 296, 315, 418, 442, 471, 485, 486
   management 6, 9, 55, 63, 176, 178, 181, 221, 254, 329, 347, 352, 355, 357, 358, 377, 378, 381, 382, 395, 399, 401, 407, 408, 413, 414, 416, 417, 418, 429, 441, 443, 444, 445, 446, 461, 463, 467, 469, 486, 492
   manure 445
   minimum tillage 90
   nitrate 221
   oats 442
   pesticide 21, 55, 89, 266, 413, 416, 417, 418, 441, 486
   soybean 442
   tillage 7, 11, 12, 56, 57, 58, 63, 90, 47, 181, 214, 215, 233, 285, 286, 287, 289, 291, 336, 339, 393, 407, 417, 436, 442, 471, 488
   wheat 57, 90, 93, 123, 47, 48, 56, 287, 442, 464, 486, 491, 494, 495, 496, 497
alluvium 21
archaeology 14

B

bedload 64, 400
bulk density 91, 181, 182, 247, 249, 251, 442

C

calibration 57, 89, 92, 93, 94, 103, 121, 122, 123, 139, 44, 45, 46, 47, 54, 55, 56, 176, 248, 339, 371, 414, 418, 420, 436, 456, 457, 469, 515, 516
   constrained 44, 45, 46, 54
carbon dioxide 7, 441, 443, 444, 445, 446
catchment 3, 6, 8, 12, 14, 16, 20, 216, 218, 219, 220, 221, 233, 286, 289, 290, 295, 305, 321, 322, 323, 329, 332, 333, 334, 336, 339, 340, 341, 346, 351, 352, 353, 354, 355, 356, 357, 391, 394, 395, 413, 414, 435, 436, 451, 464, 465, 467, 469, 481, 483, 488, 503, 504, 505, 506, 515, 527
channel 56, 58, 47, 213, 219, 220, 285, 289, 295, 297, 300, 301, 315, 326, 340, 342, 352, 353, 355, 395, 413, 414, 417, 418, 429, 431, 432, 434, 436, 441, 443, 445, 451, 452, 454, 455, 456, 457
   profile 56
climate change 6, 7, 8, 9, 10, 12, 19, 20, 123, 56, 186, 365, 371, 445, 446, 516
   greenhouse gas 6, 9
colluvium 232, 298, 299
concentrated flow 21, 285, 289, 296, 297, 298, 299, 300, 302, 303, 304, 337, 396, 471
conservation 21, 225, 259, 262, 313, 315, 320, 396, 399, 407, 436, 451, 452, 461, 462, 466, 467, 468, 469, 471, 481, 485, 488, 491, 493, 503, 504
   tillage 471

D

database 58, 249, 353, 355, 356, 357, 358, 380, 382, 403, 418, 461, 462, 463, 464, 465, 466, 467, 468, 469
deforestation 14, 175
deposition 56, 89, 123, 56, 182, 247, 252, 259, 260, 261, 262, 264, 265, 266, 295, 300, 302, 303, 331, 340, 341, 351, 389, 391, 392, 393, 395, 401, 404, 417, 433, 435, 445, 464, 481, 485
desertification 491, 495, 496, 497
digital elevation (terrain) model 354, 420, 444, 484
drainage 185, 218, 219, 221, 222, 286, 293, 294, 295, 297, 303, 304, 339, 354, 355, 368, 381, 392, 413, 429, 432, 437, 481, 503, 504, 505, 506

E

ecosystem 10, 358, 369
entrainment 260, 261, 272, 302, 404
environment 64, 259, 263, 275, 287, 356, 359, 423, 442, 467, 488
 change 9, 285, 291, 292, 302
erodibility 47, 54, 175, 182, 183, 186, 221, 235, 247, 254, 261, 266, 272, 294, 297, 299, 301, 337, 352, 370, 378, 381,
 .... 382, 393, 399, 401, 402, 403, 405, 406, 407, 408, 417, 443, 453, 456, 484, 491, 492, 494
erosion model
 ANSWERS 175, 176, 178, 186, 339, 355, 357, 358, 359, 429, 437
 APEX 441, 445
 CREAMS ... 55, 63, 64, 122, 55, 175, 176, 178, 182, 186, 299, 304, 377, 380, 381, 383, 391, 414, 423, 429, 437, 462, 486
 CSEP 4, 89, 92, 93, 103, 104, 107, 110, 115, 116, 121, 123
 DRAINMOD 185, 187
 EGEM 300, 304
 EPIC 4, 89, 92, 93, 94, 103, 104, 107, 110, 115, 116, 139, 44, 45, 46, 47, 48, 49, 50, 51, 53, 54, 55, 56, 57, 175,
 .... 176, 181, 339, 347, 377, 380, 381, 383, 441, 442, 443, 444, 445, 446, 462, 471, 481, 485, 486, 488, 489
 EUROSEM 3, 89, 92, 93, 124, 56, 253, 254, 266, 302, 339, 347, 365, 369, 370, 372, 389, 390, 391, 392, 393, 395,
 .... 396, 403, 408, 429, 432, 437
 GAMES 175, 176, 186, 352
 GLEAMS 3, 55, 58, 63, 89, 92, 93, 103, 104, 107, 110, 115, 116, 121, 122, 124, 139, 44, 45, 46, 47, 48, 49, 51, 53,
 .... 54, 55, 56, 175, 186, 300, 304, 416, 441, 446, 462
 GUEST 4, 89, 92, 93, 121, 261, 266, 399, 401, 402, 403, 404, 406, 408
 KINEROS 391, 395, 396
 LISEM 339, 340, 342, 344, 346, 347, 352, 355, 359, 429, 430, 431, 433, 435, 436, 437
 MEDRUSH 4, 89, 92, 93, 103, 104, 107, 110, 115, 116, 123, 488
 MUSLE 381, 416, 417, 442
 RUSLE 175, 186
 SWAT 413, 414, 415, 416, 417, 418, 419, 422, 423, 444, 445
 SWRRB 377, 380, 444, 445
 USLE 4, 16, 55, 56, 93, 47, 48, 175, 259, 305, 313, 315, 351, 352, 355, 377, 378, 379, 380, 381, 382, 417, 442, 461,
 .... 481, 482, 483, 484, 485, 486, 488
 WEPP 4, 89, 92, 93, 103, 104, 110, 115, 116, 121, 122, 124, 139, 44, 45, 47, 48, 49, 50, 51, 53, 54, 55, 56, 57, 175,
 .... 176, 186, 300, 339, 347, 369, 372, 377, 382, 383, 391, 403, 429, 437
 WEQ 443
erosivity 55, 56, 378, 381, 396, 485, 492
error 115, 121, 124, 176, 185, 271, 276, 277, 278, 281, 282, 356, 358, 463, 464, 466, 472, 485, 486
evapotranspiration 55, 175, 185, 186, 423, 445, 485, 486, 493
extreme event 12, 16, 17, 10, 501

F

field capacity 18, 185, 414, 417, 493
flooding 17, 19, 89, 123, 298, 299, 301, 302, 464
flume 259, 263, 264, 265, 299, 303, 396, 453, 471
fractals 57

G

Geographical Information System 4, 337, 339, 351, 352, 353, 354, 355, 356, 357, 358, 359, 417, 418, 420, 422, 423, 429,
 .... 430, 435, 436, 444, 445, 484, 488, 491
geomorphology 12, 291, 336, 355, 358, 503
geostatistics 353
GIS
 raster 339, 352, 355, 417, 429, 430, 436, 444
 vector 352
grassland 62, 90, 221, 273, 277, 282, 315
gully 4, 15, 285, 286, 287, 289, 290, 291, 292, 293, 294, 295, 297, 298, 299, 302, 303, 304, 305, 399, 451, 452, 453,
 .... 454, 455, 456, 457, 506
 bank 285, 286, 292, 295, 297, 299, 302
 development 293, 294, 295, 297, 451, 456
 ephemeral 12, 15, 219, 285, 286, 287, 289, 290, 291, 292, 293, 294, 295, 297, 298, 299, 302, 303, 304, 305, 337
 head 293, 294, 302, 456
 incision 294, 295, 305, 452

H

hillslope 12, 15, 139, 45, 213, 219, 220, 221, 222, 259, 273, 281, 282, 283, 291, 295, 355, 389, 395, 494
Hortonian overland flow...214, 233, 275, 503, 504, 505
hydraulics 271, 273, 275, 276, 278, 279, 281, 282, 283
   Chézy C 276, 281
   flow velocity...248, 272, 275, 342, 344, 389, 393, 404, 433, 451, 452, 453
   hydraulic conductivity........... 47, 54, 178, 180, 181, 182, 185, 186, 217, 219, 330, 339, 342, 352, 354, 366, 369, 393, 395, ...436, 438
   Manning's n 56, 93, 44, 46, 47, 276, 281, 302, 339, 342, 344, 346, 395, 404, 414, 417, 432, 434
hydrology............................. 4, 10, 55, 58, 64, 124, 213, 221, 222, 281, 283, 365, 369, 380, 381, 414, 416, 423, 445, 506

I

impacts of soil erosion...254, 470
infiltration.............................. 4, 9, 55, 44, 46, 47, 54, 175, 180, 181, 185, 186, 213, 214, 215, 216, 217, 218, 219, 220, 221, ... 222, 225, 228, 234, 262, 275, 276, 282, 294, 329, 331, 344, 346, 352, 365, 366, 369, 370, 371, 382, 389, ... 391, 429, 432, 435, 485, 492, 504
   Green and Ampt 213, 275, 429
interrill processes..............................55, 47, 252, 266, 271, 272, 282, 283, 285, 287, 289, 290, 291, 303, 380, 381, 382, 389, ... 391, 392, 393, 395, 396, 433
   erosion 289, 290, 291, 382, 433
   flow 282, 392, 393, 395, 396

K

kinetic energy 247, 249, 252, 346, 392, 432, 483

L

land capability 481, 485
land use................. 6, 7, 8, 9, 12, 14, 15, 18, 19, 20, 57, 175, 176, 178, 181, 182, 222, 228, 315, 326, 329, 331, 332, 336, 340, ... 351, 352, 354, 413, 417, 418, 420, 422, 423, 435, 436, 444, 461, 467, 468, 481, 482, 483, 484, 486, 516
landscape .... 3, 9, 10, 12, 13, 14, 15, 19, 20, 285, 286, 290, 292, 294, 295, 297, 302, 304, 313, 314, 315, 325, 358, ... 408, 464, 469, 471, 481, 492, 493, 495, 497, 505
   sensitivity 481
leaf area index..58, 416
legislation 380

M

mapping 233, 289, 290, 313, 352, 354, 355, 358, 481, 486, 491, 492, 495, 497
mass movement 294, 297, 302, 399, 403, 451, 503
MEDALUS...4, 11, 122, 123, 302, 488, 497
Mediterranean 90, 285, 286, 287, 291, 292, 302, 304, 305, 488, 491, 492, 495, 497, 503, 506, 507
microtopography...12, 217, 275, 504
moisture content 213, 217, 219, 297, 298, 299, 301, 356, 492
morphology 15, 44, 286, 326, 451, 467, 468

N

O

overland flow.................... 47, 49, 182, 213, 214, 217, 218, 219, 220, 221, 225, 226, 227, 228, 232, 233, 250, 259, 263, 265, ... 266, 271, 272, 275, 281, 282, 283, 285, 286, 290, 291, 294, 295, 297, 298, 339, 344, 345, 346, 347, ... 351, 392, 393, 396, 399, 408, 429, 431, 432, 433, 435, 436, 437, 491, 503, 504, 505

P

peak runoff...177, 178, 381, 395, 416, 441, 442, 443
percolation 55, 185, 371, 414, 416
piping..292
policy...8, 515
ponding...217, 220, 369, 492

porosity 181
precipitation 57, 58, 63, 175, 176, 177, 178, 221, 226, 287, 288, 351, 352, 365, 366, 369, 370, 381, 382, 420, 442,
.... 484, 485, 486, 503
probability 116, 121, 297, 452
profile
   concave 395

R

rain splash 247, 315
rainfall 6, 8, 10, 12, 18, 20, 21, 55, 91, 92, 93, 44, 45, 46, 47, 50, 51, 53, 54, 57, 177, 178, 182, 185, 186, 213,
.... 214, 217, 218, 219, 221, 225, 226, 230, 232, 233, 234, 247, 248, 252, 254, 259, 261, 262, 263, 264, 265, 266,
.... 272, 273, 275, 281, 287, 294, 297, 302, 313, 315, 320, 326, 329, 331, 332, 337, 339, 340, 346, 352, 378, 381,
.... 382, 389, 391, 392, 393, 395, 396, 399, 400, 402, 406, 407, 414, 417, 418, 429, 431, 432, 435, 436, 441, 442,
.... 451, 461, 463, 464, 465, 466, 467, 468, 471, 481, 484, 485, 492, 493, 494, 503, 504, 505, 506
   detachment 247, 261, 262, 264
   energy 389, 504
   intensity 182, 213, 214, 217, 218, 219, 272, 294, 378, 382, 414, 442, 466, 467, 492, 494, 503
   simulation 218, 248, 252, 273, 484, 504
   terminal velocity 260, 392
rainfall
   simulation 182, 217, 382, 494
rangeland 9, 285, 287, 295, 296, 382, 423
remote sensing 357, 358, 359, 429, 437
residue 46, 180, 339, 416, 417, 442, 494
rill 55, 45, 47, 49, 178, 182, 213, 219, 220, 266, 285, 286, 287, 289, 290, 291, 303, 305, 329, 332, 333, 352, 380,
.... 381, 382, 389, 391, 392, 396, 399, 402, 404, 407, 433, 437, 451, 464
   erosion 178, 219, 285, 290, 291, 303, 329, 332, 333, 352, 380, 382, 451
   flow 49, 389, 404
   initiation 392, 433
roughness 8, 181, 247, 250, 302, 329, 331, 332, 334, 339, 342, 344, 346, 352, 382, 393, 395, 396, 404, 442, 443, 492
runoff 9, 15, 17, 18, 20, 55, 57, 59, 61, 62, 63, 89, 90, 91, 92, 93, 94, 97, 103, 104, 106, 107, 110, 112, 114, 115,
.... 116, 118, 121, 122, 123, 124, 44, 45, 46, 51, 53, 54, 56, 57, 176, 177, 178, 180, 181, 182, 185, 213, 214, 218,
.... 219, 220, 221, 226, 227, 228, 230, 231, 232, 233, 234, 247, 248, 250, 252, 254, 259, 264, 266, 272, 282, 283,
.... 289, 290, 291, 297, 303, 322, 324, 329, 331, 332, 333, 334, 336, 337, 347, 366, 369, 370, 371, 378, 380, 381,
.... 382, 389, 391, 392, 395, 396, 399, 400, 401, 402, 407, 413, 414, 416, 417, 418, 419, 429, 430, 432, 436, 441,
.... 442, 443, 445, 451, 464, 471, 483, 485, 486, 491, 492, 493, 494, 495, 497, 504, 505, 506, 515
   detachment 391, 392
   plot 289, 290, 291, 399, 400, 407

S

sediment
   concentration 259, 260, 261, 263, 264, 291, 346, 370, 391, 392, 401, 404, 405, 406, 407, 417, 433, 443, 452, 453,
.... 454, 455, 503, 504, 505, 506
   transport 55, 176, 219, 221, 271, 272, 274, 275, 276, 278, 281, 282, 382, 392, 396, 408, 451, 505, 506
   yield 13, 63, 91, 177, 178, 234, 266, 291, 303, 383, 396, 416, 417, 441, 442, 443, 503, 504, 505, 506
sensitivity analysis 47, 56, 271, 339, 340, 342, 347, 436, 467, 469, 471
shear strength 182, 186, 250, 254, 298, 299, 301, 365
shear stress 47, 260, 271, 272, 275, 276, 297, 298, 299, 300, 301, 302, 382, 401, 403, 404, 408, 451, 452, 453, 456
simulation 7, 10, 55, 89, 90, 92, 93, 122, 123, 139, 175, 176, 182, 185, 218, 220, 248, 252, 273, 340, 344, 355, 365,
.... 366, 369, 370, 377, 380, 382, 389, 393, 396, 413, 414, 417, 419, 429, 437, 441, 442, 444, 445, 469, 484,
.... 485, 486, 494, 504, 515
slope 15, 56, 58, 45, 175, 178, 181, 182, 218, 219, 230, 234, 249, 250, 251, 252, 254, 264, 272, 275, 289, 291,
.... 293, 294, 295, 297, 298, 302, 304, 321, 322, 323, 324, 325, 329, 333, 340, 344, 345, 346, 351, 352, 354,
.... 355, 377, 378, 381, 382, 393, 395, 399, 400, 404, 407, 417, 420, 443, 451, 452, 453, 455, 456, 464, 465,
.... 467, 468, 482, 483, 485, 486, 492, 493, 494
   angle 295, 298, 381, 395, 486
   morphology 467, 468
   profile 56, 58
slumping 285, 302
snow 365, 366, 367, 368, 369, 370, 371, 372
snowmelt 4, 122, 226, 285, 365, 366, 367, 368, 369, 370, 371, 451, 456
soil
   crusting 4, 9, 12, 15, 139, 47, 54, 180, 181, 182, 214, 217, 220, 221, 225, 247, 332, 334, 351, 357, 484, 515
soil surface 175, 326

spatial scale 6, 8, 9, 89, 175, 176, 225, 304, 352, 391, 445, 461, 469
statistics 274, 275, 281, 313, 354, 356, 469
steady-state 182, 253, 263, 404

T

talweg 15, 213, 219, 298, 333, 527
temporal scale 7, 12, 351
terrace 56, 286
terraced land 292
testing set 89, 92, 93, 94, 97, 102, 103, 107, 110, 115, 121
threshold 9, 219, 261, 285, 286, 293, 294, 295, 302, 304, 305, 344, 346, 368, 404, 405, 443, 444, 463, 464, 465, 492, 493
throughfall 392, 396, 429, 431, 432
tillage erosion 287, 291
timescale 110, 115, 121, 122, 214, 217
topography 58, 181, 213, 218, 220, 273, 297, 332, 351, 352, 358, 366, 395, 413, 417, 418, 420, 429, 441, 445, 491, 492, 494
training set 89, 92, 93, 94, 103, 107, 121, 463, 464

V

validation 3, 5, 123, 139, 56, 176, 339, 413, 420, 429, 436, 437, 465, 515, 516, 527
vegetation 6, 10, 57, 90, 91, 277, 282, 287, 291, 294, 297, 315, 331, 366, 396, 432, 433, 452, 455, 491, 492, 493, 496, 503, 504, 506

W

water quality 175, 260, 266, 413, 423, 441, 444
watershed 55, 57, 62, 63, 124, 56, 176, 177, 186, 220, 300, 358, 359, 372, 378, 383, 396, 413, 414, 418, 441, 443, 444, 445, 446, 483, 484, 488, 527
weather generator... 46, 382
 CLIGEN 44, 46, 54, 382
wilting point 93, 185, 414, 493
wind erosion 3, 9, 10, 441, 443, 485

The ASI Series Books Published as a Result of Activities of the Special Programme on Global Environmental Change

This book contains the proceedings of a NATO Advanced Research Workshop held within the activities of the NATO Special Programme on Global Environmental Change, which started in 1991 under the auspices of the NATO Science Committee.

The volumes published as a result of the activities of the Special Programme are:

Vol. 1: **Global Environmental Change.**
Edited by R. W. Corell and P. A. Anderson. 1991.
Vol. 2: **The Last Deglaciation: Absolute and Radiocarbon Chronologies.**
Edited by E. Bard and W. S. Broecker. 1992.
Vol. 3: **Start of a Glacial.** Edited by G. J. Kukla and E. Went. 1992.
Vol. 4: **Interactions of C, N, P and S Biogeochemical Cycles and Global Change.**
Edited by R. Wollast, F. T. Mackenzie and L. Chou. 1993.
Vol. 5: **Energy and Water Cycles in the Climate System.**
Edited by E. Raschke and D. Jacob. 1993.
Vol. 6: **Prediction of Interannual Climate Variations.**
Edited by J. Shukla. 1993.
Vol. 7: **The Tropospheric Chemistry of Ozone in the Polar Regions.**
Edited by H. Niki and K. H. Becker. 1993.
Vol. 8: **The Role of the Stratosphere in Global Change.**
Edited by M.-L. Chanin. 1993.
Vol. 9: **High Spectral Resolution Infrared Remote Sensing for Earth's Weather and Climate Studies.**
Edited by A. Chedin, M.T. Chahine and N.A. Scott. 1993.
Vol. 10: **Towards a Model of Ocean Biogeochemical Processes.**
Edited by G. T. Evans and M.J.R. Fasham. 1993.
Vol. 11: **Modelling Oceanic Climate Interactions.**
Edited by J. Willebrand and D.L.T. Anderson. 1993.
Vol. 12: **Ice in the Climate System.** Edited by W. Richard Peltier. 1993.
Vol. 13: **Atmospheric Methane: Sources, Sinks, and Role in Global Change.**
Edited by M. A. K. Khalil. 1993.
Vol. 14: **The Role of Regional Organizations in the Context of Climate Change.**
Edited by M. H. Glantz. 1993.
Vol. 15: **The Global Carbon Cycle.**
Edited by M. Heimann. 1993.
Vol. 16: **Interacting Stresses on Plants in a Changing Climate.**
Edited by M. B. Jackson and C. R. Black. 1993.
Vol. 17: **Carbon Cycling in the Glacial Ocean: Constraints on the Ocean's Role in Global Change.**
Edited by R. Zahn, T. F. Pedersen, M. A. Kaminski and L. Labeyrie. 1994.
Vol. 18: **Stratospheric Ozone Depletion/UV-B Radiation in the Biosphere.**
Edited by R. H. Biggs and M. E. B. Joyner. 1994.
Vol. 19: **Data Assimilation: Tools for Modelling the Ocean in a Global Change Perspective.**
Edited by P. O. Brasseur and J. Nihoul. 1994.

Vol. 20: **Biodiversity, Temperate Ecosystems, and Global Change.**
Edited by T. J. B. Boyle and C. E. B. Boyle. 1994.
Vol. 21: **Low-Temperature Chemistry of the Atmosphere.**
Edited by G. K. Moortgat, A. J. Barnes, G. Le Bras and J. R. Sodeau. 1994.
Vol. 22: **Long-Term Climatic Variations – Data and Modelling.**
Edited by J.-C. Duplessy and M.-T. Spyridakis. 1994.
Vol. 23: **Soil Responses to Climate Change.**
Edited by M. D. A. Rounsevell and P. J. Loveland. 1994.
Vol. 24: **Remote Sensing and Global Climate Change.**
Edited by R. A. Vaughan and A. P. Cracknell. 1994.
Vol. 25: **The Solar Engine and Its Influence on Terrestrial Atmosphere and Climate.**
Edited by E. Nesme-Ribes. 1994.
Vol. 26: **Global Precipitations and Climate Change.**
Edited by M. Desbois and F. Désalmand. 1994.
Vol. 27: **Cenozoic Plants and Climates of the Arctic.**
Edited by M. C. Boulter and H. C. Fisher. 1994.
Vol. 28: **Evaluating and Monitoring the Health of Large-Scale Ecosystems.**
Edited by D. J. Rapport, C. L. Gaudet and P. Calow. 1995.
Vol. 29: **Global Environmental Change Science: Education and Training.**
Edited by D. J. Waddington. 1995.
Vol. 30: **Ice Core Studies of Global Biogeochemical Cycles.**
Edited by R. J. Delmas. 1995.
Vol. 31: **The Role of Water and the Hydrological Cycle in Global Change.**
Edited by H. R. Oliver and S. A. Oliver. 1995.
Vol. 32: **Atmospheric Ozone as a Climate Gas.**
Edited by W.-C. Wang and I. S. A. Isaksen. 1995.
Vol. 33: **Carbon Sequestration in the Biosphere.**
Edited by M. A. Beran. 1995.
Vol. 34: **Climate Sensitivity to Radiative Perturbations: Physical Mechanisms and Their Validation.**
Edited by H. Le Treut. 1996
Vol. 35: **Clouds, Chemistry and Climate.**
Edited by P. J. Crutzen and V. Ramanathan. 1996.
Vol. 36: **Diachronic Climatic Impacts on Water Resources.**
Edited by A. N. Angelakis and A. S. Issar. 1996.
Vol. 37: **Climate Change and World Food Security.**
Edited by T. E. Downing. 1996.
Vol. 38: **Evaluation of Soil Organic Matter Models.**
Edited by D. S. Powlson, P. Smith and J. U. Smith. 1996.
Vol. 39: **Microbiology of Atmospheric Trace Gases.**
Edited by J. C. Murrell and D. P. Kelly. 1996.
Vol. 40: **Forest Ecosystems, Forest Management and the Global Carbon Cycle.**
Edited by M. J. Apps and D. T. Price. 1996.
Vol. 41: **Climate Variations and Forcing Mechanisms of the Last 2000 Years.**
Edited by P. D. Jones, R. S. Bradley and J. Jouzel. 1996.
Vol. 42: **The Mount Pinatubo Eruption: Effects on the Atmosphere and Climate.**
Edited by G. Fiocco, D. Fuà and G. Visconti. 1996.
Vol. 43: **Chemical Exchange Between the Atmosphere and Polar Snow.**
Edited by E. W. Wolff and R. C. Bales. 1996.

Vol. 44: **Decadal Climate Variability: Dynamics and Predictability.**
Edited by D. L. T. Anderson and J. Willebrand. 1996.

Vol. 45: **Radiation and Water in the Climate System: Remote Measurements.**
Edited by E. Raschke. 1996.

Vol. 46: **Land Surface Processes in Hydrology: Trials and Tribulations of Modeling and Measuring.** Edited by S. Sorooshian, H. V. Gupta, and J. C. Rodda. 1997.

Vol. 47: **Past and Future Rapid Environmental Changes: The Spatial and Evolutionary Responses of Terrestrial Biota.**
Edited by B. Huntley, W. Cramer, A. V. Morgan, H. C. Prentice, and J. R. M. Allen. 1997.

Vol. 48: **The Mathematics of Models for Climatology and Environment.**
Edited by J. I. Díaz. 1997.

Vol. 49: **Third Millennium BC Climate Change and Old World Collapse.**
Edited by H. N. Dalfes, G. Kukla, and R. Weiss. 1997.

Vol. 50: **Gravity Wave Processes:**
Their Parameterization in Global Climate Models.
Edited by K. Hamilton. 1997.

Vol. 51: **Sediment Records of Biomass Burning and Global Change.**
Edited by J. S. Clark, H. Cachier, J. G. Goldammer, and Brian Stocks. 1997.

Vol. 52: **Solar Ultraviolet Radiation. Modelling, Measurements and Effects.**
Edited by C. S. Zerefos and A. F. Bais. 1997.

Vol. 53: **Atmospheric Ozone Dynamics.**
Observations in the Mediterranean Region.
Edited by C. Varotsos. 1997.

Vol. 54: **The Stratosphere and Its Role in the Climate System**
Edited by G. P. Brasseur. 1997.

Vol. 55: **Modelling Soil Erosion by Water.**
Edited by J. Boardman and D. Favis-Mortlock. 1998.

Printing: Druckhaus Beltz, Hemsbach
Binding: Buchbinderei Schäffer, Grünstadt